PULMONARY IMMUNOTOXICOLOGY

PULMONARY IMMUNOTOXICOLOGY

Edited by

Mitchell D. Cohen

Judith T. Zelikoff

Richard B. Schlesinger

New York University School of Medicine
Department of Environmental Medicine
Tuxedo, New York, U.S.A.

Kluwer Academic Publishers
Boston / Dordrecht / London

Distributors for North, Central and South America:
Kluwer Academic Publishers
101 Philip Drive
Assinippi Park
Norwell, Massachusetts 02061 USA
Telephone (781) 871-6600
Fax (781) 681-9045
E-Mail <kluwer@wkap.com>

Distributors for all other countries:
Kluwer Academic Publishers Group
Distribution Centre
Post Office Box 322
3300 AH Dordrecht, THE NETHERLANDS
Telephone 31 78 6392 392
Fax 31 78 6546 474
E-Mail <services@wkap.nl>

 Electronic Services <http://www.wkap.nl>

Library of Congress Cataloging-in-Publication Data

Pulmonary immunotoxicology / edited by Mitchell D. Cohen, Judith T. Zelikoff, Richard
B. Schlesinger.
   p.; cm.
  Includes bibliographical references and index.
  ISBN 0-7923-7843-1 (alk. paper)
  1.Pulmonary toxicology. 2. Immunotoxicology. 3. Respiratory allergy. I. Cohen,
  Mitchell D., 1959- II. Zelikoff, Judith T. III. Schlesinger, Richard B.
   [DNLM: 1. Lung—drug effects. 2. Air Pollutants—adverse effects. 3. Immune
  System—drug effects. 4. Lung—immunology. WF 600 P98349 2000]
  RC720 .P85 2000
  616.2'407—dc21

                                                   00-030161

Printed on acid-free paper.   Printed in the United States of America

*The Publisher offers discounts on this book for course use and bulk purchases.
For further information, send email to <molly.taylor@wkap.com> .*

DEDICATION

I dedicate this book to the memory of my companions Vaska, Whitey, and Bebe.

M.D.C.

This book is dedicated to the research assistants and graduate students who helped build my laboratory. I also dedicate this book to my Mother and Sister who have always believed in me and provided support and encouragement during difficult times.

J.T.Z.

To my grandson Jack and the next generation of toxicologists.

R.B.S.

TABLE OF CONTENTS

Section I. The Respiratory Tract

Chapter 1
Comparative Structure of the Respiratory Tract: Airway Architecture
Jack R. Harkema, Charles G. Plopper, and Kent E. Pinkerton

Chapter 2
Robert W. Lange and Meryl H. Karol

Chapter 3
Richard B. Schlesinger

Section II. Adverse Effects of Altered Pulmonary Immunity

Chapter 4
M. Ian Gilmour

Chapter 5
David J. P. Bassett and Deepak K. Bhalla

Chapter 6
Thomas M. Jeitner and David Lawrence

Section III. Immunotoxicants

Chapter 7
Robert L. Sherwood

Chapter 8
Kathleen E. Rodgers

Section IV. Risk Assessment

LIST OF CONTRIBUTORS

David J.P. Bassett, Department of Occupational and Environmental Health Sciences, Wayne State University, Detroit, MI 42807

Deepak K. Bhalla, Department of Occupational and Environmental Health Sciences, Wayne State University, Detroit, MI 42807

Leigh Ann Burns-Naas, Toxicology and Product Safety, Health and Environmental Sciences, Dow Corning Corporation, Midland, MI 48611

Lung-Chi Chen, New York University School of Medicine, Department of Environmental Medicine, 57 Old Forge Road, Tuxedo, New York 10987

Mitchell D. Cohen, New York University School of Medicine, Department of Environmental Medicine, 57 Old Forge Road, Tuxedo, New York 10987

Gregory L. Finch, Inhalation Toxicology Laboratory, Lovelace Respiratory Research Institute, PO Box 5890, Albuquerque, NM 87185

Donald E. Gardner, Inhalation Toxicology Associates, Inc., P.O. Box 97605, Raleigh, NC 27624-7605

M. Ian Gilmour, Immunotoxicology Branch, Experimental Toxicology Division, National Health and Environmental Effects Research Laboratory, USEPA, Research Triangle Park, NC 27711

Jack R. Harkema, Department of Pathology, College of Veterinary Medicine, Institute for Environmental Toxicology, Michigan State University, East Lansing, MI 48824

Andrea K. Hubbard, Department of Pharmaceutical Sciences, School of Pharmacy, University of Connecticut, Storrs, CT 06269-2092

Thomas M. Jeitner, Laboratory of Environmental and Clinical Immunology, Wadsworth Center, New York State Department of Health, Empire State Plaza, Albany, NY 12201-509

Meryl H. Karol, Department of Environmental and Occupational Health, University of Pittsburgh, 260 Kappa Drive, Pittsburgh, PA 15238

Robert W. Lange, Department of Environmental and Occupational Health, University of Pittsburgh, 260 Kappa Drive, Pittsburgh, PA 15238

David Lawrence, Laboratory of Environmental and Clinical Immunology, Wadsworth Center, New York State Department of Health, Empire State Plaza, Albany, NY 12201-509

Kent E. Pinkerton, Department of Anatomy, Physiology, and Cell Biology, School of Veterinary Medicine, Institute of Toxicology and Environmental Health, University of California, Davis, CA 95616

Charles G. Plopper, Department of Anatomy, Physiology, and Cell Biology, School of Veterinary Medicine, Institute of Toxicology and Environmental Health, University of California, Davis, CA 95616

Kathleen E. Rodgers, University of Southern California, Livingston Research Institute, 1321 North Mission Road, Los Angeles, CA 90033

Lisa K. Ryan, Immunotoxicology Branch, National Health and Environmental Effects Research Laboratory, USEPA, Research Triangle Park, NC 27711

Richard B. Schlesinger, New York University School of Medicine, Department of Environmental Medicine, 57 Old Forge Road, Tuxedo, New York 10987

MaryJane K. Selgrade, National Health and Environmental Effects Research Laboratory, USEPA, Research Triangle Park, NC 27711

Robert L. Sherwood, Microbiology and Immunology Division, IIT Research Institute, 10 West 35th Street, Chicago, IL 60616

David B. Warheit, DuPont Haskell Laboratory, 1090 Elkton Road, Newark, DE 19714

Judith T. Zelikoff, New York University School of Medicine, Department of Environmental Medicine, 57 Old Forge Road, Tuxedo, New York 10987

PREFACE

Within the toxicological sciences, pulmonary toxicology has been recognized as a specialized field of study for more than forty years, and immunotoxicology has been expanding since its inception in the early 1970s as a separate area of research. With increasing efforts in both areas, it is not unexpected that there would be a convergence of the two in the researchers' quests to understand how inhaled toxins and toxicants can affect human health. The area of pulmonary immunotoxicology has been very active over the past decade in attempting to elucidate how workplace and general environmental agents can produce changes in immunological function in the respiratory tract that may allow for alterations in health status.

The purpose of this book is to provide the Reader, in a single voluume, information concerning the effects of various inhaled materials upon the immune system of the respiratory tract. The book will be useful both to investigators in the field of pulmonary toxicology and immunotoxicology, as well as to those involved in administration and regulation of matters related to inhaled materials. It can also serve as a textbook for a course in pulmonary immuno-toxicology at the graduate or advanced undergraduate level.

The book is comprised of four sections. The first provides basic background concepts essential for understanding pulmonary immuno-toxicology. This includes discussions of the normal structure and function of the respiratory system, its basic immunology, and the manner by which inhaled particles and gases are removed from the air and deposit upon respiratory tract surfaces. The second section provides an overview of the major types of pathological consequences which can arise from immunomodulation within the respiratory tract. These include hypersensitivity and asthma, inflammation and fibrosis, and immunosuppression and autoimmunity. The third section, which comprises the largest portion of the book, deals specifically with major classes of airborne agents that are known to alter immune function of the respiratory tract. These are arranged into major classes, i.e., organic agents, metals, gases, particles, biologics, and complex mixtures. The fourth and final section explores the area of risk assessment. This includes discussions of the basic concepts of risk assessment as they apply specifically to immunotoxicologic effects upon the lungs, and the use of biomarkers as indices of potential pulmonary immunotoxic responses to inhaled materials.

This is the first book which attempts to consolidate information evolving from the rapidly expanding field of pulmonary immuno-toxicology. As such, it is a valuable resource to those actively engaged in research in this area, as well as to those who may be thinking of entering the field. Because the book provides current information about many diverse agents in one readily accessible site, it may allow for some heretofore unseen linkages, such as mechanisms of action, to be made between these varied agents. Thus, the book is not only a review of what has been seen in the past, but also provides a starting point for what could be research agendas for the future.

Mitchell D. Cohen

Judith T. Zelikoff

Richard B. Schlesinger

Tuxedo, New York

EDITORS' PROFILES

Mitchell D. Cohen, Ph.D., is an Assistant Professor in the Department of Environmental Medicine of the New York University School of Medicine. Dr. Cohen received his B.S. in Chemistry and Physics from the State University of New York at Albany in 1981. He received his M.S. and Ph.D. degrees in Toxicology/Food Science and Human Nutrition in 1984 and 1988, respectively, from the University of Florida. Dr. Cohen then undertook two post-doctoral fellowships at the Nelson Institute for Environmental Medicine at the New York University School of Medicine; the first in the molecular toxicology of metals in the Laboratory for Molecular Toxicology and the second in the pulmonary immunotoxicology of inhaled environmental pollutant gases/metals/mixtures in the Laboratory of Pulmonary Biology and Toxicology.

Dr. Cohen is a member of several scientific societies including the Society of Toxicology, the Society for Leukocyte Biology, and American Association for Cancer Research. He has served on the editorial board for two toxicology journals, as well as on several society committees in the areas of Immunotoxicology and Inhalation Toxicology. Dr. Cohen has been involved in inmmunotoxicology research since 1981 and his research has involved investigations into the role(s) by which ambient and occupational pollutants may alter host resistance through modifications of immune cell cytokine receptor-cytokine interactions and biochemistry .

Judith T. Zelikoff, Ph.D., is an Associate Professor in the Department of Environmental Medicine of the New York University School of Medicine. She is currently President of the Immunotoxicology Specialty Section and Chairman of the Continuing Education Committee of the Society of Toxicology; she is also a member of the Inhalation and Metals Specialty Sections. Dr. Zelikoff has previously been an Editor on books in the areas of metal immunotoxicology and in ecotoxicology. Dr. Zelikoff is an Associate Editor for the journal Biomarkers and serves on editorial boards of several others including Toxicology, Toxicology and Applied Pharmacology, Journal of Toxicology and Environmental Health, and Aquatic Toxicology. She has organized several international conferences and workshops and has, as visiting scholar, lectured in many developing nations. Dr. Zelikoff has been involved in the area of pulmonary immunotoxicology for over 10 years. Her research interest is in discerning the role of indoor and outdoor airborne particulate mixtures in altering pulmonary host defense mechanisms against infectious disease-producing organisms.

Richard B. Schlesinger, Ph.D., is a Professor in the Department of Environmental Medicine of the New York University School of Medicine, where he is Director of the Systemic Toxicology Program. Dr. Schlesinger has served as President of the Inhalation Specialty Section of the Society of Toxicology, and was recepient of the Career Achievement Award from the Section in 1999. He has been recipient of an NIEHS Research Career Development Award and of the ILSI Kenneth Morgareidge Award for contributions to the field of inhalation toxicology. He has served on numerous national scientific committees, including those of EPA and NRC. His research involves investigations into the role which ambient and occupational air contaminants may play in the development of chronic nonneoplastic pulmonary disease.

1. COMPARATIVE STRUCTURE OF THE RESPIRATORY TRACT: AIRWAY ARCHITECTURE IN HUMANS AND ANIMALS

Jack R. Harkema, D.V.M., Ph.D., Dipl. A.C.V.P.
Department of Pathology
College of Veterinary Medicine
Institute for Environmental Toxicology
Michigan State University
East Lansing, MI 48824

Charles G. Plopper, Ph.D and Kent E. Pinkerton, Ph.D
Department of Anatomy, Physiology, and Cell Biology
School of Veterinary Medicine
Institute of Toxicology and Environmental Health
University of California
Davis, CA 95616

INTRODUCTION

The mammalian respiratory system is a structurally complex arrangement of organs designed principally for the intake of oxygen and the elimination of carbon dioxide (i.e., respiration). Though its main function is gas exchange, the respiratory system is composed of specialized tissues and cells that have other important functions such as the production of proteins and lipids, the activation and inactivation of hormones, and the metabolism of xenobiotic compounds entering the body through inhalation or other routes. Another important function of the respiratory system is defense against inhaled infectious (e.g., bacteria, viruses, fungi) and non-infectious agents (e.g., respirable dusts and gaseous air pollutants). The respiratory tract comprises the largest mucosal surface of the body with an internal surface area that is 25 times greater than the external surface of the body covered by skin. In contrast to the other mucosa-lined organs of the body (e.g., alimentary and reproductive), that are only periodically exposed to the external environment, the respiratory organs are constantly being exposed to large amounts of inhaled air. An adult human at rest takes in 10,000 - 15,000 L of ambient air through the nasal passages each day. Therefore the respiratory tract serves as an important interface between the environment and the host and plays a crucial role in maintaining the immune status of the body.

Species differences in respiratory tract anatomy and physiology have long concerned biomedical scientists who wish to extrapolate results from laboratory animal studies to humans. Claude Bernard, the famous 19th Century physiologist, remarked in "An Introduction to the Study of Experimental Medicine" (Bernard, 1957) that the choice of animals for experimentation is "..so important that the solution of a physiological or pathological problem often depends solely on the appropriate choice of the animal for the experiment.." Today, concerns about species differences in structure and function continue and are not limited only to those at the gross and subgross levels of analysis, but also at the cellular and molecular levels.

This Chapter will provide a brief overview of the structure of the respiratory tract from the nose to the lungs, emphasizing the key tissues and cells that comprise this unique system of the body. Throughout, we will highlight: 1) similarities and differences in macroscopic and microscopic structure of the respiratory organs between humans and mammalian animals commonly used in respiratory research (e.g., mice, rats, dogs, and monkeys), and 2) tissues and cells located in the upper and lower respiratory tract that participate in the host immune response to various inhaled toxic, infectious, and antigenic agents. For more detailed and comprehensive reviews of specific areas of respiratory structure and function, the reader is referred to several excellent publications found in the scientific literature (Tyler et al., 1983; Parent, 1992; Jeffery, 1995; Pinkerton et al., 1997; Weibel and Taylor, 1998).

The respiratory tract can be divided into two portions based on gross anatomy and physiology: 1) the proximal conducting (nonrespiratory) airways that include the nasopharyngeal airways (nose and pharynx), larynx, and tracheobronchial airways (trachea, bronchi, and bronchioles), and 2) the distal respiratory portion comprised of the respiratory bronchioles, alveolar ducts, and alveolar sacs. Gaseous exchange between air and blood is restricted to the latter portion located in the lung paren-chyma. This division of the respiratory tract can be identified in all mammalian species even though there is tremendous species variability in the architecture of individual airway segments. Figure 1 serves as a diagrammatic overview of the human respiratory tract.

NASOPHARYNGEAL AIRWAYS

The Nose

The nose is a structurally and functionally complex organ in the upper respiratory tract. It is the primary site of entry for inhaled air in the respiratory system of mammals and therefore has many important and diverse functions. The nose not only serves as principal organ for the sense of smell (olfaction), but also functions as an efficient filter, heater, and humidifier of the inhaled air to protect the delicate structures in the lung. The nose has been described as the "scrubbing tower" that removes inhaled chemicals potentially harmful to the lower airways and lung

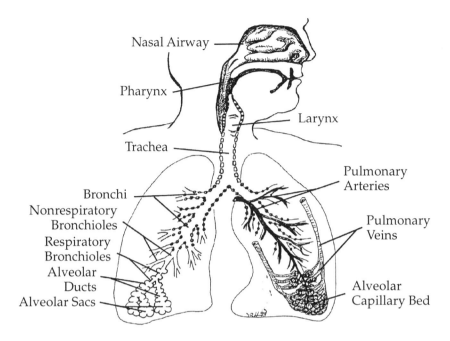

Figure 1. Diagrammatic overview of the human lung and upper respiratory tract. The lung lobe on the left side of the diagram illustrates the branching pattern of the intrapulmonary airways, from the bronchi to the alveolar sacs. The pattern of the pulmonary vasculature is illustrated in the lung lobe on the right.

parenchyma (Brain, 1970). The mucous membranes (nasal mucosa) lining this proximal airway have been shown to metabolize and detoxify many commonly-inhaled toxicants (Dahl and Hadley, 1991). The nasal airways and associated para-nasal sinuses may be afflicted by many diseases. The majority of these conditions are a consequence of viral or bacterial infections, allergic reactions, or aging. However, exposure of humans to toxic agents may also cause or exacerbate certain nasal diseases. In recent years there has been a marked increase in the study of nasal toxicology and in assessing the human risk of nasal injury from inhaled toxicants (Miller, 1995).

The nasal airway is divided into two passages by the nasal septum. Each nasal passage extends from the nostrils to the nasopharynx. The nasopharynx is defined as the airway posterior to the termination of the nasal septum and proximal to the termination of the soft palate. In adult humans, the distance from the opening of the nostril (i.e., naris) to the nasopharynx is ≈10 - 14 cm (Proctor and Chang, 1983). Inhaled air flows through the nostril openings, or nares, into the vestibule, which is a slight dilatation just inside the nares and before the main chamber of the nose. Unlike the more distal main nasal chamber that is surrounded by bone, the nasal vestibule is surrounded primarily by cartilage. The luminal surface is lined by a squamous epithelium similar to that of external skin and bears varying numbers of hairs in some species. After passing through the nasal vestibule, inhaled air courses through the narrowest part of the entire respiratory tract, the nasal valve (ostium internum), into the main nasal chamber. Each nasal passage of the main chamber is defined by a lateral wall, a septal wall, a roof, and a floor. The lumen of the main chamber is lined by well-vascularized and innervated mucous membranes that are covered by a continuous layer of mucus. The nasal mucous layer is moved distally, by underlying cilia to the oropharynx where it is swallowed into the esophagus.

Three distinct surface epithelia (ciliated respiratory, non-ciliated transitional, and olfactory) are located in specific positions within the main nasal chamber of most mammalian species (Harkema, 1991). Turbinates, bony structures lined by well-vascularized mucosal tissue, protrude into the airway lumen from the lateral walls into the main chamber of the nose. Nasal turbinates increase the inner surface area of the nose, which is important in the filtering, humidification, and warming of inspired air. Although the turbinated main chamber of the human nose is only about 5 - 8 cm long, the surface area is 160 cm^2, about four times that of the trachea (Guilmette et al., 1989).

Though there are some general gross similarities in the nasal passages of most mammalian species, there are also striking interspecies differences in nasal architecture (Table 1). From a comparative viewpoint, humans have relatively simple noses with breathing as the primary function, while other mammals have more complex noses with olfaction as the primary function. In addition, the nasal and oral cavities of humans and some non-human primates are arranged in a manner to allow for both

TABLE 1. INTERSPECIES COMPARISON OF NASAL CAVITY CHARACTERISTICS

	Rat (Sprague Dawley)	Guinea Pig	Beagle Dog	Rhesus Monkey	Human
Body weight	0.25 kg	0.60 kg	10 kg	7 kg	~70 kg
Naris cross-section (mm^2)	0.7	2.5	16.7	22.9	140
Bend in naris	40°	40°	30°	30°	
Length (cm)	2.3	3.4	10	5.3	7-8
Greatest vertical diameter (mm)	9.6	12.8	23	27	40-45
Surface area (both sides of nasal cavity) (cm^2)	10.4	27.4	220.7	61.6	181
Volume (both sides) (cm^3)	0.4	0.9	20	8	16-19*
Bend in naso-pharynx	15°	30°	30°	80°	~90°
Turbinate complexity	Complex scroll	Complex scroll	Very complex membranous	Simple scroll	Simple scroll

*Does not include sinuses

Source: From Proctor DF. and Chang JCF., in *Nasal Tumors in Animals and Man,* Vol. III, *Experimental Nasal Carcinogenesis,* Reznik G. and Stinson S.F, Eds., CRC Press, Boca Raton, FL, 1983, pp. 1-26. With permission.

nasal and oronasal breathing. Conversely, laboratory rodents (e.g., rats, mice, hamsters, guinea pigs) are obligate nose breathers due to the close apposition of the epiglottis to the soft palate. Interspecies variability in nasal gross anatomy has been emphasized in previous reviews (Negus, 1958; Harkema, 1991).

Marked differences in airflow patterns among mammalian species are primarily due to variation in the shape of nasal turbinates. The human nose has three turbinates: the superior, middle, and inferior. These structures are relatively simple in shape compared to turbinates in most non-primate species (e.g., dog, rat, mouse, rabbit) that have complex folding and branching patterns (Figure 2). In laboratory rodents (e.g., rat, mouse, hamster, guinea pig), evolutionary pressures concerned chiefly with olfactory function and dentition have defined the shape of the turbinates and the type and distribution of the cells lining the turbinates. In the proximal nasal airway, the complex maxilloturbinates of small laboratory rodents and rabbits provide far better protection of the lower respiratory tract, by better filtration, absorption, and disposal of airborne particles and gases, than do the simple middle and inferior turbinates of the human nose. The highly complex shape of the ethmoid turbinates, lined predominantly by olfactory neuroepithelium, in the distal half of the nasal cavity of small laboratory animals is suitably designed for acute olfaction. Monkeys have turbinate structures that are more similar to those of humans than those of other laboratory animals such as the dog, rat, or rabbit (Figures 2 and 3).

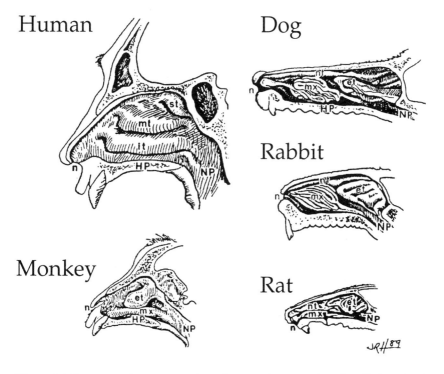

Figure 2. Diagrammatic representation of exposed mucosal surface lining the lateral wall of the nasal cavity of human, monkey, dog, rabbit, and rat. HP = hard palate; n = naris; st = superior turbinate; mt = middle turbinate; it =inferior turbinate; NP = nasopharynx; nt = nasoturbinate; mx = maxilloturbinate; et = ethmoturbinate.

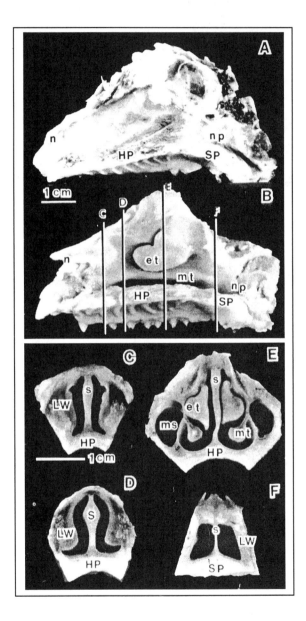

Figure 3. Dissected nasal cavity of bonnet monkey (*Macaca radiatus*) exposing the mucosal surface of septum (A) and lateral wall (B): In (B), vertical lines labeled C-F represent levels of the anterior surfaces of transverse blocks shown below: n = naris; np = nasopharynx; HP = hard palate; SP = soft palate; et = ethmoturbinate; mt = maxilloturbinate; ms = maxillary sinus; s = septum; LW = lateral wall. (From Harkema et al., Am. J. Pathol. 180:266-279, 1987).

The Nasal Mucociliary Apparatus

The luminal surface of the nasal mucosa (with the exception of the most proximal regions of the nose lined by squamous epithelium) are covered by a watery, sticky glycoprotein called mucus. Its physical and chemical properties are well suited for it role as an upper airway defense mechanism — filtering the inhaled air by trapping inhaled particles and various gases or vapors. Mucus is produced by mucous secretory cells in the surface epithelium and subepithelial glands in the lamina propria. Synchronized beating of surface cilia propel the mucus at different speeds and directions depending on the intranasal location. Mucus covering the olfactory mucosa moves very slowly, with a turnover time of probably several days. In contrast, the mucus covering of the transitional and respiratory epithelium is driven along rapidly (1 to 30 mm/min) by the ciliary activity with an estimated turnover time of about 10 min in the rat (Morgan et al., 1984). The mucus with the entrapped materials ultimately is carried to the nasopharynx and is swallowed into the gastrointestinal tract. The nasal mucociliary apparatus (i.e., mucus and cilia) exhibits a range of responses to inhaled xenobiotic agents and can be a sensitive indicator of toxicity (Morgan et al., 1986). Since this upper airway apparatus is one of the first lines of defense against inhaled pathogens, dusts, and irritant gases, toxicant-induced compromises in its defense capabilities could lead to increased nasal infections and increased susceptibility to lower respiratory tract diseases.

Nasal Blood Vessels and Blood Flow

The subepithelial connective tissue (lamina propria) of the nasal mucosa has a rich and complex network of blood vessels, with each of the epithelial regions receiving blood from a separate arterial supply (Sorokin, 1988). The vascular system in the nose is composed of resistance and capacitance vessels. Resistance vessels are small arteries, arterioles, and arteriovenous anatomoses. Blood flow to the mucosa is regulated by constriction and dilation of these vessels (Cauna, 1982).

A unique feature of the vasculature of the nose is the large venous sinusoids (i.e., capacitance vessels, venous erectile tissue, or swell bodies) that are distributed throughout the mucosa. In humans and laboratory animals these blood vessels are well developed in specific sites of the anterior or proximal aspects of the nasal passages. Capacitance vessels have dense adrenergic innervation, and the congestion and constriction of these vascular structures are regulated by the sympathetic nerve supply to the nose (Olsson and Bende, 1986). Congestion of blood in these vessels can change nasal airflow patterns and increase nasal resistance.

Though the rodent nose receives <1% of the cardiac output, the vascular uptake of nonreactive gases is strongly dependent on nasal blood perfusion rates (Morris et al., 1993). Little is known, however, about the influence of nasal blood supply on local deposition and removal of reactive gases in the nose, and the potential impact of the blood supply on nasal toxicity. It is believed that nasal blood flow may be important in removing certain toxic materials from the nose and protecting it from toxicant-induced injury.

The vascular system in the nose is also important in the delivery of non-inhaled materials or their metabolites to the nose. A wide range of chemicals have been administered to rodents by non-inhalation routes with subsequent nasal damage (e.g., nitrosamines, some herbicides) (Brittebo et al., 1991; Genter et al., 1992).

Nasal Innervation and Nasal Reflexes

At the entrance of the respiratory tract, the nose is in the ideal position to detect chemical and physical irritants in the inhaled air that could potentially damage the lower respiratory airways. The nasal mucosa is primarily innervated by olfactory and trigeminal nerves. The olfactory nerves enter from the cribiform plate and terminate in the olfactory epithelium as sensory cells. Detecting odorants is the chief function of these chemoreceptor cells (see below).

The trigeminal nerves in the head provide the sense of touch, pain, hot, cold, itch, and the sensation of nasal airfow. These nasal nerve endings detect irritating inhaled chemicals, such as ammonia and sulfur dioxide, and a range of organic substances, such as methanol, acetone, and pyridine. Stimuli from inhaled chemical or physical irritants may initiate respiratory and cardiovascular reflexes via the trigeminal nerves resulting in apnea and bradycardia. Concentration-dependent reductions in respiratory rate in rodents have been demonstrated after exposure to a number of sensory irritants (Alarie, 1981; Buckley et al., 1984).

The nasal mucosa is also innervated by both sympathetic and parasympathetic nerve fibers (Eccles, 1982). Parasympathetic fibers supply nasal glands and regulate their secretion. Sympathetic fibers innervate the blood vessels in the lamina propria of the mucosa. Stimulation of these fibers causes nasal vasoconstriction, reduction in blood flow, decongestion of capacitance vessels, and subsequent decrease in nasal airway resistance.

Cellular Composition of Nasal Surface Epithelium

Besides the differences in the gross architecture of the nose among different species, there are also species differences in the surface epithelial populations of the mucosal tissue lining the nasal passages. These differences among species are found in the distribution of nasal epithelial populations and in the types of nasal cells within these populations. There are, however, four distinct nasal epithelial populations in most mammalian species (Figures 4-6). These include the squamous epithelium, which is primarily restricted to the nasal vestibule; ciliated respiratory epithelium in the main chamber and nasopharynx; non-ciliated transitional epithelium, lying between squamous epithelium and the respiratory epithelium in the proximal or anterior aspect of the main chamber; and olfactory epithelium, located in the dorsal or dorsoposterior aspect of the nasal cavity. Figure 5 illustrates the distribution of these distinct epithelial cell populations in the nasal cavity of the F-344 rat and bonnet macaque monkey.

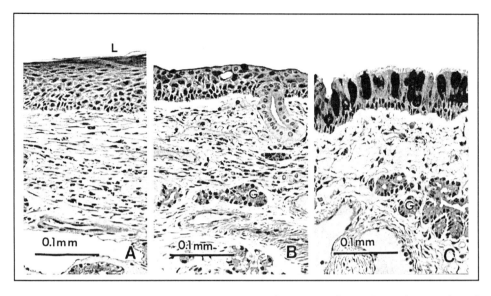

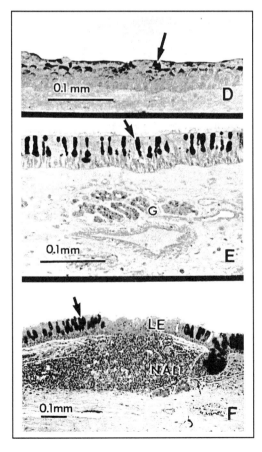

Figure 4. Light photomicrographs of surface epithelium in various locations of the nasal cavity and nasopharynx of a bonnet monkey. G = subepithelial glands. A) Stratified squamous epithelium from the lateral wall of the nasal vestibule, stained with toluidine blue. L = nasal lumen. B) Transitional epithelium from the anterior nasal septum, stained with toluidine blue. C) respiratory epithelium from the maxilloturbinate, stained with toluidine blue. D) Transitional epithelium from lateral wall, stained with Alcian Blue (pH = 2.5)/Periodic Acid Schiff (AB/PAS) demonstrating the presence of scant amounts of acidic and neutral mucosubstances within epithelium (arrow). E) Nasal respiratory epithelium from the septum containing AB/PAS-stained mucosubstances (arrow). F) Respiratory epithelium and lymphoepithelium (LE) overlying nasal associated lymphoid tissue (NALT) from the mucosa lining the nasopharynx. Tissue was stained with toluidine blue. Notice the presence of mucosubstances in the respiratory epithelium (arrow), but the absence of this material in the lymphoepithelium. (From Harkema et al., Am. J. Pathol. 180:266-279, 1987)

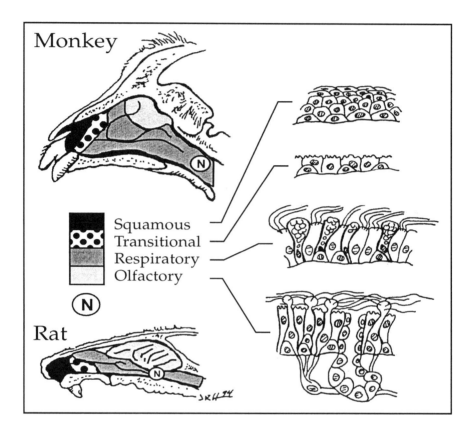

Figure 5. Distribution of the four principal surface epithelia in the nasal airways of the monkey and rat. Approximately half of the nasal cavity of the rat is covered by olfactory epithelium. Nasal associated lymphoid tissue (NALT) covered by a lymphoepithelium is located at the opening of the nasopharynx.

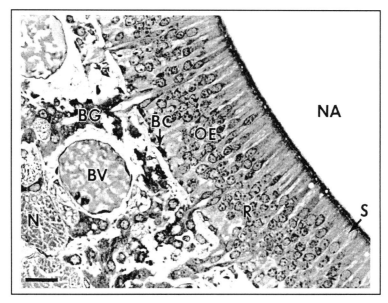

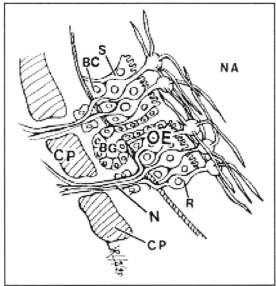

Figure 6. A) Olfactory epithelium (OE) in the nasal cavity of a F-344 rat. B) Caricatural representation of the OE and underlying lamina propria. S = sustentacular (support) cells; R = olfactory sensory receptor cells; BC = basal cells; N = olfactory nerve in the lamina propria. BG = Bowman's gland; CP = cribiform palate between nasal and cranial cavities; BV = blood vessel; NA = nasal airway, Bar = 25 μm (From Harkema and Morgan. 1996. "Normal Morphology of the Nasal Passages in Laboratory Rodents." In *Respiratory System, Monographs on Pathology of Laboratory Animals*, eds. Jones, Dungworth and Mohr, Berlin: Springer-Verlag, p. 12.).

Olfactory Epithelium

The major difference in nasal epithelium among animal species is the percentage of the nasal airway that is covered by olfactory epithelium. For example, the olfactory epithelium covers a much greater percentage of nasal cavity in rats, which have a more acute sense of smell, compared to that in monkeys or humans. Approximately 50% of the nasal cavity surface area in F-344 rats is lined by this sensory neuroepithelium (Gross et al., 1982). Olfactory epithelium in humans is limited to an area of about 500 mm^2, which is only 3% of the total surface area of the nasal cavity (Sorokin, 1988). Mice, rabbits, and dogs are much closer to rats than humans or monkeys in respect to the relative amount of olfactory epithelium within their nasal passages.

Three basic cell types compose the olfactory epithelium (sensory, sustentacular, and basal cells) (Figure 6). The olfactory sensory cells are bipolar neurons interposed between the sustentacular cells. The dendritic portions of these neurons extend above the epithelial surface and terminate into a bulbous olfactory knob from which protrude 12 or more immotile cilia. These cilia are enmeshed with each other and with microvilli in the surface fluid and provide an extensive surface area for reception of odorants. The axon of the olfactory sensory cell originates from the base of the cell and passes through the basal lamina to join axons from other sensory cells forming nonmyelinated nerves in the lamina propria. These axons perforate the cribiform plate to synapse with neurons in the olfactory bulb of the brain. Unlike other neurons in the body, the olfactory sensory cell can regenerate, having a 28- to 30-day turnover rate in the rat (Graziadei and Monti-Graziadei, 1977). Basal cells are generally considered the progenitor or stem cells for the regenerating olfactory epithelium. Regeneration of olfactory epithelium after experimental injury has been found to be an excellent model for the study of neurogenesis and axon regeneration in mammals.

Sustentacular cells in olfactory epithelium have been considered as support cells for the sensory cells. These cells also have abundant smooth endoplasmic reticulum (SER) and xenobiotic-metabolizing enzymes (e.g., esterases, cytochrome P_{450}). The metabolism in these cells may be important in detoxification of inhaled xenobiotics and in the function of smell (Dahl and Hadley, 1991).

Other important sites of xenobiotic metabolism associated with olfactory epithelium are the Bowman's glands (Figure 6). These structures, located in the underlying lamina propria and interspersed among the olfactory nerve bundles, are simple tubular-type glands composed of small compact acini. Ducts from these glands transverse the basal lamina at regular intervals and extend through the olfactory epithelium to the luminal surface. Bowman's glands contain copious amounts of neutral and acidic mucosubstances that contribute to the mucus layer covering the luminal surface of the olfactory epithelium.

With few exceptions, the olfactory epithelium has greater xenobiotic metabolizing activity than does the respiratory epithelium in most animal species (Dahl and Hadley, 1991). Immunohistochemical analyses suggest that sustentacular cells in the olfactory epithelium and Bowman's glands in the underlying lamina propria tend to have especially high concentrations of xenobiotic-metabolizing enzymes. The presence of these enzymes in the olfactory epithelium readily explains numerous observations of nasal toxicity in animals exposed to certain inhaled toxicants. With its generally higher enzyme activity, the olfactory mucosa is more sensitive to the toxic effects of many metabolizable materials.

Squamous Epithelium

The naris and nasal vestibule are completely lined by a lightly keratinized, stratified, squamous epithelium (Figure 4A). It is composed of basal cells along the basal lamina and several layers of squamous cells, which become progressively flatter toward the luminal surface. Only 3.5% of the entire nasal cavity of the F-344 rat is lined by squamous epithelium. This region of the nasal mucosa probably functions like the epidermis in the skin, to protect the underlying tissues from potentially harmful atmospheric agents.

Transitional Epithelium

Distal to the stratified squamous epithelium and proximal to the ciliated respiratory epithelium is a narrow zone of non-ciliated, microvilli-covered surface epithelium, which has been referred to as nasal, non-ciliated, respiratory epithelium or nasal transitional epithelium (Figure 4B). Common, distinctive features of this nasal epithelium in all laboratory animal species and humans include: 1) anatomical location in the proximal aspect of the nasal cavity between the squamous epithelium and the respiratory epithelium; 2) the presence of non-ciliated cuboidal or columnar surface cells and basal cells; 3) a scarcity of mucous (goblet) cells and a paucity of intraepithelial mucosubstances; and 4) an abrupt morphological border with squamous epithelium, but a less abrupt border with respiratory epithelium.

In rodents, this surface epithelium is thin (i.e., one to two cells thick), pseudostratified, and composed of three distinct cell types (basal, cuboidal, and columnar) (Monteiro Riviere and Popp, 1984). In contrast, transitional epithelium in monkeys is thick (i.e., four to five cells), stratified, and composed of at least five different cell types. The luminal surfaces of transitional epithelial cells lining the nasal airway possess numerous microvilli. Luminal, non-ciliated cells in the transitional epithelium of rodents have no secretory granules but do have abundant SER in their apices (Harkema et al., 1987) (Figure 6). SER is an important intracellular site for xenobiotic-metabolizing enzymes, including cytochromes P_{450}. Though the transitional epithelium normally contains few or no secretory cells in monkeys and rodents, exposure to high ambient concentrations of irritating pollutants, like ozone, can cause a rapid appearance of numerous mucous-secreting cells (i.e., mucous cell metaplasia) (Harkema et al., 1987, 1998).

Nasal Respiratory Epithelium

The majority of the nonolfactory nasal epithelium of laboratory animals and humans is ciliated respiratory epithelium (Figure 7). Approximately 46% of the

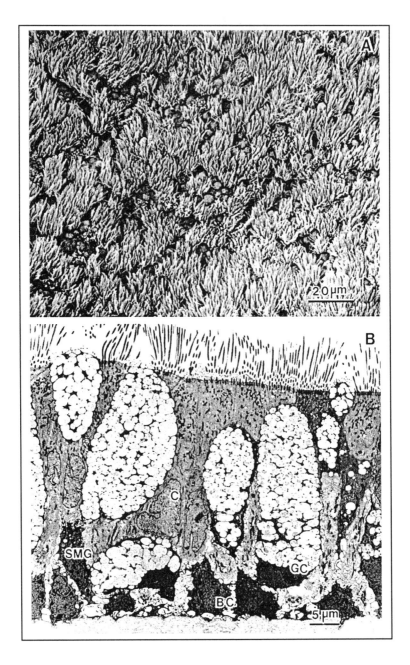

Figure 7. A) Scanning electron micrograph of ciliated respiratory epithelium from the anterior nasal septum of a bonnet monkey. B) transmission electron micrograph of pseudostratified respiratory epithelium from the anterior nasal septum. C = ciliated cell; GC = mucous (goblet) cell; BC = basal cell; SMG = small mucous granule cell.

nasal cavity in an F-344 rat is lined by respiratory epithelium (Gross et al., 1982). Although this pseudostratified nasal epithelium is similar to ciliated epithelium lining other proximal airways (i.e., tracheobronchial airways; see below), it also has unique features. Nasal respiratory epithelium in the rat is composed of six morphologically distinct cell types: mucous, ciliated, non-ciliated columnar, cuboidal, brush, and basal. These cells are unevenly distributed along the rat mucosal surface (Popp and Martin, 1984). There is a proximal-to-distal increase in ciliated cells along the lateral walls of the rat.

Like the ciliated cells, the mucus-secreting cells are unevenly distributed in the respiratory epithelium of the nasal cavity. In the normal rat, mucous (goblet) cells are predominantly located in respiratory epithelium lining the proximal septum and the nasopharynx (Harkema et al., 1989). Serous cells are the primary secretory cells in the remainder of the respiratory epithelium in rodents. Secretory cells in the respiratory epithelium of rodents have abundant SER (Yamamoto and Masuda, 1982), suggesting that these cells, like the non-ciliated cell in the transitional epithelium, have metabolic capacities for certain xenobiotic agents. Many xenobiotic enzymes have been identified in the nasal respiratory and olfactory epithelium (e.g., carboxylesterase, aldehyde dehydrogenase, cytochrome P_{450}, epoxide hydrolase, and glutathione S-transferases) (Bogdanffy, 1990). The distribution of these enzymes appears to be cell type-specific, and the presence of the enzyme may predispose particular cell types to enhanced susceptibility or resistance to chemical-induced injury.

The Pharynx

The pharynx connects the nasal and oral airways with the laryngeal airway. In human and non-human primates (e.g., rhesus monkey), the pharynx is situated posterior to the nasal cavity, mouth, and larynx (Figure 1). In many other laboratory mammals (e.g., rodent and dog), the pharynx is distal to most of the nasal airway (except for the anterior portion of the nasopharynx that lies ventral to the distal aspect of the main nasal cavity) and dorsal to the oral cavity and larynx. The pharynx is a musculomembranous tube, ≈130 mm in the adult human, ≈115 mm in the adult male beagle dog, 35 mm in the adult male rhesus monkey, and ≈22 mm in the adult Sprague-Dawley rat (Schreider and Raabe, 1981). It can be anatomically divided into nasal, oral, and laryngeal regions. The nasopharynx is lined with ciliated respiratory epithelium with mucous goblet cells, while the oropharynx and laryngo-pharynx are lined by nonkeratinized, squamous epithelium.

One of the more conspicuous anatomical differences in the pharyngeal region of the upper airway among mammals is the angle of the dorsoventral bend in the nasopharynx (Table 1). Rats have the slightest bend in the nasopharynx (15°), monkeys have the sharpest nasopharyngeal bend (80°), and the bends in the nasopharyngeal airways of dogs are in between those of rats and monkeys (45°) (Schreider and Raabe, 1981). The large bend of the nasopharynx in rhesus monkeys is similar to that of humans (≈90°) and is due to the more erect posture of this species. Because the right and left portions of the nasopharynx join at this bend, it could be an important site of inhaled particle and gas deposition. Interestingly, it is also the site where mucosal lymphoid tissues are found and where streams of mucus from the nasal cavity converge before being swallowed (See description of nasal associated lymphoid tissue below).

Immune Tissues of the Nasopharyngeal Airways

Scattered throughout the lamina propria of the nasal mucosa are lymphocytes, plasma cells and mast cells that are important participants in the immune response. Antibodies produced by plasma cells protect the nasal airways against inhaled antigens and invasion of infectious agents. In addition to these widely scattered individual or small aggregates of lymphoid cells, there are larger focal sites of discrete lymphoid tissue, designated as nasal-associated lymphoid tissue (NALT). They are located in the nasopharyngeal mucosa of humans and animals (Harkema et al., 1987; Spit et al., 1989; Kuper et al., 1990; Heritage et al., 1997). A lympho-epithelium covers the luminal side of NALT and is composed of lymphoid cells and non-ciliated, cuboidal cells with luminal microvilli similar to membranous cells (i.e., M cells) found in the gut- and bronchus-associated lymphoid tissues (GALT and BALT, respectively) (Figure 4F). There are few if any mucous cells or ciliated cells in this specialized airway epithelium. The non-ciliated cuboidal cells are thought to be involved in the uptake and translocation of inhaled antigen from the nasal lumen to the underlying lymphoid structures. In rodents, NALT is restricted to the ventral aspects of the lateral walls at the opening of the nasopharyngeal duct (Figure 5).

NALT has also been described in the nasopharyngeal airways of the monkey, but these lymphoid structures are more numerous than those in rodents and are located on both the lateral and septal walls of the proximal nasopharynx (Harkema et al., 1987). The correlate of NALT in humans is Waldeyer's ring, the orophayngeal lymphoid tissues composed of the adenoid, and the bilateral tubular, palatine, and lingual tonsils (Brandtzaeg, 1984).

The location of NALT at the entrance of the nasopharyngeal duct is a very strategic position as most of the nasal secretions and inhaled air, both presumably laden with antigenic material, pass over this area. Though the function of NALT and its place in the general mucosal-associated lymphoid system are not fully under-stood, these mucosal lymphoid tissues presumably have an important function in regional immune defense of the upper airways. NALT has been studied primarily in rat and mouse models. Immunohistochemical characterization of rat NALT has demonstrated that B- and T-lymphocytes are distributed in distinct areas with a high CD4:CD8 T-lymphocyte ratio and a predominance of B- over T-lymphocytes (Koornstra et al., 1993). Initial studies in mice suggest the NALT is distinct from that found in rats and, if examined solely on immune cell content and subset ratios, more closely resembles the spleen and not the Peyers Patches located in the intestinal mucosa (Heritage et al., 1997). However, the capability of NALT to elicit specific IgA responses locally suggests that this structure might represent a unique mucosal lymphoid tissue that is capable of expressing both mucosal and systemic immune responses.

THE LARYNX

The larynx is part of the upper respiratory tract between the pharynx and the trachea (Figure 1). It is a bilaterally symmetrical structure that is framed by cartilages and bound by ligaments and muscles. It functions as: (1) a pathway for inhaled and exhaled air during breathing; (2) a valve, with the epiglottis, to prevent swallowed food from entering the lower respiratory tract during eating and drinking; and, (3) a tone-producing structure. The airway lumen in the human larynx is lined, for the most part, by pseudostratified, ciliated respiratory epithelium, except for a small area on the vocal folds where it is stratified squamous epithelium. In rats and mice, a transitional zone of non-ciliated, cuboidal epithelium has also been identified between the squamous and ciliated epithelium (Nakano and Muto, 1987). Overlying the laryngeal airway epithelia is a layer of mucus that is propelled, by the under-lying cilia, toward the oropharynx where it is swallowed into the esophagus.

In the adult human, the larynx is about 3 cm in length in females and 4 cm in males (Zrunek et al., 1988). Morphometric studies of human laryngeal cartilage from cadavers have also demonstrated clear sexual dimorphism in the length of the vocal cords. The vocal cords are over two times longer in adult males than in adult females (Kahane, 1978). There are also many species differences in the structure and function of the larynx, and these have been described in detail by Negus (Negus, 1962). For example, the adult male beagle dog has a larynx that is $\approx$35 mm in length with the narrowest cross-sectional area being about 90 mm^2, while the larynx of an adult male Sprague-Dawley rat is $\approx$4 mm in length with its narrowest cross-sectional area being about 2 mm^2 (Schreider and Raabe, 1981).

The laryngeal cross-sectional area varies and depends on the airflow rate passing through it. The rapid expansion and contraction of this organ of phonation undoubtedly produce some degree of turbulence that could lead to significant deposition of inhaled particles. In addition, the larynx is a major resistive element to airflow and has been shown to create an inspiratory air jet that leads to particle impaction on the wall of the trachea (Schlesinger and Lippmann, 1976).

TRACHEOBRONCHIAL AIRWAYS

Trachea

The trachea is continuous with the larynx in the neck and extends distally into the thoracic cavity to the carina where it bifurcates to form two primary, or "main-stem", bronchi (Figure 1). The trachea lies close to the ventral or anterior aspect of the neck where the surrounding tissue gives little structural support. The tracheal airway remains patent on inspiration due to U-shaped cartilage rings that lie within the ventral and lateral walls. In humans, the dorsal wall of the trachea is composed of flexible fibroelastic and smooth muscle tissue (trachealis muscle). There are usually 20 cartilaginous rings in the human trachea, which is $\approx$10 - 12 cm in length with an internal diameter of 2 cm (Yeh and Schum, 1980).

The airway wall of the trachea and the more distal conducting airways (bronchi and bronchioles) is composed of three distinct layers of tissue: the mucosa, the submucosa, and the adventitia. The mucosa lining the airway lumen is composed of a pseudostratified, ciliated epithelium (respiratory epithelium) and a rather thin sub-epithelial lamina propria. A fine smooth muscle layer divides the mucosa from the submucosa. These airways are connected to the surrounding tissue by a loose connective tissue, the adventitia.

Bronchi and Bronchioles

The distal end of the trachea divides, at the carina within the thoracic cavity, into two extrapulmonary mainstem bronchi. These bronchial airways further divide into smaller bronchi. In the human respiratory tract, there are seven generations of bronchi, and the tracheobronchial branching system is relatively symmetrical (i.e., the parent airway divides into two smaller airways with relatively equal diameter). This is in contrast to most other mammals (e.g., monkey, dog, rat, mouse), that have a monopodial branching pattern, consisting of daughter branches of unequal diameter (Tyler, 1983).

As the bronchi divide into smaller and smaller diameter airways within the human lungs and the lungs of large laboratory animal species (e.g., dogs, cats, and monkeys), there is a point (airway diameter of ≈ 1 mm in the human respiratory tract) when cartilage and submucosal glands are no longer present. The walls of intrapulmonary conducting airways contain smooth muscle and loose connective tissue, but no cartilage or glands. The presence of intramural cartilage and sub-mucosal glands distinguishes bronchial airways from the more distal and smaller in diameter bronchioles in the lungs of dogs, cats, non-human primates and humans. In contrast, no intrapulmonary conducting airways (i.e., airways distal to the hilus of the lung lobe) of laboratory rodents (e.g., rats and mice) contain intramural cartilage or submucosal glands (Tyler, 1983). Therefore, according to conventional definition, these rodents have only large and small diameter bronchioles within their lungs. Cartilage, but no submucosal glands are present in the extrapulmonary, mainstem bronchi. Subepithelial secretory glands are present in the lamina propria of the nose, larynx and proximal trachea of rodents, but the remainder of conducting airways in these laboratory animals are free of intramural glands.

The human respiratory tract contains several generations of nonrespiratory bronchioles (Weibel, 1963). Considerably fewer generations of these small conduct-ing airways are present in other mammalian species (e.g., dog, cat, monkey). The number of generation of nonrespiratory bronchioles in small laboratory mammals (e.g., rat, mouse, guinea pig, hamster, rabbit) is more similar to those in humans (Tyler, 1983).

The most distal nonrespiratory bronchiole is defined as the terminal bronchi-ole. The distal end of the conducting airways connects to the aveolarized (respiratory) airways. This focal area where conducting and respiratory airways join is called the centriacinus, a common site of injury from a variety of inhaled gases and particles. In humans and some laboratory mammals (e.g., monkey, dog, cat), the terminal bronchioles end into several generations of respiratory bronchioles that are defined as bronchioles with a few, widely scattered, intramural alveoli (or alveolar outpocket-ings) (Figure 8). This is in contrast to several small laboratory mammals (e.g., rat, mouse, rabbit, hamster) whose terminal bronchioles end directly into one short segment of respiratory bronchiole or into airways with walls completely covered by alveoli (i.e., alveolar ducts) (Tyler, 1983) (Figure 8).

Epithelial Cells of the Tracheobronchial Airways

The respiratory epithelium lining the lumens of the tracheobronchial airways varies from a tall columnar ciliated epithelium in the proximal trachea to a low cuboidal ciliated epithelium in the distal bronchioles, and a simple squamous epithelium lining the alveolar outpocketings in the terminal respiratory bronchioles (Figure 9). Surface epithelial cell type and abundance varies among species and airway generations (Table 2). Thickness of the epithelium directly correlates with the size of the animal species. For example, the thickness of the tracheal epithelium in a rat is approximately half of that found in humans (24 µm compared to 43 µm). In contrast, there is an indirect correlation with the airway generation number and the thickness of airway epithelium, independent of the species. The distal bronchioles have a high airway generation number and a thin luminal epithelium compared to that of the trachea and mainstem bronchi with the lowest airway generation numbers, but the thickest surface epithelium of the tracheobronchial airways.

At least eight types of cells have been identified in surface epithelium lining the tracheobronchial conducting airways of humans and laboratory animals (Harkema et al., 1991). These epithelial cells differ in morphology, function and response to injury. The eight principal cells types are ciliated, basal, mucous (goblet), neuroendocrine, Clara, serous, small mucous granule, and brush cells. In addition, there are a variety of intermediate cell forms. Similar epithelial cells are present in the respiratory epithelium in the nasal airways as previously described above.

Species differences in the composition of the surface epithelium lining the trachea have also been demonstrated (Harkema et al., 1991; Pinkerton et al., 1997). The composition of tracheal epithelial cell populations in the tracheobronchial airways of humans, monkeys and rats are compared in Table 2. Species variability exists not only in the numbers of cells in the entire population but also in the relative percentages of specific cell types within the population. Only the proportions of ciliated cells are relatively constant among species. Relative proportions of basal cells per millimeter of basal lamina vary among species by a factor of almost 10, and the number of total epithelial cells per millimeter of basal lamina can vary by as much as a factor of 4. The most striking interspecies differences in tracheal epithelium are related to the numbers and types of secretory cells. In humans, monkeys, dogs and cats the mucous (goblet) cell or small mucous granule cell is the predominant cell type, whereas in mice, hamsters and rabbits, the Clara cell predominates. In rats, the predominant secretory cell in the trachea is the serous cell.

Ciliated Cells

Ciliated cells are the most abundant epithelial cells that line the conducting airways from the nose to the bronchioles (Figures 7-10). Along with the secretory

TABLE 2. CELL (%) COMPOSITION OF THE TRACHEA AND PULMONARY BRONCHI FOR SELECTED SPECIES

Cell type	Airway level	Rat[a,b]	Monkey[c]	Human[d,e]
Ciliated				
	Trachea	32	33	49
	Primary bronchi	35	44	
	Lobar bronchi	53	47	
	Segmental bronchi		49	
Mucous				
	Trachea	2	17	9
	Primary bronchi	<1	15	
	Lobar bronchi	<1	15	
	Segmental bronchi		14	
Serous				
	Trachea	42	0	0
	Primary bronchi	21	0	
	Lobar bronchi	20	0	
	Segmental bronchi		0	
Clara				
	Trachea	0	0	0
	Primary bronchi	0	0	
	Lobar bronchi	0	0	
	Segmental bronchi		0	
Basal				
	Trachea	21	42	33
	Primary bronchi	27	32	
	Lobar bronchi	14	32	
	Segmental bronchi		29	
Other				
	Trachea	1	4	9
	Primary bronchi	16	5	
	Lobar bronchi	12	5	
	Segmental bronchi		3	

Source: [a]Nikula et al., 1988; [b]Jeffery and Reid, 1975; [c]Plopper et al., 1989; [d]Pavelka et al., 1976; [e]Gilljam et al., 1987.

Mean ± SD, data not available if space is blank, based on Mariassy, 1992, N is the number of animals; cells classified as "other" are typically seromucus or non-ciliated epithelial cells.

cells, they comprise the mucociliary defense apparatus that is important in clearing various inhaled foreign agents (e.g., dusts, bacteria) that deposit on a moving layer of mucus (sometimes referred to as the "mucociliary escalator"). In human airways, these cells are columnar ≈20 μm long and 7 μm thick (Breeze et al., 1977). They are attached to the basal lamina and extend to the airway lumen. The primary function of the cells is to generate the flow of airway mucus with the synchronized beating of numerous motile cilia (200 - 300/cell) that line the apical surfaces of these dynamic cells. The mechanisms involved in propulsion of airway mucus have been reviewed

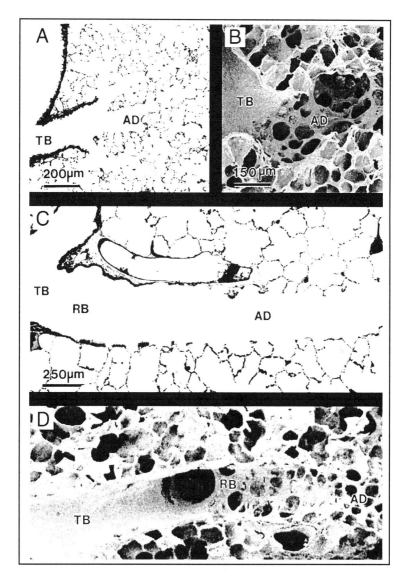

Figure 8. A light microscopic and scanning electron microscopic comparison of the centriacinar regions of species with minimal (A,B) and extensive (C,D) respiratory bronchioles. In most animals, such as the mouse (A) and rat (rat), the transition from terminal bronchiole (TB) to alveolar duct (AD) is abrupt. In contrast, in some species, such as the cat (C,D), the transition from terminal bronchiole (TB) to alveolar duct is extensive and includes a long respiratory bronchiole (RB). (From Harkema et al. 1991. "Epithelial Cells of the Conducting Airways." In *The Airway Epithelium: Physiology, Pathophysiology, and Pharmacology*, eds. Farmer and Hay, Marcel Dekker, Inc., New York, p. 29.).

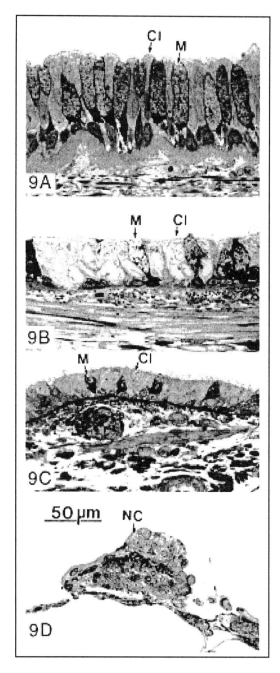

Figure 9. Light microscopic comparison of airway epithelium lining different airway generations in the rhesus monkey: A) trachea (airway generation 0); B) mainstem bronchus (airway generation 1); C) sixth generation, intrapulmonary bronchus; and D) respiratory bronchiole. CI = ciliated cell; M = mucous cell; .NC = nonciliated bronchiolar (Clara) cells.

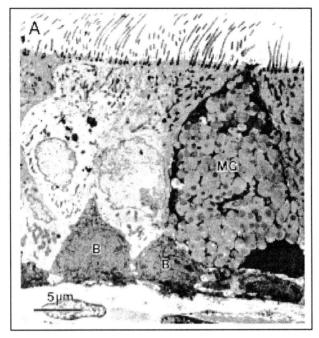

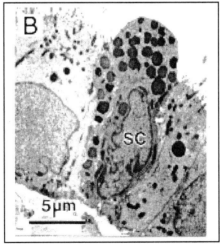

Figure 10. Transmission electron photomicrographs comparing the tracheal epithelium in the rhesus monkey (A) and the rat (B). The epithelium is shorter in the rat compared with the monkey, and the rodent epithelium contains fewer cells per length of basal lamina. More basal cells are present in the tracheal epithelium of the monkey, compared with the rat. Mucous goblet cells are the primary secretory cells in the monkey. Serous cells (SC) are the primary secretory cell in the rat. (From Harkema et al. 1991. "Epithelial Cells of the Conducting Airways." In *The Airway Epithelium: Physiology, Pathophysiology, and Pharmacology*, eds. Farmer and Hay, Marcel Dekker, Inc., New York, p. 16.).

in detail (Sleigh et al., 1988; Lee and Forrest, 1997). Cilia beat at a frequency of 12 – 15 beats/second depending on both the site and method of measurement. The beating cilia move the overlying mucus in a specific direction, depending on the location within the tracheobronchial airways. In the nasopharyngeal airways, mucus flows distally towards the orifice of the esophagus. In the tracheobronchial airways, the cilia propel the mucus proximally, but again towards the opening of the esophagus where it is swallowed into the gastrointestinal tract.

Cilia are unique cytoplasmic extensions of the cell and are covered by the same outer plasma membrane as the rest of the cell. Each cilium is ≈ 6 μm long and 0.3 μm thick, though in mice it has been shown that cilia get shorter the more distal the pulmonary airway. Ultrastructurally, the shaft of each cilium is composed of a set of characteristic microtubules with structures and arrangements that vary over its length. In the shaft of the cilium, the complex of longitudinal microtubules (axoneme) is most complete and comprises a central pair of singlet tubules surrounded by nine doublet tubules, with a complex of fine radial spokes binding them together. In the outer doublets, the microtubules have two extensions, or "arms," composed of a specific protein called dynein. Dynein contains ATPase and provides motility to the cilium. A basal body anchors the axoneme to the cytoplasm in the apical portion of the cell.

Long microvilli are interspersed among the cilia along the apical surface of the cell. The numerous microvilli and cilia provide the cell with added luminal surface area that may be important for ion exchange between the extracellular airway fluid and the intracellular cytoplasm. The cytoplasm of ciliated cells is more electron-lucent, compared with that in other surface epithelial cells, and the apical region of the cytoplasm contains numerous mitochondria. This cell also contains a well-developed supranuclear golgi apparatus.

Ciliated cells are generally considered to be terminally differentiated or "end-stage" cells meaning that they are unable to divide and proliferate. They are believed to originate from basal or secretory epithelial cells (Inayama et al., 1989; Johnson and Hubbs, 1990). Ciliated cells are particularly susceptible to injury from inhaled irritants. Which ciliated cells are injured and the degree of that injury depends on the airway level reached by the specific inhaled toxicant and sensitivity of the cells at that level.

Mucous (Goblet) Cells

Secretory epithelial cells are interspersed among the ciliated cells in the surface epithelium of conducting airways. The three principal secretory cells in the conducting airways of mammals include mucous (goblet) cells, serous cells and Clara cells. All of these cells contain unique spherical, membrane-bound secretory granules in the cytoplasm of the apical portion, or theca, of the cell. These secretory granules, along with those from submucosal gland cells, are the source of the fluid material that blankets the surfaces of the conducting airways.

Normal airway mucus is a hydrated gel that consists of about 1% salts and other small dialyzable components, 1 - 4% lipids, 0.5 - 1% free protein, 0.2 - 1% of carbohydrate-rich glycoproteins or "mucins," and 95% or more water. Mucins are derived from the secretory granules of mucous cells. These airway secretory cells often have expanded theca filled with a large number of secretory granules and narrow basal portions, or "stems", giving them the classical shape of a goblet (Figure 10). The height and shape of these cells, however, vary depending on the amount of intracellularly stored mucosubstances. The nucleus and other organelles in mucous cells are basally located. This region of the cell contains a rich network of granular endoplasmic reticulum, a well-developed Golgi apparatus, and varying numbers of mitochondria.

Ultrastructurally, the secretory granules in airway mucus cells vary in electron density and often coalesce near the apex of the cells (Figure 10). Histochemically, the stored mucosubstances within these granules stain with Alcian Blue and Periodic Acid Schiff (AB/PAS) indicating the presence of acid and neutral mucosubstances, respectively. The granules in human mucous goblet cells are ≈800 nm in diameter and generally contain acidic mucins, owing to the presence of sulfate or sialic acid. In the dog (Spicer et al., 1971) and the monkey (St.George et al., 1984), sulfomucins predominate in the mucous cell. In humans, either sulfomucins or sialomucins predominate, depending on the airway level at which the cell is found.

Production and secretion of the correct amount of airway mucus from mucous cells and other secretory cells within the surface epithelium and underlying sub-mucosal glands are important for maintaining the mucociliary apparatus, which traps potentially deleterious, inhaled particulates and gases, and clears these agents from the airways. Hypersecretion of mucins from airway mucous cells can be induced by chemical irritants (e.g., cigarette smoke, ozone, sulfur dioxide) and proteases from bacteria or human neutrophils (Adler et al., 1986; Christensen et al., 1989). Laboratory animal studies have indicated that neutrophils migrating through respiratory airway epithelium trigger hypersecretion of mucus from mucous cells (Harkema et al., 1988), though the specific mechanisms for this response is yet unknown. In addition, the number of mucous cells in conducting airways increases (i.e., mucous cell hyperplasia/metaplasia) in human airway diseases, such as chronic bronchitis and asthma (Reid, 1954; Dunnill et al., 1969; Aikawa et al., 1992). It also increases in the airways of laboratory animals following experimental inhala-tion exposure to irritating air pollutants (e.g., ozone, sulfur dioxide) (Lamb and Reid, 1968; Schlesinger et al., 1992; Harkema et al., 1998) or with an allergenic response to inhaled antigens (e.g., ovalbumin) (Blyth et al., 1998).

Clara Cells

Clara cells, the non-ciliated cell population in the epithelial lining of bronchioles, have been called the most heterogeneous and multifunctional cell types in the mammalian lung (Plopper and Hyde, 1992). These cells are cuboidal to columnar epithelial cells that extend from the basal lamina to the lumen and are free of cilia, of long apical microvilli, and of dense-cored neurosecretory granules (Figures 11 and 12). The proximal, or central, portion of the pulmonary acinus is the principal location of these airway epithelial cells. The density of differentiated Clara cells in the epithelial populations of terminal and respiratory bronchioles in adult mammals are species-specific (Table 3).

**TABLE 3. COMPARISON OF CELLULAR COMPOSITION OF
CENTRIACINAR BRONCHIOLAR EPITHELIUM IN
RATS, MONKEYS, AND HUMANS**

Species	Cell Types in Terminal Bronchiole				Reference
	Clara	Ciliated	Goblet	Basal	
Rat	>50%	<50%	-	-	Chang et al., 1988 Young et al., 1986
Macaque monkey	-	± 50%	± 20%	± 10%	Plopper et al., 1989 Tyler and Plopper, 1985
Human	-	+	+	+	Cutz and Conen, 1971 Jarkovska, 1970 Plopper et al., 1989

Note: - = not present; + = present in variable amounts.

There are also species differences in the ultrastructural characteristics of Clara cells (Plopper and Hyde, 1992). In humans and other primates, Clara cells are ultra-structurally low cuboidal cells with only slight apical projections and with little polarity to their organelles (Figure 11). These cells have a large central nucleus, little smooth endoplasmic reticulum (SER), more rough endoplasmic reticulum (RER) and small amounts of glycogen. A few electron-dense secretory granules are present in the apices of these cells. In contrast, the Clara cell of most mammalian species, including laboratory rodents (rats and mice) is a distinctly polarized epithelial cell with a basally located nucleus, apically located electron-dense secretory granules, a prominent apical projection into the airway lumen, and abundant amounts of SER (Figure 12). Greater than 40% of the cytoplasmic volume of the Clara cell of a rat or mouse is occupied by SER, while in the Clara cell of human or non-human primates the SER occupies <10% of the cell's cytoplasmic volume. The apical projections found in many animal species are the result of a specific organization of microfilaments (Sasaki et al., 1988). In all mammalian species, the Clara cell has lateral cytoplasmic interdigitations and complex junctional complexes that bind these cells together at the luminal aspect of the basolateral membranes. Some species, like the dog and cat, have large amounts of cytoplasmic glycogen and few, if any, secretory granules (Plopper and Hyde, 1992).

There are also species differences in the distribution of Clara cells throughout the respiratory tract. For example, the rabbit has Clara cells in tracheal, bronchial, and bronchiolar epithelium, whereas the monkey has Clara cells only in the epithelium lining the terminal and respiratory bronchioles (Plopper et al., 1989). Clara cells have been reported in the airway epithelium of all of the conducting airways of the mouse, including the nasal airways (Matulionis and Parks, 1973).

The Clara cell's primary functions in most mammalian species are: 1) to provide secretory material for the fluid lining the lumenal surface of bronchioles; 2) to metabolize xenobiotic compounds through the cytochrome P_{450} monooxygenase system; and, 3) to serve as the progenitor cell for ciliated cells and other Clara cells in the bronchiolar epithelium. Cytochrome P_{450} monooxygenase activity has been identified in Clara cells by immunocytochemical analysis using antibodies specific for cytochromes. Ultrastructural studies in rabbits have shown that P_{450} protein is

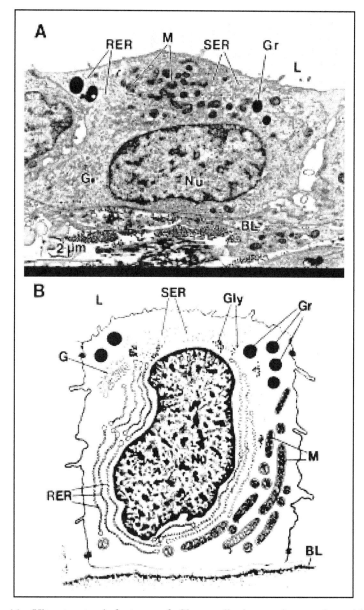

Figure 11. Ultrastructural features of Clara cells in respiratory bronchioles of primates are illustrated in a transmission electron photomicrograph (top figure) and an accompanying drawing (bottom figure). There is little smooth endoplasmic reticulum (SER) or rough endoplasmic reticulum (RER), or glycogen (Gly). Mitochondria (M) and Golgi apparatus (G) are als opresent and, in most primate species, secretory granules (Gr) are also evident in these epithelial cells. Nu= nucleus; BL = basal lamina. (From Harkema et al. 1991. "Epithelial Cells of the Conducting Airways." In *The Airway Epithelium: Physiology, Pathophysiology, and Pharmacology*, eds. Farmer and Hay, Marcel Dekker, Inc., New York, p. 11.).

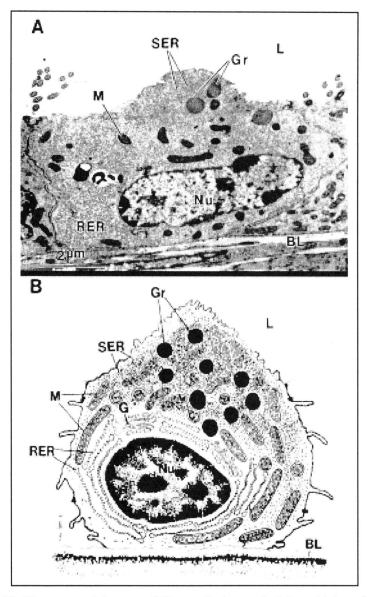

Figure 12. Ultrastructural features of Clara cells in terminal bronchioles of mice, rats, rabbits, guinea pigs, hamsters, pigs, and horses are illustrated in a transmission electron photomicrograph (top figure) and an accompanying drawing (bottom figure). Smooth endoplasmic reticulum (SER) is abundant (40% of the cytoplasmic volume), whereas rough endoplasmic reticulum (RER), mitochondria (M) and Golgi apparatus (G) are present in minimal amounts. Secretory granules (Gr) have been observed in all of these species. Nu = nucleus; BL = basal lamina. (From Harkema et al. 1991. "Epithelial Cells of the Conducting Airways." In *The Airway Epithelium: Physiology, Pathophysiology, and Pharmacology*, eds. Farmer and Hay, Marcel Dekker, Inc., New York, p. 9.).

found associated with the plasma membrane and SER-rich zones of the Clara cell (Serabjit-Singh et al., 1988). Spectrophotometric comparison of cytochrome P_{450} proteins in microsomal preparations from rabbit Clara cells, alveolar type II cells, and pulmonary alveolar macrophages demonstrated that Clara cells have approximately four times the amount of P_{450} protein as either other cell type (Devereux et al., 1989). By measuring cytochrome P_{450}-dependent metabolic activity in isolated Clara cells, using a number of specific substrates and, subsequently, identifying these substances within Clara cells (Plopper and Dungworth, 1987), investigators have been able to substantiate the metabolism of xenobiotic agents by these cells.

The secretory content of Clara cells has also been characterized using a variety of histochemical methods for glycoprotein and lipids, by autoradiographic characterization of precursor uptake, and by immunocytochemistry using antibodies developed against characterized proteins (Widdicombe and Pack, 1982; Plopper, 1983; Plopper and Dungworth, 1987). Clara cells do not contain acidic or sulfated glycoproteins, but some species have Clara cells that contain glycoproteins with vicinal hydroxyl groups. Ultrastructural studies of Clara cells from rats and rabbits demonstrate that periodic acid Schiff (PAS)-positively stained material is restricted to matrix immediately adjacent to the membranes.

A secretory protein that has homology with the uteroglobin produced by rabbit uterus has been identified in both rat and rabbit Clara cells (Singh et al., 1988). The function of this secretory protein is unknown, but it has some antiprotease activity suggesting airway protection from the injurious effects of inflammatory cell influx (e.g., protease releasing neutrophils). In addition, the human Clara cell appears to be the source of some pulmonary surfactant apoproteins (Plopper and Dungworth, 1987), and antileukoproteases found in the surface lining material in the peripheral airways (DeWater et al., 1986; Willems et al., 1989).

Clara cells also serve as progenitor cells for the maintenance of the bronchiolar epithelium. This role has been identified through laboratory animal studies (Plopper and Dungworth, 1987). In the steady-state condition, the turnover rate of epithelial populations in the bronchiolar region of the adult lung is very low, <1% of the epithelial cells are normally in the S-phase of the cell cycle and undergoing DNA synthesis in rats (0.2%), mice (0.3%), and monkeys (0.1%). However, in response to oxidant injury from NO_2 exposure, the numbers of Clara cells undergoing DNA synthesis may increase by as much as 10 - 20-fold in rats (Evans et al., 1976). During the repair process, some of the Clara cells differentiate into ciliated cells to restore the normal epithelium. *In vitro* studies have also shown that Clara cells isolated from rabbit lungs can repopulate the epithelium of denuded rat tracheas and, in so doing, express their ability to serve as progenitor cells (Brody et al., 1987).

Serous Cells

Serous cells are the primary secretory cell type in the surface respiratory epithelium lining the tracheobronchial airways of pathogen-free rats (Jeffery and Reid, 1975). Serous cells are also present in the subepithelial lateral and septal glands in the nasal airways of rats and mice (Klaassen et al., 1981), and in the submucosal glands in human airways (Meyrick and Reid, 1970). A small population of serous-like cells has been recently identified in the bronchioles and small bronchi of humans (Rogers et al., 1993). The serous cell is not found in the surface epithelium lining the conducting airways of other mammalian species (e.g., primates, mice, dogs, cats, rabbits) (Harkema et al., 1991) (Table 2) .

Discrete, electron-dense, membrane-bound granules of about 600 nm in diameter are a distinguishing ultrastructural features of the serous cell (Figure 10). These secretory granules are located in the apical cytoplasm and contain neutral mucosubstances. In contrast to Clara cells, these secretory cells have abundant amounts of RER, rather than SER, and a prominent Golgi apparatus. Besides the neutral secretory product, secretory granules in serous cells in the surface epithelium lining the rat trachea have been shown to contain calcitonin gene-related peptide (Baluk et al., 1993). Glandular serous cells in human airways also produce two important antibacterial products, lysozyme and lactoferrin (Bowes et al., 1981; Dohrman et al., 1994). These cells also participate in the transport of plasma-cell-derived IgA across the glandular epithelium (Brandtzaeg, 1974). Through immuno-cytochemical and *in situ* hybridization techniques, it also been observed that the cystic fibrosis transmembrane conductance receptor (CFTR) and the cystic fibrosis gene are present in the serous component of human bronchial submucosal glands (Engelhardt et al., 1992).

It has been noted in several studies that inhalation exposures to bacterial endo-toxin, a potent inducer of neutrophilic airway inflammatory, or to elastases derived from human inflammatory cells (i.e., neutrophils), cause a marked and rapidly occur-ring mucous cell metaplasia in the tracheobronchial airways of rats (Breuer et al., 1985; Harkema and Hotchkiss, 1992). The endotoxin-induced metaplastic response is characterized by a marked reduction in the numeric density of serous cells and a corresponding increase in mucous (goblet) cells within 48 hours after the end of the exposure (Harkema and Hotchkiss, 1992). This rapid change in secretory cells with the appearance of intermediate "seromucous" cells containing both serous- and mucus-like secretory granules and insignificant amounts of cell turnover (cell death and reparative proliferation) suggests that the new mucous cells were differentiated from pre-existing serous cells. A similar loss of serous cells with a corresponding increase in mucous cells is a common histopathologic feature in the submucosal glands of humans with chronic bronchitis and in human nasal explants *in vitro* (Ali et al., 1996). Interestingly, rats sensitized to ovalbumin and then exposed by inhalation to the same antigen develop a similar mucous cell metaplasia. The cellular mechanisms for this shift in secretory cell populations from serous cells to mucous cells have not been fully determined, but may be mediated in part by epidermal growth factor receptor and its ligand (Takeyama et al., 1999).

Basal Cells

The primary support cells in the upper conducting airways are the basal cells. These epithelial cells are very small, flattened cells that are closely attached to the basal lamina and do not extend to the airway lumen. It is the presence of these epithelial cells located in the larger conducting airways (i.e., nose, trachea, and bronchi) of all mammalian species that accounts for the pseudostratified classification given to the respiratory epithelium (Figures 7 and 10). Basal cells have a small cytoplasm/nucleus ratio, and a organelle-poor cytoplasm that is filled primarily with intermediate filaments. Numerous desmosomes attach the basal cell to the surrounding epithelial cells.

It has been suggested that the primary role of basal cells in the tracheobronchial airways is to attach columnar cells to the basal lamina. Evans et al. (1989), studying the cellular ultrastructure of airway epithelium in various species, have determined the percentage of the columnar cell attachment surface in direct contact with the basal lamina and the basal cells in the trachea of various mammalian species, each of which have epithelia of different height. They found that in tall epithelia most columnar cell attachments to the basal lamina occur indirectly through the basal cells. In shorter epithelium, there are very few basal cells, and most of the columnar cell attachment surface is directly in contact with the basal lamina.

Basal cells have long been considered to be the principal progenitor cell of the airway epithelium (Nettesheim et al., 1990). Several studies, however, have now demonstrated that basal cells are not the only progenitor cell type in the tracheobronchial airways, and that secretory cells are more likely the primary stem cells responsible for both development and renewal of the airway surface epithelium (Ayers and Jeffery, 1982; Johnson and Hubbs, 1990; Liu et al., 1994).

Neuroendocrine Cells

Airway epithelial cells with abundant round-to-ovoid neurosecretory-like granules in an electron-lucent cytoplasm are called neuroendocrine cells. The distinctive secretory granules are 70 and 150 nm with a dense core that is separated from the outer limiting membrane by a clear, electronlucent halo. The granules are argyrophilic and located in the basal portion of the cell (Figure 13).

Neuroendocrine cells are relatively rare in the airway epithelium of most species. These cells can occur individually or in small clusters of four to ten cells (i.e., neuroepithelial bodies; Figure 13). Some investigators have reported a close association of intraepithelial nerve axons with neuroepithelial bodies (Lauweryns and van Lommel, 1983). These cells increase in number from the main bronchi to the bronchioli, but are infrequently found in terminal bronchioles (Tateishi, 1973).

Neuroendocrine cells can be immunocytochemically located with the neuroendocrine marked enzyme, neuron-specific enolase (Wharton et al., 1981) and several other peptide hormones (e.g., calcitonin, bombesin, calcitonin gene-related peptide) (Tsutsumi et al., 1983; Wharton et al., 1978). There appears to be a marked heterogeneity of these cells both within a species and among species, as demonstrated by differences in the size and character of the dense-core granules within the cells (Sorokin et al., 1983).

Because neuroendocrine cells degranulate under hypoxic conditions and are sometimes associated with sensory nerves, it has been suggested that these cells have a chemoreceptor function. The morphologic and immunohistologic heterogeneity of these cells, however, suggest that they probably have multiple physiologic functions.

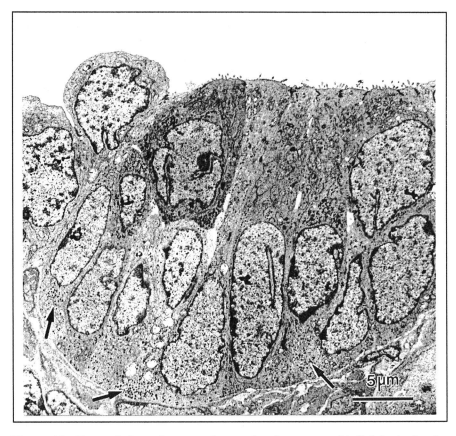

Figure 13. Transmission electron micrograph of a neuroepithelial body in the bronchial epithelium of a rabbit. Notice the small electron dense secretory granules located in the basal portion of the neuroendocrine cells (arrows).

Hyperplasia of pulmonary neuroendocrine cells has been reported in hamsters exposed to hyperoxia and the carcinogen diethylnitrosamine (Schuller et al., 1988; Nylen et al., 1990). Proliferation of neuroendocrine cells has also been observed in animals exposed to asbestos (Sheppard et al., 1982), naphthalene (Stevens et al., 1997), and sidestream cigarette smoke (Joad et al., 1995).

Brush Cells

Other relatively rare epithelial cells in the mammalian airways are brush cells. Their name comes from the border of closely packed microvilli that is present on their luminal surface. These microvilli are unusually long (2 μm) and have axial filaments that continue into the cell without ending in a terminal web. These cells have been identified in the conducting airways of a number of animal species. They are present not only in the epithelium of tracheobronchial airways (Meyrick and Reid, 1968; Jeffery and Reid, 1975; Chang et al., 1986), but also in the nasal respiratory epithelium of some species (Monteiro Riviere and Popp, 1984). The function of these epithelial cells is unknown. It has been speculated that they may play a role in any or all of the following: absorption of pericilliary fluid, chemoreception, and ciliogenesis.

Bronchus-Associated Lymphoid Tissues (BALT)

In some mammalian species well-organized aggregates or follicles of lymphoid tissue are found in the bronchial mucosa between the pulmonary artery and airway lumen, predominantly at airway bifurcations. Bienenstock and colleagues labeled these distinctive mucosal lymphoid tissues as bronchus-associated lymphoid tissue (BALT) (Bienenstock et al., 1973a and b). BALT morphologically resembles gut-associated lymphoid tissue (GALT) or Peyers Patches in the intestinal tract and the nasal-associated lymphoid tissue (NALT) in the nasopharyngeal airways (previously described above). BALT is prominent in rats and rabbits (Bienenstock et al., 1973b; Gregson et al., 1979), and less well-developed in mice and guinea pigs. Plesch reported that there can be as many as 50 BALT aggregates in adult specific-pathogen free rat lung (Plesch, 1982). Very little, if any, BALT is present in bronchial airways of healthy (respiratory disease-free) hamsters, dogs, cats, non-human primates, and humans (Brownstein et al., 1980; Pabst and Gehrke, 1990; Haley, 1993).

Figure 14 illustrates the histologic features of BALT. Like NALT and GALT, BALT is covered on its luminal side by a specialized non-ciliated lymphoepithelium (LE) that is composed of non-ciliated cuboidal epithelial cells with microvilli along with numerous infiltrating lymphocytes. The LE is devoid of ciliated or mucous cells. The absence of a mucociliary barrier and the phagocytic ability of the cuboidal epithelial cells, facilitates the transport of inhaled infectious agents, particulate antigens and pollutants across the LE to the underlying BALT.

The BALT is organized in a dome region directly beneath the LE, a follicular region beneath the dome, and a parafollicular region in the abluminal periphery of the lymphoid tissue. BALT follicles are composed predominantly of small lymphocytes and lesser numbers of larger lymphoblastic cells, reticular cells, dendritic cells, macrophages and occassionally basophil-like cells. In the parafollicular region there is an extensive network of blood vessels, most notably highly endothelial venules through which bloodborne lymphocytes enter the BALT. Lymphocytes are mainly B-lymphocytes, a large number of which express the IgA isotype (Bienenstock et al., 1973a), but distinct B-lymphocyte and T-lymphocyte regions have been described in the BALT of the rat (Plesch, 1982).

The rate and degree of BALT development in mammalian airways appears to be related to the degree of local antigenic stimulation. BALT has been shown to be poorly-developed in germ-free animals (Bienenstock et al., 1973b), while lymphoid hyperplasia occurs in the BALT of the laboratory rats and rabbits experimentally exposed to inhaled particulate antigens (Murray and Driscoll, 1992). Though the presence of BALT in humans has been questioned (Pabst and Gehrke, 1990), its presence in human airways appears to be environmentally- and antigenically-stimulated like that in animals. BALT has been reported in the airways of humans with a history of chronic pulmonary disease (Meuwissen and Hussain, 1982) and in

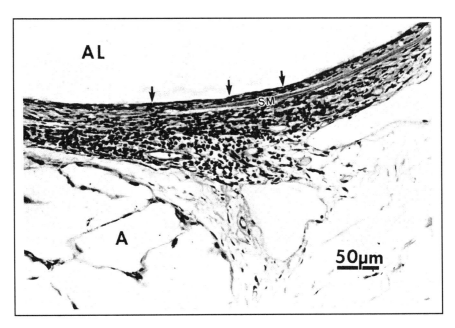

Figure 14. Light photomicrograph of bronchus-associated lymphoid tissue (BALT) in wall of the axial airway from the left lung lobe of a specific pathogen-free F-344 rat. A thin, non-ciliated lymphoepithelium (arrows) overlies the intramural lymphoid tissue. AL = airway lumen; SM = smooth muscle. A = adjacent alveolar air space.

a high percentage of smokers (Richmond et al., 1993). BALT is believed to develop at local sites in the tracheobronchial airways that are preferentially exposed to high concentrations of inhaled antigens or pollutants due to intraluminal airflow patterns, and rates of local deposition and mucociliary clearance of respirable antigenic agents (e.g., airway bifurcations).

GAS-EXCHANGE REGIONS OF THE LUNG

Respiratory Bronchioles

Bronchioles that have a few alveolar outpockets arising from their walls are defined as respiratory bronchioles (Tyler, 1983). These poorly-alveolarized bronchioles are the transitional airways between the conducting (nonrespiratory) bronchioles and the alveolar ducts that are completely covered by alveoli (Figure 8). They are present in some but not all mammalian species. Several generations of respiratory bronchioles are present in the lungs of humans, dogs, cats and monkeys. In contrast, the lungs of mice, rats, hamsters, guinea pigs and rabbits have no respiratory bronchioles (Tyler, 1983). In these latter species there is an abrupt transition from the nonrespiratory terminal bronchial to the alveolar duct (Figure 8). Interestingly, the lung of the horse, like the rat, is devoid of respiratory bronchioles (McLaughlin et al., 1961; Tyler and Julian, 1992).

The luminal surface of the non-alveolarized portion of respiratory bronchioles is lined by ciliated cells and Clara cells, but normally lacks mucous cells and basal cells. The cellular and acellular structures of the alveolarized out-pocketings of the respiratory bronchiole resemble those of the alveolar ducts and of the alveoli that are described below.

Ventilatory Units and the Pulmonary Acini of the Lung Parenchyma

In 1962, Storey and Staub proposed a basic unit of ventilation in the mammalian lung that consists of the alveoli and the alveolar ducts distal to the transition from one bronchiole to an alveolar duct system (Figure 15). Mercer and Crapo (1992) used serial section reconstructions of the lung to demonstrate that the alveoli and the alveolar ducts of these basic "ventilatory units" occur in a highly interconnected pattern that forms distinct subdivisions of the alveolar parenchyma which do not overlap. As described by these investigators, "The ventilatory unit is functionally important because it is the smallest common denominator in determining the distribution of inspired gas to the gas-exchange surfaces of the lungs. Thus the ventilatory unit has a central role when considering what effect variations in size and architecture have in determining the dose of inhaled pollutants delivered to the gas-exchange region." Though there are distinctive species-related differences in the branching patterns of the small airways and in the presence or absence of respiratory bronchioles, all mammalian species have the interconnected bronchiole to alveolar duct junctions that are the entrance ways for the distinct ventilatory units.

The "pulmonary acinus" consists of all the gas-exchange structures distal to a single terminal bronchiole and is composed of two or more ventilatory units (Mercer and Crapo, 1992) (Figure 15). The human lung contains 30,000 - 40,000 terminal bronchioles and therefore, by definition, an equal number of pulmonary acini. Each acinus in the human lung contains 10,000 - 12,000 alveoli. The average number of alveoli per acinus is relatively constant among mammalian species. The ventilatory unit size and alveolar size are in direct proportion to the size of the species (Mercer and Crapo, 1992). Therefore, the alveoli and alveolar duct structures in the human lung are scaled-up versions of those in smaller mammalian species. For example, human lungs have ≈50-fold greater numbers of acini, a 170-fold greater acinar volume, and a 7-fold greater acinar diameter compared to acini in the lungs of mice.

As mentioned above, the most conspicuous species-related difference in acinar structure is the arrangement of the bronchioles supplying the gas-exchange regions in the lung parenchyma (i.e., the presence or absence of respiratory bronchioles). In addition to this structural difference, the amount of smooth muscle that occupies the wall of the alveolar ducts varies considerably among mammalian species. In humans, guinea pigs, and baboons, smooth muscle fibers extend into the first few generations of alveolar ducts (Tyler and Julian, 1992). In contrast, smooth muscle does not extend past the bronchiole-alveolar duct junction in rats and mice.

Using serial section techniques on *in situ* perfusion-fixed lungs of rats, investigators have found that there is considerable variation in the volume of ventilatory units (Mercer and Crapo, 1989). There can be as much as a 15-fold size variation among ventilatory units distal to a single bronchus. Therefore, the ventilation to

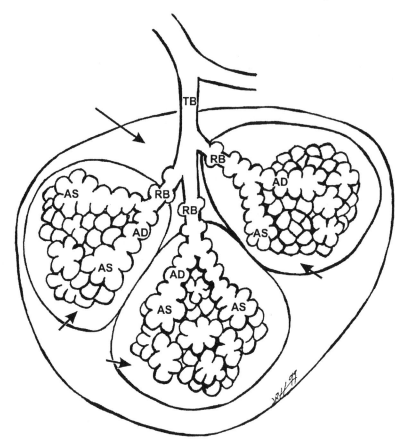

Figure 15. Diagrammatic representation of components of the pulmonary acinus. The terminal bronchiole (TB) branches into two or more respiratory bronchioles (RB) and ventilatory units dividing the alveolar parenchyma. Each ventilatory unit within the acinus consists of alveolar sacs (AS) and alveolar ducts (AD) distal to a single bronchiole-alveolar duct junction. (Modified from Mercer and Crapo. 1992. "Architecture of the Acinus". In *Comparative Biology of the Normal Lung*, ed. R.A. Parent, CRC Press, Inc., Boca Raton, FL, p. 111.).

each acinus and its ventilatory units throughout the mammalian lung, regardless of species, is not constant. This is particularly important in understanding the severity and distribution of pulmonary damage caused by inhalation of toxic agents (e.g., ozone, particulate matter) in laboratory animals and predicting the likelihood of similar responses in human lungs.

Alveolar Duct and Alveoli

The pulmonary parenchyma (gas-exchange region of the lung) composes 80 - 90% of the total lung volume in most mammalian species. The most prominent structures in the parenchyma are the alveolar ducts and alveoli. Alveolar ducts arise from the most distal (terminal) conducting airways (respiratory or nonrespiratory bronchioles, depending on the animal species). The walls of alveolar ducts are completely composed of a linearized arrangement of alveoli (Figure 8). The proximal alveolar duct branches into secondary alveolar ducts each of which end in a blind outpouching composed of two or more small clusters of alveoli called alveolar sacs.

The alveoli are the primary functional and structural units of the lung parenchyma where a very thin tissue barrier separates the surface of the airspace from the luminal surface of the capillaries. There are $\approx$500 million alveoli in the human lung with a total surface area of 100 m^2. The ratio of capillary surface area to the surface area of the alveolar air space in most mammalian species is slightly less than 1. The mean distance between the surface of the alveolar air space to the blood surface of the capillaries located in the interalveolar wall (septa) is only 0.4 μm in the Sprague Dawley rat and 0.6 μm in the human. This extremely thin air-blood barrier in mammalian lungs allows for efficient gas transfer between the inspired air and the circulating blood. Table 4 provides some additional morphometric data on the comparative anatomy of the lung parenchyma.

TABLE 4. COMPARATIVE ANATOMY OF THE LUNG PARENCHYMA AND AIR-BLOOD TISSUE BARRIER[a]

Species	N	Body Weight (kg)	Lung Volume	Alveolar Surface Area (both lungs) (cm^2)
[b]White mouse (*Mus musculus*)	5	0.023 ± 0.002	0.74 ± 0.07	680 ± 85
[c]White rat (F-344; Male (5-mo-old)	4	0.289 ± 0.013	8.60 ± 0.31	3915 ± 390
[d]Baboon (*Papio papio*)	5	29 ± 3	2393 ± 100	496,000 ± 77,000
[e]Man (*Homo sapiens*)	8	74 ± 4	4341 ± 285	1,430,000 ± 120,000

Species	N	Capillary Surface Area (both lungs) (cm^2)	T$_h$ Tissue (mm)
[b]White mouse	5	590 ± 60	0.032 ± 0.01
[c]White rat (F-344)	4	3830 ± 395	0.38 ± 0.03
[d]Baboon	5	386,000 ± 95,000	0.67 ± 0.06
[e]Man	8	1,260,000 ± 120,000	0.62 ± 0.04

[a]All values are mean ± SEM.
[b]Geelhaar and Weibel, 1971; [c]Pinkerton et al., 1982; [d]Crapo et al., 1980;
[e]Gehr et al., 1978.

The interconnected capillaries within the alveolar wall form a single vascular bed that is separated from the alveolar air space by a thin structural barrier of epithelial, interstitial, and endothelial tissues (Weibel and Taylor, 1998). The integrity of the alveolar septa is maintained by interstitial connective tissues (extracellular matrix) composed of collagen and elastin fibers that interweave around the

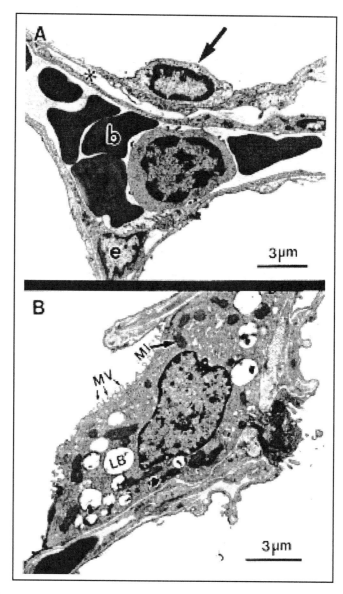

Figure 16. Transmission electron photomicrographs of alveolar type I (arrow) and II epithelial cells (A and B, respectively). Asterisk = single basement membrane between the epithelial and endothelial cells in thin portion of the alveolar wall (i.e., minimal barrier between alveolar air/capillary blood) b = red blood cell in alveolar capillary. E = endothelial cell nucleus. Alveolar type II cell in (B) contains luminal microvilli (MV), mitochondria (MI), and lamellar bodies (LB).

capillaries, forming what are called "thick" and "thin" regions within the wall. The thick regions of the septa contain the extracellular matrix of collagen and elastin fibers intermixed with interstitial cells (e.g., fibroblasts, myofibroblasts) all of which lie between the epithelial lining of the alveolar air space and the endothelial lining of the capillary lumen. The thin portions of the septa represent the air-blood barrier and are formed by epithelial and endothelial single cell layers that are separated only by a single basal membrane (Figure 16A).

Epithelial Cells of the Alveolus

Table 5 provides some cellular characteristics of the alveolar region of normal mammalian lungs. There are only two types of epithelial cells that line the alveolar lumen—alveolar type I and type II cells (also called type I and type II pneumocytes) (Figure 16A). The type I cells are large, flat (in most areas <0.2 μm thick) squamous cells with a central ovoid nucleus. Type I cells function primarily as a gas permeable membrane that lines ≈93 - 97% of the alveolar air space, depending on the species. These cells are terminally-differentiated epithelial cells and are derived from the other alveolar epithelial cells, the type II cells (described below).

TABLE 5. CHARACTERISTICS OF CELLS FROM ALVEOLAR REGION OF NORMAL MAMMALIAN LUNGS[a]

	F-344 Rat[b]	Baboon	Human[c]
Total no. cells per lung, x 10^9	$0.67 \pm 0.02_a$	$99 \pm 9_b$	230 ± 25
Total lung cells (%)			
Alveolar type I	$8.1 \pm 0.3_a$	$11.8 \pm 0.6_b$	$8.3 \pm 0.6_a$
Alveolar type II	$12.1 \pm 0.7_a$	7.7 ± 1.0	$15.9 \pm 0.8_b$
Endothelial	$51.1 \pm 1.7_a$	36.3 ± 2.4	30.2 ± 2.4
Interstitial	$24.4 \pm 0.7_a$	41.8 ± 2.7	36.1 ± 1.0
Macrophage	$4.3 \pm 1.0_a$	$2.3 \pm 0.7_a$	9.4 ± 2.2
Alveolar surface covered (%)			
Alveolar type I	$96.4 \pm 0.5_a$	$96.0 \pm 0.6_a$	92.9 ± 1.0
Alveolar type II	$3.6 \pm 0.3_a$	$4.0 \pm 0.6_a$	$7.1 \pm 1.0_a$
Average cell volume (mm^3)			
Alveolar type I	$1530 \pm 121_a$	$1224 \pm 136_a$	1764 ± 155
Alveolar type II	$455 \pm 108_a$	$539 \pm 184_a$	889 ± 101
Endothelial	$275 \pm 25_a$	$365 \pm 61_a$	632 ± 64
Interstitial	$427 \pm 55_a$	$227 \pm 30_b$	637 ± 26
Macrophage	$639 \pm 131_a$	$1059 \pm 287_a$	2492 ± 167
Average cell surface area (mm^2)			
Alveolar type I	$7287 \pm 755_b$	$4004 \pm 383_a$	$5098 \pm 659_a$
Alveolar type II	$185 \pm 56_{a,b}$	$285 \pm 85_b$	$183 \pm 14_{a,b}$
Endothelial	$1121 + 95_a$	$1040 + 209_a$	$1353 + 67_a$

[a]All data are mean ±SEM. For comparisons between species, letter subscripts indicate values that are not different from other values having same letter subscript.
[b]Data from Pinkerton KE et al., Am. J. Anat., 1982;164;155-174.
[c]Data from Crapo JD et al., Am. Rev. Respir. Dis., 1983;128:542-546.

The type I cell has relatively few cytoplasmic organelles (few mitochondria, some endoplasmic reticulum, microfilaments, and a Golgi apparatus) that are predominantly centered around the small nucleus. This alveolar epithelial cell is two to four times larger in volume and covers 20 - 40 times more surface area than the type II cell. The human alveolus, with a luminal surface area of 200,000 μm^2 , is covered on average by 32 type I cells and 51 type II cells (Crapo et al., 1983). In contrast, the rat alveolus has a surface area of $\approx 14,000$ μm^2 and is covered by two type II cells and a single type I cell (Pinkerton et al., 1992). Interestingly, this suggests that the size and shape of type I and II cells are not dependent on the size of the animal and are relatively similar among mammalian species.

In contrast, the alveolar type II cell (Figure 16B) is a small (370 μm^3 in the rat and 900 μm^3 in the human) cuboidal epithelial cell that covers only 3 - 4% of the alveolar lumen and is normally positioned at the corners of the alveolus away from the thin portions (air-blood barrier) of the interalveolar septa (Weibel and Taylor, 1998). The main function of these cells is to produce and secrete pulmonary surfactant, a complex mixture of lipids and proteins that lines the luminal surface of the alveolus, reducing the alveolar surface tension and preventing the collapse of the alveolus during respiration (Pattle, 1965). The primary lipids in surfactant are phosphatidylcholine and phosphatidylglycerol. Intracellular surfactant is packaged in unique membrane-bound lamellar bodies located in the apical portion of the cell. Besides containing the phospholipids, these lamellar bodies contain surfactant proteins A, B, and C (SP-A, SP-B, and SP-C), lysozyme, and lysosomal enzymes (Singh and Katyal, 1992). SP-A, -B, and -C are important for the spreading of surfactant on the alveolar surface. SP-D is thought to function in antibacterial defense. Various anti-inflammatory cytokines and antioxidants (e.g., superoxide dismutase) are also produced and secreted by these epithelial cells. Nitric oxide has been shown to be secreted through the activation of nitric oxide synthase.

Type II cells also proliferate in response to alveolar epithelial injury and then differentiate into type I cells to restore the normal air-blood barrier (Bowden, 1981). It is not known what regulates these progenitor cells to differentiate into type I cells, but one hypothesis is that the signal for transformation is regulated in part by the extracellular matrix (Sannes, 1984).

Type II cells also act as ionic pumps that move fluid from the alveolar lumen to the adjacent interstitial compartment of the interalveolar septa (Lubman et al., 1997). These cells move sodium from the lumen to the interstitium via a cyclic AMP-dependent sodium channel. Subsequently, the water passively follows the movement of sodium preventing excessive buildup of aqueous fluid in the airspace.

MOBILE CELLS IN THE LUNG

Alveolar Macrophages

Alveolar macrophages are mobile, mononuclear, phagocytic cells that reside in the lung parenchyma and play crucial roles in homeostasis, host defense, and the removal of inhaled particles from the alveolar lumen (Figure 17). Extensive reviews have been published on structure and function of pulmonary macrophages to which

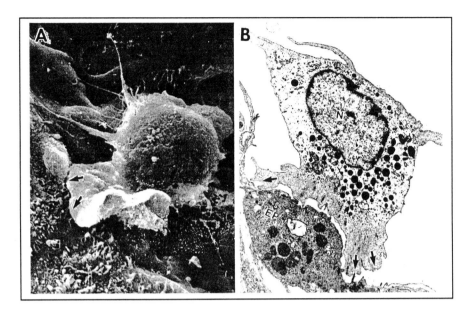

Figure 17. Scanning and transmission electron photomicrographs (A and B, respectively) of an alveolar macrophage lying on the alveolar epithelium of a rat lung. Cytoplasmic lamella (arrows) represent the leading edge of this mobile phagocytic cell. N = nucleus; EP = alveolar epithelium.

the reader is referred for more details (Bezdicek and Crystal, 1997; Lipscomb and Russell, 1997). These cells have a highly convoluted plasma membrane that has proteins specialized for attachment and ingestion of foreign particles. Because of their phagocytic and locomotive properties, these cells have numerous lysosomes and cytoskeletal fibers (e.g., microfilaments) throughout their cytoplasm. These professional phagocytes are submerged below the surfactant layer in the hypophase of the alveolar lining fluid and move across the alveolar epithelial surface by way of pseudopodia. They normally keep the alveolar surface clean and sterile by engulfing inhaled foreign material such as dust and bacteria that first land on the surfactant lining. Alveolar macrophages are the first line of cellular defense in the lung parenchyma. Macrophages may also phagocytize degenerating and apoptotic neutrophils to help resolve pulmonary inflammatory responses (Cox et al., 1995). After the phagocytic process, the engulfed material is packaged in a cytoplasmic vesicle that fuses with a lysosome to form a phagolysosome. The material is then degraded by enzymatic processes. Macrophages may also remove the phagocytized material from the alveolar air space by migrating up the bronchial tree along the moving mucous layer towards the oropharynx where they are swallowed. In addition, particle-laden macrophages may reenter the interstitial space often at the entrance of the acinus around conducting airways where they can enter lymphatic vessels and are transported to regional tracheobronchial lymph nodes (Harmsen et al., 1985).

The numbers of macrophages in the lung are quite variable and dependent on environmental and health conditions. For example, human smokers have more alveolar macrophages in their lungs than nonsmokers (Plowman, 1982). In addition, human lungs have significantly more macrophages per volume of lung tissue than laboratory rodents (Pinkerton et al., 1992). This may be due to the daily exposures of humans to various airborne pollutants to which rats and mice are not routinely exposed in their pristine, air-filtered laboratory environments.

Alveolar macrophage numbers increase after inhalation exposure to a particulate or gaseous air pollutant or an airborne infectious agent. Numbers can increase by division of pre-existing alveolar macrophages (Coggle and Tarling, 1984) or by an influx of peripheral monocytes from the bone marrow via the blood that then differentiate into these specialized phagocytic cells (Blusse van Oud Alblas and van Furth, 1979).

Besides the alveolar air spaces, pulmonary macrophages are also present in the conducting airways and are actually more numerous per unit of surface area in these airways than in alveoli (Brain et al., 1984). However, 99% of the total air space macrophages reside in the alveoli, because of the huge surface area of the alveolar parenchyma. In addition, interstitial macrophages are found in the connective tissue surrounding conducting airways and blood vessels and occasionally in interalveolar septa. Another population of lung macrophages is found in the pulmonary vasculature in some species including sheep, pigs, cats and humans, but not rodents (Warner and Brain, 1990). These intravascular macrophages protect the lung by ingesting potentially harmful bacteria, neutrophils or particles circulating in the pulmonary blood.

Alveolar macrophages are also able to secrete more than 100 substances (e.g., proteases, nitric oxide, lipid metabolites, cytokines, growth factors), many of which are important in mediating both acute and chronic inflammatory responses in the lung (Murray and Driscoll, 1992; Kunkel et al., 1997; Peters-Golden, 1997). On stimulation from bacterial agents (e.g . endotoxin) or inorganic particulates, the alveolar macrophage will secrete early-response cytokines, such as interleukin-1 (IL-1) and tumor necrosis factor (TNF), that act in either an autocrine or paracrine fashion to stimulate the production of more distal mediators of inflammation (e.g., chemokines) from various non-immune cells (e.g., interstitial fibroblasts, airway epithelial cells).

Compared to blood monocytes and macrophages from other tissues, alveolar macrophages function poorly as accessory cells for fostering immune B- or T-lymphocyte responses. The results of several *in vitro* and *in vivo* studies demonstrate that alveolar macrophages actually play an important role in downregulating the immune responses in the lung (Kraal et al., 1997).

Dendritic Cells

Though alveolar macrophages are not equipped to perform as antigen-presenting cells, professional antigen-presenting dendritic cells (DC) are abundantly present throughout the airways. These mononuclear cells have high expression of major histocompatibility complex (MHC) Class II and a high antigen-presenting capacity for immune T and naïve cells (Holt, 1995). DC are derived from bone-marrow and reside in lymphoid organs and a variety of other tissues, including the respiratory tract. They are distinguished from other mononuclear cells by their bean-shaped or somewhat lobulated nucleus, long cytoplasmic extensions, large amounts of MHC Class II (Ia) antigen expression on their plasma membranes, and an absence of acid phosphatase (Van Voorhis et al., 1982).

In the respiratory tract of humans (van Haarst et al., 1994) and rodents (Breel et al., 1988), DC are mainly located in the conducting airways with lesser numbers also found in the alveolar parenchyma and visceral pleura. DC are normally present in the epithelial and subepithelial regions of the tracheobronchial airways, as well as in the BALT. The tracheal mucosa of rats has been estimated, using immuno-histochemistry and whole mount techniques, to contain 1405 ± 140 DC/mm^2 (Brokaw et al., 1998). The highly branched cytoplasmic processes of DC are intimately associated with airway epithelial cells.

Airway populations of DC increase in response to inhaled antigen (McWilliam et al., 1996) and decrease after glucocorticoid treatments (Brokaw et al., 1998). The glucorcoticord-induced decrease in airway DC may be mediated through a decrease in emigration of these cells to the respiratory tract or through apoptosis. In contrast, antigen-stimulated increases in DC populations are thought to be due to local cell proliferation, increased emigration to the airway site from blood and bone marrow, and/or an enhanced residence time in the airway tissue. Antigen-induced upregulation of the DC network in the lung has been shown to be accompanied by an increase in DC-progenitor cell numbers in the bone marrow of sensitized rats (Lambrecht et al., 1999). In addition, the pulmonary vasculature of rats has been shown to be enriched with populations of DC precursors compared to the circulating blood in the vena cava (Suda et al., 1998).

Lymphocytes

Under disease-free conditions, most mobile cells that migrate to the pulmonary tissues from the blood and bone marrow are mononuclear cells (e.g., monocytes, lymphocytes). Only a few, if any, polymorphonuclear leukocytes (i.e., neutrophils and eosinophils) migrate into the normal lung tissue. In addition to the large numbers of lymphocytes comprising the tracheobronchial lymph nodes and BALT, individual or small aggregates of these immune cells can be found in the pulmonary airways and alveolar parenchyma. Most of the lymphocytes in the normal lung are T-lymphocytes with considerably fewer numbers of B-lymphocytes and NK cells (Richeldi et al., 1997). Lung T-lymphocytes express specific mucosal and memory phenotypes enabling them to respond to recall antigens with quick and strong proliferation.

Mast Cells

Mast cells are also present in the normal mammalian lung. These mononuclear cells, with distinctive metachromatic granules, can be found in or beneath the airway epithelium, around blood vessels and glands in the airway submucosa, associated with smooth muscle bundles, and in the interalveolar septa. It has been estimated that mast cells comprise 2% of the tissue in the human alveolar wall (Fox et al.,

1981). In monkey lungs, ≈83% of the mast cells are located around airways and associated blood vessels and the remainder in the parenchyma (Guerzon et al., 1979). These cells have important roles in allergic airway reactions. IgE binds to mast cell receptors via its F_c portion and initiates a series of membrane events that result in the granular discharge of various mediators associated with allergic airway reactions (e.g., histamine).

Neutrophils

A very large marginated pool of polymorphonuclear neutrophils resides in the pulmonary capillary bed (Hogg, 1987). Under normal conditions, most of these leukocytes remain in the microvessels of the lung, with only a few entering into the pulmonary interstitium or alveolar airspace. Following toxicant-induced lung injury or pulmonary infections of bacterial or fungal agents, neutrophils quickly adhere to the endothelial cells and migrate out of the vascular space into the interstitial or air spaces to kill the invading microorganisms or help to repair damaged tissue. The principal function of neutrophils in the lung is to phagocytose foreign material. Phagocytosis stimulates increased oxygen consumption in neutrophils which results in the generation of oxygen-derived free radicals that are important in microbial killing, but may also secondarily injure surrounding lung tissue. Neutrophils also have secretory granules containing numerous enzymes and other proteins designed to kill the engulfed organisms (Holland and Gallin, 1997).

Eosinophils

Like neutrophils, eosinophils are infrequently observed in normal lung tissue. These bone marrow-derived polymorphonuclear leukocytes, however, invade the respiratory tract in response to various parasitic infections and allergic reactions. Production and tissue recruitment of eosinophils is primarily controlled by T-helper lymphocytes and enhanced by various cytokines (e.g., IL-5). Eosinophilic infiltration is also a principal cellular feature of the airway inflammation associated with bronchial asthma (Barnes, 1989). Though eosinophils are capable of phagocytosis, their primary function is to produce basic granule proteins (e.g., major basic protein) and oxidant species used to damage invading helminths (Allen and Davis, 1997).

PULMONARY INTERSTITIUM

The pulmonary interstitium is the connective tissue space interspersed among the airway epithelium, the vascular endothelium, and the outer visceral pleura of the lung. It has been described as a well-designed "mechanical scaffold" that binds the functional components of the lung from the hilus to the visceral pleura via a "continuous system of fibers" (Weibel and Bachofen, 1997; Weibel and Crystal, 1997; Weibel and Taylor, 1998). It is composed of an integrated system of fibers (i.e., collagen and elastic fibers), specialized mesenchymal cell types (e.g., fibroblasts, myofibroblasts, and pericytes) and two basement membranes to which epithelial and endothelial cells are attached. The interstitial space in the lung parenchyma also serves as a conduit for excess fluid to drain from the alveoli to the juxtaalveolar regions and subsequently out of the lungs via the lymphatics. The interstitial compartment of the lung parenchyma is substantially greater in human lungs (56% of alveolar tissue) than in other mammalian species, including laboratory rodents (29 - 47% of alveolar tissue). Though the total number of interstitial cells in the parenchyma are several fold higher in the human lung compared to the rodent lung, the proportion of specific interstitial cell types are very similar (Pinkerton et al., 1992).

The mechanical integrity of the lung is based on three interconnected, regional fiber sytems; axial, peripheral and septal (Weibel and Gil, 1977). The axial fiber system is composed of a strong sheath of fibers that enwrap the conducting airways from the bronchi entering the hilus to the distal terminal and respiratory bronchioles and alveolar ducts. The peripheral fiber system is made up of interstitial fibers in the visceral pleura and extensions of these fibers into the interlobular connective tissue septa that subdivide lobes into various parenchymal units. The axial and peripheral fiber systems are linked by the septal fiber network consisting of a delicate, but strong, network of fibers coarsing through the interalveolar septa, as noted above.

The fiber systems are comprised of collagen and elastic fibers (Weibel and Crystal, 1997). Bundles of collagen fibrils bound by proteoglycans form the collagen fibers. These fibers have high tensile strength, but low extensibility giving them a twine-like consistency. In contrast, elastic fibers are composed of a amorphorous elastin core surrounded and pervaded by microfibrils. Elastic fibers have high extensibility and low tensile strength giving it a rubber-like consistency. Collagen and elastic fibers occur in the lung parenchyma and visceral pleura in volume ratios of 2.5:1 and 10:1, respectively. Besides the collagen and elastic fibers, the extracellular matrix of the interstitium also contains various amounts of fibronectin and proteoglycans and basement membranes of endothelium and epithelium.

In the normal lung, the major mesenchymal cells in the alveolar wall are fibroblasts, myofibroblasts, and pericytes (Weibel and Crystal, 1997). The fibroblasts are closely associated with the interstitial fibers and are engaged in the synthesis and secretion of collagen, proteoglycans and fibronectin. Myofibroblasts have contractile properties and are thought to be regulators of alveolar capillary blood flow and possibly the flow of interstitial fluids within the septa. Elastin is produced by myofibroblasts. Pericytes are closely associated with alveolar capillaries and also have contractile properties. Smooth muscle cells are not present in the interalveolar septa, but are present at the free edges of the alveolar septa of some mammalian species, including humans.

PULMONARY VASCULATURE

The pulmonary blood vessels are mechanically supported by the lung's interstitial scaffold (Weibel and Taylor, 1998). The pulmonary artery branches along side the conducting airways and within the axial fiber system. The main pulmonary artery and the next several generations of pulmonary arteries with diameters >0.5 cm are called elastic arteries. The walls of these arteries contain multiple concentric layers of elastic lamina, as well as a layer of smooth muscle and a layer of collagen. Smaller arteries (0.1 - 0.5 cm in diameter), called muscular pulmonary arteries, follow the small bronchi and the nonrespiratory and respiratory bronchioles in the human lung and contain circular smooth muscle located between the internal and external elastic lamina. These pulmonary arteries enter the acini and divide into the capillary network spreading out as a continuous sheet of vessels within the inter-connected walls of the alveoli, as previously described above.

In contrast to the pulmonary arteries, the pulmonary veins are associated with the peripheral fiber system that is located between pulmonary acini. Therefore, the oxygen-deficient arterial blood enters the acinus along side of the small airways, then moves outward through the capillary sheet in the interalveolar walls to the peri-phery where pulmonary veins collect the well-oxygenated blood from multiple acini.

A separate blood supply arising from the systemic circulation feeds the walls of the bronchial airways and the pulmonary arteries. These blood vessels originate from the aorta or the internal mammary, subclavian, or intercostal arteries and branch into smaller arterial vessels traveling in the walls of bronchi and bronchioles. Bronchial veins are restricted to the more central bronchi and empty into the azygos and hemizygos veins. The remainder of the arterial bronchial circulation empties into pulmonary veins and moves to the left atrium of the heart.

Comparative morphologic features of the pulmonary vasculature in mammals have been reviewed (Kay, 1983). There are interspecies differences in the quantity and arrangement of collagen, smooth muscle, and elastic tissue in the pulmonary vessels. Though primates have similar muscular pulmonary arteries, smooth muscle is found in much smaller pulmonary arterial vessels in most other species. Like in humans, the pulmonary veins have thin fibrous walls in the cat, dog, monkey and rabbit. However, the pulmonary veins are muscular in rats, mice and guinea pigs. In humans and most animals, the intrapulmonary veins are devoid of cardiac muscle. In contrast, the intrapulmonary veins of the rat and mouse have a thick medial layer of striated cardiac muscle that is continuous with the left atrium (Kay, 1983).

Endothelial cells are flat, thin cells with an ovoid nucleus that line the lumen of the pulmonary vessels. In contrast to the endothelial cells lining the arteries and veins, the alveolar endothelial cells have few organelles. They contain some mitochondria, free ribosomes, rough endoplasmic reticulum, microtubules and microfilaments and a few Weibel Palade bodies.

Besides their involvement in gas exchange, the pulmonary capillary endothelium has a number of metabolic functions (Silverman et al., 1997). One of these functions is the conversion of angiotension I to angiotension II (a potent vaso-constrictive molecule) via angiotension-converting enzyme (ACE) that takes place on the surface of the cell. In addition, ACE inactivates circulating bradykinin, a potent vasodilator. Endothelial cells in the lung also metabolize other mediators, such as serotonin and norepinephrine, as well as adenonucleotides and various prostaglandins. These cells produce prostacyclin, thrombomodulin, and tissue plasminogen activator. Nitric oxide, a vasorelaxant, and enodothelins, vasoconstric-tor peptides, are also secreted by endothelial cells.

LYMPHATIC VESSELS IN THE LUNG

The pulmonary lymphatic vessels provide a one way drainage system of excess fluid and proteins (lymph) from interstitial areas in the lung back to the blood stream (Leak and Jamuar, 1983; Leak and Ferrans, 1997). This prevents the fluids and plasma proteins that are leaked from the alveolar blood capillaries to build up in the interstitial space in the interalveolar walls and subsequently leak into the alveolar airspaces. There is a set of lung lymphatics that accompanies the conducting airways and pulmonary vessels deep into the lungs along the axial fiber system. Another set of lymphatics lies in the interstitium of the visceral pleura covering the outer surface of the lung lobes. These two sets of lymphatics meet at the hilus of the lung lobe where they enter the tracheobronchial lymph nodes. In contrast to the pulmonary vessels, the lymphatic vessels extend only to the terminal and respiratory bronchioles where they end as blind saccules (Lauweryns, 1970). In humans and common laboratory mammals, lymphatics are not present in the interalveolar septa. The plexus of lymphatic capillaries circumscribing the distal airways serves as a sump for fluids arising from the alveolar septal interstitium. Lymphatic capillaries are extremely thin lacking both a smooth muscle layer (tunica media) and a continuous basal lamina. The intima of these small vessels are composed of overlapping endothelial cells and are attached to the surrounding interstitium by anchoring filaments. Lymphatic capillaries are also present in the peripheral interlobular interstitium that connects to the outer pleura of the lungs. Lymphatic capillaries drain into larger collecting lymphatic vessels that have a thicker wall composed of intimal, medial, and adventitial layers and intraluminal valves that prevent backflow. The medial layer of these larger vessels is composed of contracting smooth muscle that facilitates the movement of lymph centripetally from the deep parenchymal tissue and outer pleura to the hilus of the lung lobe. As mentioned previously, alveolar macrophages containing phagocytized material (e.g., inhaled antigens, dust particles) can enter the lymphatics and be transported in the lymph to the regional lymph nodes.

NEURAL SUPPLY TO THE LUNG

Excellent reviews of comparative innervation of the lungs of mammals have been previously published, to which the reader is referred for more details not covered in this Chapter (Richardson, 1983; Zorychta and Richardson, 1991; Barnes, 1997). The mammalian lungs are innervated by both motor efferent neurons (parasympathetic and sympathetic) and afferent sensory nerves. Motor nerves regulate, in part, the flow of air in the conducting airways and the flow of blood in the pulmonary vessels via the stimulation or inhibition of smooth muscle tone. In addition, efferent innervation modulates the luminal secretion of mucus from submucosal glands in the walls of the tracheobronchial airways. Stimulation of sensory nerves modulate various airway reflexes (e.g., the cough reflex, the Hering-Breur reflex) and airway responses to irritating particles and gases in the inhaled air. There is disparity, however, in the innervation of the human airways with large-diameter central, bronchial airways being highly innervated and small-diameter peripheral airways having little or no innervation (Doidge and Satchell, 1982). The innervation of the airways and neural control of the pulmonary airways have been well studied and a brief review is provided below. The reader is referred to the other previously mentioned publications for a review of the innervation and neural control of the pulmonary vasculature.

The prominent neuropathways in the human lung are cholinergic excitatory fibers and nonadrenergic, noncholinergic (NANC) inhibitory fibers (Richardson and Beland, 1976; Davis et al., 1982). Though terminals of sympathetic (adrenergic) nerve fibers have been morphologically identified near the smooth muscle and submucosal glands in human bronchial airways, their physiologic function has not been determined. There is little direct effect of sympathetic nervous system on bronchial airway caliber. The airways of non-human primates are similarly innervated with a cholinergic excitatory pathway and a NANC inhibitory pathway. In contrast, dogs have cholinergic excitatory and adrenergic inhibitory fibers in all airways, both large and small. Guinea pigs have a cholinergic excitatory pathway and two inhibitory pathways, adrenergic and nonadrenergic.

The primary motor neurons in the pulmonary airways are the parasympathetic (cholinergic) motor fibers that originate in the dorsal motor vagal nuclei (nucleus ambiguus) in the brain stem (Bennett et al., 1981). These efferent nerve fibers course through the vagus nerve (cranial nerve X) to synapse in ganglia within the walls of the bronchial airways. In contrast, the sympathetic (adrenergic) motor neurons which store and release norepinephrine in the bronchial airways originate from the thoracic chain ganglia. Stimulation of the vagus nerve causes the release of acetylcholine (Ach) from parasympathetic nerve terminals resulting in broncho-constriction and increased secretion of airway mucus. There is a certain degree of resting bronchomotor tone caused by tonic parasympathetic activity that can be abolished by atropine and increased by an inhibitor of acetylcholinesterase (which normally inactivates Ach release from nerve terminals). Interestingly, the vascular intestinal peptide (VIP), the primary neural inhibitor of bronchoconstriction, is also stored and released from these nerve fibers, as well as sensory nerve endings, and may moderate the effects of acetylcoline (Barnes, 1997). This is an example of the complex interactions of the various neurotransmittors that can occur in the lung.

The nonadrenergic, noncholinergic (NANC) neurons in the pulmonary airways also travel via the vagus and have both excitatory and inhibitory influences over the tone of airway smooth muscle. The terminal endings of these fibers release neuro-peptides such as substance P and VIP, which contract and relax the airway smooth muscle cells, respectively. Substance P-releasing neurons also participate in local reflexes where there is irritation of sensory nerve endings in the airway epithelium and airway smooth muscle. VIP is a potent bronchodilator and the NANC system appears to be the most potent inhibitor of smooth muscle tone in the human airways.

Sensory afferent nerve fibers are present in the airway surface epithelium, submucosal glands and smooth muscle. Stimulation of these sensory fibers by inhaled irritants can activate cholinergic reflexes that result in bronchoconstriction, increased blood flow, and increased secretion of airway mucus through the activation of muscarinic receptors on the airway smooth muscle cells and submucosal glands. Cholinergic reflex-mediated bronchoconstriction is an important airway defensive mechanism that prevents the irritating agent from penetrating further down the airways and increases the removal of the agent by way of increased mucociliary clearance.

VISCERAL PLEURA

The outside surface of the mammalian lung is covered by a visceral pleura that is efficiently designed to prevent leakage of air into the thoracic cavity. Its roles in normal function and its responses in disease have been extensively reviewed (Chretien et al., 1985). The visceral pleura is composed of an outer, single cell layer of mesothelium lying on a thin, dense layer of fibrous tissue, and an underlying thick layer of connective tissue, rich in elastic fibers, that joins the peripheral interlobular fiber system of the lung. The low cuboidal to squamoid mesothelial cells have complex intercellular junctions and a well-developed apical brush border of microvilli (Mariassy and Wheeldon, 1983). The visceral pleura of humans and other large mammals (e.g., humans) are thick, while those of smaller mammals (e.g., dogs, cats, rats, and mice) are thin, reflecting principally the amount of connective tissue (Tyler and Julian, 1992). Animals with thick pulmonary pleura usually have extensive lymphatic networks (Leak and Jamuar, 1983). Laboratory rodents (e.g., mouse, rat) and rabbits have a thin pleura containing few lymphatic vessels. In addition, the blood to the thick pleura of human lungs is supplied by the bronchial artery, while the blood supply to thin pleura found in most laboratory animals is supplied by the pulmonary artery (Tyler and Julian, 1992).

LUNG LOBES

Besides obvious differences in the size of the lung among mammalian species, there are gross species differences in lobular structure of the lungs (Tyler, 1983). In humans, the left lung is divided into two lobes, the superior and inferior, and the right lung is divided into three lobes. Since other mammalian species are quadrapeds, the left lung lobes are referred to as cranial and caudal, rather than superior and inferior. In most domestic animal species (e.g., dogs, cats) the cranial lobe of the left lung is further divided into a cranial and a caudal part. The right lungs in these animals are divided into four lobes (cranial, middle, caudal, and accessory). The right lungs of laboratory rodents and non-human primates are also divided into four lobes. The left lungs of laboratory rodents (e.g., rats, mice, hamsters), however, are not divided. In larger laboratory mammals (e.g., rabbits, guinea pigs), the left lungs are divided like those of domestic animals.

Another conspicuous difference in lung anatomy among mammals is the amount of interlobular connective tissue in the lung lobes. Human lungs have extensive amounts, while laboratory mammalian species, including non-human primates, have little, if any (Tyler and Julian, 1992).

SUMMARY

In this Chapter, we have provided a comparative overview of the structure of the mammalian respiratory tract at the organ (gross), tissue (subgross) and cellular (microscopic) levels. Though complex in nature, the mammalian respiratory system is elegantly designed to efficiently provide its major function, gas exchange. The upper respiratory tract serves as a thorough conditioning chamber for cleaning, warming and humidifying the inhaled air before it reaches the delicate gas-exchange regions in the distal lung parenchyma. In the deep lung, the cellular architecture of the alveolar septa with its large surface area and ultra-thin air-blood barrier allows for rapid exchange of oxygen and carbon dioxide.

We have illustrated the diversity in structure that exists not only among the different regions of the respiratory tract, but also among mammalian species. Though there are many similarities in respiratory structure and function between laboratory animals and humans, there are also important species-related differences that exist at all anatomic and physiologic levels. For the toxicologist and immunologist who use animal models in human health-related respiratory research, it is critical to recognize these species differences before designing experiments or interpreting results.

In humans and animals, there are cellular defense mechanisms (including those involving mobile phagocytic and immune cells) that protect the airways from inhaled infectious or toxic agents and thus help to maintain airway homeostasis and normal respiratory function. The epithelial surfaces of the airways and alveoli are the initial cellular sites for inhaled foreign agents to interact with the host. The absence or presence of adverse effects from inhaled toxic, infectious or antigenic agents is primarily determined by the airway epithelial and immune/inflammatory responses.

REFERENCES

Adler KB, Hendley DD, Davis GS. Bacteria associated with obstructive pulmonary disease elaborate extracellular products that stimulate mucin secretion by explants of guinea pig airways. Am. J. Pathol., 1986;125:501-514.

Aikawa T, Shimura S, Sasaki H, Ebina M, Takishima T. Marked goblet cell hyperplasia with mucus accumulation in the airwaya of patients who died of severe acute asthma attack. Chest, 1992;101:916-921.

Alarie Y. Bioassay for evaluating the potency of airborne sensory irritants and predicting acceptable levels of exposure in man. Food Cosmet. Toxicol., 1981;19:623-626.

Ali M, Maniscalco J, Baraniuk JN. Spontaneous release of submucosal gland serous and mucous cell macromolecules from human nasal explants *in vitro*. Am. J. Physiol., 1996;270:L595-L600.

Allen JN, Davis WB. "Eosinophils." In *The Lung: Scientific Foundations*, RG Crystal, JB West, ER Weibel, PJ Barnes, eds. Philadelphia: Lippincott-Raven Publishers, 1997, pp. 905-915.

Ayers M, Jeffery PK. "Cell Division and Differentiation in the Respiratory Tract." In *Cell Biology and the Lung*, G Cumming, G Bonsignore, eds. New York: Plenum Press, 1982, pp. 33-60.

Baluk P, Nadel JA, McDonald DM. Calcitonin gene-related peptide in secretory granules of serous cells in the rat tracheal epithelium. Am. J. Respir. Cell Mol. Biol., 1993;8:446-453.

Barnes PJ. Drug-therapy - A new approach to the treatment of asthma. N. Engl. J. Med., 1989;321:1517-1527.

Barnes PJ. "Neural Control of Airway Smooth Muscle." In The Lung: Scientific Foundations, RG Crystal, JB West, ER Weibel, PJ Barnes, eds. Philadelphia: Lippincott-Raven Publishers, 1997, pp. 1269-1285.

Bennett JA, Kidd C, Latif AB, McWilliam PN. A horseradish peroxidase study of vagal motoneurons with axons in cardiac and pulmonary branches of the cat and dog. Quart. J. Exp. Physiol., 1981;66:145-154.

Bernard C. (ed.) An Introduction to the Study of Experimental Medicine. New York: Dover Publications, Inc., 1957.

Bezdicek P, Crystal RG. "Pulmonary Macrophages." In The Lung: Scientific Foundations, RG Crystal, JB West, ER Weibel, PJ Barnes, eds. Philadelphia: Lippincott-Raven Publishers, 1997, pp. 859-875.

Bienenstock J, Johnston N, Perey DY. Bronchial lymphoid tissue. Part I. Morphologic characteristics. Lab. Invest., 1973b;28:686-692.

Bienenstock J, Johnston N, Perey DY. Bronchial lymphoid tissue. Part II. Functional characterisitics. Lab. Invest., 1973a;28:693-698.

Blusse van Oud Alblas A, van Furth R. Origin, kinetics, and characteristics of pulmonary macrophages in the normal steady-state. J. Exp. Med., 1979;149:1504-1518.

Blyth DI, Pedrick MS, Savage TJ, Bright H, Beesley JE, Sanjar S. Induction, duration, and resolution of airway goblet cell hyperplasia in a murine model of atopic asthma: Effect of concurrent infection with respiratory syncytial virus and response to dexamethasone. Am. J. Respir. Cell Mol. Biol. 1998;19:38-54.

Bogdanffy MS. Biotransformation enzymes in the rodent nasal mucosa: The value of a histochemical approach. Environ. Health Perspect., 1990;85:177-186.

Bowden DH. Alveolar response to injury. Thorax, 1981;36:801-804.

Bowes D, Clark AE, Corrin B. Ultrastructural localization of lactoferrin and glycoprotein in human bronchial glands. Thorax, 1981;36:108-115.

Brain JD. The uptake of inhaled gases by the nose. Ann. Otol. Rhinol. Laryngol., 1970;79:529-539.

Brain JD, Gehr P, Kavet RI. Airway macrophages. The importance of the fixation method. Am Rev Respir Dis 1984;129:823-826.

Brandtzaeg P. Mucosal and glandular distribution of immunoglobulin components. Immuno-histochemistry with a cold ethanol-fixation technique. Immunology, 1974;26:1101-1114.

Brandtzaeg P. "Immune Function of Human Nasal Mucosa and Tonsils in Health and Disease." In Immunology of the Lung and Upper Respiratory Tract, J Bienenstock, ed. New York: McGraw-Hill, 1984, pp. 28-95.

Breel M, Vanderende M, Sminia T, Kraal G. Subpopulations of lymphoid and non-lymphoid cells in bronchus-associated lymphoid-tissue (BALT) of the mouse. Immunology, 1988;63:657-662.

Breeze RG, Aalberse RC, Wheeldon EB. The cells of the pulmonary airways. Am. Rev. Respir. Dis., 1977;116:705-777.

Breuer R, Christensen TG, Lucey EC, Stone PJ, Snider GL. Quantitative study of secretory cell metaplasia induced by human neutrophil elastase in the large bronchi of hamsters. J. Lab. Clin. Med., 1985;105:635-640.

Brittebo EB, Eriksson C, Feil V, Bakke J, Brandt I. Toxicity of 2,6-dichlorothiobenzamide (chlorthiamid) and 2,6-dichlorobenzamide in the olfactory nasal mucosa of mice. Fundam. Appl. Toxicol., 1991;17:92-102.

Brody AR, Hook GE, Cameron GS, Jetten AM, Butterick CJ, Nettesheim P. The differentiation capacity of Clara cells isolated from the lungs of rabbits. Lab. Invest., 1987;57:219-229.

Brokaw JJ, White GW, Baluk P, Anderson GP, Umemoto EY, McDonald DM. Glucocorticoid-induced apoptosis of dendritic cells in the rat tracheal mucosa. Am. J. Respir. Cell Mol. Biol., 1998;19:598-605.

Brownstein DG, Rebar AH, Bice DE, Muggenburg BA, Hill JO. Immunology of the lower respiratory tract. Serial morphologic changes in the lungs and tracheobronchial lymph nodes of dogs after intrapulmonary immunization with sheep erythrocytes. Am. J. Pathol., 1980;98:499-514.

Buckley LA, Jiang XZ, James RA, Morgan KT, Barrow CS. Respiratory tract lesions induced by sensory irritants at the RD_{50} concentration. Toxicol. Appl. Pharmacol., 1984;74:417-429.

Cauna N. "Blood and Nerve Supply of the Nasal Lining." In *The Nose: Upper Airway Physiology and the Atmospheric Environment*, DF Proctor, I Anderson, eds. Amsterdam: Elsevier Biomedical Press, 1982, pp. 45-69.

Chang LY, Mercer R, Crapo RD. Differential distribution of brush cells in rat lung. Anat. Rec., 1986;216:49-54.

Chang LY, Mercer RR, Stockstill BL, Miller FJ, Graham JA, Ospital JJ, Crapo JD. Effects of low levels of NO_2 on terminal bronchiolar cells and its relative toxicity compared to O_3. Toxicol. Appl. Pharmacol., 1988;96:451-464.

Chretien J, Bignon J, Hirsch A. (eds.) *The Pleura in Health and Disease*. New York: Marcel Dekker, 1985.

Christensen TG, Breuer R, Lucey EC, Stone PJ, Snider GL. Regional difference in airway epithelial response to neutrophil elastase: Tracheal secretory cells discharge and recover in hamsters that develop bronchial secretory-cell metaplasia. Exp. Lung Res., 1989;15:943-959.

Coggle JE, Tarling JD. The proliferation kinetics of pulmonary alveolar macrophages. J. Leukocyte Biol., 1984;35:317-327.

Cox G, Crossley J, Xing Z. Macrophage engulfment of apoptotic neutrophils contributes to the resolution of acute pulmonary inflammation *in vivo*. Am. J. Respir. Cell Mol. Biol., 1995;12:232-237.

Crapo JD, Barry BE, Foscue HA, Shelburne J. Structural and biochemical changes in rat lungs occurring during exposures to lethal and adaptive doses of oxygen. Am. Rev. Respir. Dis., 1980;122:123-143.

Crapo JD, Young SL, Fram EK, Pinkerton KE, Barry BE, Crapo RO. Morphometric characteristics of cells in the alveolar region of mammalian lungs. Am. Rev. Respir. Dis., 1983;128:S42-S46.

Cutz E, Conen PE. Ultrastructure and cytochemistry of Clara cells. Am. J. Pathol., 1971;62:127-141.

Dahl AR, Hadley WM. Nasal cavity enzymes involved in xenobiotic metabolism: Effects on the toxicity of inhalants. CRC Crit. Rev. Toxicol., 1991;21:345-372.

Davis C, Kannan MS, Jones TR, Daniel EE. Control of human airway smooth muscle: *In vitro* studies. J. Appl. Physiol., 1982;53:1080-1087.

Devereux TR, Domin BA, Philpot RM. Xenobiotic metabolism by isolated pulmonary cells. Pharmacol. Ther., 1989;41:243-256.

DeWater R, Willems LNA, van Muijen GNP, Franken C, Fransen JAM, Dijkman JH, Kramps JA. Ultrastructural localization of bronchial antileukoprotease in central and peripheral human airways by a gold-labeling technique using monoclonal antibodies. Am. Rev. Respir. Dis., 1986;133:882-890.

Dohrman A, Tsuda T, Escudier E, Cardone M, Jany B, Gum J, Kim Y, Basbaum C. Distribution of lysozyme and mucin (MUC2 and MUC3) mRNA in human bronchus. Exp. Lung Res., 1994;20:367-380.

Doidge JM, Satchell DG. Adrenergic and non-adrenergic inhibitory nerves in mammalian airways. J. Auton. Nerv. Syst., 1982;5:83-99.

Dunnill MD, Massarella GR, Arderson JA. A comparison of the quantitative anatomy of the bronchi in normal subjects, in status asthmaticus, in chronic bronchitis, and in emphysema. Thorax, 1969;24:176-179.

Eccles R. "Neurological and Pharmacological Considerations." In *The Nose: Upper Airway Physiology and the Atmospheric Environment*, DF Proctor, IB Anderson, eds. Amsterdam: Elsevier Biomedical Press, 1982, pp. 191-214.

Engelhardt JF, Yankaskas JR, Ernst SA, Yang Y, Marino CR, Boucher RC, Cohn JA, Wilson JM. Submucosal glands are the predominant site of CFTR expression in the human bronchus. Nat. Genet., 1992;2:240-248.

Evans MJ, Cox RA, Shami SG, Wilson B, Plopper CG. The role of basal cells in attachment of columnar cells to the basal lamina of the trachea. Am. J. Respir. Cell Mol. Biol., 1989;1:463-469.

Evans MJ, Johnson LV, Stephens RJ, Freeman G. Renewal of the terminal bronchiolar epithelium in the rat following exposure to NO_2 or O_3. Lab. Invest., 1976;35:246-257.

Fox B, Bull TB, Guz A. Mast cells in the human alveolar wall - an electron-microscopic study. J. Clin. Path., 1981;34:1333-1342.

Geelhaar A, Weibel ER. Morphometric estimation of pulmonary diffusion capacity. 3. The effect of increased oxygen consumption in Japanese Waltzing mice. Respir. Physiol., 1971;11:354-366.

Gehr P, Bachofen M, Weibel ER. The normal human lung: Ultrastructure and morphometric estimation of diffusion capacity. Respir. Physiol., 1978;32:121-140.

Genter MB, Llorens J, O'Callaghan JP, Peele DB, Morgan KT, Crofton KM. Olfactory toxicity of β,β'-iminodipropionitrile in the rat. J. Pharmacol. Exp. Ther., 1992;263:1432-1429.

Gilljam H, Motakefi AM, Robertson B, Strandvik B. Ultrastructure of the bronchial epithelium in adult patients with cystic fibrosis. Eur J Respir Dis 1987;71:187-194.

Graziadei PPC, Monti-Graziadei GA. "Continuous Nerve Cell Renewal in the Olfactory System." In *Handbook of Sensory Physiology*, M Jacobson, ed. New York: Springer-Verlag, 1977, pp. 55-82.

Gregson RL, Davey MJ, Prentice DE. Bronchus-associated lymphoid tissue (BALT) in the laboratory-bred and wild rat, *Rattus norvegicus*. Lab. Anim., 1979;13:239-243.

Gross EA, Swenberg JA, Fields S, Popp JA. Comparative morphometry of the nasal cavity in rats and mice. J. Anat., 1982;135:83-88.

Guerzon GM, Pare PD, Michoud MC, Hogg JC. Number and distribution of mast cells in monkey lungs. Am. Rev. Respir. Dis., 1979;119:59-66.

Guilmette RA, Wicks JD, Wolff RK. Morphometry of human nasal airways *in vivo* using magnetic resonance imaging. J. Aerosol Med., 1989;2:365-377.

Haley PJ. "Immunological Responses Within the Lung After Inhalation of Airborne Chemicals." In *Toxicology of the Lung, 2 Edition*, DE Gardner, JD Crapo, RO McClellan, eds. New York: Raven Press, 1993, pp. 389-416.

Harkema JR. Comparative aspects of nasal airway anatomy: Relevance to inhalation toxicology. Toxicol. Pathol., 1991;19:321-336.

Harkema JR, Hotchkiss JA. *In vivo* effects of endotoxin on intraepithelial mucosubstances in rat pulmonary airways. Quantitative histochemistry. Am. J. Pathol., 1992;141:307-317.

Harkema JR, Hotchkiss JA, Barr EB, Bennett CB, Gallup M, Lee JK, Basbaum C. Long-lasting effects of chronic ozone exposure on rat nasal epithelium. Am. J. Respir. Cell Mol. Biol., 1998;19:1-13.

Harkema JR, Hotchkiss JA, Harmsen AG, Henderson RF. *In vivo* effects of transient neutrophil influx on nasal respiratory epithelial mucosubstances. Quantitative histochemistry. Am. J. Pathol., 1988;130:605-615.

Harkema JR, Hotchkiss JA, Henderson RF. Effects of 0.12 and 0.80 ppm ozone on rat nasal and nasopharyngeal epithelial mucosubstances: Quantitative histochemistry. Toxicol. Pathol., 1989;17:525-535.

Harkema JR, Mariassy A, St.George JA, Hyde DM, Plopper CG. "Epithelial Cells of the Conducting Airways: A Species Comparison." In *The Airway Epithelium: Physiology, Pathophysiology, and Pharmacology*, SG Farmer, DW Hay, eds. New York: Marcel Dekker, Inc., 1991, pp. 3-39.

Harkema JR, Plopper CG, Hyde DM, St.George JA, Dungworth DL. Effects of an ambient level of ozone on primate nasal epithelial mucosubstances. Quantitative histochemistry. Am. J. Pathol., 1987;127:90-96.

Harkema JR, Plopper CG, Hyde DM, Wilson DW, St.George JA, Wong VJ. Nonolfactory surface epithelium of the nasal cavity of the bonnet monkey: A morphologic and morphometric study of the transitional and respiratory epithelium. Am. J. Anat., 1987;180:266-279.

Harmsen AG, Muggenburg BA, Snipes MB, Bice DE. The role of macrophages in particle translocation from lungs to lymph nodes. Science, 1985;230:1277-1280.

Heritage PL, Underdown BJ, Arsenault AL, Snider DP, McDermott MR. Comparison of murine nasal-associated lymphoid tissue and Peyer's patches. Am. J. Respir. Crit. Care Med., 1997;156:1256-1262.

Hogg JC. Neutrophil kinetics and lung injury. Physiol. Rev. 1987;67:1249-1295.

Holland SM, Gallin JI. "Neutrophils." In *The Lung: Scientific Foundations*, RG Crystal, JB West, ER Weibel, PJ Barnes, eds. Philadelphia: Lippincott-Raven Publishers, 1997, pp. 877-890.

Holt PG. "Macrophage and Dendritic Cell Populations in the Respiratory Tract." In *Immunopharmacology of the Respiratory System*, ST Holgate, ed. London: Academic Press, 1995, pp. 1-12.

Inayama Y, Hook GE, Brody AR, Jetten AM, Gray T, Mahler J, Nettesheim P. *In vitro* and *in vivo* growth and differentiation of clones of tracheal basal cells. Am. J. Pathol., 1989;134:539-549.

Jarkovska D. Ultrastructure of the epithelium of the respiratory bronchioles in man. Folia Morphol. Praha, 1970;18:352-358.

Jeffery PK. "Structural, Immunologic, and Neural Elements of the Normal Human Airway Wall." In *Asthma and Rhinitis*, WW Busse, ST Holgate, eds. Oxford: Blackwell Scientific Publications, 1995, pp. 80-106.

Jeffery PK, Reid L. New observations of rat airway epithelium: A quantitative and electron microscopic study. J. Anat., 1975;120:295-320.

Joad JP, Ji C, Kott KS, Bric JM, Pinkerton KE. *In utero* and postnatal effects of sidestream cigarette smoke exposure on lung function, hyperresponsiveness, and neuroendocrine cells in rats. Toxicol. Appl. Pharmacol., 1995;132:63-71.

Johnson NF, Hubbs AF. Epithelial progenitor cells in the rat trachea. Am. J. Respir. Cell Mol. Biol., 1990;3:579-585.

Kahane JC. A morphological study of the human prepubertal and pubertal larynx. Am. J. Anat., 1978;151:11-19.

Kay JM. Comparative morphologic features of the pulmonary vasculature in mammals. Am. Rev. Respir. Dis., 1983;128:S53-S57.

Klaassen AB, Kuijpers W, Denuce JM. Morphological and histochemical aspects of the nasal glands in the rat. Anat. Anz., 1981;149:51-63.

Koornstra PJ, Duijvestijn AM, Vlek LF, Marres EH, van Breda Vriesman PJ. Immunohistochemistry of nasopharyngeal (Waldeyer's ring equivalent) lymphoid tissue in the rat. Acta Otolaryngol. Stockh., 1993;113:660-667.

Kraal G, Broug E, van Iwaarden JF, Persoons JHA, Thepen T. "The Role of Alveolar Macrophages in Pulmonary Immune Function." In *Lung Macrophages and Dendritic Cells in Health and Disease*, MF Lipscomb, SW Russell, eds. New York: Marcel Dekker, Inc., 1997, pp. 203-220.

Kunkel SL, Chensue SW, Lucas NW, Strieter RM. "Macrophage-Derived Cytokines in Lung Inflammation." In *Lung Macrophages and Dendritic Cells in Health and Disease*, MF Lipscomb, SW Russell, eds. New York: Marcel Dekker, Inc., 1997, pp. 183-202.

Kuper CF, Hameleers DM, Bruijntjes JP, van der Ven I, Biewenga J, Sminia T. Lymphoid and non-lymphoid cells in nasal-associated lymphoid tissue (NALT) in the rat. An immuno- and enzyme-histochemical study. Cell Tissue Res., 1990;259:371-377.

Lamb D, Reid L. Mitotic rates, goblet cell increase and histochemical changes in mucus in rat bronchial epithelium during exposure to sulphur dioxide. J. Pathol. Bacteriol., 1968;96:97-111.

Lambrecht BN, Carro-Muino I, Vermaelen K, Pauwels RA. Allergen-induced changes in bone marrow progenitor and airway dendritic cells in sensitized rats. Am. J. Respir. Cell Mol. Biol., 1999;20:1165-1174.

Lauweryns JM. The juxta-alveolar lymphatics in the human adult lung. Histologic studies in 15 cases of drowning. Am. Rev. Respir. Dis., 1970;102:877-885.

Lauweryns JM, van Lommel A. The intrapulmonary neuroepithelial bodies after vagotomy: Demonstration of their sensory neuroreceptor-like innervation. Experientia, 1983;39:1123-1124.

Leak LV, Ferrans VJ. "Lymphatics and Lymphoid Tissue." In *The Lung: Scientific Foundations*, RG Crystal, JB West, ER Weibel, PJ Barnes, eds. Philadelphia: Lippincott-Raven Publishers, 1997, pp. 1129-1137.

Leak LV, Jamuar MP. Ultrastructure of pulmonary lymphatic vessels. Am. Rev. Respir. Dis., 1983;128:S59-S65.

Lee RMK, Forrest JB. "Structure and Function of Cilia." In *The Lung: Scientific Foundations*, RG Crystal, JB West, ER Weibel, PJ Barnes, eds. Philadelphia: Lippincott-Raven Publishers, 1997, pp. 459-478.

Lipscomb MF, Russell SW. (eds.) *Lung Macrophages and Dendritic Cells in Health and Disease.* New York: Marcel Dekker, Inc., 1997.

Liu JY, Nettesheim P, Randell SH. Growth and differentiation of tracheal epithelial progenitor cells. Am. J. Physiol., 1994;266:L296-L307.

Lubman RL, Kim KJ, Crandall ED. "Alveolar Epithelial Barriers Proteins." In *The Lung: Scientific Foundations*, RG Crystal, JB Went, ER Weibel, PJ Barnes, eds. Philadelphia: Lipincott-Raven Publishers, 1997, pp. 585-602.

Mariassy AT, Wheeldon EB. The pleura: A combined light microscopic, scanning, and transmission electron microscopic study in the sheep. Part I. Normal pleura. Exp. Lung Res., 1983;4:293-314.

Matulionis DH, Parks HF. Ultrastructural morphology of the normal nasal respiratory epithelium of the mouse. Anat. Rec., 1973;176:64-83.

McLaughlin R, Tyler W, Canada R. A study of the subgross pulmonary anatomy in various mammals. Am. J. Anat., 1961;108:149-165.

McWilliam AS, Napoli S, Marsh AM, Pemper FL, Nelson DJ, Pimm CL, Stumbles PA, Wells TNC, Holt PG. Dendritic cells are recruited into the airway epithelium during the inflammatory response to a broad spectrum of stimuli. J. Exp. Med., 1996;184:2429-2432.

Mercer RR, Crapo JD. "Architecture of the Lung." In *Comparative Biology of the Normal Lung*, RA Parent, ed. Boca Raton, FL: CRC Press, 1992, pp. 109-119.

Mercer R, Crapo JD. "Anatomical Modeling of Microdosimetry of Inhaled Particles and Gases in the Lung." In *Extrapolation of Dosimetric Relationships for Inhaled Particles and Gases*, JD Crapo, ED Smolko, FJ Miller, JA Graham, AW Hayes, eds. New York: Academic Press, 1989, pp. 69-78.

Meuwissen HJ, Hussain M. Bronchus-associated lymphoid tissue in human lung: Correlation of hyperplasia with chronic pulmonary disease. Clin. Immunol. Immunopathol., 1982;23:548-561.

Meyrick B, Reid L. The alveolar brush cell in rat lung - a third pneumonocyte. J. Ultrastruct. Res., 1968;23:71-80.

Meyrick B, Reid L. Ultrastructure of cells in the human bronchial submucosal glands. J. Anat., 1970;107:281-299.

Miller FJ. (ed.) *Nasal Toxicology and Dosimetry of Inhaled Xenobiotics: Implications for Human Health.* Washington, DC: Taylor & Francis, 1995.

Monteiro Riviere NA, Popp JA. Ultrastructural characterization of the nasal respiratory epithelium in the rat. Am. J. Anat., 1984;169:31-43.

Morgan KT, Jiang XZ, Patterson DL, Gross EA. The nasal mucociliary apparatus. Correlation of structure and function in the rat. Am. Rev. Respir. Dis., 1984;130:275-281.

Morgan KT, Patterson DL, Gross EA. Responses of the nasal mucociliary apparatus of F-344 rats to formaldehyde gas. Toxicol. Appl. Pharmacol., 1986;82:1-13.

Morris JB, Hassett DN, Blanchard KT. A physiologically-based pharmacokinetic model for nasal uptake and metabolism of nonreactive vapors. Toxicol. Appl. Pharmacol., 1993;123:120-129.

Murray MJ, Driscoll KE. "Immunology of the Respiratory System." In *Comparative Biology of the Normal Lung*, RA Parent, ed. Boca Raton, FL: CRC Press, 1992, pp. 725-746.

Nakano T, Muto H. The transitional zone in the epithelium lining the mouse epiglottis. Acta Anat. Basel, 1987;130:285-290.

Negus VE. (ed.) *Comparative Anatomy and Physiology of the Nose and Paranasal Sinuses.* Edinburgh: E&S Livingstone, 1958.

Negus VE. (ed.) *The Comparative Anatomy and Physiology of the Larynx.* London: Hafner, 1962.

Nettesheim P, Jetten AM, Inayama Y, Brody AR, George MA, Gilmore LB, Gray T, Hook GE. Pathways of differentiation of airway epithelial cells. Environ. Health Perspect., 1990;85:317-329.

Nikula KJ, Wilson DW, Giri SN, Plopper CG, Dungworth DL. The response of the rat tracheal epithelium to ozone exposure. Injury, adaptation, and repair. Am. J. Pathol., 1988;131:373-384.

Nylen ES, Becker KL, Joshi PA, Snider RH, Schuller HM. Pulmonary bombesin and calcitonin in hamsters during exposure to hyperoxia and diethylnitrosamine. Am. J. Respir. Cell Mol. Biol., 1990;2:25-31.

Olsson P, Bende M. Sympathetic neurogenic control of blood flow in human nasal mucosa. Acta Otolaryngol. Stockh., 1986;102:482-487.

Pabst R, Gehrke I. Is the bronchus-associated lymphoid tissue (BALT) an integral structure of the lung in normal mammals, including humans? Am. J. Respir. Cell Mol. Biol., 1990;3:131-135.

Parent RA (ed.) *Comparative Biology of the Normal Lung.* Boca Raton, FL: CRC Press, Inc., 1992.

Pattle RE. Surface lining of the lung aveoli. Physiol. Rev., 1965;45:48-79.

Pavelka M, Ronge HR, Stockinger G. [Comparative study of tracheal epithelium of different mammals] Vergleichende Untersuchungen am Trachealepithel verschiedener Sauger. Acta Anat. Basel, 1976;94:262-282.

Peters-Golden M. "Lipid Mediator Synthesis by Lung Macrophages." In *Lung Macrophages and Dendritic Cells in Health and Disease*, MF Lipscomb, SW Russell, eds. New York: Marcel Dekker, Inc., 1997, pp. 151-182.

Pinkerton KE, Barry BE, O'Neil JJ, Raub JA, Pratt PC, Crapo JD. Morphologic changes in the lung during the lifespan of Fischer 344 rats. Am. J. Anat., 1982;164:155-174.

Pinkerton KE, Gehr E, Crapo JD. "Architecture and Cellular Composition of the Air-Blood Barrier." In *Comparative Biology of the Normal Lung*, RA Parent, ed. Boca Raton, FL: CRC Press, 1992, pp. 121-128.

Pinkerton KE, Plopper CG, Hyde DM, Harkema JR, Tyler WS, Morgan KT, St.George JA, Kay M, Mariassy AT. "Structure and Fuction of the Respiratory Tract." In *Handbook of Human Toxicology*, EJ Massaro, ed. Boca Raton, FL: CRC Press, 1997, pp. 469-491.

Plesch BE. Histology and immunohistochemistry of bronchus-associated lymphoid tissue (BALT) in the rat. Adv. Exp. Med. Biol., 1982;149:491-497.

Plopper CG. Comparative morphologic features of bronchiolar epithelial cells. The Clara cell. Am. Rev. Respir. Dis., 1983;128:S37-S41.

Plopper CG, Dungworth DL. "Structure, Function, Cell Injury and Cell Renewal of Bronchiolar and Alveolar Epithelium." In *Lung Carcinoma*, EM McDowell, ed. London: Churchill-Livingstone, 1987, pp. 29-44.

Plopper CG, Heidsiek JG, Weir AJ, George JA, Hyde DM. Tracheobronchial epithelium in the adult rhesus monkey: A quantitative histochemical and ultrastructural study. Am. J. Anat., 1989;184:31-40.

Plopper CG, Hyde DM. "Epithelial Cells of Bronchioles." In *Comparative Biology of the Normal Lung*, RA Parent, ed. Boca Raton, FL: CRC Press, 1992, pp. 85-92.

Plopper CG, Weir AJ, George JA, Tyler NK, Mariassy A, Wilson D, Nishio SJ, Cranz DL, Heidsiek JG, Hyde DM. "Species Differences in Airway Cell Distribution and Morphology." In *Extrapolation of Dosimetric Relationships for Inhaled Particles and Gases*, JD Crapo, ED Smolko, FJ Miller, JA Graham, AW Hayes, eds. New York: Academic Press, Inc., 1989, pp. 19-34.

Plowman PN. The pulmonary macrophage population of human smokers. Ann. Occup. Hyg., 1982;25:393-405.

Popp JA, Martin JT. Surface topography and distribution of cell types in the rat nasal respiratory epithelium: Scanning electron microscopic observations. Am. J. Anat., 1984;169:425-436.

Proctor DF, Chang JCF. "Comparative Anatomy and Physiology of the Nasal Cavity." In *Nasal Tumors in Animals and Man*, G Reznik, S Stinson, eds. Boca Raton, FL: CRC Press, 1983, pp. 1-33.

Reid LM. Pathology of chronic bronchitis. Lancet, 1954;1:275-276.

Richardson JB. Recent progress in pulmonary innervation. Am. Rev. Respir. Dis., 1983;128:S65-S68.

Richardson JB, Beland J. Nonadrenergic inhibitory nervous system in human airways. J. Appl. Physiol., 1976;41:764-771.

Richeldi L, Franchi A, Rovatti E, Cossarizza A, du Bois RM, Saltini C. "Lymphocytes." In *The Lung: Scientific Foundations*, RG Crystal, JB Went, ER Weibel, PJ Barnes, eds. Philadelphia: Lippincott-Raven Publishers, 1997, pp. 803-820.

Richmond I, Pritchard GE, Ashcroft T, Avery A, Corris PA, Walters EH. Bronchus-associated lymphoid tissue (BALT) in human lung: Its distribution in smokers and non-smokers. Thorax, 1993;48:1130-1134.

Rogers AV, Dewar A, Corrin B, Jeffery PK. Identification of serous-like cells in the surface epithelium of human bronchioles. Eur. Respir. J., 1993;6:498-504.

Sannes PL. Differences in basement membrane-associated microdomains of type I and type II pneumocytes in the rat and rabbit lung. J. Histochem. Cytochem., 1984;32:827-833.

Sasaki J, Watanabe S, Nomura T, Wada T, Tanaka Y, Kanda S, Otsuka N. Presence of filaments in the non-ciliated bronchiolar epithelial (Clara) cell of mammalian lung. Okajimas Folia Anat. Japan, 1988;65:155-169.

Schlesinger RB, Gorczynski JE, Dennison J, Richards L, Kinney PL, Bosland MC. Long-term intermittent exposure to sulfuric acid aerosol, ozone, and their combination: Alterations in tracheobronchial mucociliary clearance and epithelial secretory cells. Exp. Lung Res., 1992;18:505-534.

Schlesinger RB, Lippmann M. Particle deposition in the trachea: *In vivo* and in hollow casts. Thorax, 1976;31:678-684.

Schreider JP, Raabe OG. Anatomy of the nasal-pharyngeal airway of experimental animals. Anat. Rec., 1981;200:195-205.

Schuller HM, Becker KL, Witschi HP. An animal model for neuroendocrine lung cancer. Carcinogenesis, 1988;9:293-296.

Serabjit-Singh CJ, Nishio SJ, Philpot RM, Plopper CG. The distribution of cytochrome P_{450} monooxygenase in cells of the rabbit lung: An ultrastructural immunocytochemical characterization. Mol. Pharmacol., 1988;33:279-289.

Sheppard MN, Johnson NF, Cole GA, Bloom SR, Marangos PJ, Polak JM. Neuron specific enolase (NSE) immunostaining detection of endocrine cell hyperplasia in adult rats exposed to asbestos. Histochemistry, 1982;74:505-513.

Silverman ES, Gerritsen ME, Collins T. "Metabolic Functions of the Pulmonary Endothelium." In *The Lung: Scientific Foundations*, RG Crystal, JB West, ER Weibel, PJ Barnes, eds. Philadelphia: Lippincott-Raven Publishers, 1997, pp. 629-651.

Singh G, Katyal SL. "Secretory Proteins of Clara Cells and Type II Cells." In *Comparative Biology of the Normal Lung*, RA Parent, ed. Boca Raton, FL: CRC Press, 1992, pp. 93-108.

Singh G, Katyal SL, Brown WE, Phillips S, Kennedy AL, Anthony J, Squeglia N. Amino-acid and cDNA nucleotide sequences of human Clara cell 10 kDa protein. Biochim. Biophys. Acta, 1988;950:329-337.

Sleigh MA, Blanchard JD, Liron M. The propulsion of mucus by cilia. Am. Rev. Respir. Dis., 1988;137:726-731.

Sorokin SP. "The Respiratory System." *In Cell and Tissue Biology: A Textbook of Histology*, L Weiss, ed. Baltimore: Urban & Schwarzenberg, Inc., 1988, pp. 751-814.

Sorokin SP, Hoyt-RF J, Pearsall AD. Comparative biology of small granule cells and neuroepithelial bodies in the respiratory system. Short review. Am. Rev. Respir. Dis., 1983;128:S26-S31.

Spicer SS, Chakrin LW, Wardell J Jr., Kendrick W. Histochemistry of mucosubstances in the canine and human respiratory tract. Lab. Invest., 1971;25:483-490.

Spit BJ, Hendriksen EG, Bruijntjes JP, Kuper CF. Nasal lymphoid tissue in the rat. Cell Tissue Res 1989;255:193-198.

St.George JA, Nishio SJ, Plopper CG. Carbohydrate cytochemistry of rhesus monkey tracheal epithelium. Anat. Rec., 1984;210:293-302.

Stevens TP, McBride JT, Peake JL, Pinkerton KE, Stripp BR. Cell proliferation contributes to PNEC hyperplasia after acute airway injury. Am. J. Physiol., 1997;272:L486-L493.

Storey WF, Staub NC. Ventilation of terminal air units. J. Appl. Physiol., 1962;17:391.

Suda T, McCarthy K, Vu Q, McCormack J, Schneeberger EE. Dendritic cell precursors are enriched in the vascular compartment of the lung. Am. J. Respir. Cell Mol. Biol., 1998;19:728-737.

Takeyama K, Dabbagh K, Lee HM, Agusti C, Lausier JA, Ueki IF, Grattan KM, Nadel JA. Epidermal growth factor system regulates mucin production in airways. Proc. Natl. Acad. Sci., USA, 1999;96:3081-3086.

Tateishi R. Distribution of argyrophil cells in adult human lungs. Arch. Pathol., 1973;96:198-202.

Tsutsumi Y, Osamura RY, Watanabe K, Yanaihara N. Immunohistochemical studies on gastrin-releasing peptide- and adrenocorticotropic hormone-containing cells in the human lung. Lab. Invest., 1983;48:623-632.

Tyler NK, Plopper CG. Morphology of the distal conducting airways in rhesus monkey lungs. Anat. Rec., 1985;211:295-303.

Tyler WS. Small airways and terminal units: Comparative subgross anatomy of lungs. Am. Rev. Respir. Dis., 1983;128:S32-S36.

Tyler WS, Coalson JJ, Stripp B. Comparative biology of the lung. Am Rev Respir Dis 1983;128:S1-S91.

Tyler WS, Julian MD. "Gross and Subgross Anatomy of Lungs, Pleura, Connective Tissue Septa, Distal Airways, and Structural Units." In *Comparative Biology of the Normal Lung*, RA Parent, ed. Boca Raton, FL: CRC Press, 1992, pp. 37-48.

van Haarst JM, de Wit HJ, Drexhage HA, Hoogsteden HC. Distribution and immunophenotype of mononuclear phagocytes and dendritic cells in the human lung. Am. J. Respir. Cell Mol. Biol., 1994;10:487-492.

van Voorhis WC, HAIR LS, Steinman RM, Kaplan G. Human dendritic cells - Enrichment and characterization from peripheral blood. J. Exp. Med., 1982;155:1172-1187.

Warner AE, Brain JD. The cell biology and pathogenic role of pulmonary intravascular macrophages. Am. J. Physiol., 1990;258:L1-L12.

Weibel ER. (ed.) *Morphometry of the Human Lung*. New York: Academic Press, 1963.

Weibel ER, Bachofen H. "The Fiber Scaffold of Lung Parenchyma." In *The Lung: Scientific Foundations*, RG Crystal, JB West, ER Weibel, PJ Barnes, eds. Philadelphia: Lippincott-Raven Publishers, 1997, pp. 1139-1146.

Weibel ER, Crystal RG. "Structural Organization of the Pulmonary Interstitium." In *The Lung: Scientific Foundations*, RG Crystal, JB West, ER Weibel, PJ Barnes, eds. Philadelphia: Lippincott-Raven Publishers, 1997, pp. 685-695.

Weibel ER, Gil J. "Structure-Function Relationships at the Alveolar Level." In *Bioengineering Aspects of the Lung*, JB West, ed. New York: Marcel Dekker, 1977, pp. 1-81.

Weibel ER, Taylor CR. "Functional Design of the Human Lung for Gas Exchange." In *Fishman's Pulmonary Diseases and Disorders, 3rd Edition*, AP Fishman, ed. New York: McGraw-Hill, 1998, pp. 21-71.

Wharton J, Polak JM, Bloom SR, Ghatei MA, Solcia E, Brown MR, Pearse AG. Bombesin-like immunoreactivity in the lung. Nature, 1978;273:769-770.

Wharton J, Polak JM, Cole GA, Marangos PJ, Pearse AG. Neuron-specific enolase as an immunocytochemical marker for the diffuse neuroendocrine system in human fetal lung. J. Histochem. Cytochem., 1981;29:1359-1364.

Widdicombe JG, Pack RJ. The Clara cell. Eur. J. Respir. Dis., 1982;63:202-220.

Willems LN, Kramps JA, Stijnen T, Sterk PJ, Weening JJ, Dijkman JH. Antileukoprotease-containing bronchiolar cells. Relationship with morphologic disease of small airways and parenchyma. Am. Rev. Respir. Dis., 1989;139:1244-1250.

Yamamoto T, Masuda H. Some observations on the fine structure of the goblet cells in the nasal respiratory epithelium of the rat, with special reference to the well-developed agranular endoplasmic reticulum. Okajimas Folia Anat. Japan, 1982;58:583-594.

Yeh HC, Schum GM. Models of human lung airways and their application to inhaled particle deposition. Bull. Math. Biol., 1980;42:461-480.

Young SL, Fram EK, Randell SH. Quantitative three-dimensional reconstruction and carbohydrate cytochemistry of rat non-ciliated bronchiolar (Clara) cells. Am. Rev. Respir. Dis., 1986;133:899-907.

Zorychta E, Richardson JB. "Innervation of the Lung." In *Comparative Biology of the Normal Lung*, RA Parent, ed. Boca Raton, FL: CRC Press, 1991, pp. 157-161.

Zrunek M, Happak W, Hermann M, Streinzer W. Comparative anatomy of human and sheep laryngeal skeleton. Acta Otolaryngol. Stockh., 1988;105:155-162.

2. IMMUNOLOGY OF THE RESPIRATORY TRACT

Robert W. Lange, Ph.D. and Meryl H. Karol, Ph.D.
Department of Environmental & Occupational Health
University of Pittsburgh
260 Kappa Drive
Pittsburgh, PA 15238

INTRODUCTION

The pulmonary immune system is composed of innate and acquired components. Innate (natural) immunity includes physical barriers, cells, and blood-borne macromolecules. These elements are present prior to exposure to a foreign substance (xenobiotic) and do not discriminate among (recognize) xenobiotics. The inflammatory response represents one form of natural immunity.

By contrast, acquired (specific) immunity consists of elements, both cellular and soluble, that are induced by exposure to foreign substances. Acquired immunity is characterized by an increase in number and defensive capability of distinct macromolecules that recognize and respond to the xenobiotic, with recognition increasing with repeated exposure to the same foreign material. This memory of the past ensures that future encounters with the same agent will be increasingly effective.

An intimate link exists between the innate and acquired immune systems. Cellular and soluble components of the innate system influence antigen processing, selection and presentation, as well as the magnitude of the response of the acquired immune system. This Chapter will describe the structure and operation of the integrated immune system that functions following inhalation of a foreign agent.

CELLS AND TISSUES OF THE IMMUNE SYSTEM

The soft tissue in hollow bones contains hematopoietic stem cells, the precursors of all blood cell populations. During fetal life, hematopoiesis occurs in the yolk sac and eventually in the liver and spleen. By puberty, hematopoiesis occurs mostly in the sternum, vertebrae, iliac bones and ribs. Mature blood cells exit via a dense network of vascular sinuses to become part of the circulatory system. Cells of the immune system originate from a common progenitor stem cell that becomes committed to differentiate along a particular lineage, i.e., myeloid, lymphocytic, or null cell line (see Figure 1).

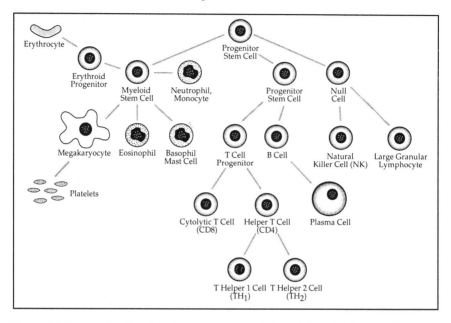

Figure 1 Lineage of cells of the immune system (adapted from Weigle, 1997).

The mononuclear phagocyte system is a major cell population of the immune cells. These cells develop from the myeloid stem cell in the bone marrow and enter the bloodstream as monocytes. Once settled within a tissue they become macrophages. Since they are found in all tissues, they are designated by their locations. For example, those in the pulmonary airways are called "alveolar macrophages."

Macrophages function by phagocytizing foreign particles, and secreting enzymes, reactive oxygen intermediates, and lipid-derived mediators such as prostaglandins. By doing such, they are able to kill microbes and control the spread of infection Occasionally however, they injure normal tissue. They frequently affect distant sites via the production of cytokines, which recruit other inflammatory cells, especially neutrophils (see Table 1) (Sibille and Reynolds, 1990).

Granulocytes (including neutrophils, eosinophils, and basophils) comprise another cellular component of innate immunity that develops from the myeloid stem cell (Sibille and Marchandise, 1993; Kroegel et al., 1994; Warner and Kroegel, 1994). Frequently termed inflammatory cells, granulocytes respond to chemotactic stimuli and function to eliminate microbes and dead tissues.

Several granulocytic cells are important in the defense against organisms. Neutrophils are the primary cell type in the acute inflammation response. They contain receptors for antibody molecules and for complement components. These cells perform many of the functions of macrophages and act as effector cells of humoral immunity. Parasites stimulate production of high levels of IgE. Eosinophils express receptors for IgE and are abundant at sites of allergic inflammation. Encounter of the cell-surface bound IgE with antigen results in release of granular contents, leading to immediate-type hypersensitivity reactions. Basophils, circulating cells that are counterparts to tissue mast cells, also express IgE receptors.

Natural killer (NK) cells comprise another cell population associated with innate immunity. These cells develop from a null stem cell (Figure 1) and are capable of lysing tumor- and virus-infected cells. Although developmentally related to T-lymphocytes, they lack expression of the T-lymphocyte receptor. Human NK cells are defined by the expression of the surface markers (CD antigens). CD antigens function to promote cell-cell interactions and adhesion, and transduce signals that lead to lymphocyte activation (Abbas et al., 1997). Clusters of CDs can identify the lineage or stage of differentiation of lymphocytes and thus distinguish one class from another. For example, T helper (T_H) cells express CD4 and cytotoxic T-lymphocytes (T_{CTL}) express CD8 on their surface. Such markers can be identified with the use of antibodies and serve to identify T-lymphocytes, B-lymphocytes and other leukocytes that participate in immune and inflammatory responses. Human NK cells express CD16 and CD56 (Nagler et al., 1989). Rodent NK cells share similar functions and are defined by unique surface markers (Warren, 1996).

Lymphocytes, a central component of acquired immunity, develop from a lymphocyte stem cell. Cytokines stimulate the proliferation and maturation of various precursors. These "colony stimulating factors," including interleukin (IL)-3 and granulocyte-macrophage colony stimulating factor (GM-CSF), are produced by T-lymphocytes (Gabrilove and Jakubowski, 1990) (Table 1). Macrophages and marrow stromal cells produce IL-1 and IL-6, which enhance colony stimulating activity. Marrow stromal-derived IL-7 preferentially stimulates the maturation of B-lymphocytes from marrow precursors. Little is known about the nature of the uncommitted stem cell or the mechanisms that regulate its commitment to specific lineages.

T-Lymphocytes arise in the bone marrow, then migrate to the thymus where they mature. T-Lymphocytes are subdivided into T_{CTL} and T_H cells (see Figure 1), both of which recognize antigen when associated with proteins encoded by the major histocompatibility complex (MHC) expressed on the surface of accessory cells (antigen-presenting cells; APC). In response to antigenic stimulation, T_H cells secrete cytokines that may promote proliferation and differentiation of T- and B-lymphocytes and macrophages, or recruitment and activation of inflammatory leukocytes (see Table 1). T_{CTL} lyse cells that produce foreign antigens (Agostini et al., 1993).

B-Lymphocytes were demonstrated first in the avian bursa of Fabricus. These cells are responsible for humoral immune responses through their production of antibodies. Functionally distinct lymphocytes can be distinguished by the products they secrete, such as antibodies or cytokines, and by the expression of membrane-associated markers (CD).

The pulmonary interstitial pool of lymphocytes amounts to the estimated pool of those in circulation (10×10^9) In the bronchoalveolar space the majority of cells are macrophages; about 10% are T-lymphocytes.

TABLE 1. CYTOKINES AFFECTING THE LUNG

Cytokine	Cell Source	Target Cell	Effect on Target Cell
Interleukin-1 (IL-1)	Monocyte, alveolar macrophage, neutrophil, endothelial cell, fibroblast, smooth muscle cell, T- and B-lymphocyte NK cell, dendritic cell, epithelial cell	T_H lymphocyte (CD4$^+$)	Secrete IL-2, express IL-2 receptor, proliferation, adhesion, chemotaxis
		B-lymphocyte	Proliferation, Ig synthesis, adhesion, responsive to IL-5, chemotaxis
		NK cell, neutrophil, fibroblast, thymocyte	Proliferation, activation, adhesion
		Endothelial cell	Inhibit growth, alter function, promote thrombotic and anti-coagulatory mechanisms, increase ICAM expression
		Macrophage	Chemotaxis
		Myeloid progenitor cell	Potentiate CSF
		Monocyte	Adhesion
		Tumor cell	Monocyte-mediated cytotoxicity, regression
		All cells	Radioprotection
Interleukin-2 (IL-2)	CD4$^+$ T-lymphocyte NK cell	T-lymphocyte (CD4$^+$, CD8$^+$)	Proliferation, cytokine production, long-term growth of primed clones
		B-lymphocyte	Proliferation, antibody synthesis
		NK cell	Proliferation, growth
Interleukin-3* (IL-3)	T-lymphocyte (CD4$^+$,CD8$^+$), monocyte, NK cell, mast cell, endothelial cell	Stem cell	Proliferation and differentiation
		Eosinophil	Chemotaxis
		Mast cell	Proliferation, inhibit MHC Class II expression
		Macrophage	Proliferation, phago-cytosis, IL-1, IL-6, TNFα secretion
		Basophil	Responsive to further activation by IL-8

TABLE 1. CYTOKINES AFFECTING THE LUNG (cont'd)

Cytokine	Cell Source	Target Cell	Effect on Target Cell
Interleukin-4* (IL-4)	CD4+ T-lymphocyte mast cell basophil	B-lymphocyte	Proliferation, differentiation, isotype switching to IgG_1 and IgE, inhibit IgM, IgG_3, IgG_{2a}, IgG_{2b} synthesis, enhance MHC Class II and IgE receptors expression
		T-lymphocyte	Proliferation
		Alveolar macrophage	Upregulation of MHC Class II, inhibit synthesis of IL-6, increase phagocytosis
		Smooth muscle cell	Proliferation
		Thymocyte	Differentiation without proliferation
		Mast cell	Stimulate growth
		NK cell	Inhibit activation
Interleukin-5 (IL-5)	T-lymphocyte, mast cell	Eosinophil	Proliferation, differentiation, activation
		B-lymphocyte	Proliferation, differentiation, production and secretion IgM, IgA
		Progenitor cell	Proliferation of BFU, differentiation/growth inhibition of CFU-E
Interleukin-6 (IL-6)	T- and B-lymphocyte monocyte, macrophage, endothelial cell, epithelial cell, stem cell, mast cells, granulocyte, smooth muscle cell, eosinophil, fibroblast,	B-lymphocyte	*Major physiological mediator of acute phase reaction* Differentiation into plasma cell, secretion antibody in IL-4-activated B-cells
		T-lymphocyte	Activation, differentiation into CD8+ T-lymphocyte in presence of IL-2
		Thymocyte	Proliferation
		Human lung cancer lines	Block growth
Interleukin-7 (IL-7)	Epithelial cell, bone marrow, thymic stromal cell	B-lymphocyte	Proliferation of immature B-lymphocytes only
		T-lymphocyte	Proliferation, expression & secretion of IL-3, GM-CSF
		Thymocyte	Proliferation
		Monocyte	Synthesis IL-1, IL-6, MIP
		Macrophage	Downregulate TGF-β expression

TABLE 1. CYTOKINES AFFECTING THE LUNG (cont'd)

Cytokine	Cell Source	Target Cell	Effect on Target Cell
Interleukin-8 (IL-8)	NK cell, fibroblast, monocyte, endothelial cell, T-lymphocyte, macrophage, tumor cell lines	Neutrophil	Activation, chemotaxis, adhesion and extravasation into tissue
		B-lymphocyte	Inhibition IgE production, alter production of IgM, IgG, and IgA, chemotaxis
		T-lymphocyte	Chemotaxis
		Endothelial cell	Inhibit leukocyte adhesion resulting in anti-inflammatory effect
		Erythrocyte	Strongly bind and regulate inflammatory reactions (bound IL-8 cannot activate neutrophil)
Interleukin-9 (IL-9)	CD4$^+$ T-lymphocyte	Stromal cell, T-lymphocyte	Proliferation in absence of antigen (acts as mitogen)
		B-lymphocyte	Enhance IgG, IgM, and IgE production by IL-4 - induced B-lymphocytes
		Mast cell	Secretion of IL-6, proliferation in bone marrow in presence of IL-3
Interleukin-10 (IL-10)	T-lymphocyte, CD5$^+$ B-lymphocyte macrophage	Macrophage	Inhibit IL-1, IL-6, TNFα production, inhibit antigen presentation
		B-lymphocyte	Regulate proliferation and differentiation, MHC Class II expression, IgG, IgM, IgA secretion
		Mast cell	Proliferation in presence of IL-3 and/or IL-4
		T-lymphocyte	Inhibit IFN-γ, IL-2, TNFα production, downregulate function, chemotaxis (CD8$^+$)
		Monocyte	Inhibit MHC Class II expression
		Thymocyte	Co-stimulate with IL-2, IL-4, and IL-7 differentiation into CD8$^+$
Interleukin-11 (IL-11)	Stromal fibroblast	Plasmacytoma, progenitor B-lymphocyte, megakaryocyte	Proliferation
		All cells	Prevent apoptosis

TABLE 1. CYTOKINES AFFECTING THE LUNG (cont'd)

Cytokine	Cell Source	Target Cell	Effect on Target Cell
Interleukin-12 (IL-12)	Monocyte T-lymphocyte	NK cell	Activation (CD56[+]), proliferation
		T-lymphocyte	Co-stimulate differentiation into CD8[+] T-lymphocyte with IL-2
		Lymphoblast	Proliferation
		LAK (lympho-kine-activated killer) cell	Co-stimulate proliferation with IL-2
		Bone marrow progenitor cell	Co-stimulate myelopoiesis with SCF, Co-stimulate proliferation with CSF
Interleukin-13 (IL-13) *Not expressed in the lungs*	T-lymphocyte	LAK cell	Enhance killer activity
		Macrophage	Modulates activity, inhibit IL-1, IL-6, IL-8, IL-10, IL-12, MIP-1, MCP synthesis
		B-lymphocyte	Differentiation, proliferation, isotype switching to IgE and IgG, induce IgM and IgG production, induce CD23, CD72 and MHC Class II expression
Interleukin-14 (IL-14, BCGF)	T- and B-lymphocyte	B-lymphocyte	Proliferation and immunoglobulin secretion
Interleukin-15 (IL-15)	Mast cell, epithelial cell, stromal cell	T-lymphocyte	Proliferation, differentia-tion to CD8[+] T-lympho-cyte, inhibit apoptosis
		LAK cell NK cell	Activate cytolytic function
		Mast cell	Proliferation in absence of IL-2 receptor
Interleukin-16 (IL-16)	CD8[+] T-lymphocyte	T-lymphocyte (CD4[+]), eosinophil, macrophage	Chemotaxis
Interleukin-17 (IL-17)	CD4[+] T-lymphocyte	Fibroblast	ICAM-1 expression, secretion of IL-6, IL-8, G-CSF, prostaglandin E_2, maturation into neutrophil
		T cell	Proliferation
		Epithelial and endothelial cell	Secretion of IL-6, IL-8, G-CSF, prostaglandin E_2

TABLE 1. CYTOKINES AFFECTING THE LUNG (cont'd)

Cytokine	Cell Source	Target Cell	Effect on Target Cell
Interleukin-18 (IL-18)	Macrophage	T-lymphocyte	Synthesis of IFN-γ, proliferation, cytotoxicity
		NK cell	Enhance activity
Granulocyte macrophage colony forming cells GM-CFC	T-lymphocyte, endothelial cells, fibroblasts, phagocytes	Granulocyte and phagocyte colony forming units	Differentiation and colony formation
Leukocyte migration inhibitory factor (LIF)	Lung fibroblast, alloreactive T-lymphocyte, spleen cell, monocyte, thymic epithelial cell, bone marrow stromal cell	Hematopoietic progenitor cell	Regulation of proliferation and differentiation
Colony stimulating factor-1 (CSF-1, macrophage colony stimulating factor, MCSF)	Macrophage, endothelial cell fibroblast T-lymphocyte	Hematopoietic progenitor cell	Proliferation, differentiation into macrophage colony
		Macrophage	Maintain membrane integrity, prolong life
Gamma interferon (IFN-γ)	T- and B-lymphocyte NK cell	Transformed cell	Inhibit proliferation
		Normal cell	Inhibit viral replication
		T-lymphocyte	Expression Il-2 receptor, regulate cytotoxicity, modulate proliferation, and differentiation
		B-lymphocyte	Inhibit IgG$_1$, IgE in IL-4 - stimulated cells, costimulate proliferation with anti-Ig, regulate expression MHC class II
		Neutrophil, monocyte	Expression of IgG receptors
		Monocyte, macrophage	Secretion of TNFα, synthesis of G-CSF, M-CSF, activation
		Endothelial cell, smooth muscle cell	Inhibit proliferation
		Myofibroblast	Inhibit collagen synthesis

TABLE 1. CYTOKINES AFFECTING THE LUNG (cont'd)

Cytokine	Cell Source	Target Cell	Effect on Target Cell
Tumor necrosis factor-alpha (TNF-α)	Macrophage, monocyte, neutrophil, T-lymphocyte, LPS-stimulated NK cell	Tumor cell	Cytolysis, cytostasis
		All cells	Inhibit cell-cell contact
		Endothelial cell	Inhibit anti-coagulation, promote thrombosis, angiogenesis
		Neutrophil	Chemotaxis, adhesion to endothelial cells
		Fibroblast	Synthesis collagenase and prostaglandin
		Macrophage	Synthesis IL-1 and prostaglandin, phagocytosis, synthesis superoxide dismutase
		Progenitor cell	Expression MHC Class I, Class II, differentiation antigens, synthesis IL-1, CSF, IFN-γ, arachidonic acid metabolism
Tumor necrosis factor-beta (TNF-β)	T-lymphocyte, leukocyte, fibroblast, endothelial cell, epithelial cell	Fibroblast	Synthesis GM-CSF, G-CSF, IL-1, collagenase, prostaglandin
		Tumor cell	Cytolysis, cytostasis
		Monocyte	Differentiation, synthesis G-CSF
		B-lymphocyte	Mitogen
		Neutrophil	Synthesis reactive oxygen species, chemotaxis, phagocytosis, adhesion of endothelial cells
		Endothelial cell	Proliferation
Lymphotoxin	T-lymphocyte	Neutrophil, endothelial cell	Activation

* Species specific; ref: Ibelgauft, 1998.

Abbreviations: **BCGF**, B-cell growth factor; **BFU**, B-cell forming unit; **CD**, cluster of differentiation; **CFU**, colony-forming unit; **GM-CSF**, granulocyte-macrophage colony stimulating factor; **ICAM**, intracellular adhesion molecule; **IGIF**, interferon-gamma inducing factor; **LCF**, lymphocyte chemoattractant factor; **LPS**, lipopolysaccharide; **MCP**, monocyte chemoattractant protein; **MIP**, macrophage inflammatory protein; **NK**, natural killer; **SCF**, stem cell factor; **TGF**, T-cell growth factor.

HUMORAL IMMUNITY

Antibodies are responsible for the acquired humoral component of immunity. These immunoglobulins are produced by B-lymphocytes and plasma cells, and are composed of two identical light chains (about 24 kD) and two identical heavy chains (about 55 or 70 kD). Antibodies are divided into classes (isotypes), i.e., IgA, IgG, IgE, IgD, and IgM, based on differences in size, charge and solubility. As indicated in Table 2, the isotypes also separate antibodies by function since different isotypes possess different functions.

TABLE 2. EFFECTOR FUNCTIONS OF HUMAN IMMUNO-GLOBULIN ISOTYPES

Antibody	Secretory Form	Function
IgA	Monomer, dimer, trimer	Neutralize toxins
IgD	–	Unknown
IgE	Monomer	Immediate hypersensitivity reactions
IgG	Monomer	Opsonize bacteria Activate complement Immune complex-mediated injury
IgM	Pentamer	Activate complement Immune complex-mediated injury

The ability of antibodies to bind antigen is attributed to a region of the immunoglobulin structure known as the variable (V) region and in particular to the hypervariable portion of the region. Antibody recognition of antigen involves reversible, non-covalent binding, the strength of which increases with repeated immunization. This phenomenon is known as affinity maturation. Antibodies are found on the surface of B-lymphocytes, the cells that synthesize antibodies and on the surface of cells, such as mononuclear phagocytes, mast cells, and NK cells, that possess receptors for antibodies. Antibodies are also found in the plasma, in interstitial fluid, and in some secretory fluids, such as milk and mucus.

B-Lymphocytes may undergo isotype switching resulting in progeny that express different Ig isotypes but the same antigen-binding characteristics. The earliest cell that synthesizes an Ig gene product is the pre-B-lymphocyte. This cell does not produce light chains and does not express surface immunoglobulin. It therefore does not respond to external antigen. Immature B-lymphocytes express IgM on their surface. Encounter of these cells with antigen leads to either cell death or functional unresponsiveness.

B-Lymphocytes continue to recognize the same antigen throughout their life-time. However, repeated encounter with antigen usually results in changes in antibody affinity. The average affinity of the antibody population increases with antigen exposure. This affinity maturation results from mutations that occur in the DNA that codes for the variable region of the antibody molecules, followed by selection of the cells that respond with the highest affinity to antigen.

B-Lymphocytes migrate out of the bone marrow into the circulation and lymphoid tissues. These mature B-lymphocytes possess IgM and IgD and are responsive to antigens. Once stimulated by antigen, they are termed activated B-lymphocytes. Some of the progeny of activated B-lymphocytes undergo heavy chain (isotype) switching and express heavy chain classes that include γ, α, or ε, in addition to μ and δ.

Exposure of the lung to antigen results in recruitment of antibody-forming cells from the blood (Bice and Muggenburg, 1996). Immune responses to antigens delivered to the lung are produced in the lung-associated lymph nodes B-lymphocytes produced in the lymph nodes are released into the blood. These cells do not enter the lung in an antigen-specific manner, but rather as a result of pulmonary inflammation. Furthermore, the cells are not amplified by contact with antigen in the lung. Amplification results from continued cell migration to sites of antigen localization.

ORGANS OF THE IMMUNE SYSTEM

The thymus is located dorsal to the cranial part of the sternum in the thorax, and functions to generate immunocompetent T-lymphocytes from cells released from the bone marrow. It is the site of selection of antigen-specific cells and the deletion of self-reactive T-lymphocytes. The thymus undergoes involution with age, and by puberty is difficult to locate Since T-lymphocyte selection occurs into adulthood, either thymus remnants accomplish this task, and/or extrathymic sites of T-lymphocyte selection exist.

In the thymus, immature antigen-committed T-lymphocytes come into contact with macrophages, epithelial cells and dendritic cells. The cells mature, and are released into the blood, lymph and peripheral lymphoid tissues as $CD4^+$ (T_H) and $CD8^+$ (T_{CTL}) T-lymphocytes.

The spleen is an abdominal, lymphoid organ with a primary function to filter blood. Non-specific phagocytosis occurs in the sinusoidal red pulp of the organ, resulting in the removal of senescent erythrocytes. The white pulp of the spleen is composed of dense lymphoid tissues arranged in periarteriolar sheets, and is the site of the initiation of immunological reactions.

Lymph nodes, located throughout the body at the junction of lymph and blood vessels, function to filter lymph. The major lymph nodes associated with the respiratory tract are indicated in Figure 2. Encounter of pathogens in the lymph with lymphocytes results in the initiation of antigen-specific immunological reactions.

IMMUNE STRUCTURES OF THE RESPIRATORY TRACT

The expansive surface of the lung is protected from airborne contaminants by the persistent surveillance of circulating immune cells. Vast numbers of these cells are prominent in the vascular and interstitial regions of the lung.

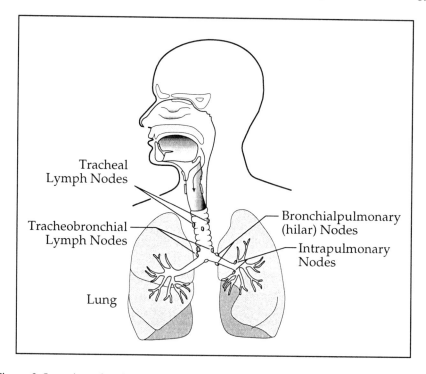

Figure 2. Location of major pulmonary lymph nodes (adapted from Netter, 1980).

Nasal Region

Large diameter materials are removed in the nasal region by filtration and impaction, thus preventing them from entering the deeper regions of the respiratory tract. Further protection is provided by non-specific factors, including mucus, ciliary activity, phagocytic cells, and secretions of antimicrobial molecules, such as lactoferrin and lysozyme.

Associated with many mucosal surfaces throughout the body are organized lymphoid nodules coated with a specialized epithelium. These structures constitute the "associated lymphoid tissue" of the nose (NALT), larynx (LALT), bronchus (BALT), and intestine (Peyer's patches) and are believed to play a central role in induction of immune reactions against particulate and microbial antigens (see Tables 3A and 3B) (Hiller et al., 1998).

The Waldeyer's ring of tonsils and adenoids constitute the laryngeal-associated lymphoid tissue (LALT) in humans (Brandtzaeg and Halstensen, 1992). Although rodents and other animals do not possess tonsils, the upper respiratory tract frequently contains an equivalent structure, nasal-associated lymphoid tissue (NALT) (Spit et al., 1989). This structure is well organized in rats and mice, and is comprised of bilateral strips of non-encapsulated lymphoid tissue underlying the epithelium on the ventral aspect of the posterior nasal passages (Kuper et al., 1992). NALT is a first line of defense against inhaled antigens and is also effective for immunizing against respiratory pathogens.

TABLE 3A. OCCURRENCE OF BRONCHUS- AND NASAL-ASSOCIATED LYMPHOID TISSUE IN NON-HUMAN MAMMALS

Species	% Animals with BALT	% Animals with NALT	Reference
Mice	43	100	Pabst and Gehrke, 1990 van Der Ven and Sminia, 1993
Rats	100	100	Pabst and Gehrke, 1990 Spit, Hendricksen et al., 1989
Rabbits	100	NE	Pabst and Gehrke, 1990
Guinea Pigs	50	NE	Pabst and Gehrke, 1990
Cats	0	NE	Pabst and Gehrke, 1990
Pigs	50	NE	Pabst and Gehrke, 1990

Abbreviations: **BALT**, bronchus-associated lymphoid tissue; **NALT**, nasal-associated lymphoid tissue; **NE**, not examined.

TABLE 3B. OCCURRENCE OF BRONCHUS- AND LARYNGEAL-ASSOCIATED LYMPHOID TISSUE IN HUMANS

Age	% Patients Expressing BALT	% Patients Expressing LALT	Reference
<20 years	40	80	Hiller, Tschernig et al., 1998
>20 years	<10	56	Hiller, Tschernig et al., 1998

Abbreviations: **BALT**, bronchus-associated lymphoid tissue; **LALT**, laryngeal-associated lymphoid tissue; **NE**, not examined.

Cell- and humoral-mediated immune responses are prominent in upper airway defenses Resident immune cells include abundant populations of CD3[+], CD4[+], CD8[+] T-lymphocytes in the superficial lamina propria. The Vδ1-Jδ1-bearing T-lymphocytes predominate and appear to be specific to intraepithelial lymphocytes of the mucosa (Takeuchi et al., 1997). B-Lymphocytes, monocytes, and macrophages are distributed more sparsely throughout the mucosa (Igarashi et al., 1993). Differentiated B-lymphocyte populations are distinguished by surface antigen expression CD22 antigens are present on B-lymphocytes in circulation and those found in the lymph node. B-Lymphocytes located around submucosal glands are J-chain and Ig[+], predominantly IgA[+]. IgG-bearing cells in the nasal mucosa are diffuse and likely arise from the vasculature. The majority of cells staining positive for IgE are mast cells (Igarashi et al., 1993).

B- and T-lymphocyte populations of NALT function through the expression of cytokines that may initiate and perpetuate immune responses (Table 1). Responses may be further exacerbated by release of cytokines and other immune mediators from the nasal epithelium. Proinflammatory interleukins IL-1α, -6, and -8 are produced by human nasal epithelium and contribute to the pathophysiology of allergic respiratory disorders (Kenney et al., 1994).

Bronchus-Associated Lymphoid Tissue (BALT)

The hierarchy of lung lymphatics include nodes, nodules, aggregates and small clusters of cells BALT, an organized accumulation of lymphoid cells situated near branch points throughout the bronchial tree, is commonly restricted to the latter aggregate and cluster structures (Bienenstock et al., 1973). BALT is covered with a layer of flattened epithelial cells, termed M cells (Neutra et al., 1996). BALT has been recognized in animals and in developing humans (Table 3). The constitutive occurrence of BALT varies in mammals and is most extensively found in the rabbit, rat and guinea pig (Table 3A) (Bienenstock, 1984). Chronic stimulation of the airways by cigarette smoking, microbial infection or disease results in hyperplasia of BALT mainly at bifurcations where antigen deposition easily occurs (Sato et al., 1992; Richmond et al., 1993). Whether BALT is necessary after continuous contact with environmental antigens is controversial. Dendritic and bronchial epithelial cells may assume the role of antigen uptake and presentation (Holt, 1993).

Acting as sentinels at a portal of entry, these specialized cells transport inhaled antigens to the adjoining nodules of lymphoid tissue. Within BALT there is a preponderance of IgA precursor cells; these cells are responsible for the production of secretory IgA (sIgA). IgA acts as an immunological barrier of the bronchial tree by preventing the adherence and absorption of antigen (Tomasi, 1992).

A common mucosal immune system has been proposed (Bienenstock, 1984). Lymphocytes obtained from the mucosa are capable of repopulating other mucosal organs as well as the tissue of origin. However, cells obtained from BALT preferentially return to the lung more frequently than cells from the largest mucosal immune system, the intestines (Bienenstock and Befus, 1984).

MUCOSAL IMMUNITY

Mucosa-associated lymphoid tissue (MALT) is the lymphoid tissue or organs in immediate connection with the mucosal layer in nasopharynx, airways and the intestinal tract MALT serves as the immunological defense at secretory surfaces and to some extent is independent of the systemic response. Although IgA constitutes less than one fourth of the immunoglobulin in plasma, it plays a key role in mucosal immunity. It is the only isotype that can be transported across mucosal barriers into the lumens of mucosal-lined organs. Receptors for IgA are expressed on epithelial cells. Dimeric IgA bound on the basal surface of the cell is transported to the mucosal surface where the dimer, and a portion of its receptor (secretory piece), is released into the secretions. Mucosal IgA therefore is larger than serum IgA since it is dimeric and carries an extra peptide component.

The clinical importance of mucosal IgA relates to protection against pathogenic microorganisms and viruses, and perhaps to protection against allergies. Secretory antibodies (both sIgA and sIgM) inhibit uptake of soluble antigen and block epithelial colonization by microorganisms. IgA-deficient patients may compensate with sIgM, or are prone to recurrent airway infections.

Efforts have been directed to the design of effective mucosal vaccines. Induction of mucosal IgA is known to be dependent on CD4+ T-lymphocytes in MALT. The role of IL-6 in regulating production of IgA was demonstrated in mice rendered deficient in IL-6 production as a result of disruption of the gene that encodes IL-6 (Ramsay et al., 1994). Such mice fail to mount an optimal response to injury or infection. Mice inoculated intranasally with virus failed to develop large numbers of antibody-secreting cells, unless restored with the ability to make IL-6.

Interferon-γ (IFNγ) is also important in mucosal immunity. Its role in allergic inflammation was studied by mucosal immunization with IFNγ. Gene transfer of IFNγ to the mucosa of mice was performed by treating mice intratracheally with plasmid-liposome containing gene constructs (Li et al., 1996). Expression of the gene transcripts was detected only in the lungs (epithelium), not in the spleens or livers of the mice. The IFNγ inhibited both antigen and T$_H$2 cell-induced pulmonary eosinophilia, as well as airway hyperreactivity to acetylcholine.

ORCHESTRATION OF THE IMMUNE RESPONSE

The lymph system collects fluid from interstitial spaces and transports it through dedicated vessels to the ductus thoracicus and the blood via the lymph nodes. Two major systems of lymphatic vessels drain the lung. The first begins at the respiratory bronchioles, proceeds centrally and drains at the hilum. The second travels a peripheral course along the pleural surface and drains at the hilum. Negative fluid pressure from the blood capillaries toward the interstitium and lymphatics assures removal of fluid and macromolecules (Guyton et al., 1976). Migration into secondary lymphoid tissues, such as peripheral and mesenteric lymph nodes, occurs by extravasation of lymphocytes through the endothelium of post-capillary, high endothelial venules (HEV).

The concept that "naive" and "memory" T-lymphocytes migrate into lymph and nonlymphoid organs was introduced by Charles Mackay (Mackay et al., 1990). Lymphocytes continually recirculate from blood through tissue, into lymph and back to blood. Continuous circulation throughout the body increases the likelihood that specific immunocompetent cells (especially T-lymphocytes) will encounter antigen. Upon first encounter with antigen, lymphocytes acquire a predilection to home to the same tissue. The continuous recirculation of cells is one of the most characteristic features of the immune system. Highly-regulated trafficking of lymphocytes is responsible for the integration and control of systemic immune responses. This circuit is completed twice per day, increasing the likelihood that naive cells will encounter antigen throughout the body. Although granulocytes and monocytes emigrate from the bloodstream in response to molecular signals, they do not recirculate.

Naive lymphocytes circulate through secondary lymph tissues (lymph nodes, Peyer's patches, tonsils, and spleen). Mucosa-associated lymphocytes home primarily to mucosal tissue. Lymphocytes home to the mucosa-associated lymphoid tissues by way of interaction of adhesion molecules on their surface with those on the surface of high endothelial venules (Burchiel and Davila, 1997). After extravasation, memory B-lymphocytes are regulated by local APCs and T-lymphocytes. *In vitro* studies have shown Class II$^+$-epithelial cells can activate CD8$^+$-suppressor cells (Brandtzaeg et al., 1996).

Differentiation of lymphocytes into memory/effector cells is driven by antigen presentation and occurs in the lymph nodes associated with the lung. These mature lymphocytes traffic through lymphoid organs, while also accessing and recirculating through extralymphoid sites (pulmonary interstitium, inflamed skin and intestinal lamina propria) (Williams and Butcher, 1997; Pitzalis et al., 1991; Schweighoffer et al., 1993; Saltini et al., 1990).

Lymphocytes within the lung most likely arise from transmigration from the vasculature. Tissue-specific homing is regulated in large part by lymphocyte recognition of post-capillary venular endothelial cells (EC) via interaction of lymphocyte receptors and their EC ligands. The nature of the emigration stimulus determines the class of immune cell that will predominate. The specificity of the niche is controlled by a complex array of variables including endothelial wall changes, and local chemokine release (Meeusen et al., 1996).

Several stages are likely involved in lymphocyte-endothelial cell recognition and recruitment of lymphocytes. The initial contact of a flowing leukocyte at the endothelial wall occurs via the selectin family of adhesion molecules (Hogg and Berlin, 1995). This unstable, easily reversible adhesion of leukocytes permits tethering and rolling in the direction of blood flow. P-selectin, in the endothelial cell, is mobilized to the plasma surface to bind neutrophils and monocytes. L-selectin is expressed on the surface of all circulating lymphocytes. E-selectin is induced on the endothelial cells by IL-1, lipopolysaccharide (LPS), or tumor necrosis factor-α (TNFα) (Wellicome et al., 1990). In bacterial-induced pulmonary inflammation, L-selectin maintains the sequestration of neutrophils within capillaries and the accumulation of neutrophils within non-capillary vessels. However, in uninflamed lungs of L-selectin-deficient mice, margination of neutrophils was comparable to that of wild type, implying the contribution of other mediators (Doyle et al., 1997).

The subsequent, activation-dependent adhesion mechanisms require integrins. Seconds after lymphocyte-endothelial contact, adhesion to the vasculature occurs as a result of integrin-mediated activation of receptors on leukocytes. The GTP binding protein, RhoA, has been implicated in this process by acting as an intracellular control point in signaling from chemoattractants and other receptors to lymphocyte integrins (Bacon et al., 1998). The next step in the process, stable arrest, involves the β2 integrins. The final step is diapedesis.

Lymphocytes recruited to different tissues home to specialized microenvironments that may be geographically distinguishable, or have specialized stromal elements. Tissue-specific homing and retention is characterized by a combinatorial display of adhesive ligands and regulatory factors (Westerman and Pabst, 1996). Adhesive function is regulated by signaling through cell surface receptors. Human, pulmonary memory/effector T-lymphocytes express low levels of cutaneous lymphocyte antigen (CLA), L-selectin and moderate levels of $\alpha\epsilon\beta7$-integrin. This contrasts with skin lymphocytes that are CLA and L-selectin positive, but nearly lack the expression of $\alpha\epsilon\beta7$-integrin (Picker et al., 1994). $\alpha\epsilon\beta7$-integrin expression has been implicated in T-lymphocyte binding to mucosal epithelial cells (Cepek et al., 1993). Other factors likely contribute to the pulmonary retention of these cells given the inconsistency of ligand expression on lung T-lymphocytes.

Although many cytokines and growth factors have been implicated in the regulation of lymphocyte trafficking, the chemokines receive particular attention because of their widespread tissue expression and their specificity for distinct lymphocyte subsets. Virtually all nucleated blood and tissue cells produce chemokines (Table 4). Chemokines are small proteins (8-10 kD) and comprise a family containing greater then 40 members (Luster, 1998). Two large subfamilies are distinguished by the arrangement of the first two cysteines which are either separated by one amino acid (CXC chemokines) or are adjacent (CC chemokine). CXC chemokines act mainly on neutrophils; CC chemokines activate monocytes, basophils, eosinophils and T-lymphocytes (Baggiolini et al., 1994). A novel membrane-bound chemokine (CX_3C) "fractalkine" that is a potent chemoattractant for T-lymphocytes and monocytes has been described (Bazan et al., 1997). Its inducible expression on endothelial cells suggests a role in cell trafficking. Another unique chemokine, lymphotactin, is a lymphocyte specific chemokine (Kelner et al., 1994) and has only two cysteines in the mature protein. Antagonists of chemokine activity have potential as anti-inflammatory agents (McFadden and Kelvin, 1997).

TABLE 4. ORGANIZATION OF CHEMOKINE SUPERFAMILY

Subfamily	Chemokine	Target Cell(s)
C	Lymphotactin	Resting T-lymphocyte
CC	Eotaxin, MCP, MIP, RANTES	Activated T-lymphocyte, basophil, eosinophil, monocyte
CXC	ENA-78, GCP, GRO, IL-8, LIX, NAP-2	Neutrophil
CXXXC	Fractalkine	Monocyte, T-lymphocyte

Abbreviations: **MCP**, monocyte chemoattractant protein; **MIP**, macrophage inflammatory protein; **RANTES**, regulated upon activation normal T-lymphocytes expressed and secreted; **ENA-78**, epithelial-cell-derived neutrophil-activating peptide 78; **GCP**, granulocyte chemotactic protein; **GRO**, growth-regulated oncogene; **IL-8**, interleukin-8; **LIX**, lipopolysaccharide-induced CXC chemokine; **NAP-2**, neutrophil-activating peptide 2 (adapted from Luster, 1998).

ANTIGEN PRESENTATION/DENDRITIC CELLS (DC)

As a primary barrier to the environment, the respiratory tract continuously encounters antigen. For encounter to result in initiation of an immune response, the antigen must be effectively presented to naive T-lymphocytes. This process involves dendritic cells (DC) functioning as APCs in lymphoid and nonlymphoid organs. In the lung, the density of the DC network is likely determined by the degree of inhaled antigen, as evidenced by a correlative decrease in DC density with airway diameter (Schon-Hegrad et al., 1991).

Dendritic cells differentiate from monocytes under the influence of the local microenvironment (Lu et al., 1995). GM-CSF is a critical growth factor for DC. Synergistically, GM-CSF and TNFα increase GM-CSF receptor expression on DC (Santiagoschwarz et al., 1993). Species differences exist in the function and phenotype of DC precursors. In rodents, the precursor in the bone marrow is MHC Class II⁻ (Bowers et al., 1983; Inaba et al., 1992); in humans, it expresses CD34 (Reid et al., 1990).

Antigen processing begins with the endocytosis of antigen by the APC, partial degradation, then binding to an MHC molecule. The complex is then delivered to the surface of the cell. Endogenous antigens, processed within cells, are presented by MHC Class I molecules (Rotzschke and Kalk, 1991), and recognized by CD8[+] T-lymphocytes. Exogenous antigens are bound to MHC Class II molecules and recognized by CD4[+] T-lymphocytes (Cresswell, 1994). Class II MHC-bearing DCs make up less than 1% of lung cells and are found within the bronchoalveolar lumen, airway epithelium, alveolar septa, and in connective tissue surrounding pulmonary veins and airway vessels (Sertl et al., 1986; van Haarst et al., 1994). Intramucosal DCs turnover in less than 72 hr, allowing rapid sampling and reaction to antigen, as well as rapid return to baseline upon clearance (Holt et al., 1994). The kinetics of DC response resemble those of the neutrophil response (McWilliam et al., 1994).

Antigen uptake and presentation by DC is influenced by cytokines. *In vitro*, GM-CSF + IL-4 enhances, whereas IFNγ treatment reduces, uptake of antigen by a DC cell line. *In vitro*, IL-2 is a potent DC chemoattractant; intratracheal administration of IL-2 in rodents results in DC accumulation (Kradin et al., 1996).

Phagocytes in the lung influence the manner in which DC handle soluble vs. insoluble antigens. In the rat, inhaled soluble antigen is trapped by airway DC, transported to lymph nodes and presented to naive T-lymphocytes (Xia et al., 1995). By contrast, the pulmonary immune response to inhaled particulates is generated only when the protective phagocytic capacity of the lung is exceeded. Excess antigen interacts with DC and provokes an antigen-specific immune response (MacLean et al., 1996).

Pulmonary dendritic cells possess a portion of the receptor for IgE. Asthmatics display both a greater number of DC in the airways compared with non-asthmatics, and a higher proportion of DC bearing an IgE receptor (Tunon-de Lara et al., 1996). Expression of the IgE receptor may increase the ability of DC to capture, process and present antigens to T-lymphocytes, suggesting an important function of DC in pulmonary disorders such as asthma. The integral role of DC in the perpetuation of airway inflammation has been demonstrated using mice in which airway DCs were chemically ablated (Lambrecht et al., 1998).

Neuropeptides have been implicated in the regulation of DC. In the normal rat lung DC are located proximal to unmyelinated nerve fibers that contain the tachykinin neuropeptide substance P (SP). Depletion of SP by capsaicin inhibits the pulmonary response to inhaled antigen including perivascular accumulation of DC and other inflammatory cells (Kradin et al., 1997). Alteration of DC function may offer novel therapeutic strategies to control chronic diseases of the lung such as asthma.

ACTIVATION/ACUTE AND CHRONIC ACTIVATION

The frequent exposure of the respiratory epithelium to inhaled, environmental antigens implies that T-lymphocyte responses are under strict control. Antigen-specific mechanisms are required to suppress T-lymphocyte locomotion and increase APC contact time. T-Lymphocytes must interact with APCs for several hours to initiate cytokine production and enter the cell cycle. Engagement of the T-lymphocyte receptor results in an essential stop signal for altered locomotion and lymphocyte activation (Dustin et al., 1997).

The T-lymphocyte population in the lung exhibits post-activation characteristics and expression of IL-2 receptor and CD2 below those expected of activated populations (Strickland et al., 1996). Sensitized quiescent T-lymphocytes, sequestered in the lung respond rapidly to secondary antigen encounter leading to specific T-lymphocyte activation and to the generation of pulmonary cell-mediated immunity (Marathias et al., 1991).

Resolution of airway inflammation and prevention of tissue damage is incompletely understood. Airway epithelial cells secrete immunoregulatory products such as adhesion molecules, cytokines, enzyme inhibitors, inducible nitric oxide, lipid mediators, MHC molecules, metalloendopeptidases, and regulatory peptides (Levine, 1995). A chronic inflammatory state is exemplified by eosinophil accumulation and activation in the airways of asthmatics. Apoptosis appears to be involved in suppression of T- and B-lymphocyte immunity and inflammation in humans (Mariana and Krammer, 1998). *In situ*, allergen-challenged murine eosinophils express *Fas* receptor and are subject to apoptosis regardless of co-culturing in the presence of the eosinophil hematopoietins IL-3, -5, and GM-CSF (Tsuyuki et al., 1995).

SUMMARY

The immune system of the lung is composed of both innate and acquired elements. Through soluble mediators, such as cytokines and chemokines, the innate system has been found to have a profound effect on development and maintenance of acquired immunity. Various lymphoid structures are associated with the lung including lymph nodes, nodules, and lymphocyte aggregates. The number and complexity of the structures decrease with reduced airway diameter. This correlation probably reflects the reduced contact of the lower airways relative to the larger airways to inhaled foreign agents. The lung possesses mucosal immune tissue similar to that of other organs, such as the intestine. This lymphoid structure has the potential for efficient vaccination against pathogens.

Acquired immunity is composed of antigen-specific T- and B-lymphocytes, and of antibodies. Development of acquired immunity relies on a sequence of events initiated by contact of the inhaled material with dendritic cells. Migration of the cells to local lymphoid tissue enables contact and presentation of the processed material to naive T-lymphocytes.

Mature cells circulate to other lymphoid organs. Through a complex process involving release of soluble chemoattractants, cytokines, cell surface ligands and receptors, lymphocytes home to native lymphoid areas and establish an immune response. Increased understanding of the activation, inactivation, and regulatory control elements of the immune system is necessary to prevent chronic immunologic lung disease.

ACKNOWLEDGMENTS

The authors thank Timothy C. Mitchell for technical research, and Donna Stoliker for assistance with preparation of the manuscript.

REFERENCES

Abbas AK, Lichtman AH, Pober JS. "Appendix: Principal Features of Known CD Molecules." In *Cellular and Molecular Immunology, 3rd Edition*, AH Lichtman, JS Pober, eds. Philadelphia: W.B. Saunders Company, 1997, pp. 463-469.

Agostini C, Chiloshi M, Zambello R, Trentin L, Semenzato G. Pulmonary immune cells in health and disease: Lymphocytes. Eur. Respir. J., 1993;6:1378-1401.

Bacon KB, Schall TJ, Dairaghi DJ. RANTES activation of phospholipase D in Jurkat T-cells: Requirement of GTP-binding proteins ARF and RhoA. J. Immunol., 1998;160:1894-1900.

Baggiolini M, Dewald B, Moser B. Interleukin-8 and related chemotactic cytokines-CXC and CC chemokines. Adv. Immunol., 1994;55:97-179.

Bazan JF, Bacon KB, Hardiman G, Wang W, Soo K, Rossi D, Greaves DR, Zlotnik A, Schall TJ. A new class of membrane-bound chemokine with a CX3C motif. Nature, 1997;385:640-644.

Bice DE, Muggenburg BA. Pulmonary immune memory: Localized production of antibody in the lung after antigen challenge. Immunology, 1996;88:191-197.

Bienenstock J. "Bronchus-Associated Lymphoid Tissue." In *Immunology of the Lung and Upper Respiratory Tract*, J Bienenstock, ed. New York: McGraw Hill, 1984, pp. 96-118.

Bienenstock J, Befus D. Gut- and bronchus-associated lymphoid tissue. Am. J. Anat., 170;1984:437-445.

Bienenstock J, Johnston N, Perey DYE. Bronchial lymphoid tissue I. Morphological characteristics. Lab. Invest., 28;1973:686-692.

Bowers WE, Berkowitz MR, La Bade JH, Klinkert WEF. Differentiation of dendritic cells from rat bone marrow. Transplant. Proc., 1983;15:1611-1612.

Brandtzaeg P, Halstensen TS. Immunology and immunopathology of tonsils. Acta Otorlaryngol., 1992;47:64-75.

Brandtzaeg P, Jahnsen FL, Farstad IN. Immune functions and immunopathology of the mucosa of the upper respiratory pathways. Acta Otorlaryngol., 1996;116:149-159.

Burchiel SW, Davila DR. "Mucosal Immunology." In *Comprehensive Toxicology*, IG Sipes, CA McQueen, AJ Gandolphi, eds. *Volume 5: Toxicology of the Immune System*, DA Lawrence, ed. Cambridge: Pergamon, 1997, pp. 143-51.

Cepek KL, Parker CM, Madara JL, Brenner MB. Integrin $\alpha\epsilon\beta7$ mediates adhesion of T-lymphocytes to epithelial cells. J. Immunol., 1993;150:3459-3470.

Cresswell P. Antigen presentation. Getting peptides into MHC Class II molecules. Curr. Biol., 1994;4:541-543.

Doyle NA, Bhagwan SD, Meek BB, Kutkoski GJ, Steeber DA, Tedder TF, Doerschuk CM. Neutrophil margination, sequestration, and emigration in the lungs of L-selectin-deficient mice. J. Clin. Invest., 1997;99:526-533.

Dustin ML, Bromley SK, Kan ZK, Peterson DA, Unanue ER. Antigen receptor engagement delivers a stop signal to migrating T-Lymphocytes. Proc. Natl. Acad. Sci. USA, 1997;94:3909-3913.

Gabrilove JL, Jakubowski A. Hematopoietic growth factors: Biology and clinical application. J. Natl. Cancer Inst. Monogr., 1990;10:73-77.

Guyton AC, Taylor AE, Brace RA. A synthesis of interstitial fluid regulation and lymph formation. Fed. Proc. Am. Soc. Exp. Biol., 1976;35:1881-1885.

Hiller AS, Tschernig T, Kleeman WJ, Pabst R. Bronchus-associated lymphoid tissue (BALT) and larynx-associated lymphoid tissue (LALT) are found at different frequencies in children, adolescents and adults. Scand. J. Immunol., 1998;47:159-162.

Hogg N, Berlin C. Structure and function of adhesion receptors in leukocyte trafficking. Immunol. Today, 1995;16:327-330.

Holt PG. Regulation of antigen-presenting cell function(s) in lung and airway tissues. Eur. Respir. J., 1993;6:120-129.

Holt PG, Haining S, Nelson DJ, Sedwick JD. Origin and steady-state turnover of Class II MHC-bearing dendritic cells in the epithelium of the conducting airways. J. Immunol., 1994;153:256-261.

Ibelgauft H. Cytokine Online Pathfinder Encyclopaedia. http://bioinfo.weizmann.ac.il/cgi-bin/cope. Laboratory of Molecular Biology - Gene Center, Ludwig - Maximilians - University of Munich, FRG: Munich, FRG.

Igarashi Y, Kaliner MA, Hausfeld JN, Irani A-MA, Schwartz LB, White MV. Quantification of resident inflammatory cells in the human nasal mucosa. J. Allergy Clin. Immunol., 1993;91:1082-1093.

Inaba K, Inaba M, Romani N, Aya H, Deguchi M, Ikehara S, Muramatsu S, Steinman RM. Generation of large numbers of dendritic cells from mouse bone marrow cultures supplemented with granulocyte/macrophage-colony stimulating factor. J. Exp. Med., 1992;176:1693-1702.

Kelner GS, Kennedy J, Bacon KB, Kleyensteuber S, Largaespada DA, Jenkins NA, Copeland NG, Bazan JF, Moore KW, Schall TJ, Zlotnik A. Lymphotactin: A cytokine that represents a new class of chemokine. Science, 1994;266:1395-1399.

Kenney JS, Baker C, Welch MR, Altman LC. Synthesis of interleukin-1α, interleukin-6 and interleukin-8 by cultured human nasal epithelial cells. J. Allergy Clin. Immunol., 1994;93:1060-1067.

Kradin RL, Xia W, Pike M, Randolph Byers H, Pinto C. Interleukin-2 promotes the motility of dendritic cells and their accumulation in lung and skin. Pathobiology, 1996;64:180-186.

Kradin R, MacLean J, Duckett S, Schneeberger EE, Waeber C, Pinto C. Neuroimmune interactions promote the recruitment of dendritic cells to the lung and the cellular immune response to inhaled antigen. Am. J. Pathol., 1997;150:1735-1743.

Kroegel C, Warner JA, Virchow JC, Matthys H. Pulmonary immune cells in health and disease: The eosinophil leucocyte (part II). Eur. Respir. J. 1994;7:743-760.

Kuper CF, Koornstra PJ, Hameleers DM, Biewenga J, Spit BJ, Duijvestijn AM, van Breda Vriesman PJ, Sminia T. The role of nasopharyngeal lymphoid tissue. Immunol. Today, 1992;13:219-224.

Lambrecht BN, Salomon B, Klatzmann D, Pauwels RA. Dendritic cells are required for the development of chronic eosinophilic airway inflammation in response to inhaled antigen in sensitized mice. J. Immunol., 1998;160:4090-4097.

Levine SJ. Bronchial epithelial cell-cytokine interactions in airway inflammation. J Invest. Med., 1995;43:241-249.

Li XM, Chopra RK, Chou TY, Schofield BH, Wills-Karp M, Huang SK. Mucosal IFN-γ gene transfer inhibits pulmonary allergic responses in mice. J. Immunol., 1996;157:3216-3219.

Lu L, Rudert WA, Qian S, McCaslin D, Fu F, Rao AS, Trucco M, Fung JJ, Starzi TE, Thomson AW. Growth of donor-derived dendritic cellls from the bone marrow of murine liver allograft recipients in response to granulocyte/macrophage colony-stimulating factor. J. Exp. Med., 1995;182:379-387.

Luster AD. Chemokines - Chemotactic cytokines that mediate inflammation. N. Engl. J. Med., 1998;338:436-445.

Mackay CR, Marston WL, Dudler L. Naive and memory T-cells show distinct pathways of lymphocyte recirculation. J. Exp. Med., 1990;171:801-817.

MacLean JA, Xia W, Pinto CE, Zhao L, Lui HW, Kradin RL. Sequestration of inhaled particulate antigens by lung phagocytes. A mechanism for the effective inhibition of pulmonary cell-mediated immunity. Am. J. Pathol., 1996;148:657-666.

Marathias KP, Preffer FI, Pinto C, Kradin RL. Most human pulmonary infiltrating lymphocytes display the surface immune phenotype and functional responses of sensitized T-cells. Am. J. Respir. Cell Mol. Biol., 1991;5:470-476.

Mariana SM, Krammer PH. Differential regulation of TRAIL and CD95 ligand in transformed cells of the T- and B-lymphocyte lineage. Eur. J. Immunol., 1998;28:973-982.

McFadden G, Kelvin D. New strategies for chemokine inhibition and modulation. Biochem. Pharmacol., 1997;54:1271-1280.

McWilliam AS, Nelson D, Thomas JA, Holt PG. Rapid dendritic cell recruitment is a hallmark of the acute inflammatory response at mucosal surfaces. J. Exp. Med., 1994;179:1331-1336.

Meeusen NT, Premier RR, Brandon MR. Tissue-specific migration of lymphocytes: A key role for T_H1 and T_H2? Immunol. Today, 1996;17:421-424.

Nagler A, Lanier LL, Cwirla S, Phillips JH. Comparative studies of human F_cRIII-positive and negative natural killer cells. J. Immunol., 1989;143:3183-3191.

Neutra MR, Frey A, Kraehenbuhl J-P. Epithelial M cells: Gateways for mucosal infection and immunization. Cell, 1996;86:345-348.

Netter, FH. The CIBA Collection of Medical Illustrations. Vol. 7, Respiratory System, MB. Divertie, A Brass, eds. Rochester, NY:CIBA-GEIGY, 1980, p.32.

Picker LJ, Martin RJ, Trumble A, Newman LS, Collins PA, Bergstresser, Leung DYM. Differential expression of lymphocyte homing receptors by human memory/effector T-cells in pulmonary versus cutaneous immune effector sites. Eur. J. Immunol., 1994;24:1269-1277.

Pitzalis C, Kingsley GH, Covelli M, Meliconi R, Markey A, Panayi GS. Selective migration of the human helper-inducer memory T-cell subset: Confirmation by in vivo cellular kinetic studies. Eur. J. Immunol., 1991;21:369-376.

Ramsay AJ, Husband AJ, Ramshaw IA, Bao S, Matthaei KI, Koehler G, Kopf M. The role of interleukin-6 in mucosal IgA antibody responses in vivo. Science, 1994;264:561-563.

Reid CD, Fryer PJ, Clifford C, Kirk A, Tikerpae J, Knight SC. Identification of hematopoietic progenitors of macrophages and dendritic Langerhans cells (DL-CFU) in human bone marrow and peripheral blood. Blood, 1990;76:1139-1149.

Richmond I, Pritchard GE, Ashcroft T, Avery A, Corris PA, Walters EH. Bronchus-associated lymphoid tissue (BALT) in human lung: Its distribution in smokers and non-smokers. Thorax, 1993;48:1130-1134.

Rotzschke O, Falk K. Naturally-occurring peptide antigens derived from the MHC Class-I-restricted processing pathway. Immunol. Today, 1991;12:447-455.

Saltini C, Kirby M, Trapnell BC, Tamura N, Crystal RG. Biased accumulation of T-lymphocytes with 'memory'-type CD45 leukocyte common antigen gene expression on the epithelial surface of the human lung. J. Exp. Med., 1990;171:1123-1140.

Santiagoschwarz F, Divaris N, Kay C, Carsons SE. Mechanisms of tumor necrosis factor-granulocyte-macrophage-colony stimulating factor-induced dendritic cell development. Blood, 1993;82:3019-3028.

Sato A, Chida K, Iwata M, Hayakawa H. Study of bronchus-associated lymphoid tissue in patients with diffuse pan-bronchiolitis. Am. Rev. Respir. Dis., 1992;146:473-478.

Schon-Hegard MA, Oliver J, McMenamin PG, Holt PG. Studies on the density, distribution and surface phenotype of intraepithelial Class II major histocompatibility complex antigen (Ia)-bearing dendritic cells (DC) in the conducting airways. J. Exp. Med., 1991;173:1345-1356.

Schweighoffer T, Tanaka Y, Tidswell M, Erle DJ, Horgan KJ, Ginther Luce GE, Lazarovits AI, Buck D, Shaw S. Selective expression of integrin $\alpha4\beta7$ on a subset of human CD4$^+$ memory T-cells with hallmarks of gut-tropism. J. Immunol., 1993;151:717-729.

Sertl K, Takemura T, Tschachler E, Ferrans VJ, Kaliner MA, Shevach EM. Dendritic cells with antigen presenting capability reside in airway epithelium, lung parenchyma, and visceral pleura. J. Exp. Med., 1986;163:436-451.

Sibille Y, Marchandise FX. Pulmonary immune cells in health and disease: Polymorphonuclear neutrophils. Eur. Respir. J., 1993;6:1529-1543.

Sibille Y, Reynolds HY. Macrophages and polymorphonuclear neutrophils in lung defense and injury. Am. Rev. Respir. Dis., 1990;141:471-501.

Spit BJ, Hendricksen EGJ, Bruijntjes JP, Kuper CF. Nasal lymphoid tissue in the rat. Cell Tissue Res., 1989;255:193-198.

Strickland D, Kees UR, Holt PG. Regulation of T-cell activation in the lung: Isolated lung T-cells exhibit surface phenotype characteristics of recent activation including downmodulated T-cell receptors, but are locked into the G_0/G_1 phase of the cell cycle. Immunology, 1996;87:242-249.

Takeuchi K, Hirata N, Ukai K, Sakakura Y. Analysis of $\gamma\delta$ T-cell receptor repertoire in the human nasal mucosa. J. Allergy Clin. Immunol., 1997;99:251-253.

Tomasi TB. The discovery of secretory IgA and the mucosal immune system. Immunol. Today, 1992;13:416-418.

Tsuyuki S, Bertrand C, Erard F, Trifilieff A, Tsuyuki J, Wesp M, Anderson GP, Coyle AJ. Activation of the *Fas* receptor on lung eosinophils leads to apoptosis and the resolution of eosinophilic inflammation of the airways. J. Clin. Invest., 1995;96:2924-2931.

Tunon-de Lara JM, Redington AE, Bradding P, Church MK, Hartley JA, Semper AE, Holgate ST. Dendritic cells in normal and asthmatic airways: Expression of the α subunit of the high affinity immunoglobulin E receptor ($F_{c\epsilon}RI-\alpha$). Clin. Exp. Allergy, 1996;26:648-655.

van Haarst JMW, deWit HJ, Drexhage HA, Hoogsteden HC. Distribution and immunophenotype of mononuclear phagocytes and dendritic cells in the human lung. Am. J. Respir. Cell Mol. Biol., 1994;10:487-492.

Warner JA, Kroegel C. Pulmonary immune cells in health and disease: Mast cells and basophils. Eur. Respir. J., 1994;7:1326-1341.

Warren HS. NK cell proliferation and inflammation. Immunol. Cell Biol., 1996;74:473-480.

Weigle WO. "Overview of the Immune System." In *Comprehensive Toxicology*, IG Sipes, CA McQueen, AJ Gandolphi, eds. *Volume 5: Toxicology of the Immune System*, DA Lawrence, ed. Cambridge: Pergamon, 1997, p. 19.

Wellicome SM, Thornhill MH, Pitzalis C, Thomas DS, Lanchbury JSS, Panayi GS, Haskard DO. A monoclonal antibody that detects a novel antigen on endothelial cells that is induced by tumor necrosis factor, IL-1, or lipopolysaccharide. J. Immunol., 1990;144:2558-2565.

Westermann J, Pabst R. How organ-specific is the migration of 'naive' and 'memory' T-cells? Immunol. Today, 1996;17:278-282.

Williams MB, Butcher EC. Homing of naive and memory T-lymphocyte subsets to Peyer's patches, lymph nodes, and spleen. J Immunol 1997;159:1746-1752.

Xia W, Pinto CE, Kradin RL. The antigen-presenting activities of Ia+ dendritic cells shift dynamically from lung to lymph node after an airway challenge with soluble antigen. J. Exp. Med., 1995;181:1275-1283.

3. DISPOSITION OF INHALED PARTICLES AND GASES

Richard B. Schlesinger, Ph.D.
New York University School of Medicine
Department of Environmental Medicine
57 Old Forge Road
Tuxedo, New York 10987

INTRODUCTION

Biological responses following the inhalation of chemical toxicants are dependent upon the disposition of these materials within the respiratory system. This, in turn, depends upon patterns of deposition, i.e., the sites where material carried in the airstream initially contacts epithelial surfaces, and clearance, i.e., the routes via which deposited material is translocated within the respiratory tract. For chemicals which exert their action upon surface contact, initial deposition is the main predicator of toxic response. In many other cases, however, it is the net result of deposition and clearance, namely retention, i.e., the amount of toxicant remaining in the respiratory tract at specific times after exposure, which influences response.

Immunotoxic chemicals may be inhaled in the gaseous or particulate phases; these latter occur as aerosols, suspensions of finely dispersed liquid or solid particles in air. This Chapter provides an overview of the fate of inhaled particles and gases within the respiratory tract.

DISPOSITION OF INHALED PARTICLES

Characterization of Inhaled Particles

Airborne particles consist of material having diverse physicochemical characteristics, but the one having the greatest influence upon particle behavior within the respiratory tract is size. However, specific chemical makeup and other physical characteristics, such as surface area, are factors in the production of biological responses.

An understanding of particle deposition processes requires some background related to the size characterization of aerosols. It is important that size be expressed in such a manner that particle behavior within an airstream can be properly described. In many cases, such behavior is a function not of the actual physical or geometric size of the particle, such as would be obtained with a microscopic measurement, but rather of what is termed an "aerodynamic" size. Since particles having the same physical size but having different shapes and densities may behave differently once inhaled into the respiratory tract, the use of this parameter allows for normalization in terms of the airborne behavior of particles having diverse physical characteristics. One common descriptor in this regard is aerodynamic equivalent particle diameter (D_{ae}), defined as the diameter of a spherical shaped particle having a unit density (1 g/mL) which has the same settling velocity within a gravitational field as the particle of interest. In general, the aerodynamic behavior and deposition characteristics for particles with actual physical diameters $\geq$ 0.5 μm is governed largely by their aerodynamic diameter. On the other hand, the behavior of smaller inhaled particles is governed largely by actual physical diameter.

All particles within an aerosol are rarely the same size, and aerosols can be classified by the size uniformity of their constituent particles relative to some central tendency. Because particle sizes within aerosols generally show a log-normal distribution, the statistical description of central tendency is the median diameter. The specific measurement basis for this median can, however, vary. It may be reflective of the actual number of particles in an aerosol, in which case it is termed the count median diameter (CMD), it can reflect particle mass, in which case it is the mass median diameter (MMD), or it can be based upon the aerodynamic diameter, such as the mass median aerodynamic diameter (MMAD).

Aerosols are also classified in terms of the size uniformity of their constituent particles relative to the median. This measure of variability is the geometric standard deviation (s_g). An aerosol is characterized as monodisperse if the constituent particles are essentially all the same size; in such a case, $s_g = 1$. A polydisperse aerosol contains particles of different sizes, and $s_g > 1$. If the s_g of a polydisperse aerosol is <2, the total amount of deposition within the respiratory tract will probably not differ substantially from that for a monodisperse aerosol having the same median size. However, size distribution is critical in determining the regional pattern of deposition, since this depends upon the sequential removal of particles within each successive region of the respiratory tract through which the particles pass following inhalation and which, in turn, depends upon the actual particle sizes present within the aerosol. Thus, while the total respiratory tract deposition of a monodisperse and a polydisperse aerosol having the same median size may be comparable, regional deposition patterns may show substantial differences.

Particle Deposition

Mechanisms of Deposition

Inhaled particles may deposit within the respiratory tract by five mechanisms. These are impaction, sedimentation, Brownian diffusion, electrostatic precipitation, and interception. While one mechanism may predominately control the deposition of specific particles within specific respiratory tract regions, they generally act in a competitive manner.

Impaction is the inertial deposition of a particle onto an airway surface. It occurs when the particle's momentum prevents it from changing direction of movement when there is a rapid change in the direction of bulk airflow. Impaction is the main mechanism by which particles having $D_{ae} \geq 0.5$ µm deposit in the upper respiratory tract, and at or near branching points (bifurcations) of the tracheobronchial tree. The probability of impaction increases with increasing air velocity, rate of breathing, and particle aerodynamic diameter.

Sedimentation is deposition due to gravity. When the gravitational force on an airborne particle is balanced by the total of forces due to air buoyancy and air resistance, the particle will fall out of the inhaled air stream at a constant rate. In order for deposition to occur, this rate must be attained before the particle exits the airway. The probability of sedimentation increases with the particle's residence time in the airway and with aerodynamic diameter, and decreases with increasing breathing rate. Sedimentation is an important deposition mechanism for particles with $D_{ae} \geq 0.5$ µm which penetrate to airways where air velocity is relatively low, e.g., mid to small bronchi and bronchioles.

Submicrometer-sized particles, especially ultrafines, which are those having actual physical diameters <0.1 µm, acquire a random motion due to bombardment by surrounding air molecules; this motion may then result in particle contact with the airway wall. The displacement sustained by the particle is a function of a parameter known as the diffusion coefficient, which is inversely related to actual particle size (specifically cross-sectional area) but is independent of particle density and, therefore, of aerodynamic diameter. The probability of deposition by diffusion increases with increasing particle residence time within the airway, and diffusion is a major deposition mechanism where the bulk air flow is low or absent, e.g., in bronchioles and the alveolated airways. However, ultrafine particles can show significant deposition in the upper respiratory tract, the trachea and larger bronchi; this likely occurs by turbulent diffusion.

Airborne particles can be electrically charged and such particles may exhibit enhanced deposition over that expected based upon size alone. Deposition can be due to image charges induced on the surface of the airway by these particles, and/or to space-charge effects, whereby repulsion of particles containing like charges results in increased migration towards the airway wall. The effect of charge on deposition is inversely proportional to actual particle size and airflow rate.

Interception is a significant deposition mechanism for fibrous particles, which are those having length-to-diameter ratios >3:1. While fibers are also subject to all of the same deposition mechanisms as are more spherical or compact particles, they have the additional possibility of deposition when an edge contacts, or intercepts, an airway wall. The probability of interception increases as airway diameter decreases, but it can also be fairly significant in both the upper respiratory tract (larynx and proximal airway) and upper tracheobronchial tree. While interception probability increases with increasing fiber length, the aerodynamic behavior of a fiber and the probability of impaction or sedimentation is influenced largely by fiber diameter.

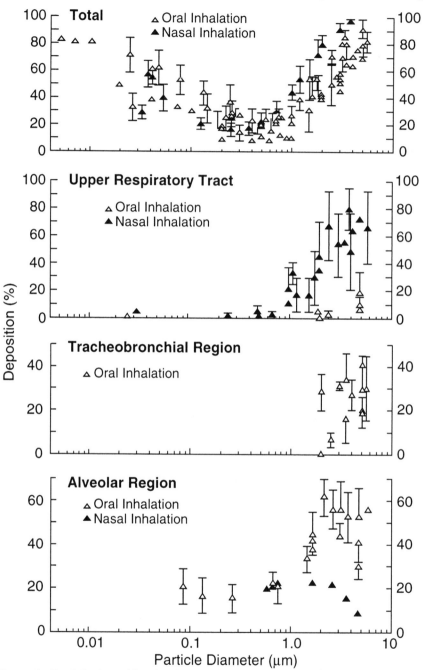

Figure 1. Particle deposition in the human respiratory tract. The % deposition represents the deposition efficiency, i.e., the percentage deposition of the amount originally inhaled into the respiratory tract. This is plotted as a function of the median particle size of the inhaled aerosol. Particle size is aerodynamic equivalent (MMAD) for particles ≥0.5 μm and diffusion equivalent for those <0.5 μm. Modified from Schlesinger (1995).

Deposition Patterns Within the Respiratory Tract

The net result of the deposition mechanisms described above is shown in Figure 1 which depicts experimentally-determined deposition patterns within the human respiratory tract as a function of median size of the inhaled aerosol. The top panel shows the pattern of total respiratory tract deposition. There is a deposition minima around the 0.1 - 0.5 µm size range, with increasing deposition with increasing size for larger particles and with decreasing size for smaller ones. As previously noted, particles with diameters ≥0.5 µm are subject to impaction and sedimentation, while deposition of those is ≤0.1 µm is diffusion dominated. Particles with diameters between these values are not strongly influenced by any of these mechanisms and tend to have relatively prolonged suspension times in the inhaled air. They undergo minimal deposition after inhalation and most are carried out of the respiratory tract in the exhaled air.

Humans can breath either through the nose and mouth, and mode of breathing clearly affects total deposition. Inhalation via the nose results in greater total deposition than does oral inhalation for particles with diameters >0.5 µm; this is due to enhanced collection in the upper respiratory tract with the former. On the other hand, there is little apparent difference in total deposition between nasal or oral breathing for those particles with diameters between 0.02 - 0.5 µm. For particles below this size range, an increase in total deposition with nose compared to mouth breathing also occurs, but the difference is relatively small.

In many instances, production of a toxicologic response depends upon deposition of inhaled particles within specific areas of the respiratory tract. The respiratory tract is commonly divided into three regions for describing regional deposition; the upper respiratory tract, the tracheobronchial tree, and the alveolar region.

The second panel shows deposition in the upper respiratory tract. It is evident that nasal inhalation results in enhanced deposition compared to oral; this accounts for differences in total respiratory tract deposition related to breathing mode which were noted above in reference to total respiratory tract deposition. The greater the deposition in the upper respiratory tract, the less is the amount available for removal in the lungs.

The third panel depicts deposition in the tracheobronchial tree. There appears not to be as well a defined relationship between deposition and particle size as in other regions; fractional tracheobronchial deposition is relatively constant over a wide particle size range.

Deposition in the alveolar region is shown in the bottom panel. With oral inhalation, deposition increases with particle size after a minimum at ~0.5 µm. With nasal breathing, on the other hand, deposition tends to decrease with increasing particle size. The removal of particles in more proximal airways determines the shape of these curves. For example, increased upper respiratory and tracheobronchial deposition would be associated with a reduction of alveolar region deposition; thus, nasal breathing results in less penetration of larger particles in the deep lung, and a lesser fraction of deposition for entering aerosol than does oral inhalation. Thus, in the latter case, the peak for pulmonary deposition shifts upwards to a larger sized particle, and is more pronounced. On the other hand, with nasal breathing, there is a relatively constant alveolar region deposition over a wider particle size range.

Deposition of ultrafine particles is of special interest in toxicology since these particles present a large surface area for potential adsorption of other materials for delivery to the respiratory tract. From Figure 1 it can be seen that total respiratory tract deposition increases as particle size decreases <0.2 μm. In terms of regional patterns, as particle size decreases <0.2 μm, deposition in the upper respiratory tract and tracheobronchial tree increases substantially, while deposition within the alveolar region decreases. Deposition of ultrafine particles in the upper respiratory tract, especially the nasal passages, can be quite high.

The deposition profiles presented in Figure 1 are for spherical or compact particles. Due to the potential toxicity of fibrous particles, experimental fiber deposition data in humans is not available. However, studies in animals and with mathematical and physical models (e.g., Hammad et al., 1982; Asgharian and Yu, 1988, 1989; Sussman et al., 1991) note that long fibers (>10 μm) tend to show enhanced deposition in the tracheobronchial tree, and reduced deposition in the pulmonary region, compared to shorter fibers. But fibers which are very long (e.g., >50 μm) and thin (e.g., <0.5 μm) can reach distal conducting airways and significant amounts of such particles can deposit in the alveolar region. However, the deposition of fibers is much more complex than that for spherical particles. For example, shape of the former is important, as straight fibers penetrate more distally than do curly ones.

Many inhaled particles have a characteristic which may dynamically alter size after inhalation. This is hygroscopicity. Hygroscopic particles will absorb water in the humid atmosphere of the respiratory tract and will, thus, grow while they are still airborne within the airways. The deposition characteristics of such particles is a function of their hydrated size rather than of their initial size when inhaled.

Localized Patterns of Particle Deposition

Particle deposition is not homogeneous along airway surfaces. Specific patterns of enhanced local deposition are important in determining initial dose, which depends on the surface density of deposition. Nonuniformity implies that this dose delivered to specific sites may be greater than that occurring if a uniform density of surface deposit is assumed. This may be a factor in the site selectivity of certain diseases (e.g., Schlesinger and Lippmann, 1972).

In the upper respiratory tract, enhanced deposition occurs at areas characterized by constrictions, directional changes and high air velocities, such as the larynx, oropharyngeal bend, and nasal turbinates. Likewise, deposition in the tracheobronchial tree is not homogeneous. Airflow turbulence produced by the larynx results in enhanced localized deposition in the upper trachea and larger bronchi, while deposition is also greatly enhanced at bronchial bifurcations, especially along the carinal ridges. This occurs for particles >0.5 μm diameter due to impaction, as well as for submicrometer particles with diameters down to ≈0.1 μm due to turbulent diffusion (Schlesinger et al., 1982; Cohen et al., 1988). As particle size decreases further, effects of localized flow patterns upon particle behavior become less important, resulting in more uniform deposition along airway surfaces (Gradon and Orlicki, 1991). Nonuniform deposition also occurs within the alveolar region, at bifurcations of alveolar ducts near bronchioalveolar junctions (Brody and Roe, 1983; Warheit and Hartsky, 1990).

Factors Modulating Deposition

Various factors may alter deposition patterns compared to those occurring in normal, healthy adults. These include previous or co-exposure to airborne irritants, lung disease, and growth, all of which can affect deposition by changing its controlling parameters, namely ventilation pattern/airway geometry. For example, irritant inhalation-induced bronchoconstriction would tend to increase impaction deposition in the upper bronchial tree. Likewise, deposition may be altered due to the presence of a disease state (Bennett et al., 1997; Kim and Kang, 1997). Bronchial obstruction associated with various pulmonary conditions tends to enhance deposition within the upper respiratory tract and tracheobronchial tree, especially for particles >1 μm; on the other hand, peripheral lung deposition may be reduced. In fact, deposition may be eliminated in entire portions of the lungs due to impaired ventilation. Deposition of ultrafine particles may also increase in obstructive lung disease due to increased residence time and to flow perturbations arising from reduced airway lumen caliber.

One of the current concerns in toxicology involves differences in deposition between children and adults. A number of attempts have been made to estimate the influence upon deposition of anatomical and ventilatory changes during postnatal growth (e.g., Hofmann, 1982; Crawford, 1982; Phalen et al., 1991). They indicate that the relative effectiveness of the major deposition mechanisms likely differ at various times during growth and that this, in turn, may alter regional deposition patterns. Taking into account anatomical differences and the greater ventilation per unit body weight in children, the deposition fractions for some particle sizes, especially those >1 μm, within certain regions of the growing respiratory tract could be quite different, sometimes well above, those occurring in adults; such differences would become even more significant when deposition is expressed on a per unit surface area basis (Bennett et al., 1996; Bennett and Zeman, 1998). Thus, the initially deposited dose to specific lung compartments from inhaled particles may vary with age from newborn to adult. Anatomical changes with aging post-maturity may also affect deposition for particles >1 μm, increasing pulmonary region deposition in older compared to younger adults (Phalen et al., 1991). On the other hand, the deposition of ultrafine particles does not show dramatic differences between children and adults, nor with aging beyond adolescence (Phalen et al., 1991; Swift et al., 1992).

Interspecies Deposition Patterns

Various animals are employed in experimental inhalation toxicological studies, with the ultimate goal being extrapolation to humans. To adequately apply results to human risk assessment, however, it is essential to consider interspecies differences in total and regional deposition patterns. Although the same deposition mechanisms operate for humans and animals, different species exposed to the same aerosol at identical concentrations may not receive identical doses in comparable respiratory tract regions, due largely to anatomical and ventilatory differences. If the total amount of deposition is divided by body (or lung) weight, smaller animals would receive greater initial particle burdens per unit weight per unit exposure time than would larger ones. For example, initial deposition of 1 μm particles in the rat will be five to ten times that of humans, and the dog three times that of humans, if deposition is calculated on a per unit lung or body weight basis (Phalen et al., 1977). Furthermore, anatomy of the upper respiratory tract is a major factor influencing deposition pattern. The greater complexity of the nasal passages in rodents compared to primates results in more upper respiratory tract deposition in the former (Schreider, 1986). Thus, selection of a particular species may influence the estimated human lung, or systemic, dose as well as its relation to potential health effects when extrapolating from animal studies to human exposure. Descriptions of experimentally-determined regional deposition patterns in various animal species are available (Schlesinger et al., 1997).

Clearance of Deposited Particles

Mechanisms and Pathways of Clearance

Particles which deposit upon airway surfaces may be cleared from the respiratory tract completely, or may be translocated to other sites within this system. The timeframe over which this occurs affects dose delivered to the respiratory tract, as well as to extrapulmonary organs.

Table 1. OVERVIEW OF RESPIRATORY TRACT PARTICLE CLEARANCE AND TRANSLOCATION MECHANISMS

Extrathoracic Region

    Mucociliary transport
    Sneezing
    Nose wiping and blowing
    Dissolution (for "soluble" particles) and absorption into blood

Tracheobronchial Region

    Mucociliary transport
    Endocytosis by macrophages/epithelial cells
    Coughing
    Dissolution (for "soluble" particles) and absorption into blood

Alveolar region

    Macrophages, epithelial cells
    Interstitial
    Dissolution for "soluble" and "insoluble" particles (intra- and
      extracellular)

Clearance mechanisms are regionally distinct, as outlined in Table 1 (Schlesinger, 1995). They may be broadly classified into two groupings, namely absorptive, which is dissolution, or nonabsorptive, which involves mechanical transport of intact particles; these may occur simultaneously or with temporal variations. It should be mentioned that particle solubility in terms of clearance refers to solubility *in vivo* within the fluid lining the airway surfaces. Thus, an insoluble particle is considered to be one whose rate of clearance by dissolution is insignificant compared to its rate of clearance by mechanical processes.

The clearance of relatively insoluble particles deposited in the nasal passages occurs via mucociliary transport, and the general flow of mucus is backwards, i.e., towards the nasopharynx. The epithelium in the most anterior portion of the nasal passages is not ciliated, and mucus flow just distal to this is forward, clearing deposited material to a site where removal is by sneezing, nose wiping, or nose blowing. Soluble material deposited on the nasal epithelium will be accessible to underlying cells if it can diffuse to them through the mucus prior to removal via mucociliary transport. Since there is a rich vasculature in the nose, uptake into the blood may occur rapidly.

Mucus flow rates in the posterior nasal passages are highly nonuniform. The median rate is about 5 mm/min, with a mean transport time for insoluble particles over the entire region of ~10 - 20 min (Proctor, 1980; Rutland and Cole, 1981; Stanley et al., 1985). On the other hand, particles depositing in the non-ciliated anterior portion of the nasal passages are cleared at a much slower rate, about 1 - 2 mm/hr, by mucus moved via traction induced by more distal cilia, and such deposits are usually more effectively removed by sneezing, wiping, or nose blowing. Clearance of insoluble particles deposited in the oral passages is by swallowing into the gastrointestinal tract. Soluble particles are likely rapidly absorbed after deposition.

Like the nasal passages, insoluble particles deposited on tracheobronchial tree surfaces are cleared primarily by mucociliary transport, with net movement of fluid towards the oropharynx. Mucus transport in the tracheobronchial tree occurs at different rates in different regions; mucus movement velocity is fastest in the trachea and becomes progressively slower in more distal airways. In healthy adult humans, a mucus transport rate can range from 4 - 6 mm/min in the trachea to 0.001 mm/min in the most distal ciliated airways (Morrow et al., 1967; Yeates et al., 1975, 1981; Yeates and Aspin, 1978; Foster et al., 1980; Leikauf et al., 1981, 1984).

Clearance of particles from the tracheobronchial tree may also occur following phagocytosis by airway macrophages located on or beneath the mucus lining throughout the bronchial tree, which then move cephalad on the mucociliary blanket, or via macrophages which enter the airway lumen from the bronchial or bronchiolar mucosa. Soluble particles may be absorbed through the mucus layer directly into the blood.

Total duration of bronchial clearance (or some other time parameter) is often used as an index of mucociliary function. In healthy adult non-smoking humans, generally >90% of insoluble particles depositing on the tracheobronchial tree will be cleared by 48 hr after deposition, depending upon the individual subject and the particle size; this latter does not affect the actual rate of surface transport, but does affect the depth of particle penetration and deposition and the subsequent pathway length for clearance. A number of studies with both rodents and humans have indicated that some fraction of insoluble material may be retained for a prolonged period of time within the upper respiratory tract (specifically within the nasal passages) or tracheobronchial tree (Patrick and Stirling, 1977; Gore and Patrick, 1982; Camner et al., 1997). While the reason for such extended retention is unclear, it is known that long-term tracheobronchial retention patterns are not uniform. There appears to be an enhancement at bifurcation regions (Henshaw and Fews, 1984; Cohen et al., 1988), perhaps a result of both greater deposition and ineffectual mucus clearance within these areas. Soluble material may also undergo long-term retention in tracheobronchial airways due to binding to cells or macromolecules.

Clearance from the alveolar region occurs via a number of mechanisms and pathways which are outlined in Figure 2, but the relative importance of each is not always certain. The major pathway involves alveolar macrophages. These cells reside on the alveolar epithelium, and phagocytize and transport deposited material which they contact by random motion or via directed migration in response to local chemotactic factors.

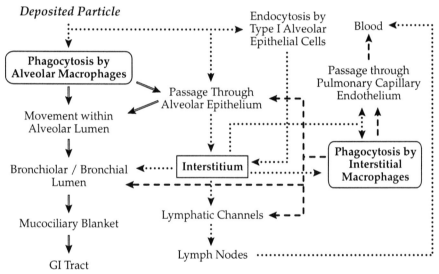

Figure 2. Diagram showing possible mechanical clearance pathways for insoluble particles depositing in the alveolar region of the lungs. From Schlesinger (1995).

Alveolar macrophages normally comprise ~3 - 5% of the total alveolar cells in healthy (non-smoking) humans and other mammals. However, the actual cell count is influenced by particle loading. While low numbers of deposited particles may not result in an increase in cell number, above some level, macrophage numbers increase until a saturation point is reached (Brain, 1971; Adamson and Bowden, 1981). Since the magnitude of this increase is related more to the number of deposited particles than to total deposition by weight, equivalent masses of an identical deposited substance may not produce the same response if particle sizes differ; thus, smaller particles tend to result in a greater elevation in cell number than do larger ones.

The primary route for clearance of particle-laden macrophages from the pulmonary region is cephalad transport via the mucociliary system after the cells reach the distal terminus of the mucus blanket. This can occur via chance encounter; passive movement along the alveolar surface due to surface tension gradients between the alveoli and conducting airways; directed locomotion along a gradient produced by chemotactic factors released by macrophages ingesting deposited material; or passage through the alveolar epithelium and the interstitium. Some cells which follow interstitial clearance pathways are likely resident interstitial macrophages which have ingested free particles transported through the alveolar epithelium (Brody et al., 1981; Bowden and Adamson, 1984). Particle-laden interstitial macrophages can also migrate across the alveolar epithelium, becoming part of the alveolar macrophage cell pool.

Macrophages which are not cleared via the tracheobronchial tree may actively migrate within the interstitium to a nearby lymphatic channel or, along with uningested particles, be carried in the flow of interstitial fluid into the lymphatic system. Lymphatic endothelium may also actively engulf particles from the surrounding interstitium. Particles within the lymphatic system may be translocated to tracheobronchial lymph nodes, which often become reservoirs of retained material. Particles penetrating the nodes and subsequently reaching the post-nodal lymphatic circulation may enter the blood.

Uningested particles or macrophages in the interstitium may traverse the alveolar-capillary endothelium, directly entering the blood; endocytosis by endothelial cells followed by exocytosis into the vessel lumen seems, however, to be restricted to particles <0.1 μm, and may increase with increasing lung burden (Oberdörster, 1988; Lee et al., 1989). Once in the systemic circulation, transmigrated macrophages, as well as free particles, can travel to extrapulmonary organs.

Free particles and macrophages within the interstitium may travel to perivenous, peribronchiolar or subpleural sites, where they become trapped, increasing lung particle burden. The migration and grouping of particles and macrophages within the lungs can lead to the redistribution of initially diffuse deposits into focal aggregates. Some particles, notably fibers, can be found in the pleural space, often within macrophages which have migrated across the visceral pleura.

Clearance by the absorptive mechanism involves dissolution in the alveolar surface lining fluid, followed by transport through the epithelium and into the interstitium, and diffusion into lymph or blood. Some soluble particles translocated to and trapped in interstitial sites may be absorbed there. Although the factors affecting the dissolution of deposited particles are poorly understood, it is influenced by the particle surface to volume ratio and other surface properties (Mercer, 1967; Morrow, 1973). Thus, materials generally considered to be relatively insoluble may have high dissolution rates and short dissolution half-times if particle size is small.

Some deposited particles may undergo dissolution after phagocytic uptake by macrophages. For example, engulfed metals may dissolve in the acidic milieu of the phagolysosomes (Lundborg et al., 1985). It is, however, not certain whether the dissolved material then emigrates from the macrophage. Finally, some particles can bind to epithelial cell components, delaying clearance from the lungs.

Fibrous particles deposited in the pulmonary region may be additionally subject to a process of disintegration, which involves the subdivision of a large particle into smaller segments. This can occur by leaching within the fibrous structure, which then fractures, or by surface etching, resulting in a change in the external dimensions of the fiber. Some fiber types breakup by length, while others will disintegrate into smaller diameter particles (Lippmann, 1992).

Particles deposited in the alveolar region generally remain in the respiratory tract longer than do those deposited in airways cleared by mucociliary transport. There are limited data on rates in humans, while within any species rates vary widely due to different properties of the particles used in the various studies. Furthermore, many reported studies using animals employed high concentrations of insoluble particles, which may have interfered with normal clearance mechanisms, producing rates different from those which would occur at lower exposure levels. Prolonged exposure to high particle concentrations is associated with what is termed particle "overload." This is a non-specific effect noted in animal studies using many different kinds of insoluble particles, including fibers, and results in clearance slowing or stasis, with an associated inflammation and aggregation of macrophages in the lungs. (Lehnert, 1990; Muhle et al., 1990). While it is, however, likely to be of little relevance for most general atmospheric exposures of humans, it is of concern in interpreting some long-term experimental animal exposure data and may also have some significance for human occupational exposures.

Clearance from the alveolar region can be described as a multiphasic process, each component considered to represent removal by a different mechanism or pathway, and characterized by increasing half-times of clearance with time post-exposure. For example, an initial fast phase, having a half-time of ~2 - 6 wks, presumably represents rapid clearance via macrophages, while a phase of prolonged clearance, with a half-time of months to years, represents removal by dissolution. This latter is extremely variable, but likely dominates the long-term clearance of relatively insoluble particles (Kreyling et al., 1988). An intermediate phase, with a half time on the order of months, may represent a slower phase of macrophage clearance via interstitial pathways.

Although the kinetics of overall clearance from the alveolar region have been assessed, much less is known concerning relative rates along specific pathways. The usual initial step in clearance, i.e., uptake of deposited particles by alveolar macrophages, is very rapid. Unless the particles are cytotoxic or very large, ingestion by macrophages occurs within 24 hr of a single inhalation (Lehnert and Morrow, 1985; Naumann and Schlesinger, 1986). But the actual rate of subsequent macrophage clearance is not certain; it appears that 5% or less of their total number is translocated from the lungs each day (Masse et al., 1974; Lehnert and Morrow, 1985). The actual time for clearance of particle-laden alveolar macrophages via the mucociliary system depends upon the site of particle uptake relative to the distal terminus of the mucus blanket at the bronchiolar level. Furthermore, clearance pathways, and subsequent kinetics, may depend to some extent upon particle size. For example, ultrafine particles <0.02 μm are less effectively phagocytosed than are larger ones (Oberdörster, 1993). But once ingestion occurs, alveolar macrophage-mediated kinetics are independent of the particle involved, as long as solubility and cytotoxicity are low.

Free particles may penetrate into the interstitium, largely by Type I cell endocytosis, within a few hours following deposition (Sorokin and Brain, 1975; Ferin and Feldstein, 1978; Brody et al., 1981). This transepithelial passage seems to increase as particle loading increases, especially to a level above the saturation point for increasing macrophage number (Ferin, 1977; Adamson and Bowden, 1981). It may also be particle size dependent, since insoluble ultrafine particles <0.05 μm diameter show increased access to and greater lymphatic uptake than do larger ones (Oberdörster et al., 1992). Similarly, a depression of phagocytosis by toxic particles or the deposition of large numbers of smaller ultrafine particles may increase the number of free particles in the alveoli, enhancing removal by other routes. In any case, free particles and alveolar macrophages may reach the lymph nodes within a few days after deposition (Harmsen et al., 1985; Lehnert, 1988). However, most translocation to the lymphatic system is very slow, on the order of 0.020 - 0.003 %/day (Snipes, 1989), and elimination from the lymph nodes is even slower, with half-times measured in tens of years (Roy, 1989).

Soluble particles depositing in the alveolar region are rapidly cleared via absorption through the epithelial surface into the blood. Actual rates depend upon the size of the particle (i.e., molecular size), with smaller ones clearing faster than larger ones. Some solubilized material may be retained in lung tissue due to binding with cellular components, preventing passage into the circulation.

Factors Modulating Clearance

A number of factors may modify normal clearance patterns, thereby affecting the dose delivered by exposure to inhaled particles. These include age, disease, and irritant inhalation.

The evidence for aging-related effects on mucociliary function in healthy individuals is equivocal, with studies showing either no change or a slowing in clearance function with age after maturity (Goodman et al., 1978; Yeates, 1981). There are few data to assess changes in clearance in the growing lung and no data to allow assessment of aging-relating changes in clearance from the pulmonary region. Although functional differences have been found between alveolar macrophages of mature and senescent mice (Esposito and Pennington, 1983), no age-related decline in macrophage function has been seen in humans (Gardner et al., 1981).

In terms of gender, no difference in nasal mucociliary clearance rate was observed between male and female children (Passali and Ciampoli, 1985), nor in tracheal transport rates in adults (Yeates et al., 1975). Slower bronchial clearance has been noted in male compared to female adults, but this was attributed to differences in lung size (and resultant clearance pathway length) rather than to inherent gender related differences in transport velocities (Gerrard et al., 1986).

Various respiratory tract diseases are associated with clearance alterations. Nasal mucociliary clearance is prolonged in humans with chronic sinusitis, bronchi-ectasis, or rhinitis (Majima et al., 1983; Stanley et al., 1985), and in cystic fibrosis (Rutland and Cole, 1981). Bronchial mucus transport may be impaired in people with bronchial carcinoma (Matthys et al., 1983), chronic bronchitis (Vastag et al., 1986), asthma (Pavia et al., 1985), and in association with various acute infections (Lourenco et al., 1971; Camner et al., 1979; Puchelle et al., 1980). In certain of these cases, coughing may enhance mucus clearance, but it generally is effective only if excess secretions are present (Groth et al., 1997).

Rates of pulmonary region particle clearance appear to be reduced in humans with chronic obstructive lung disease (Bohning et al., 1982) and in experimental animals with viral infections (Creasia et al., 1973). The viability and functional activity of macrophages was found to be impaired in human asthmatics (Godard et al., 1982). Inflammation may enhance particle and macrophage penetration through the alveolar epithelium into the interstitium, perhaps by increasing epithelium and lymphatic endothelium permeabilities (Adamson and Prieditis, 1998). Neutrophils, which are phagocytic cells present in alveoli during inflammation, may contribute to the clearance of particles via the mucociliary system (Bice et al., 1990).

Inhaled irritants can have an effect upon mucociliary clearance function in both humans and experimental animals (Wolff, 1986). Single exposures to a particular material may increase or decrease the overall rate of tracheobronchial clearance, often depending upon the exposure concentration (Schlesinger, 1990). Alterations in clearance rate following single exposures to moderate levels of irritants are generally transient, lasting <24 hr. However, repeated exposures may result in an increase in intra-individual variability of clearance rate and persistently retarded clearance.

Acute and chronic exposures to inhaled irritants may also alter pulmonary region clearance (Ferin and Leach, 1977; Cohen et al., 1979; Schlesinger et al., 1986), which may be accelerated or depressed, depending upon the specific irritant and/or length of exposure. Alterations in alveolar macrophages likely underlie some of the observed changes, since numerous irritants have been shown to impair the numbers and functional properties of these cells.

Interspecies Patterns of Clearance

The basic mechanisms of clearance from the respiratory tract are similar for humans and most other mammals. However, regional clearance rates show substantial variation between species, even for similar particles deposited under comparable exposure conditions (Snipes, 1989). There are interspecies differences in rates of mechanical transport, e.g., macrophage-mediated clearance of insoluble particles from the pulmonary region (Bailey et al., 1985), transport of particles from the pulmonary region to pulmonary lymph nodes (Snipes et al., 1983; Mueller et al., 1990) and mucociliary transport in conducting airways (Felicetti et al., 1981). There are thus likely to be species-dependent rate constants for specific clearance pathways.

DISPOSITION OF INHALED GASES AND VAPORS

Gas Transport and Deposition

Gases and vapors[1] are not subject to the inertial or gravitational deposition mechanisms which can influence the transport of particles from the inhaled airstream

---

[1]The basic difference between a gas and vapor is that the latter is the gaseous state of a substance that would be a solid or liquid under normal temperature and pressure, while the former is completely in the gaseous state under these conditions. Since gases and vapors behave identically in terms of respiratory tract deposition processes, for purposes of this chapter, the generic term "gas" will be used for both.

onto airway surfaces. However, like ultrafine particles, individual gas molecules undergo kinetic motion and are, therefore, subject to random movement within the airway lumen via passive diffusion. Diffusive transport to the air-liquid surface lining of the airway wall will occur if there is a concentration gradient for the gas between the air and this interface. In the presence of such a gradient, movement of gas will be from a region of higher to one of lower concentration, a process proportional to the diffusion coefficient of the gas. The term "deposition" when applied to an inhaled gas refers to this random migration of gas molecules towards the airway wall and their subsequent contact with and absorption by (or reaction with) airway surface fluid lining. Once a gas deposits on the airway surface, it or its reaction products become accessible to underlying epithelial cells and local blood circulation.

While individual gas molecules can move via passive diffusion, inhaled gases also undergo bulk motion due to movement of the airstream, in which the material is entrained, during inspiration and expiration. This type of gas transport is termed convection. Both convection and passive diffusion play roles in the movement of gases within the respiratory tract, although one or the other may dominate. The former plays a major role in the upper respiratory tract (nasal and oral passages) and upper tracheobronchial tree, while the latter predominates in the lower tracheobronchial tree and alveolar region.

Deposition of an inhaled gas is a function largely of two physical properties of the gas, namely its water solubility and its reactivity with the surface fluid lining or cell/tissue components of the airway. This differs from the case involving particles, whereby deposition does not depend upon the chemistry of the inhaled material but, rather, solely upon its size. But as with particles, chemistry does impact upon any biological response following deposition of a gas.

There are other differences, and some similarities, between gases and particles in terms of deposition phenomena. As noted above, there is similarity in the mechanism of transport and deposition between gases and particles small enough to be influenced by passive diffusion. But, regardless of their chemical properties or solubility, all particles which contact the liquid lining of the airway surface are effectively considered to be removed from the air stream and to remain deposited on the surface, i.e., they are not re-entrained into the air. Thus, the airway surface lining acts as an infinite sink for particles. This results in a concentration gradient between the air and the surface fluid that persists until the particle concentration within the inhaled air reaches zero. In some cases, gas deposition may be similar. For example, deposition of a highly water soluble gas inhaled at a low concentration may be followed by rapid transport via the blood away from the site of contact, such that the liquid airway surface lining acts as an infinite sink in this case as well, and overall transport is limited by transport on the gaseous side of the interface. A similar situation would occur following deposition of a soluble reactive gas if the gas undergoes rapid and irreversible chemical reaction with the airway liquid lining, and the reaction products are then transported away from the site of contact. On the other hand, inhalation of a gas with low water solubility and low chemical reactivity will result in a gas concentration at the air-liquid interface that is essentially equal to the concentration of the gas in the airway lumen. In this case, movement of gas through the interface is much lower than for a soluble and/or reactive gas, and the deposition of the gas may be spread widely over a large surface area, e.g., the entire alveolar region. Furthermore, if the concentration of the gas becomes higher in the surface lining or tissue than in the airstream, the direction of gas diffusion will be from tissue to airway lumen, rather than the reverse. Thus, unlike the situation with particles, interaction of a deposited gas with the liquid lining of the airway surface and underlying epithelium can influence the rate of deposition or removal of the gas from the air stream.

The regional deposition site of an inhaled gas is also influenced by its water solubility and chemical reactivity. For example, highly soluble gases may be removed almost completely from inhaled air within the upper respiratory tract, especially with nasal breathing; this can result in little gas penetration into the tracheobronchial tree and beyond, especially when the inhaled concentration is low. This occurs because the partial pressure of the gas in the inhaled air decreases rapidly as the air moves through the upper respiratory tract and into the lungs. But with increasing concentration of even a highly soluble gas, a concentration gradient may develop, resulting in significant gas penetration and ultimate deposition beyond the upper respiratory tract. Less soluble gases may have deposition which is distributed throughout both the tracheobronchial tree and alveolar region. Soluble gases which do penetrate the upper respiratory tract would tend to be deposited in the upper tracheobronchial tree, while insoluble, nonreactive gases, or less soluble but reactive gases, may penetrate much deeper into the lungs and can become distributed throughout the lungs. Bear in mind that these relationships between deposition site and solubility/reactivity of a gas are generalizations, and other factors may influence ultimate disposition. For example, a gas may adsorb onto a particle in a mixed gas/aerosol atmosphere, and deposition of the gas species would then occur at the site of particle contact with airway surfaces. In addition, local metabolism in the airways may influence disposition of a gas once deposited, as does local vascular blood flow and tissue binding. Gases which do not deposit will leave the respiratory tract with the exhaled air. Further details on factors affecting the transport and deposition of gases may be found in Miller et al. (1993) and Ultman (1988), while total respiratory tract and regional deposition patterns for specific gases is provided in Schlesinger et al. (1997).

Clearance of Deposited Gases

Once deposited within the liquid lining of the airway, the gas, or its reaction products, can be transported by convection, i.e., bulk movement of the liquid layer, or via passive diffusion within this fluid. While the former mechanism may occur in the nasal passages and tracheobronchial tree, movement in the alveolar region is dominated by diffusion, which occurs in a direction normal to the airstream/airway liquid surface interface. Deposited gas (or its reaction products) can be translocated to other sites within the respiratory tract or completely out of the system via the flow of mucus, via the bloodstream or lymphatic system, or via exhalation (this last occurring via the same diffusion processes as deposition). Reversible (or irreversible) binding to tissue macromolecules may occur at the site of deposition if the inhaled gas is reactive. Gases having low chemical reactivity will not accumulate, and an equilibrium may be reached. This will not occur for gases which are reactive, or which have toxic metabolites.

Interspecies Patterns of Deposition

Anatomical and ventilatory differences between species result in differences in sites of regional gas deposition and differences in the extent of uptake of a particular gas within each respiratory tract region. This must be considered in interspecies extrapolation of biological responses to inhaled gaseous toxicants. Details of comparative aspects of gas uptake in the respiratory tract are available in Gerrity, (1989), Overton et al. (1989), Martonen et al. (1995), and Miller and Kimbell (1995).

CONCLUSIONS

The biological responses from inhaled particles and gases are dependent upon both the amount and pattern of deposition and the time frame of their persistence in the respiratory tract. The deposition of particles on airway surfaces is the result of specific physical mechanisms that are influenced by particle characteristics, namely size, shape and density, as well as host factors of airflow patterns, and respiratory tract anatomy. The deposition of gases is also affected by ventilatory and airway geometric factors, but depends upon specific chemical properties of the gas, namely water solubility and chemical reactivity. Biological effects from inhaled toxicants are often related more to the quantitative pattern of deposition at specific sites than to the total amount depositing in the respiratory tract. The accurate interpretation of results from inhalation toxicological studies requires an appreciation of those factors which control and affect deposition and clearance of inhaled toxicants.

REFERENCES

Adamson IYR, Bowden DH. Dose response of the pulmonary macrophagic system to various particulates and its relationship to transepithelial passage of free particles. Exp. Lung Res., 1981;2:165-175.

Adamson IYR, Prieditis H. Silica deposition in the lung during epithelial injury potentiates fibrosis and increases particle translocation to lymph nodes. Exp. Lung Res., 1998;24:293-306.

Asgharian B, Yu CP. Deposition of fibers in the rat lung. J. Aerosol Sci., 1989;20:355-366.

Asgharian B, Yu CP. Deposition of inhaled fibrous particles in the human lung. J. Aerosol Med., 1988;1:37-50.

Bailey MR, Hodgson A, Smith H. Respiratory tract retention of relatively insoluble particles in rodents. J. Aerosol Sci., 1985;16:279-293.

Bennett WD, Zeman KL, Kang CW, Schechter MS. Extrathoracic deposition of inhaled, coarse particles (4.5 μm) in children vs adults. Ann. Occup. Hyg., 1997;41(Suppl. 1):497-502.

Bennett WD, Zeman KL, Kim C, Mascarella J. Enhanced deposition of fine particles in COPD patients spontaneously breathing at rest. Inhal. Toxicol., 1997;9:1-14.

Bennett WD, Zeman KL. Deposition of fine particles in children spontaneously breathing at rest. Inhal. Toxicol., 1998;10:831-842.

Bice DE, Harmson AG, Muggenburg BA. Role of lung phagocytes in the clearance of particles by the mucociliary apparatus. Inhal. Toxicol., 1990;2:151-160.

Bohning DE, Atkins HL, Cohn SH. Long-term particle clearance in man: Normal and impaired. Ann. Occup. Hyg. 1982;26:259-271.

Bowden DH, Adamson IYR. Pathways of cellular efflux and particulate clearance after carbon instillation to the lung. J. Pathol., 1984;143:117-125.

Brain JD. "The Effects of Increased Particles on the Number of Alveolar Macrophages." In *Inhaled Particles III, Vol. 1*, WH Walton, ed., Old Woking, England: Unwin Bros, 1971, pp. 209-223.

Brody AR, Hill LH, Adkins Jr. B, O'Connor RW. Chryostile asbestos inhalation in rats: Deposition pattern and reaction of alveolar epithelium and pulmonary macrophages. Am. Rev. Respir. Dis. 1981;123:670-679.

Brody AR, Roe MW. Deposition pattern of inorganic particles at the alveolar level in the lungs of rats and mice. Am. Rev. Respir. Dis. 1983;128:724-729.

Camner P, Anderson M, Philipson K, Bailey A, Hashish A, Jarvis N, Bailey M, Svartengren M. Human bronchiolar deposition and retention of 6-, 8-, and 10-μm particles. Exp. Lung Res. 1997;23:517-535.

Camner P, Mossberg B, Philipson K, Strandberg K. Elimination of test particles from the human tracheobronchial tract by voluntary coughing. Scand. J. Respir. Dis., 1979;60:56-62.

Cohen BS, Harley NH, Schlesinger RB, Lippmann M. Nonuniform particle deposition on tracheobronchial airways: Implication for lung dosimetry. Ann. Occup. Hyg., 1988;32(Suppl 1):1045-1052.

Crawford DJ. Identifying critical human subpopulations by age groups: Radioactivity and the lung. Phys. Med. Biol., 1982;27:539-552.

Creasia DA, Nettesheim P, Hammons AS. Impairment of deep lung clearance by influenza virus infection. Arch. Environ. Health, 1973;26:197-201.

Esposito AL, Pennington JE. Effects of aging on antibacterial mechanisms in experimental pneumonia. Am. Rev. Respir. Dis. 1983;128:662-667.

Felicetti SA, Wolff RK, Muggenburg BA. Comparison of tracheal mucous transport in rats, guinea pigs, rabbits, and dogs. J. Appl. Physiol., 1981;51:1612-1617.

Ferin J, Feldstein ML. Pulmonary clearance and hilar lymph node content in rats after particle exposure. Environ. Res., 1978;16:342-352.

Ferin J. "Effects of Particle Content of Lung on Clearance Pathways." In *Pulmonary Macrophage and Epithelial Cells*, CL Sanders, RP Schneider, GE Dagle, HA Ragan, eds. Springfield, VA: NTIS, 1977, pp.414-423.

Foster WM, Langenbach E, Bergofsky EH. Measurement of tracheal and bronchial mucus velocities in man: Relation to lung clearance. J. Appl. Physiol., 1980;48:965-971.

Gardner ND, Lim STK, Lawton JWM. Monocyte function in ageing humans. Mech. Ageing Dev., 1981;16:233-239.

Gerrard CS, Gerrity TR, Yeates DB. The relationships of aerosol deposition, lung size, and the rate of mucociliary clearance. Arch. Environ. Health, 1986;41:11-15.

Gerrity TR. "Nasopharyngeal Uptake of Ozone in Humans and Animals." In *Extrapolation of Dosimetric Relationships for Inhaled Particles and Gases*, JD Crapo, ED Smolko, FJ Miller, JA Graham, AW Hayes, eds. San Diego: Academic Press, 1989, pp. 187-195.

Godard P, Chaintreuil J, Damon M, Coupe M, Flandre O, de Paulet AC, Michel FB. Functional assessment of alveolar macrophages: Comparison of cells from asthmatic and normal subjects. J. Allergy Clin. Immunol., 1982;70:88-93.

Gore DJ, Patrick G. A quantitative study of the penetration of insoluble particles into the tissue of the conducting airways. Ann. Occup. Hyg., 1982;26:149-161.

Gradon L, Orlicki D. Deposition of inhaled aerosol particles in a generation of the tracheobronchial tree. J. Aerosol Sci., 1990;21:3-19.

Groth ML, Macri K, Foster WM. Cough and mucociliary transport of airway particulate in chronic obstructive lung disease. Ann. Occup. Hyg., 1997;41(Suppl. 1):515-521.

Hammad Y, Diem J, Craighead J, Weill H. Deposition of inhaled man-made mineral fibers in the lungs of rats. Ann. Occup. Hyg., 1982;26:179-187.

Harmsen AG, Muggenburg BA, Snipes MB, Bice DE. The role of macrophages in particle translocation from lungs to lymph nodes. Science, 1985;230:1277-1280.

Henshaw DL, Fews AP. "The Microdistribution of Alpha Emitting Particles in the Human Lung." In *Lung Modelling for Inhalation of Radioactive Materials*, H Smith, G Gerber, eds. Luxemburg: Commission of the European Communities, 1984, pp. 199-208.

Hofmann W. Mathematical model for the postnatal growth of the human lung. Resp. Physiol., 1982;49:115-129.

Kim CS, Kang TC. Comparative measurement of lung deposition of inhaled fine particles in normal subjects and patients with obstructive airway disease. Am. J. Respir. Crit. Care Med., 1997;155:899-905.

Kreyling W, Ferron GA, Schumann G. Particle transport from the lower respiratory tract. J. Aerosol Med., 1988;1:351-370.

Lee KP, Trochimowicz HJ, Reinhardt CF. Transmigration of titanium dioxide (TiO_2) in rats after inhalation exposure. Exp. Mol. Pathol., 1989;42:331-343.

Lehnert BE, Morrow PE. Association of [59]iron oxide with alveolar macrophages during alveolar clearance. Exp. Lung Res., 1985;9:1-16.

Lehnert BE. Alveolar macrophages in a particle "overload" condition. J. Aerosol Med., 1990;3(Suppl 1):S9-S30.

Lehnert BE. Pulmonary and thoracic macrophage subpopulations and the clearance of particles from the lung. Environ. Health Perspect., 1992;97:17-46.

Leikauf G, Yeates DB, Wales KA, Albert RE, Lippmann M. Effects of sulfuric acid aerosol on respiratory mechanics and mucociliary particle clearance in healthy non-smoking adults. Am. Ind. Hyg. Assoc. J., 1981;42:273-282.

Leikauf GD, Spektor DM, Albert RE, Lippmann M. Dose-dependent effects of submicrometer sulfuric acid aerosol on particle clearance from ciliated human lung airways. Am. Ind. Hyg. Assoc. J., 1984;45:285-292.

Lippmann M. "Asbestos and Other Mineral Fibers." In *Environmental Toxicants: Human Exposures and Their Health Effects*, M Lippmann, ed. New York: Van Nostrand Reinhold, 1992, pp. 30-75.

Lourenco RV, Stanley ED, Gatmaitan B, Jackson GG. Abnormal deposition and clearance of inhaled particles during upper respiratory and viral infections. J. Clin. Invest., 1971;50:62a.

Lundborg M, Eklund A, Lind B, Camner P. Dissolution of metals by human and rabbit alveolar macrophages. Br. J. Ind. Med., 1985;42:642-645.

Majima Y, Sakakura Y, Matsubara T, Murai S, Miyoshi Y. Mucociliary clearance in chronic sinusitis: Related human nasal clearance and *in vitro* bullfrog palate clearance. Biorheology, 1983;20:251-262.

Martonen T, Zhang A, Yand Y. Interspecies modeling of inhaled gase. Inhalation Toxicol., 1995l7:1125-1139.

Masse R, Ducousso R, Nolibe D, Lafuma J, Chretien J. Passage transbronchique des particules metalliques. Rev. Fr. Mal. Respir., 1974;1:123-129.

Matthys H, Vastag E, Kohler D, Daikeler G, Fischer J. Mucociliary clearance in patients with chronic bronchitis and bronchial carcinoma. Respiration, 1983;44:329-337.

Mercer TT. On the role of particle size in the dissolution of lung burdens. Health Phys., 1967;13:1211-1221.

Miller FJ, Kimbell JS. "Regional Dosimetry of Inhaled Reactive Gases " In *Concepts in Inhalation Toxicology*, RO McClellan, R.F. Henderson, eds. Washington, DC: Taylor and Francis, 1995, pp. 257-287.

Miller FJ, Overton JH, Kimbell JS, Russell ML. "Regional Respiratory Tract Absorption of Inhaled Reactive Gases." In *Toxicology of the Lung, 2nd Edition*, DE Gardner, JD Crapo, RO McClellan, eds. New York, Raven Press, 1993, pp. 485-525.

Morrow PE, Gibb FR, Gazioglu K. A study of particulate clearance from the human lung. Am. Rev. Respir. Dis., 1967;96:1209-1221.

Morrow PE. Alveolar clearance of aerosols. Arch. Intern. Med., 1973;131:101-108.

Mueller HL, Robinson B, Muggenburg BA, Gillett NA, Guilmette RA. Particle distribution in lung and lymph node tissues of rats and dogs and the migration of particle-containing alveolar cells *in vitro*. J. Toxicol. Environ. Health, 1990;30:141-165.

Muhle H, Creutzenberg O, Bellmann B, Heinrich U, Mermelstein R. Dust overloading of lungs: Investigations of various materials, species differences, and irreversibility of effects. J. Aerosol Med., 1990;3(Suppl. 1):S111-S128.

Naumann BD, Schlesinger RB. Assessment of early alveolar particle clearance and macrophage function following an acute inhalation of sulfuric acid mist. Exp. Lung Res., 1986;11:13-33.

Oberdörster G, Ferin J, Gelein R, Soderholm SC, Finkelstein J. Role of the alveolar macrophage in lung injury: Studies with ultrafine particles. Environ. Health Perspect., 1992;97:193-199.

Oberdörster G. Lung clearance of inhaled insoluble and soluble particles. J. Aerosol Med., 1988;1:289-330.

Oberdörster G. Lung dosimetry: Pulmonary clearance of inhaled particles. Aerosol Sci. Technol., 1993;18:279-289.

Overton JH, Barnett AE, Graham RC. "Significances of the Variability of Tracheobronchial Airway Paths and their Air Flow Rates to Dosimetry Model Predictions of the Absorption of Gases." In *Extrapolation of Dosimetric Relationships for Inhaled Particles and Gases*, JD Crapo, ED Smolko, FJ Miller, JA Graham, AW Hayes, eds. San Diego: Academic Press, 1989, pp. 273-291.

Passali D, Ciampoli MB. Normal values of mucociliary transport time in young subjects. Int. J. Pediat. Otorhinolaryngol., 1985;9:151-156.

Patrick G, Stirling C. The retention of particles in large airways of the respiratory tract. Proc. Royal Soc. London Ser. V, 1977;198:455-462.

Pavia D. Lung "Mucociliary Clearance." In *Aerosols and the Lung*, SW Clarke, D Pavia, eds., London: Butterworths, 1984, pp. 127-155.

Phalen R, Kenoyer J, Davis J. "Deposition and Clearance of Inhaled Particles: Comparison of Mammalian Species." In *Proceedings of the Annual Conference on Environmental Toxicology, Vol. 7*, AMRL-TR-76-125, Springfield, VA: NTIS, 1977, pp. 159-170.

Phalen RF, Oldham MJ, Schum GM. Growth and ageing of the bronchial tree: Implications for particle deposition calculations. Radiat. Prot. Dosim., 1991;38:15-21.

Puchelle E, Zahm JM, Girard F, Bertrand A, Polu JM, Aug F, Sadoul P. Mucociliary transport *in vivo* and *in vitro* - relations to sputum properties in chronic bronchitis. Eur. J. Respir. Dis., 1980;61:254-264.

Proctor DF. "The Upper Respiratory Tract." In *Pulmonary Diseases and Disorders*, AP Fishman, ed. New York: McGraw-Hill, 1980, pp. 209-223.

Roy M. Lung clearance modeling on the basis of physiological and biological parameters. Health Phys., 1989;57(Suppl 1):255-262.

Rutland J, Cole PJ. Nasal mucociliary clearance and ciliary beat frequency in cystic fibrosis compared with sinusitis and bronchiectasis. Thorax, 1981;36:654-658.

Schlesinger RB, Ben-Jebria A, Dahl AR, Snipes MB, Ultman J. "Disposition of Inhaled Toxicants." In *Handbook of Human Toxicology*, EJ Massaro, ed. Boca Raton, FL: CRC Press, 1997, pp. 493-550.

Schlesinger RB, Gurman JL, Lippmann M. Particle deposition within bronchial airways: Comparisons using constant and cyclic inspiratory flow. Ann. Occup. Hyg., 1982;26:47-64.

Schlesinger RB, Lippmann M. Selective particle deposition and bronchogenic carcinoma. Environ. Res., 1978;15:424-431.

Schlesinger RB. The interaction of inhaled toxicants with respiratory tract clearance mechanisms. CRC Crit. Rev. Toxicol., 1990;20:257-286.

Schreider JP. "Comparative Anatomy and Functions of the Nasal Passages." In *Toxicology of the Nasal Passages*, CS Barrow, ed. New York: McGraw-Hill, 1986, pp. 1-25.

Snipes MB, Boecker BB, McClellan RO. Retention of monodisperse or polydisperse aluminosilicate particles inhaled by dogs, rats, and mice. Toxicol. Appl. Pharmacol., 1983;69:345-362.

Snipes MB. Long-term retention and clearance of particles inhaled by mammalian species. CRC Crit. Rev. Toxicol., 1989;20:175-211.

Sorokin SP, Brain JD. Pathways of clearance in mouse lungs exposed to iron oxide aerosols. Anat. Rec., 1975;181:581-626.

Stanley PJ, Wilson R, Greenstone MA, Mackay IS, Cole PJ. Abnormal nasal mucociliary clearance in patients with rhinitis and its relationship to concomitant chest disease. Br. J. Dis. Chest, 1985;79:77-82.

Sussman RG, Cohen BS, Lippmann M. Asbestos fiber deposition in a human tracheobronchial cast. I. Experimental. Inhal. Toxicol., 1991;3:145-160.

Swift DL, Montassier N, Hopke PK, Karpen-Hayes K, Cheng YS, Su YF, Yeh HC, Strong JC. Inspiratory deposition of ultrafine particles in human nasal replicate cast. J. Aerosol Sci., 1992;23:65-72.

Ultman, JS. "Transport and uptake of inhaled gases." In *Air Pollution, the Automobile and Public Health*, AY Watson, RR Bates, D Kennedy, eds. Washington, National Academy Press, 1988, pp. 323-366.

Vastag E, Matthys H, Zsamboki G, Kohler D, Daikeler G. Mucociliary clearance in smokers. Eur. J. Respir. Dis., 1986;68:107-113.

Warheit DB, Hartsky MA. Species comparsons of alveolar deposition patterns of inhaled particles. Exp. Lung Res., 1990;16:83-99.

Wolff RK. Effects of airborne pollutants on mucociliary clearance. Environ. Health Perspect., 1986;66:222-237.

Yeates DB, Aspin M, Levison H, Jones MT, Bryan AC. Mucociliary tracheal transport rates in man. J. Appl. Physiol., 1975;39:487-495.

Yeates DB, Aspin M. A mathematical description of the airways of the human lungs. Resp. Physiol., 1978;32:91-104.

Yeates DB, Pitt BR, Spektor DM, Karron GA, Albert RE. Coordination of mucociliary transport in human trachea and intrapulmonary airways. J. Appl. Physiol., 1981;51:1057-1064.

4. HYPERSENSITIVITY AND ASTHMA

M. Ian Gilmour, Ph.D.
Immunotoxicology Branch
Experimental Toxicology Division
National Health and Environmental Effects
Research Laboratory
United States Environmental Protection Agency
Research Triangle Park, NC 27711

INTRODUCTION

The cognate (acquired) immune system recognizes and responds to invading microbes with humoral and cellular defenses to effectively combat infection and contribute to wound healing. Under most circumstances these responses are appropriate and necessary for maintenance of healthy tissue. In some cases however, the immune system may also react against foreign or even self proteins that offer no pathogenic threat. If these responses go unchecked, or are amplified with intermittent or continued antigen exposure, inflammatory processes normally associated with beneficial host defenses may cause a variety of forms of tissue damage collectively known as hypersensitivity or allergic reactions.

A central feature of pulmonary allergy is that it is immune-mediated and acquired following exposure to antigen. Although hundreds of ordinary substances could conceivably trigger allergic reactions, certain proteins and autoantigens seem to be more commonly associated with hypersensitivity disease than others, suggesting that these substances may possess certain unique structural or bioactive properties (Lehrer et al., 1996). The most common airborne allergens are pollens, molds, household dust including dust mite and cockroach antigens, animal dander, and in occupational settings; industrial chemicals and detergent enzymes.

Mechanisms by which hypersensitivity reactions occur in some individuals but not others are not fully known, although epidemiology studies have clearly shown that both genetic and environmental components contribute to disease. Allergic asthma for example, is in part a genetic disease because the incidence is greater in families with a history of the syndrome (Marsh et al., 1981; Burrows et al., 1995; Barnes and Marsh, 1998). The risk of contracting childhood asthma doubles with concurrent exposure to second hand cigarette smoke (maternal smoking) however (Neuspiel et al., 1989; Martinez et al., 1992), indicating a significant environmental influence. Host susceptibility is also dependent upon other biological factors such as stress, diet, pre-existing disease, activity levels and a number of possible interactions among antigens, accessory (antigen-presenting) and immune cells, inflammatory cells, and soluble mediators (Husband and Gleeson, 1996). Finally, in addition to allergen exposure, environmental factors known to play a role in exacerbation, and perhaps induction of allergic lung disease include climatic conditions; co-exposure to other air pollutants (e.g., ozone [O_3], sulfates, nitrogen oxides, and particulate matter) (Wardlaw, 1993); and interactions with infectious agents (Beasley et al., 1993) and their products (Michel et al., 1991).

The purpose of this Chapter is to describe altered pulmonary immune responses which culminate in hypersensitivity reactions in the respiratory tract. Because of the broad nature of the subject, this Chapter will only provide an overview of the immunologic events that occur in IgE-mediated hypersensitivity reactions, hypersensitivity pneumonitis and allergic asthma. The reader is directed to other Chapters in this text for reviews on autoimmune hypersensitivity reactions, delayed-type hypersensitivity (DTH) reactions to beryllium, and immune responses to anhydrides and other small molecular weight compounds. More detailed information on the epidemiology, pathogenesis, and treatment of allergic lung disease can be found in the excellent textbooks of Danielle (1988), Crystal (1991), and Fishman et al. (1998), and in the recent primer on immunologic diseases by the Journal of the American Medical Association (Baker, 1997).

HYPERSENSITIVITY LUNG DISEASE

Classification

Several types of allergic reactions can occur in the lung and have been broadly defined according to their time of onset, whether the effects are mediated by cells or antibody, and if the responses are directed against extrinsic or "self" antigen (Gell and Coombs, 1968). Under this scheme (Table 1), immediate (Type I) hypersensitivity responses (as seen in hay fever [rhinoconjunctivitis]) arise from allergen-specific crosslinking of IgE antibodies on mast cells and rapid release of vasoactive amines, bronchoconstrictive agents, and proinflammatory mediators. Type II reactions, which are quite rare (3 per million), involve slower antibody- and complement-mediated autoimmune reactions against self proteins in the basement membrane (e.g., Goodpastures Syndrome). Type III reactions are associated with increased IgG/immune complex formation and T-lymphocyte activation in hypersensitivity pneumonitis (e.g., farmer's lung), and Type IV (DTH) reactions are characterized by cell-mediated activation, resulting in pulmonary inflammation, granuloma formation, and fibrosis.

DTH reactions were first characterized in the lungs of patients suffering chronic bacterial infections (e.g., tuberculosis), although responses are now also associated with inhalation of small molecular weight chemicals/certain metals, and occur via mechanisms similar to contact hypersensitivity reactions against poison ivy. Rarer forms of pulmonary hypersensitivity reactions (e.g., allergic bronchopulmonary aspergillosis) and diseases that have an immunological component but no identified causal agent (e.g., sarcoidosis and eosinophilic pneumonia) have been recognized but not further classified into disease groups because of their unique features.

TABLE 1. CLASSIFICATION OF TYPE I-IV HYPERSENSITIVITY REACTIONS IN THE RESPIRATORY TRACT.

Type of Hyper-sensitivity Reaction	Immunologic Mediation	Antigen	Method of Diagnosis
Type I (anaphylactic)	IgE antibody	Proteins such as animal allergens, pollens and dust mite	Correlation of skin tests or *in vitro* measurement of antibody with symptoms
Type II (cytolytic)	IgG antibody vs cells expressing self proteins	Basement membrane protein	Autoantibody titers, clinical symptoms
Type III (antigen-antibody)	Antigen-antibody complex and inflammatory cells	Avian proteins and fungal spores	Precipitating antibody correlated with exposures and symptoms
Type IV delayed (cell-mediated) hypersensitivity	Lymphocyte-mediated reactions in the lung against antigen	Microbial proteins and inhaled chemicals	Correlation of clinical disease with lymphocyte activation and cytokine release after antigen exposure.

The criteria for hypersensitivity diseases described above are still extremely useful from a clinical standpoint, and specific diagnostic techniques can successfully differentiate between allergic rhinitis to pollens (Type I), hypersensitivity pneumonitis to mold spores (Type III), and DTH to beryllium (Type IV). Although the timing and distinct pathologies of the four main reactions remain unchanged, it is now appreciated that distinct CD4$^+$ T-helper lymphocyte subpopulations (termed T_H1 and T_H2) regulate expression of the various disease types via differences in antibody class production and T-lymphocyte effector functions (reviewed in Abbas et al., 1996). At present it is known that T_H1 lymphocytes are responsible for stimulating B-lymphocyte secretion of complement-fixing antibodies seen in Type II and III reactions, whereas T_H2 lymphocytes signal production of IgE antibody (Type I hypersensitivity reactions). In terms of effector function, T_H1 lymphocytes promote cell-mediated reactions and macrophage activation via interferon (IFN)-γ production (Type IV reactions), while T_H2 lymphocytes secrete interleukin (IL)-5 which recruits eosinophils to mucosal surfaces (allergic asthma).

Despite clear distinctions in cytokine profile and immune phenotype by the various lymphocyte subsets, cytokines associated with both T_H1 and T_H2 responses can be produced simultaneously from other sources such as mast cells or eosinophils (Gordon et al., 1990). In addition, the pathological changes and symptoms seen in allergic disease may be reproduced by several different mediators, suggesting more than one mechanism of action. In allergic asthma for example, bronchial hyperresponsiveness can be induced by leukotriene D_4 (LTD$_4$) from mast cells (Smithy et al., 1985), major basic protein (MBP) from eosinophils (Gundel et al., 1991), or IL-5 from T-lymphocytes (van Oosterhout et al., 1993); all of which are increased during exacerbations of disease. Similarly, of the 50 or so mediators produced during an asthma attack (Barnes et al., 1998), approximately half have pro-inflammatory potential with an ability to recruit monocytes and/or granulocytes to mucosal surfaces. These many variables support the notion that asthma is a complex syndrome whose symptoms can occur via several different mediators from a number of different cell types, though the final pathology can be strikingly similar.

Prevalence of Atopy

Atopy has been defined as "the state of having one or more groups of diseases – allergic rhinoconjunctivitis (hay fever), allergic asthma, and atopic dermatitis (eczema) – that are caused by a genetic propensity to produce IgE antibodies to environmental allergens encountered through inhalation, ingestion and possibly skin contact." A broader definition sometimes used for epidemiological studies requires only the presence of total IgE antibody regardless of allergic disease (Institute of Medicine, 1993). Of the United States population, 23 - 30% are atopic although up to 40% of these individuals may remain asymptomatic despite having higher than normal IgE antibody levels. Of the 17% of the population who suffer from some sort of allergic disease, most are hypersensitive to pollens and suffer from allergic rhinitis and sinusitis (Figure 1). More than 6% of the population has physician-diagnosed asthma, the prevalence of which has increased by ≈30% over the last 15 years. Although it is true that most allergic asthmatics also experience nasal allergies, rhinitics do not necessarily have airway disease and asthma. Whether asthmatics comprise a special subset of rhinitics, or if the larger particles associated with rhinitis (pollens and spores) simply cause asthma-like reactions, but in the nose rather than the airways, remains controversial (Rowe-Jones, 1997).

A smaller fraction (5 - 15%) of rhinitic and asthmatic individuals exhibit intrinsic (non-atopic) disease with a similar pathology to the allergic phenotype, despite low levels of IgE antibody. At present no specific agent(s) has been implicated in these patients, although the individuals tend to be older, usually exhibit more severe forms of the disease with associated nasal polyps, and often have a rather unique sensitivity to aspirin (Zeitz, 1988). As opposed to atopic asthmatics who respond to inhalational challenge with the appropriate allergen, symptoms occur in intrinsic patients following a variety of relatively non-specific stimuli such as exercise, cold air, and emotional stress.

It should also be noted that the severity of both allergic and non-atopic asthma is highly variable between individuals and direct relationships between different markers of disease, such as IgE levels, number of circulating and/or pulmonary eosinophil numbers, and bronchial hyperresponsiveness are difficult to establish. Currently, asthmatics are classified as having mild, moderate, or severe disease based upon clinical symptoms and the type and frequency of required treatment (e.g., severe, steroid-dependent asthma).

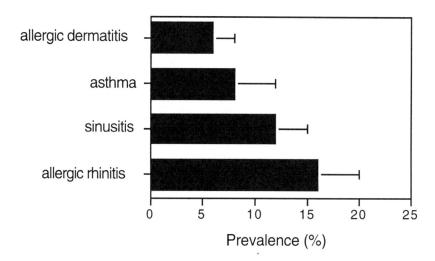

Figure 1. Estimated range of prevalence of allergic diseases in U.S. population (Institute of Medicine, 1993).

Immediate (Type 1) Hypersensitivity Reactions

Immediate (Type 1) hypersensitivity reactions are caused by overproduction of IgE antibody against certain pollens, molds, and other allergens, and this sensitivity can be confirmed by presence of a wheal and flare reaction following intradermal injections of allergen (positive skin test) or by measuring allergen-specific or total IgE levels in serum. In sensitized individuals, allergen exposure causes a variety of symptoms within 5 – 30 min including watery eyes, nasal discharge, sneezing, and blockade in both the nose and airways. These immediate allergic/asthmatic reactions (IAR) result from the binding of inhaled antigen to IgE antibodies on the surface of mast cells in the respiratory epithelium. Crosslinking of the antigen-specific (F_{ab}) regions of two neighboring IgE molecules by the relevant allergen stimulates a cascade of signal transduction events via the high affinity IgE receptor ($F_{c\varepsilon}RI$). Amongst the most prominent indicators of this activation sequence are elevated cytosolic calcium and increased phosphoinosotide turnover in the cytoplasm, which in turn stimulate the cell to secrete a variety of mediators by two major processes (reviewed in Askenase, 1977 and Costa et al., 1997).

First, preformed mediators including histamine, tryptase, and heparin are released from metachromatic granules via a process of exocytosis. Of these, histamine in particular, when topically applied to the nose, reproduces the transient symptoms of nasal blockade, watery and proteinaceous discharge, sneezing, and itching seen in the early allergic response. Histamine acts through binding to receptors on endothelial cells, vascular smooth muscle, and sensory nerves to cause plasma exudation, vasodilation and secretion, respectively, and its action can be abolished by topical treatment with H_1-receptor antagonists, or by blocking its release with mast cell stabilizers such as chromoglycate. The biological functions of tryptase and heparin in the IAR are less well-defined, because neither substance can directly elicit allergic symptoms, although tryptase is now thought to be involved in cleaving complement proteins into anaphylatoxic fragments.

A second $F_{c\epsilon}R1$ signal involves activation of phospholipase A_2 and C which stimulate the intracellular release and hydrolysis of arachidonic acid from membrane phospholipids. A number of leukotrienes (LTB_4, LTC_4, LTD_4), prostaglandins (PGD_2, PGE_2), and other lipid mediators (e.g., thromboxane A_2) are then synthesized from this substrate by lipoxygenase and cyclooxygenase and are rapidly secreted. *In toto*, release of these lipid mediators augments the early phase responses of bronchoconstriction, airway secretions, and plasma exudation (edema), as well as stimulating irritant and sensory nerves, and increasing chemoattraction of inflammatory cells. Mast cell activation also leads to transcription and synthesis of numerous cytokines including IL-1, -3, -4, -5, -6, and -10, IFNγ, TGFβ, and TNFα. Many of these latter mediators provide central control in immune and inflammatory processes and are thought to be involved in the protracted pathophysiology seen in more chronic forms of allergic disease such as asthma (Huston, 1997).

Hypersensitivity Pneumonitis

This disease, also known as extrinsic allergic alveolitis, is caused by abnormal immune reactions to antigens in inhaled organic dusts and mold spores, and is associated with high circulating levels of precipitating IgG and IgM antibodies to the offending allergen. It is usually an occupational disease affecting 1 out of 10,000 Americans (reviewed in Fink, 1987; Salvaggio and Millhollon, 1993). Acute illness begins 4 - 8 hr after exposure, and is characterized by a flu-like syndrome with fever, chills, dry cough, and dyspnea. Resolution may occur in days or weeks after expo-sure ceases, whereas repeated exposure can result in chronic illness with changes seen on chest X-ray ranging from mild infiltrate to diffuse interstitial pulmonary fibrosis. Histologically, the disease is characterized by a patchy granulomatous pneumonitis. In the early stages, neutrophils are predominant in BAL fluid, but give way to large numbers of plasma cells, lymphocytes (>50% $CD8^+$ T-lymphocytes), and multinucleated giant cells in chronic disease. Antigen-specific subtypes include farmer's lung (*Mycopolyspora faeni*), Pandora's pneumonitis, bird-fancier's lung, mushroom worker's lung, bagassosis, malt worker's lung, and maple bark disease.

Although it is true that high levels of precipitating IgG and IgM antibodies to fungal spores are found in nearly all cases of farmer's lung, disease severity does not correlate with antibody concentration (Lee et al., 1983). In addition, 10 - 40% of exposed individuals may remain asymptomatic despite high circulating titers of antibody to the offending allergen. Animal studies have shown that experimental hypersensitivity pneumonitis in mice and guinea pigs can be transferred to naive recipients via T-lymphocytes and to a far less extent by precipitating antibodies (Schuyler et al., 1987, 1990). From these reports, it would appear that immune complex formation (although involved in complement fixation and neutrophil recruitment) may play a less prominent role than T-lymphocytes in disease expression.

Asthma

In individuals who suffer from Type 1 hypersensitivity reactions without additional chronic conditions such as asthma, the immediate allergic response (IAR) usually resolves over 1 - 2 hr, is easily treated, and normally does not worsen with repeated exposure. In asthma however, as the initial symptoms of basophil and mast cell activation subside, a secondary wave of bronchoconstriction and inflammation can arise which may be of equal or even greater intensity than the immediate response, and generally lasts for several hours (Newman Taylor, 1998). This late phase allergic/asthmatic response (LAR) is characterized by a marked reduction in lung function; pronounced airway infiltration of monocytes, lymphocytes, and granulocytes including eosinophils; and increased airway responsiveness (reviewed in Djukanovic et al., 1990, Horwitz and Busse, 1995). The inflammatory process extends from the bronchi to the distal airways where plugs of mucus form and bronchial epithelium are shed. Because these reactions occur in the airways, the same kind of Type 1 hypersensitivity reactions found in rhinitis (i.e. histamine release, mucous secretion, and cellular inflammation) can have a much more profound effect on the ability to breath.

In addition to the features of the IAR described above, asthmatics also display anatomical and morphological changes that may persist for indefinite periods of time, and can worsen after antigen challenge. In the large airways, mucus secretion is increased and is associated with goblet cell hyperplasia and hypertrophy of the bronchial glands. The basement membrane around the airways also shows a marked thickening and airway smooth muscle is hypertrophied. Cellular infiltrates pack around the air passages and possibly contribute to the narrowing of the airway lumen. Finally sheets of both ciliated and non-ciliated epithelial cells are shed into the lumen and the remaining endothelial layers appear swollen and damaged. From a physiological standpoint, the disease is characterized by a marked reversibility of airflow obstruction over relatively brief periods during the pathophysiological process. Asthmatics also tend to exhibit long-lasting bronchial hyperresponsiveness to non-specific agonists, including cold air and hypertonic saline, and increased airway reactivity to cholinergic agents such as methacholine.

Allergen challenge in asthmatics causes a mixed inflammatory reaction of lymphocytes, monocytes and macrophages, and granulocytes around the airways and vasculature. Although neutrophils often appear early in this response, the principal granulocyte that persists for long periods of time is the eosinophil (Metzger et al., 1986). These cells when activated, release four principal cationic proteins (reviewed in Wardlaw et al., 1995) which are cytotoxic and may contribute to airway responsiveness possibly through exposing sensory nerve endings. Major basic protein (MBP) is the main constituent of eosinophil granules and is both cytotoxic to epithelial cells and directly causes airway hyperresponsiveness. Eosinophil cationic pro-tein (ECP) is a more potent cytotoxic agent that causes pore formation on the membranes of target cells. Eosinophil-derived neurotoxin (EDN) damages myelinated neurons, and eosinophil peroxidase (EPO) decreases degradation of bioactive leukotrienes produced by mast cells and basophils and induces mast cell histamine release.

Eosinophils can also generate a strong respiratory burst via the elaboration of oxygen radicals and secrete numerous cytokines such as IL-3, IL-5, and granulocyte-macrophage colony-stimulating factor (GM-CSF) which enhance their longevity and recruit additional eosinophils to the site, respectively (Corrigan and Kay, 1992). It has also been shown that eosinophils express both high ($F_{c\varepsilon}RI$) and low ($F_{c\varepsilon}RII$) affinity IgE receptors and major histocompatibility Class II molecules, and secrete IL-4 (Wardlaw et al., 1995). These findings would suggest that eosinophils are not simply effector cells but contribute to the regulation of specific immune responses.

The airway epithelium is usually disrupted in asthmatics and is the site of the mast cell- and eosinophil-related inflammatory response. The epithelial cells are in themselves active participants in asthma pathophysiology and can produce cytokines and other mediators (e.g., IL-1β, IL-6, IL-8, GM-CSF, prostaglandins, and platelet activating factor) which contribute to airway inflammation (Montefort et al., 1992).

The summation of these pathophysiologic features, as well as clinical history, distinguishes asthma from simple immediate hypersensitivity responses and provides a differential diagnosis between the two disease states. Though many of the mediators thought to be important in generating the LAR may be produced by activation of mast cells during the IAR, the biological mechanisms that induce transient responses in rhinitics while producing asthma in others are not known. Quite possibly, the airways of asthmatics have unique underlying phenotypic characteristics not present in rhinitics, although at present these features are not known.

Allergen Avoidance and Treatment of Symptoms

The most effective way of treating hypersensitivity reactions is through eliminating exposure to the relevant allergen. In many cases, this can be accomplished through changing the environmental exposure such as avoiding pets, removing carpets, changing occupation, relocating to a drier climate, etc. and can effectively result in a "cure" in most cases. This is especially true for those who suffer from occupational asthma or hypersensitivity pneumonitis against particular molds, spores, and specific components of organic dust. The ubiquitous nature of some common allergens such as the house dust mite, however, require more sophisticated methodologies (e.g., advanced climate control, specialized vacuum cleaners, hypoallergenic bedding, acaricides, etc.), to reduce antigen levels in households. Allergen shots (immunotherapy), which are designed to trick the immune system into downregulating specific immune responses, have been successfully used to create immunological tolerance in some rhinitic individuals against many pollens and mold, but are generally less effective for treating mite-sensitive asthmatics (Bousquet and Michel, 1995).

The most common asthma therapies employ a variety of antihistamines, mast cell stabilizers (cromolyn), smooth muscle antagonists (β_2 adrenergics and anticholinergics), non-steroidal anti-inflammatories, and lastly steroids, depending on the type and severity of disease (reviewed in Barnes et al., 1988). In mild-to-moderate asthma, cromolyn and nedocromil sodium are the preferred first treatment strategy because of their safety and efficacy in providing a degree of mast cell stability. Acute symptoms are then controlled on an as-needed basis with inhaled β agonists which relax smooth muscle via activation of adenylate cyclase with subsequent increases in cAMP and kinase-dependent inhibition of myosin phosphorylation. More severe forms of asthma are typically treated with oral or inhaled steroids that essentially inhibit immune cell expansion, mediator release, and inflammation. Although this therapy has proved effective in all but the most severe cases (steroid-resistant asthma), discontinuing their use results in a re-emergence of the disease, and the side effects of long-term therapy include decreased bone density, soft organ damage, and general immunosuppression.

Identification of the many enzyme and mediator cascades during the immediate and late-phase reactions has led to a new generation of pharmacologic agents that have a high specificity. Of note are the inhibitors of the 5-lipoxygenase pathway (Israel et al., 1993), and the receptor antagonist for LTD_4 (Spektor et al., 1994), both of which have been shown to reduce asthma symptoms. Also, antibodies and receptor antagonists against cytokines such as IL-5 have been developed and shown to reduce eosinophil numbers and bronchial hyperresponsiveness in animal models of allergy (Kung et al., 1995, Mauser et al., 1995). Such strategies for targeting the initial signaling mechanisms involved in mucosal inflammation and bronchial hyperresponsiveness may provide more effective control of airway disease than the current therapeutic agents that treat the symptoms during exacerbations of disease.

Animal Models of Pulmonary Allergy

Much of the understanding of the mediators and mechanisms involved in allergic lung disease and immunology in general has come from the use of experimental animals. A number of different species have been used to model allergy and asthma and have provided valuable information about the induction, expression and treatment of the disease (reviewed in Herz et al., 1996; Griffiths-Johnson et al., 1997). Guinea pigs were traditionally used as asthma models because of the ease in which they can be sensitized, and because their response to allergen challenge, such as histamine release, smooth muscle contraction, and eosinophilic inflammation, resembles the acute events seen in the human disease. Subsequent studies in rats and mice have achieved similar results with the added advantage of more immunologic reagents to manipulate the system, and identify key mediators in the pathogenesis of disease. The greatest drawback of these rodent models is that they lack some of the features of chronic disease such as spontaneous airflow limitation, smooth muscle hyperplasia, and goblet cell hypertrophy. Attempts to elicit these long-term pathological changes by multiple antigen exposures have, for the large part, failed because such regimes in rodents usually result in a state of immunologic tolerance or unresponsiveness (reviewed in Holt, 1993). Although asthmatic individuals may be unable to tolerize in this way because of their inherent genetic makeup, or through differences in the concentration and frequency of subsequent allergen exposures, further experimentation with different rodent strains (including transgenics) and more complex allergen exposure scenarios are needed to determine if chronic disease can indeed be achieved in mice and rats. In addition, other environmental factors such as diet, viral infections, climate variations, and exposure to indoor and outdoor air pollutants may have an impact on the severity of disease, but have not been extensively tested as co-factors in experimental systems.

Larger animal species including cats, dogs, sheep, and non-human primates are often naturally sensitized to intestinal tract parasites such as *Ascaris suum*, and pulmonary challenge with purified allergen can result in a more prolonged decrement in lung function and eosinophilic inflammation than occurs in rodents. Dogs, cats, and horses also spontaneously develop clinically important allergic lung disease that may be the domestic animal equivalent of human asthma. Although these "natural" systems are more relevant for study, and in some cases more closely resemble the anatomy and morphology of the human lung, large scale use of these species is limited by fiscal and emotional reasons, a lack of reagents for each species, and because of the large degree of variation in responses between individuals.

REGULATION OF ALLERGIC IMMUNE RESPONSES

The vast majority of people with asthma and/or rhinitis are atopic and experience disease exacerbations within minutes of challenge with the appropriate allergen. Because antigen-specific crosslinking of IgE on the surface of mast cells is the primary immunologic trigger in many allergic reactions, and overproduction of IgE antibody appears to be a heritable trait in atopic individuals, a brief review of the genetics of atopy is warranted. Further sections will describe the immunologic pathways that lead to IgE production, factors that influence polarization of immune responses toward this phenotype, and how air pollutant exposure might these events

The Genetics of Atopy

Several epidemiological studies have demonstrated that atopic diseases including asthma, have a heritable component with an estimated transmission risk of up to 70% depending upon the familial relationship (i.e., genetic similarity) (Martinez et al., 1997). Unlike single-gene diseases like cystic fibrosis and sickle cell anemia, which obey classic Mendelian laws of recessive, dominant, or co-dominant inheritance, atopy and asthma are more complex disorders that originate from co-expression (and interaction) of multiple genes (Barnes and Marsh, 1998). In addition, disease occurrence is dependent upon other environmental influences such as allergen exposure during critical periods of immune development (Husband and Gleeson, 1996; Holt et al., 1997). Despite the multifactorial nature of asthma and atopy, linkage analysis and positional cloning studies in various populations including di- and monozygotic twins have succeeded in identifying particular regions of the human genome associated with development of an allergic phenotype. Not surprisingly, many of the genes (Table 2) are associated with aspects of immune function.

TABLE 2. CHROMOSOMAL REGIONS IMPORTANT IN THE HERITABILITY OF ASTHMA.

Region	Candidate gene(s)[a]	Proposed function	Phenotype(s) implicated
5q31.1-q33	IL-2, IL-4, IL-5, IL-9, IL-13, CSF$_2$	B-lymphocyte isotype switching to IgE, cytokines upregulate eosinophil, basophil, mast cell, and IgE functions (e.g. T$_H$2 response)	Total IgE (non-cognate), asthma, BHR
	ADRB2	Membrane coupled G-protein-coupled receptor in the lung	Total IgE, BHR, asthma
	GRL	Receptor important in inflammatory modulation	Unknown
6P21.3	HLAD	Antigen presentation	Specific IgE and IgG antibodies
	TNFα	Mediates inflammatory response	Asthma
11q13	F$_{c\varepsilon}$RIB	Transduction signaling on basophils, mast cells and dendritic cells	Atopy (total and specific IgE), asthma
	FGF$_3$	Promotes cellular proliferation	"
12q14.3-q24.1	IFN-γ	Inhibits IL-4 activity	Asthma, atopy, IgE production
	SCF	Produces IL-4	"
	NFYB	Upregulates transcription of IL-4 and HLA-D genes,	"
	STAT6	essential cytokine-regulated transcription factor	"
14q11.2-13	TCRA, TCRD	Interacts with peptide-MHC complexes	Specific IgE antibody
	NF-κB	Activates immunoregulatory genes in response to inflammatory stimuli	Unknown

Abbreviations: BHR, bronchial hyperresponsiveness; MHC, major histocompatibility complex; [a]Proteins encoded by genes described in the Table: IL, interleukin; CSF, colony stimulating factor; ADRβ2, β$_2$-adrenergic receptor; GRL, glucocorticoid receptor; HLAD, human leukocyte antigen D region; TNFα, tumor necrosis factor-α; F$_{c\varepsilon}$RIB, high affinity receptor for IgE; FGF$_3$, fibroblast growth factor 3; IFNγ, interferon-γ; SCF, stem cell factor; NFYB, β-subunit of nuclear factor Y; STAT6, signal transducer and activator of transcription 6; TCRA/D, T-lymphocyte receptor α/δ chains. Adapted with permission from Barnes and Marsh, 1998.

The genetic region on chromosome 6 (6p21.3) contains a number of immune response (Ir) genes involved in antigen presentation in the context of major histocompatibility proteins. Early studies of this gene showed that 78 of 80 individuals sensitized to ragweed allergen (*Amb a*) expressed HLA-DR2 and Dw2 haplotypes in contrast to an incidence of only 22% in non-allergic individuals (Marsh et al., 1990). Further analyses using DNA sequencing and peptide-specific T-lymphocyte clones identified several Class II restriction elements associated with the production of IgE against *Amb a* and other molecules such as the group I and II allergens of house dust mite (Zwollo et al., 1991). These results offered such compelling evidence for a genetic predisposition to allergy, through the inheritance of specific proteins involved in presentation of antigen to T-lymphocytes, that this genetic region is named allergy and atopy region 1 (AA1) (Barnes and Marsh, 1998).

Another more general region of the human genome that has proven to be important in atopy and asthma is the IL-4 gene locus on chromosome 5q31.1-q33. This cluster of genes includes the cytokines IL-2, -4, -5 -9, and -13, and colony stimulating factor which are important in the differentiation of T_H2 lymphocytes and in subsequent formation of IgE antibody, and the recruitment, maturation, and activation of mast cells and eosinophils. At present, certain polymorphisms in this region are associated with elevated levels of total IgE antibody (Marsh et al., 1994) and bronchial hyperresponsiveness in asthmatic populations (Postma et al., 1995).

In addition to the examples described above, three other genetic regions are recognized to be important in the development of asthma. Region q13 of chromosome 11 (11q13) contains the high affinity receptor for IgE ($F_{c\epsilon}RI$) (Cookson et al., 1989). This molecule is known to be upregulated in cells of asthmatics and provides the signal mechanism for mast cell degranulation and the subsequent release of vasoactive amines, chemoattractive agents and numerous other pharmacologically active substances (Sutton and Gould, 1995). Region q14.3-q24.1 on chromosome 12 maps for a number of regulatory elements in the development of T_H2 lymphocyte responses including IFNγ (which downregulates IL-4 production), and STAT6 and the IL-4 promoter which activate synthesis of IL-4 (Barnes et al., 1996; Nickel et al., 1997). Finally, recent linkage analysis reports have identified certain polymorphisms for asthma in region q11.2-13 on chromosome 14, which contains the sequences for the α and δ chains of the T-lymphocyte receptor as well as the transcription factor NFκB (Moffat et al., 1994).

The epidemiological data and subsequent molecular genetic analyses strongly suggest a genetic predisposition to atopy and asthma, and identification of HLA haplotypes and distinct T-lymphocyte receptor sequences encoding for certain allergens offers firm evidence for the specific genetic expression of these responses. Principal regulatory elements of the immune response such as those in the IL-4 locus and transcription factors of inflammation are also implicated in the atopy/asthma phenotype and provide a controlling framework for ensuing responses. Although the integration of these different genetic factors seems to be required for the development of atopy, the reason(s) why some individuals progress to chronic airway disease and asthma is not known.

Antigen Presentation and Polarization of T-Lymphocyte Responses

Although one's genetic makeup is a strong controlling factor in promoting atopy, the structure and concentration of antigen, the frequency of exposure, and previous or concomitant conditioning of the immune system to other antigenic stimuli such as infections, may also profoundly affect the resulting type of immune response (Holt, 1996). The initial (sensitizing or immunizing) encounter with an allergen is a critical junction in determining the type and magnitude of the primary immune response and programming subsequent immunological memory.

After inhalation, antigenic particles deposit on the linings of the nose, airways or alveoli according to their size and aerodynamic characteristics. If the particles are not physically removed either by coughing, sneezing, or mucociliary expulsion, they are taken up by resident phagocytic macrophages or by dendritic cells which form a contiguous network throughout the pulmonary epithelium (Holt et al., 1994). Although it was originally thought that alveolar macrophages were the principal antigen-presenting cells in the lung, it is now clear that dendritic cells are much more effective at this process (reviewed by Lambrecht et al., 1996). While these cells do not bear $F_{c\gamma}$ receptors and are poorly phagocytic, recent studies have shown that they can take up soluble antigen by both pinocytosis (Sallusto et al., 1995) and mannose receptor-mediated endocytosis (Reis e Sousa et al., 1993), and may also receive processed peptides from phagocytic macrophages (Gong et al., 1994).

After internalization by dendritic cells, antigen molecules are processed intracellularly and expressed on the cell surface as peptides in the groove of the Class II antigen presentation apparatus termed the major histocompatibility complex (MHC II). During this period of antigen uptake and processing, the intraepithelial dendritic cells detach and enter afferent lymphatics in the form of veiled cells and home to regional lymph nodes where they interdigitate in the T-lymphocyte-dependent paracortical areas. At this point, circulating T-lymphocytes bearing receptors for a wide array of antigenic sites are screened and form rosettes around each dendritic cell. If the peptide is recognized with sufficient avidity in the context of the MHC/T-lymphocyte receptor complex along with appropriate accessory molecules such as CD4 and the B7.1 and B7.2/CD28 ligands, then antigen presenta-tion and T-lymphocyte activation will ensue (reviewed by Janeway and Bottomly, 1994). During this initial sensitizing phase, the peptides are presented to mature naive $CD4^+$ T-lymphocytes termed T_H0 lymphocytes which may produce a wide spectrum of T_H1 and T_H2 cytokines. Depending upon the presence of regulatory cytokines such as IFNγ, IL-4, and IL-12, the strength of the antigen signal (avidity and concentration), and the nature of co-stimulation, the T_H0 lymphocytes then polarize into T_H1 or T_H2 lymphocytes which are characterized by their contrasting cytokines and phenotypic function (reviewed by Abbas et al., 1996). Once generated, effector and memory pools of T- and B-lymphocytes circulate throughout the body or lodge in lymphocyte-rich areas of tissue in preparation for expanding into antigen-specific secondary responses. Subsequent antigen encounters produce responses of the same original type except that the reaction is stronger because of a larger responsive T-lymphocyte pool and the presence of B-lymphocytes which augment the antigen presentation processes.

The principal signal for T_H1 differentiation is IL-12 produced from macro-phages, and natural killer (NK) and dendritic cells during initial infection and antigen presentation. Once selected, T_H1 lymphocytes secrete IFNγ which has two key functions in promoting DTH reactions and defense against bacterial and viral patho-gens. First, it activates cell-mediated immune responses and enhances macrophage microbicidal activity. Second, it stimulates the production of complement-fixing IgG antibodies involved in the opsonization and phagocytosis of particulate microbes. T$_H$1 lymphocytes also produce IL-2 to stimulate further clonal expansion and promote the differentiation of CD8 lymphocytes and TNFβ, which recruits inflammatory leukocytes to the site of immune activation.

Although T$_H$1 responses may be regarded as normal protective immune responses, T$_H$2 lymphocytes are associated with IgE production and eosinophilic inflammation during helminth infections, and are also the predominant T-lymphocyte subset in Type I hypersensitivity reactions and allergic asthma. T$_H$2 lymphocytes produce IL-4, -5, -10, and -13. IL-4 and IL-13 induce class switching of B-lymphocytes from an IgG and IgM isotype to an IgE phenotype, and also promote the differentiation and maintenance of the T$_H$2 lymphocyte population. IL-4 also reduces IL-12 receptor expression which further decreases T$_H$1 differentiation. IL-5 is a chemotactic and maturation factor for eosinophils, and IL-10 serves to suppress T$_H$1-induced inflammatory processes. In addition to their differing effector functions, T$_H$1 and T$_H$2 lymphocytes are mutually antagonistic toward each other. T$_H$1-derived IFNγ potentiates growth of T$_H$1 lymphocytes while suppressing activation of T$_H$2 responses, whereas IL-4 and IL-10 fulfill similar functions for T$_H$2 lymphocytes.

Despite the well characterized distinctions between cytokine profiles of T$_H$1 and T$_H$2 lymphocytes, and their functional phenotypes, the reasons why the host at times favors a T$_H$2 response over the more normal IL-12-dependent T$_H$1 response are not known. The primary signal for the induction of T$_H$2 lymphocytes is known to be IL-4, and once operant, T$_H$2 lymphocytes may promote their own differentiation. Antigen concentration seems to be a factor in initial phenotype selection because in general it has been shown that small amounts of antigen induce T$_H$1 responses whereas greater concentrations induce a T$_H$2 response (reviewed in Murray, 1998). Experiments in the early 1970s showed that antibody formation and DTH responses (T$_H$2 and T$_H$1 responses, respectively) occurred at different immunizing concentrations of antigen. Later studies confirmed that different concentrations of the same antigen may induce either IL-4 or IFNγ in specific T-lymphocyte clones. Most recently, it has been shown that the polarization toward T$_H$1 or T$_H$2 is dependent upon the affinity of binding between the peptide and certain MHC molecules, and subsequent interaction with appropriate T-lymphocytes. It could be that certain peptide fragments from allergens have a greater capacity to induce IL-4 during the presentation process, or that the T$_H$2 phenotype is selected by exogenous IL-4 from other sources such as mast cells. Certain polymorphisms in the IL-4 promoter region or deficiencies in IFNγ gene may also be the controlling factors in this process.

ALLERGY AND IMMUNOTOXICOLOGY

The discipline of immunotoxicology is concerned with alteration of the immune response by exogenous agents which results in abnormal pathology and disease. From this definition, it could be argued that allergy is a naturally occurring immunotoxic reaction that is driven by host susceptibility and allergen exposure, and may be exacerbated by exposure to other xenobiotic agents. In the respiratory tract, this phenomenon is most commonly associated with air pollutant exposure increasing the incidence and severity of allergic lung disease. Currently, there is substantial epidemiological evidence that medication use and hospital admissions for asthma increase during episodes of high air pollution, and some of these findings have been confirmed in clinical studies (reviewed in Bascom et al., 1996a and b). It is less clear whether air pollution exposure actually increases the rate or level of initial allergic sensitization, because the development of clinical disease requires an undefined time and number of subsequent allergen challenges after the sensitization phase before symptoms may be apparent. Likewise, the question of pollutant-enhanced sensitization is also difficult to ask in clinical studies because of the risk of initiating allergic disease in human volunteers. This section will briefly review examples of air pollutant-enhanced allergic lung disease in humans and animals and discuss potential mechanisms for this effect. More detailed accounts of pollutant enhanced allergy for each agent are described in appropriate chapters of this book.

Pollutant-Enhanced Allergic Lung Disease

Air pollution exposure has long been considered a risk factor for the exacerbation of asthma attacks in asthmatic individuals. Health care statistics in Canada and the U.S.A. show that hospital admissions for asthma and bronchodilator use increase during episodes of high ozone and/or airborne particulate levels (reviewed by Pope et al., 1995). In controlled clinical exposures, O_3 and nitrogen dioxide (NO_2) have been shown to exacerbate asthmatic responses to allergen challenge (reviewed by Peden, 1997). Asthmatics are also more reactive to inhaled sulfur dioxide (SO_2), suggesting these individuals are inherently susceptible to air pollutant exposure. From an immunological perspective, asthmatics exposed to diesel exhaust particles (DEP) have increased levels of IgE-forming cells and IgE and IgG_4 antibody in nasal lavage suggesting that DEP may act as an immunologic adjuvant to enhance induction or boosting of allergic (T_H2) immune responses (Dias-Sanchez et al., 1994, Takenaka et al., 1995).

Several animal models have been used to investigate the effect of air pollutant exposure on the development or severity of IgE-mediated (Type I) hypersensitivity. Numerous studies have now demonstrated that exposure to O_3, H_2SO_4, SO_2, NO_2, and diesel particles can increase IgE antibody levels and lymphocyte activity in mice, rats, and guinea pigs. Recent experiments have also shown that co-administration of diesel particles and antigen increases IL-4 production in bronchial lymph nodes, and that pollutant-enhanced immunity results in increased allergic lung disease after subsequent antigen challenge (reviewed by Gilmour, 1995).

The mechanisms by which specific immune function can be increased following pollutant exposure are still unclear. It is known that inhalation of oxidant gases causes decrements in lung function, epithelial cell damage, increased permeability and edema, and the release of a broad spectrum of pro-inflammatory mediators. Epithelial damage and increased lung permeability could alter translocation of antigen to immunoreceptive sites in the submucosa and lymphatic tissue, whereas increased cytokine output may have profound effects on inflammatory and immune cells locally, and possibly on cells remote from the initial site of damage. Furthermore, changes in the antigen handling/presentation apparatus by defects in macrophage phagocytosis, and altered dendritic cell number and function may affect immune signaling during both initial antigen recognition and secondary clonal expansion. Finally, direct effects of air pollutant exposure on T-lymphocyte numbers and function could account for altered immune function during antigen exposure (reviewed by Gilmour and Selgrade, 1998).

SUMMARY

Pulmonary allergies are the most common form of recurrent respiratory disease and have a huge impact on quality of life and health care costs. Despite tremendous advances in understanding the genetic and immunological basis for these diseases, the incidence of atopy and asthma are increasing. Assuming the genetic pool has been relatively stable over the past 30 years, this increase must be attributed to environmental influences and/or changes in lifestyle. Increased use of climate control with less ventilation could account for changes in allergen exposure and possible influences of other indoor air pollutants. Immune function in premature babies and in vaccinated children may be different from previous generations where infectious diseases demanded survival of the fittest. The increased use of television and computers with less emphasis on outdoor activities may also impact on indoor exposures as well as reduced athletic ability. Finally, co-factors such as infections, exposure to outdoor air pollutants, chemicals in food and water, and nutritional status may alter immunological sensitization.

This Chapter has attempted to summarize some of the basic mechanisms that govern allergic reactions in Type I hypersensitivity reactions and asthma. In doing so, other important aspects of allergic lung disease such as neural regulation of smooth muscle and bronchial hyperreactivity, adhesion molecules and chemotaxis in inflammation, and the role of complement factors were not discussed but have been reviewed elsewhere (Busse and Holgate, 1995; Baker, 1997). Despite these omissions, it is clear that the regulatory pathways involved in the development of specific immune responses can be profoundly affected by toxicant exposure. Integration of exposure data in the human population and further mechanistic studies in animals will identify immunotoxic agents that may alter the incidence and severity of allergic lung disease and asthma.

ACKNOWLEDGEMENTS

I am grateful to Drs. P Bromberg, DL Costa, DB Peden, and MJK Selgrade for review of the manuscript, and to Mrs. B Culpepper for editorial assistance. This article has been reviewed by the National Health and Environmental Effects Research Laboratory, U.S. Environmental Protection Agency and approved for publication. Approval does not signify that the contents necessarily reflect the views and policies of the Agency, nor does mention of trade names or commercial products constitute endorsement or recommendation for use.

REFERENCES

Abbas AK, Murphy KM, Sher A. Functional diversity of helper T-lymphocytes. Nature, 1996;383:787-793.

Askenase PW. Role of basophils, mast cells and vasoamines in hypersensitivity reactions with a delayed time course. Prog. Allergy, 1977;23:199-320.

Baker JR. Primer on allergic and immunologic diseases-fourth edition. JAMA, 1997;278:1799-1937.

Barnes KC, Neely JD, Duffy DL, Freidhoff LR, Breazeale DR, Schou C, Naidu RP, Levett PN, Renault B, Kucherlapati R, Iozzino S, Ehrlich E, Beaty TH, Marsh DG. Linkage of asthma and total serum IgE concentration to markers on chromosome 12q: Evidence from Afro-Caribbean and Caucasian populations. Genomics, 1996;37:41-50.

Barnes KC, Marsh, DG. The genetics and complexity of allergy and asthma. Immunol. Today, 1998;19:325-332.

Barnes PJ, Fan Chung K, Page CP. Inflammatory mediators of asthma: An update. Pharmacol. Rev., 1998;50:515-561.

Barnes PJ, Rodger IW, Thomson NC. (eds.) *Asthma: Basic Mechanisms and Clinical Management, 3rd Edition*. San Diego: Academic Press, 1988.

Bascom R, Bromberg PA, Costa DL, Devlin R, Dockery D, Frampton MW, Lambert W, Samet JM, Speizer FE, Utell M. Health effects of outdoor air pollution. Part 1. Am. J. Respir. Crit. Care Med., 1996a;153:3-50.

Bascom R, Bromberg PA, Costa DL, Devlin R, Dockery D, Frampton MW, Lambert W, Samet JM, Speizer FE, Utell M. Health effects of outdoor air pollution. Part 2. Am. J. Respir. Crit. Care Med., 1996b;153:477-498.

Beasley R, Coleman ED, Hermon Y, Hoist PE, O'Donnel TV. Viral respiratory tract infections and severe acute exacerbations of asthma in adult patients. Thorax, 1993;43:479-483.

Bousquet J, Michel FB. "Specific Immunotherapy in Allergic Rhinitis and Asthma." In *Asthma and Rhinitis*, WW Busse, ST Holgate, eds. Cambridge, MA: Blackwell Scientific Publications, 1995, pp. 1309-1324.

Burrows BFD, Martinez FD, Cline MA, Lebowitz MD. The relationship between parental and children's serum IgE and asthma. Am. J. Respir. Crit. Care. Med., 1995;152:1497-1500.

Busse WW, Holgate ST. (eds) *Asthma and Rhinitis*. Cambridge, MA: Blackwell Scientific Publications, 1995.

Cookson W, Sharp PA, Fauz JA, Hopkin JM. Linkage between immunoglobulin E responses underlying asthma and rhinitis and chromosome 11q. Lancet, 1989;8650:1292-1295.

Corrigan CJ, Kay AB. T-Cells and eosinophils in the pathogenesis of asthma. Immunol. Today, 1992;13:501-507.

Costa JJ, Weller PF, Galli SJ. The cells of the allergic response: Mast cells, basophils, and eosinophils. JAMA, 1997;278:1815-1822.

Crystal RG. (ed.) *The Lung. Scientific Foundations*. New York: Raven Press, 1991.

Danielle RP. (ed.) *Immunology and Immunologic Diseases of the Lung*. Cambridge, MA: Blackwell Scientific Publications, 1988.

Diaz-Sanchez D, Dotson AR, Takeaka H, Saxon A. Diesel exhaust particles induce local IgE production *in vivo* and alter the pattern of messenger RNA isoforms. J. Clin. Invest., 1994;94:1417-1425.

Djukanovic R, Roche WR, Wilson JW, Beasley CR, Twentyman OP, Howarth RH, Holgate ST. Mucosal inflammation in asthma. Am. Rev. Respir. Dis., 1990;142:434-457.

Fink JN. Epidemiologic aspects of hypersensitivity pneumonitis. Monog. Allergy, 1987;21:59-69.

Fishman AP, Elias JA, Fishman JA, Grippi MA, Kaiser LR, Senior R (eds.). *Fishman's Pulmonary Diseases and Disorders - 3rd Edition.* New York: McGraw Hill, 1998.

Gell PG, Coombs RR. (eds.) *Clinical Aspects of Immunology – 2nd Edition.* Philadelphia: Davis, 1968.

Gilmour MI. Interaction of air pollutants and pulmonary allergic responses in experimental animals. Toxicology, 1995;105:335-342.

Gilmour MI, Selgrade MJK. "Modulation of T-Lymphocyte Responses by Air Pollutants." In *T-Lymphocyte Subpopulations*, I Kimber, MJK Selgrade, eds. Chichester, England: John Wiley and Sons, 1998, pp. 253-272.

Gong JL, McCarthy KM, Rogers RA, Schneeberger E. Interstitial lung macrophages interact with dendritic cells to present antigenic peptides derived from particulate antigens to T-cells. Immunology, 1994;81:342-351.

Gordon JR, Burd PR, Galli SJ. Mast cells as a source of multifunctional cytokines. Immunol. Today, 1990;11:458-464.

Griffiths-Johnson DA, Collins PD, Jose PJ, Williams TJ. Animal models of asthma: Role of chemokines. Meth. Enzymol., 1997;288:241-267.

Gundel RH, Letts LG, Gleich GJ. Human eosinophil major basic protein induces airway constriction and airway hyperresponsiveness in primates. J. Clin. Invest., 1991;87:1470-1473.

Herz U, Lumpp U, Da Palma JC, Enssle K, Takatsu K, Schnay N, Daser A, Kottgen E, Wahn U, Renz H. The relevance of murine animal models to study the development of allergic bronchial asthma. Immunol. Cell Biol., 1996;74:209-217.

Holt PG. Regulation of antigen presenting cell(s) in lung and airway tissues. Eur. Respir. J., 1993;6:120-129.

Holt PG. Infections and the development of allergy. Toxicol. Lett., 1996;86:205-210.

Holt PG, Yabahara A, Prescott S, Venaille C. "Allergen Recognition in the Origin of Asthma." In *The Rising Trend of Asthma*, Ciba Foundation Symposium 206, Chichester, England: John Wiley and Sons, 1997, pp. 35-55.

Holt PG, Haining S, Nelson DJ, Sedgwick JD. Origin and steady state turnover of Class II MHC-bearing dendritic cells in the epithelium of the conducting airways. J. Immunol., 1994;153:256-621.

Horwitz RJ, Busse WW. Inflammation and asthma. Clinics Chest Med., 1995;16:583-602.

Husband AJ, Gleeson M. Ontogeny of mucosal immunity: Environmental and behavioral influences. Brain Behav. Immunol., 1996;10:188-204.

Huston DP. The biology of the immune system. JAMA, 1997;278:1804-1814.

Institute of Medicine Report on Indoor Allergens, AM Pope, R Patterson, H Burge, eds. Washington, DC: National Academy Press, 1993.

Israel E, Rubin P, Kemp JP, Grossman J, Pierson W, Siegel SC, Tinkelman D, Murray JJ, Busse W, Segal AT. The effect of inhibition of 5-lipoxygenase by zileuton in mild-to-moderate asthma. Ann. Intern. Med., 1993;119:1059-1066.

Janeway CA, Bottomly K. Signals and signs for lymphocyte responses. Cell, 1994;76:275-285.

Kung TT, Stelts DM, Zurcher JA, Adams GK, Egan RW, Kreutner W, Watnick AS, Jones H, Chapman RW. Involvement of IL-5 in a murine model of allergic pulmonary inflammation: Prophylactic and therapeutic effect of an anti-IL-5 antibody. Am. J. Respir. Cell Mol. Biol., 1995;13:360-365.

Lambrecht BN, Pauwels RA, Bullock GR. The dendritic cell: Its potent role in the respiratory immune response. Cell Biol. Intl., 1996;20:111-120.

Lee TH, Wraith DG, Bennet CO, Bentley AP. Budgerigar fanciers lung: The persistence of budgerigar precipitans and the recovery of lung function after cessation of avian exposure. Clin. Allergy, 1983;13:197-202.

Lehrer SB, Horner WE, Reese G. Why are some proteins allergenic? Implications for biotechnology. Crit. Rev. Food Sci. Nutr., 1996;36:553-564.

Marsh DG, Meyers DA, Bias WB. The epidemiology and genetics of atopic allergy. N. Eng. J. Med., 1981;305:1551-1559.

Marsh DG, Zwollo, Huang SK, Ghosh B, Ansari, AA. Molecular studies of human response to allergens. Cold Spring Harbor Symp. Quant. Biol., 1990;54:459-470.

Marsh DG, Neely JD, Breazeale DR, Ghosh B, Freidhoff LR, Ehrlich-Kautzky E, Schou C, Krishnaswamy G, Beaty TH. Linkage analysis of IL-4 and other chromosome 5q31.1 markers and total serum immunoglobulin E concentrations. Science, 1994;264:1152-1156.

Martinez FD, Cline M, Burrows B. Increased incidence of asthma in children of smoking mothers. Pediatrics, 1992;89:21-26.

Martinez FD. Complexities of the genetics of asthma. Am. J. Respir. Crit. Care. Med., 1997;156:S117-S122.

Mauser PJ, Pitman AM, Fernandez X, Foran SK, Adams GK, Kreutner W, Egan RW, Chapman RW. Effects of an antibody to interleukin-5 in a monkey model of asthma. Am. J. Respir. Crit. Care Med., 1995;152:467-472.

Metzger WJ, Richerson HB, Worden K, Monick M, Hunninghake GW. Bronchoalveolar lavage of allergic asthmatic patients following allergen bronchoprovocation. Chest, 1986;89:477-483.

Michel O, Ginanni R, Duchateau J, Vertongen F, Le Bon B, Sergysels R. Domestic endotoxin exposure and clinical severity of asthma. Clin. Exp. Allergy, 1991;21:441-448.

Moffatt MF, Hill MR, Cornelis F, Schou C, Faux JA, Young RP, James AL, Ryan G, Le Souef P, Musk AW. Genetic linkage of T-cell receptor γ/δ complex to specific IgE responses. Lancet, 1994;343:1597-1600.

Montefort S, Herbert CA, Robinson C, Holgate ST. The bronchial epithelium as a target for inflammatory attack in asthma. Clin. Exp. Allergy, 1992;22:511-520.

Murray JS. How the MHC selects T_H1/T_H2 immunity. Immunol. Today, 1998;19:157-163.

Neuspiel DR, Rush D, Butler N, Golding J, Bijur PE, Kurzon M. (1989). Parental smoking and post-infancy wheezing in children: A prospective cohort study. Am. J. Public Health, 1989;79:168-171.

Newman-Taylor AJ. Asthma and allergies. Br. Med. J., 1998;316:997-1000.

Nickel R, Wahn U, Hizawa N, Maestri N, Duffy DL, Barnes KC, Beyer K, Forster J, Bergmann R, Zepp F, Wahn V, Marsh DG. Evidence for linkage of chromosome 12q15-q24.1 markers to high total serum IgE concentrations in children of the German Multicenter Allergy Study. Genomics, 1997;46:159-162.

Peden DB. Mechanisms of pollution-induced airway disease: In vivo studies. Allergy, 1997;52 (suppl 58):37-44.

Pope CA, Dockery DW, Schwartz J. Review of epidemiological evidence of health effects of particulate air pollution. Inhal. Toxicol., 1995;7:1-18.

Postma DS, Bleecker ER, Amelung PJ, Holroyd KJ, Xu J, Panhuysen CI, Meyers DA, Levitt RC. Genetic susceptibility to asthma - bronchial hyperresponsiveness coinherited with a major gene for atopy. N. Eng. J. Med., 1995;333:894-900.

Reis e Sousa C, Stahl PD, Austin JM. Phagocytosis of antigens by Langerhans cells in vitro. J. Exp. Med., 1993;178:509-519.

Rowe-Jones JM. The link between the nose and the lung, perennial rhinitis, and asthma - is it the same disease? Allergy, 1997;52 (Supp 36):20-28.

Sallusto F, Cella M, Danieli C, Lanzavecchia A. Dendritic cells use macropinocytosis and the mannose receptor to concentrate macromolecules in the major histocompatibility complex Class II compartment: Downregulation by cytokines and bacterial products. J. Exp. Med., 1995;182:389-400.

Salvaggio JE, Millhollon BW. Allergic alveolitis: New insights into old mysteries. Respir. Med., 1993;87:495-501.

Schuyler M, Subramanyan S, Hassan M. Experimental hypersensitivity pneumonitis: Transfer with cultured cells. J. Lab. Clin. Med., 1987;109:623-630.

Schuyler M, Gott K, Shopp G, Crooks L. Experimental hypersensitivity pneumonitis: Influence of donor sensitization. J. Lab. Clin. Med., 1990;115:621-628.

Smithy LJ, Greenberger PA, Patterson R, Krell RD, Bernstein PR. The effect of inhaled leukotriene D_4 in humans. Am. Rev. Respir. Dis., 1985;315:368-372.

Spector SL, Smith LJ, Glass M. Accolate asthma trialists group: Effects of 6 weeks of therapy with oral doses of ICI 204,219, a leukotriene D4 receptor antagonist, in subjects with bronchial asthma. Am. J. Respir. Crit. Care Med., 1994;150:618-623.

Sutton BJ, Gould HJ. The human IgE network. Science, 1995;366:421-428.

Takenaka H, Zhang K, Diaz-Sanchez D, Tsien A, Saxon A. Enhanced human IgE production results from exposure to the aromatic hydrocarbons from diesel exhaust: Direct effects on B-cell IgE production. J. Allergy Clin. Immunol., 1995;95:103-115.

van Oosterhout AJ, Ladenius AR, Savelkoul HF, van Ark I, Delsman KC, Nijkamp FP. Rudolf A. Effect of anti-IL-5 and IL-5 on airway hyperreactivity and eosinophils in guinea pigs. Am. Rev. Respir. Dis., 1993;147:548-552.

Wardlaw AJ. The role of air pollution in asthma. Clin. Exp. Allergy, 1993;23:81-96.

Wardlaw AJ, Moqbel R, Kay AB. Eosinophils: Biology and role in disease. Adv. Immunol., 1995;60:151-265.

Zeitz HJ. Bronchial asthma, nasal polyps and aspirin sensitivity: Samter's syndrome. Clin. Chest Med., 1988;9:967-976.

Zwollo P, Ehrlich-Kautzky E, Scharf SJ, Ansari AA, Ehrlich HA, Marsh DG. Molecular studies of human immune response genes for the short ragweed allergen Amb a V. Sequencing of HLA-D second exons in responders and nonresponders to short ragweed allergen, Amb a V. Immunogenetics, 1991;33:141-151.

5. INFLAMMATION AND FIBROSIS

David J.P. Bassett, Ph.D. and Deepak K. Bhalla, Ph.D.
Department of Occupational and
Environmental Health Sciences
Wayne State University
Detroit, MI 42807

INTRODUCTION

Inflammatory cells represent an important component of the pulmonary defenses, but they also play a critical role in the pathogenesis of lung disorders such as adult respiratory distress syndrome (Patterson et al., 1989), asthma (Bigby and Nadel, 1988), silicosis (Lugano et al., 1984; Sjostrand et al., 1991; Li et al., 1992), interstitial pulmonary fibrosis, and asbestosis (Sibille and Reynolds, 1990; Rochester and Elias, 1993; Gee and Mossman, 1995). Because of the generally recognized adverse effects of airway inflammation and contribution of inflammatory cells to lung fibrosis, the factors involved in inflammatory reactions, mechanisms of toxicity and development of chronic lung disease have received considerable attention. The initial steps in the cascade of events, that ultimately result in the development of chronic lung disease, include stimulation of resident cells, release of chemotactic agents and recruitment of inflammatory cells. These events can be set in motion by an intrapulmonary insult in the form of an acute or chronic inhalation exposure to either environmental or occupational pollutants.

Over the last decade there has been a rapid expansion of understanding of the molecular and cellular mechanisms by which such exposures result in lung inflammation. However, there still remain many unanswered questions concerning factors that determine whether such lesions resolve to a normal structure or progress from a sustained aveolitis to an irreversible potentially life-threatening fibrotic condition. Pulmonary fibrosis is an interstitial disease of the parenchymal tissue, characterized by a failure of normal repair processes of the lung that results in the collapse of alveolar structure, derangement of the epithelial and basement membrane architecture, and by the accumulations of inflammatory cells, fibroblasts and the structural protein collagen (Kuhn, 1995). In many cases, a diffuse pulmonary fibrosis can develop over a period of several years with no clear indication of the etiological cause (idiopathic pulmonary fibrosis - IPF). In other cases, the lesions can be very site specific and can be directly related to either a major acute lung injury or prolonged low level exposures to fibrogenic agents. The major experimental models that have been used to examine the inflammatory determinants of fibrogenesis have included the use of the anti-neoplastic drug bleomycin (Thrall and Scalise, 1995), radiation (Pickrell and Abdel-Mageed, 1995), and animal inhalation exposures to oxygen, ozone, and such mineral dusts as silica and asbestos (LeMaire, 1995). These exposures have provided methods to scparately examine the effects of acute injuries to endothelium, airway epithelium, and the results of direct macrophage activation. In addition, continued exposure and sensitization to antigenic dusts from bacteria, fungi, animal proteins, organic chemicals and beryllium can cause delayed hypersensitivity pneumonitis that is characterized by a sustained mononuclear cell infiltration and the formation of granulomas. Failure of these lesions to resolve can lead to scar tissue formation or extensive interstitial fibrosis (Rose and Newman, 1993). Recent advances in molecular and cellular biology have also provided a greater understanding of the mechanisms and risk factors associated with fibroproliferative disorders, that are required for improving current methods of diagnosis, prevention, and clinical management (Ward and Huninghake, 1998).

This Chapter details the mechanisms of inflammation, discusses various factors that have been identified in the development of fibrosis in the lung, and highlights some of the cells and their respective functions in the fibrogenic process.

INFLAMMATION

Since chronic inflammation presents a risk of tissue damage through the release of toxic mediators, i.e., proteases and oxygen free radicals, by activated inflammatory cells, an understanding of the mechanisms responsible for the recruitment of inflammatory cells and their activation constitutes an important subject for the appreciation of injury process. The mechanistic studies implicate cell activation, cellular mediators, chemotactic factors, extracellular matrix components, cytoskeleton and cell adhesion molecules in inflammation and lung injury. The flowchart in Figure 1 shows the interactive nature of these various components and provides an overview of the sequence of inflammatory reactions which involve stimulation of resident cells, release of chemotactic agents and toxic mediators, cytoskeleton-dependent mobility of neutrophils and adhesion molecule-regulated neutrophil-endothelial cell interactions, followed by neutrophil entry into the lung. This sequence of events takes into account that: 1) the airway epithelial and inflammatory cells, upon activation, release chemokines and products of arachidonic acid metabolism; 2) extracellular matrix components, such as fibronectin, play a critical role in the inflammatory process by providing a surface for attachment and spreading of inflammatory cells, an event that causes stimulated release of chemokines and other cellular mediators; 3) the cellular mediators can cause a wide range of pathophysiologic changes, including activation of vascular neutrophils, which is associated with upregulation of cell adhesion molecules and reorganization of cytoskeletal components so as to facilitate neutrophil motility and interaction with endothelial cells prior to their entry into the lung; and 4) activated inflammatory cells contribute, through the release of toxic products, to lung injury and so amplify the direct deleterious effects of toxic agents of gaseous, particulate, and agents of fibrous or microbial origin.

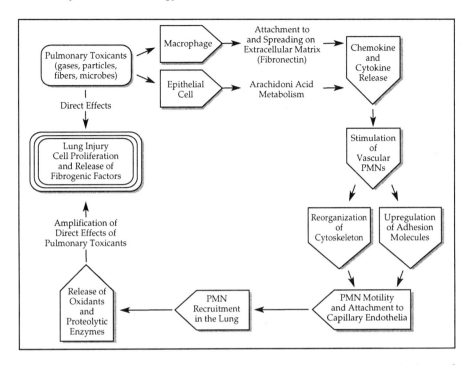

Figure 1. Sequence of inflammatory reactions leading to lung injury and release of fibrogenic factors.

In the context of our current understanding, the inflammatory response represents the end result of a set of complex reactions involving cellular components and mediators that are functionally interrelated and act in concert. The studies discussed below recognize sequential processes, multiple pathways and cytokine/adhesion molecule specificities in the induction of inflammation.

Macrophage Activation

Alveolar macrophages represent a resident cell type in the lung and compose the majority of the inflammatory cell population in healthy individuals. These cells, along with neutrophils, constitute an alveolar source of degradative enzymes, including a large number of lysosomal enzymes and oxidants capable of inducing lung injury. The slow release of these secretory products of macrophages is contained by active detoxification processes in the lung. However, under abnormal conditions, excessive release of degradative secretions results in pathophysiologic processes leading to chronic lung disease, as discussed later in this Chapter.

In addition to their role in the generation of toxic mediators, microbial killing and tissue damage, macrophages are a source of a number of cytokines and chemokines that serve to recruit and activate other inflammatory cells, including neutrophils, lymphocytes and fibroblasts. It is, thus, apparent that the inflammatory response is the outcome of a series of events involving stimulation of resident cells, release of chemotactic agents and toxic mediators, and mobility and adhesion of inflammatory cells leading to their recruitment from blood into the lung. The initial stages of this process involving macrophages are accompanied by changes in the cell surface properties relevant to their role in the inflammatory processes. Macrophage functions affecting the release of proinflammatory cytokines and the development of inflammation are stimulated as the macrophages adhere to various surfaces. The adhesion of these cells to plastic or other matrices results in the stimulated release of a neutrophil chemotactic factor (Merrill et. al., 1980), enhanced phagocytosis of opsonized bacteria (Newman and Tucci, 1990) and induction of gene expression of cytokines, interleukin-1 (IL-1), IL-8, and tumor necrosis factor-α (TNFα) (Fullbrigge et. al., 1987; Haskill et. al., 1988; Standiford et. al., 1991). These studies illustrate the early changes in the macrophage behavior following cell stimulation and suggest a relationship between cell adhesion and induction of immunological defenses in the lung. Similar changes could be produced following macrophage stimulation as a consequence of exposure to pulmonary irritants. Ozone exposure has been shown to augment macrophage adhesion to epithelial cells in culture, to stimulate macrophage motility towards a chemotactic gradient (Bhalla, 1996) and to reduce their recovery by bronchoalveolar lavage, perhaps due to their increased adhesion to alveolar walls following oxidant exposure (Pearson and Bhalla, 1997). Macrophages are inherently adhesive cells. A further increase in their adhesive potential is likely to promote their attachment to the airway lining for anchorage dependent functions, to other cell types for cell damage and to inhaled particles or microbes for phagocytosis. The importance of macrophage adhesion in lung injury relates to the suggested anchorage requirement for the induction of full effector functions (Mulligan et. al., 1993). The studies demonstrating an involvement of the integrins in macrophage activation (Arnaout, 1990) and cell accumulation in chronic inflammatory disorders (Malizia et. al., 1991; Argenbright and Barton, 1992) support the suggestion that the cell surface molecules modulate macrophage functions. The absence of regular respiratory burst of the phagocytic cells from patients genetically deficient in CD11/CD18 epitope and a concomitant increase in the frequency of bacterial infections (Hogg, 1989) further suggests that the release of toxic mediators and phagocytic functions may be cell adhesion molecule-dependent processes.

Fibronectin

Fibronectins, dimeric glycoproteins of 500 kDa molecular weight, are produced in the lung by a variety of cell types, including macrophages, fibroblasts, endothelial and epithelial cells, and neutrophils, which accumulate in the lung during inflammation and injury (Ruoslahti, 1988; Limper and Roman, 1992). These are regarded as multifunctional proteins having binding sites for various cell types and molecules, including collagen, fibrin and heparin (Dean, 1989). They are found in most body fluids and as a component of extracellular matrix (McKeown-longo, 1987; Ruoslahti, 1988). Fibronectins play an important role in a variety of biological processes, including cell adhesion, motility, and cell stimulation (Stanislawski et al., 1990). A widely recognized function of fibronectins is their role in tissue repair resulting from recruitment of cells to sites of tissue injury and promotion of wound healing. The cell recruitment and tissue repair functions of fibronectins are supported by the studies demonstrating their ability to recruit fibroblasts *in vivo* and cause enhanced migration of monocytes and neutrophils *in vitro* (Rennard et al., 1981; Denholm et al., 1989; Everitt et al., 1996).

The plasma contains a large amount of fibronectin and serves as a source of some fibronectin in tissues, where it promotes tissue repair following injury (Deno et al., 1983). In the adult lung, plasma fibronectin is found in association with the basal lamina, in alveolar lining fluid and in the interstitial spaces (Torikata et al., 1985). By providing a substrate for cell attachment, it promotes cell adhesion, cell migration and wound repair (Yamada, 1989; Hynes, 1990a and b), and mediates cell-cell and cell-substratum interactions (LaFleur et al., 1987). The heterodimer integrin $\alpha5\beta1$ serves an important role in binding monocytes to receptors on endothelial cells and to extracellular matrix fibronectin in the lung. The increased fibronectin secretion during the repair phase following lung injury is believed to influence the macrophage ability to attract neutrophils through an increased release of inflammatory cytokines. The baseline expression of fibronectin is fairly restricted in the adult lung, but its expression increases after tissue injury and inflammation. Increased concentrations of fibronectin in the bronchoalveolar lavage fluid is reflective of enhanced inflammatory activity in inflammatory lung diseases (Rennard et al., 1981). As discussed below, the inflammatory cell recruitment is mediated by several cytokines, and the products released by recruited inflammatory cells serve as chemotactic agents for migration and influx of additional mononuclear cells and fibroblasts. As reviewed by Limper and Roman (1992), lung injury in fibroproliferative disorders is accompanied by disruption of the alveolar epithelial barrier, and influx of plasma proteins, inflammatory cells, mesenchymal cells and fibroblasts. The fibroblast influx is accompanied by excessive deposition of extracellular matrix proteins, including fibronectin. These events are accompanied by alveolar epithelial type II cell proliferation and migration, leading to septal thickening and development of fibrosis.

Increased fibronectin levels have been found in the BAL of patients with interstitial and inflammatory lung disorders such as asthma, chronic bronchitis, idiopathic pulmonary fibrosis and sarcoidosis (Rennard and Crystal, 1981; Vignola et al., 1996), in the BAL of cigarette smokers (Villiger et al., 1981), in the BAL of sheep and humans exposed to asbestos (Begin et al., 1986) and in the lungs of animals exposed to ozone (O_3), an environmental pollutant (Pendino et al., 1994; Gupta et al., 1998). Increased fibronectin secretion has also been observed in rat parietal pleural mesothelial cells stimulated by asbestos fibers (Kuwahara et al., 1994) and in macrophages following exposure to nickel and cobalt chloride (Roman et al., 1990). The increased fibronectin could subsequently mobilize a set of events leading to lung inflammation. An ability of fibronectin to induce interleukin-1β (IL-1β) production (Graves and Roman, 1996), stimulate neutrophil migration (Everitt et al., 1996), and polymerize actin-cytoskeleton in neutrophils (Yang et al., 1994), represent functions that favor neutrophil recruitment in the lung. It is thus apparent that the extracellular matrix protein, fibronectin promotes conditions conducive to cell adhesion and recruitment of inflammatory cells. Extracellular matrix proteins are also important for their role in fibroblast migration, proliferation and collagen production. An understanding of the functions of fibronectin, therefore, offers an opportunity for delineation of inflammatory mechanisms and recognition of its role in the development of fibrosis.

Cytoskeleton

Cytoskeletal elements are critical to a variety of cellular functions, including maintenance of cell shape, phagocytic uptake of particles, cytoplasmic transport of molecules and cell motility in response to chemotactic signals. Neutrophil alignment in response to a chemotactic gradient involves increased assembly of actin (Sheterline and Rickard, 1989) and a concomitant change in cellular morphology from a round cell with evenly distributed surface microprojections in a resting cell to an elongated shape with microvillar concentration at the trailing end and membrane ruffling at the advancing end characteristic of a motile cell. These changes represent a behavioral response to a chemotactic stimulus. While little actin is polymerized in resting neutrophils (Sheterline et al., 1984, 1986), the motility of neutrophils is accompanied by actin assembly at the advancing ruffled end, its contraction as a dorsal sheath and its disassembly at the trailing end, which often contains a network of contractile filaments. The reorganization of contractile proteins in motile neutrophils involves their preferential distribution into specific regions of the cells so as to promote cell motility. Cytoskeletal changes could also impact cell adhesion to and migration on extracellular matrix in the lung. The reorganization and regional distribution of these cytoskeletal proteins in response to a chemotactic signal, therefore, appear critical for cell motility as the cells infiltrate lung parenchyma.

Cell motility also involves an active interplay between various cytoskeletal elements and cell adhesion molecules expressed at the cell surface. Cells are generally attached to extracellular matrix by α, β-integrins and to each other by calcium-dependent adhering molecules, cadherins. The surface distribution of adhesion molecules, receptor mobility in the plane of cell membrane and transmembrane interactions constitute events that play a vital role in cellular functions such as phagocytosis, activation, attachment and mobility of leukocytes. It is suggested that CD11/CD18 expressed on the leukocyte surface is generally inactive and requires activation for binding to its ligand (Hynes, 1992; Lollo et al., 1993). This activation is associated with mobility of cell surface receptors and their redistribution into aggregates or patches, which in turn facilitates ligand recognition and binding. Consistent with this suggestion are the studies demonstrating an increase in CD11/CD18 binding to ICAM-1 when the receptors are clustered (van Kooyk et al., 1994) and an increase in the diffusion rate of CD11/CD18 following phorbol myristate acetate (PMA)-induced activation of leukocytes (Kucik et al., 1996). Cytoskeleton indirectly plays a key role in this activation process. Both cadherins and β-integrins are attached by their COOH-terminal to cytoplasmic proteins. Prominent among these linkages are associations with the cytoskeletal proteins, including talin, vinculin, filamin, tensin (or α-actinin) and actin (Burridge et al., 1988). It is suggested that cell activation is associated with a brief dissociation of receptor-cytoskeleton linkage, permitting receptor clustering and cell adhesion (Kucik et al., 1996; Lub et al., 1997). A disruption of this linkage, formation of receptor clusters and reformation and redistribution of receptor-cytoplasmic linkages appear essential for cell motility and tissue infiltration.

Cell Adhesion Molecules

The lung is regarded as an active site of neutrophil margination (Doerschuk et al., 1990). As blood flows through pulmonary microvasculature, the leukocytes roll and stick to the endothelial cells and become static for a period of less than one to several seconds (Kuhnle et al., 1995). These conditions permit extended contact of leukocytes with endothelial cells and facilitate leukocyte entry into the lung in response to stimuli generated by airway irritation and initial injury. Leukocyte rolling on the capillary endothelium is mediated by cell adhesion molecules belonging to the selectin family including L-selectin expressed on the surface of neutrophils and E- and P-selectins on endothelial cells (von-Adrian et al., 1991; Springer, 1994). Subsequent attachment and tighter binding of neutrophils to endothelia is mediated by neutrophil LFA-1 (CD11a/CD18) and MAC-1 (CD11b/CD18) β_2-integrins and counter receptor intercellular adhesion molecule-1 (ICAM-1) on endothelial cells. While this simplified model describes the early events of inflammatory cell recruitment from systemic circulation, the mechanism of neutrophil recruitment from the pulmonary microcirculation and the precise role of cell adhesion molecules in lung inflammation constitute an area of active investigation.

Mulligan et al. (1992, 1993a and b, 1998) have investigated the role of specific integrins in inflammatory reactions under different conditions. In these studies endothelial injury in pulmonary capillaries and neutrophil accumulation following complement activation by cobra venom factor was dependent on CD11a/CD18, CD11b/CD18, and ICAM-1, but there was no requirement for E-selectin, TNFα, IL-1, or IL-8. On the other hand, lung injury and alveolar accumulation of neutrophils in the IgG immune complex model was dependent on CD11a/CD18, but not on CD11b/CD18. In the IgA immune complex lung injury model, endothelial injury involved ICAM-1 and macrophage chemotactic protein-1, but not TNFα, IL-1 or IL-8. Lung injury produced by anti-glomerular basement membrane antibody was subject to the presence of neutrophils and required β_1- and β_2-integrins, L- and E-selectins and ICAM-1 for emergence of injury. These studies serve to demonstrate differential integrin and cytokine specificities for different inducers of lung injury and inflammation. Doerschuk et al. (1990) and Burns et al. (1994a) reported that *Streptococcus pneumoniae* caused an increased expression of CD11/CD18 in neutrophils in pulmonary capillaries. However, *Escherichia coli* endotoxin-induced neutrophil migration resulted in a decrease in L-selectin and an increase in CD11/CD18 expression only after the neutrophils had moved across endothelial cells. These studies further characterize β_2-integrin specificity and dynamics, reveal their vital contributions, and emphasize unique roles of neutrophil subsets involved in specific inflammatory reactions. Besides their role in inflammatory reactions, β-integrins seem to play a more direct role in development of pulmonary fibrosis. Monoclonal antibodies to CD11a and CD11b were effective in preventing silica or bleomycin-induced collagen deposition in mice (Piguet et al., 1993a).

Intercellular adhesion molecule-1 (ICAM-1) is constitutively-expressed on endothelial cells, epithelial cells, fibroblasts and mesenchymal cells (Dustin et al., 1986; Wegner et al., 1990, 1996). Its expression is substantially increased by cytokines, following microbial infections, in inflammatory lung disorders and upon exposure to pulmonary toxicants. Increased ICAM-1 levels have been reported in the BAL of patients with sarcoidosis (Shijubo et al., 1994). An adherence of neutrophils to fibroblasts in culture by β_2-integrins (Schock and Laurent 1991) and the release of fibroblast replication factors by neutrophils (Schock and Laurent, 1989) suggest a role for these adhesion molecules in cellular interactions and the role of neutrophils in fibrosis. Studies demonstrating increased levels of ICAM-1 the BAL of O_3-exposed rats (Gupta and Bhalla, 1998), in lungs by hyperoxia (Piedboeuf et al., 1996), and in cultured type II epithelial cells (A549) by respiratory syncytial virus (Patel et al., 1995), provides the basis for a role of this molecule in directing neutrophil mobility within the lung. A redistribution of ICAM-1 along the intercellular contacts of type II epithelial cells in culture following hyperoxia (Kang et al., 1993) and increased expression of ICAM-1 on epithelial and endothelial cells in response to inflammatory stimuli (Tosi et al., 1992; Burns et al., 1994b) support the idea that ICAM-1 plays a role in directing neutrophils. Another adhesion molecule, platelet-endothelial cell adhesion molecule-1 (PECAM-1), which is structurally related to ICAM-1 and belongs to Ig gene superfamily (Newman et al., 1990), is concentrated at the junctions between endothelial cells (Muller et al., 1989). PECAM-1 appears to play a role in neutrophil transmigration and therefore it seems to constitute a molecule responsible for neutrophil "traffic control" functions as opposed to the role of ICAM-1 in "neutrophil directionality". Antibodies to PECAM-1 have been effectively used in inhibiting stimulated neutrophil entry into the lungs, peritoneum and skin (Vaporciyan et al., 1993).

Cytokines and Chemokines

Inflammatory and epithelial cells are capable of producing a wide range of mediators with multiple functions and potential for progressive inflammation and excessive tissue injury. Macrophages activated *in vitro* by cytokines or following exposure to pollutant gases and particles, as well as macrophages recovered from patients with pulmonary diseases such as sarcoidosis, have been shown to release increased amounts of chemotactic mediators, suggestive of their role in inflammation and fibrosis (Hunninghake, 1984; Pierce et al., 1996; Stringer et al., 1996; Bhalla and Gupta, submitted for publication). Although IL-1 and TNFα have been studied for their chemotactic activity and inflammatory response in the lung (Leff et al., 1993), there is evidence to suggest that the neutrophil chemotactic activity of macrophage supernatant may be dissociated from IL-1 and TNFα activity (Georgilis et al., 1987; Schroeder et al., 1987), and antibodies to IL-1 and TNFα were not effective in blocking O_3-induced inflammatory response or airway permeability (Pearson and Bhalla, 1997). However, the potency of these antibodies in blocking O_3-induced *in vitro* adherence of macrophages to alveolar epithelial cells in the same study (Pearson and Bhalla, 1997) represents an observation consistent with the suggested role of TNFα and IL-1 in triggering the upregulation of cell adhesion molecules (Bochner et. al., 1995; Mulligan et al., 1993c; Schleimer at al., 1992). Therefore, TNFα and IL-1 appear to play a role in neutrophil margination and attachment to endothelia through cell adhesion molecules, but other cytokines, especially IL-8 in humans, and cytokine-induced neutrophil chemoattractant (CINC) and macrophage inflammatory protein-2 (MIP-2) in rats could present major stimuli for neutrophil motility and entry into the lung. The importance of IL-8 is supported by the studies demonstrating IL-8 release by epithelial cells at inflammatory sites and in patients with chronic lung diseases such as idiopathic pulmonary fibrosis and ARDS (Carre et al., 1991; McElvaney et al., 1992; Miller et al., 1992). TNFα and IL-1 may be involved indirectly, as they both induce production of proinflammatory IL-8 (Strieter et al., 1989; Smart and Casale, 1994; Liu et al., 1996) and cause upregulation of adhesion molecules (Albelda, 1991; Tosi et al., 1992). IL-8, in turn, causes shedding of LECAM, upregulation of β2 integrins and migration of neutrophils (Huber et al., 1991). Anti-IL-8 antibodies have been found to be effective in inhibiting neutrophil migration across cultured monolayers (Bittleman and Casale, 1995) and blocking IgG immune complex-induced inflammatory reactions in the rat lung (Mulligan et al., 1993d), thereby suggesting a role for the pro-inflammatory cytokines in lung inflammation.

Although IL-8 has not been cloned in rat, newly emerged chemokines, MIP-1, MIP-2 and CINC have been studied for their chemotactic potential. Studies in the last few years have revealed an induction of MIP-2 and KC mRNA expression in rat macrophages and lungs by lipopolysaccharide and vanadium compounds (Watanabe et al., 1991; Huang et al., 1992; Gupta et al., 1996; Pierce et al., 1996), MIP-1 and MIP-2 mRNA expression in rat alveolar macrophages by mineral dust (Driscoll et al., 1993a), and MIP-2 and CINC mRNA expression in mice and rat lungs by O_3 (Driscoll et al., 1993b; Koto et al., 1997). In the recent studies of Yuen et al. (1996), MIP-2 and KC mRNA expression in rats exposed to silica correlated with neutrophilic inflammation and chemotactic activity, even though the cytokine expression was only transient while the inflammation persisted for 10 days. While these results support the role of MIP-2 and CINC as important cytokines, they also recognize a complex mechanism involving multiple cytokines in the induction of lung inflammation and tissue injury.

A large number of cytokines are constitutively-expressed in the lung, but their production is upregulated in connective tissue disorders. It is suggested that an abnormal production of cytokines maintains connective tissue buildup preceding tissue remodeling (Kovacs, 1991; Kovacs and Dipietro, 1994). Interleukin-2 (IL-2)-stimulated cytokines were found to cause fibroblast proliferation, and leukocyte supernatant effectively induced increased expression of type 1 procollagen and fibronectin mRNA (Kovacs et al., 1993). Other cytokines have been implicated in activation and proliferation of mesenchymal cells leading to increased secretion and deposition of extracellular matrix. Transforming growth factor-β (TGFβ) is regarded as a fibrogenic cytokine involved in the development of pulmonary fibrosis. An increase in the production of TGFβ has been reported both in an animal model of pulmonary fibrosis (Khalil and Greenberg, 1991; Westergren-Thorsson et al., 1993) and in lung tissues from patients with idiopathic pulmonary fibrosis (Khalil et al., 1991; Broekelmann et al., 1993). TNFα and IL-1 are believed to play important roles in silica-induced fibrosis. Higher levels of TNFα mRNA were found in lungs from patients with idiopathic pulmonary fibrosis than in normal lungs (Piguet et al., 1993b). Furthermore, an anti-TNFα antibody, a TNFα antagonist and an IL-1 receptor antagonist prevented silica-induced hydroxyproline increase and collagen deposition in mice (Piguet et al., 1990, 1993c; Piguet and Vesin, 1994).

PROGRESSION TO FIBROSIS

Following the onset of toxic exposure, lung inflammatory reactions, and endothelial, epithelial and interstitial repair processes usually proceed in an orderly fashion and under normal circumstances would be expected to result in resolution of the initial lesion. Understanding the pathogenesis of interstitial fibrosis is therefore dependent on the identification of factors that alter these normal repair mechanisms. The site and extent of the initial interaction, the characteristics of the associated injury and inflammation, and the persistence of the offending agent can all influence the pathogenic outcome (Doherty et al. 1993). Continuing exposures to the same or different agents have the potential to continue to cause injury, perpetuate the inflammatory condition, and as illustrated by studies with oxidants, interfere with endothelial, epithelial, and interstitial repair processes (Haschek and Witschi, 1979; Haschek et al., 1983; Last et al., 1993; Choi at al., 1994; Riley and Poiani, 1995). A sustained level of inflammatory cells within the lung together with high levels or altered patterns of mediator production, oxidant metabolism, and proteinase release all appear to be common determinants of fibrogenesis. Extensive damage to extra-cellular matrix components, either as a result of the initial or continuing injury, has also been associated with the loss of a suitable template for normal epithelial replicative repair, and subsequent infiltration and proliferation of fibroblasts can lead to the excessive collagen synthesis and deposition associated with fibrogenesis.

Fibrogenesis is clearly a very complex process that involves every cell type within the lung alveolar septum and within the extracellular matrix in which these cells and resident and infiltrating inflammatory cells operate. These different cell-types are subject to a series of mechanisms that control, stimulate, and/or inhibit molecular, cellular, and biochemical processes. It is therefore not surprising that in addition to a wide range of occupational and environmental risk factors associated with the development of fibrosis, studies have also been conducted to seek evidence of genetic factors that might render individuals more susceptible to fibrogenesis (Marshall et al., 1997).

FIBROGENESIS

Fibrogenesis can proceed from a wide range of different inflammatory scenarios (Figure 2). In the case of acute damage to either lung endothelium or airway epithelium, brought about for example by septic shock or by accidental exposure to reactive substances such as phosgene or methyl mercaptan, an exudative phase associated with edema fluid and neutrophil accumulation is followed by a period of proliferative repair activity. It is during this latter period that continuing toxic exposures might either interfere with epithelial and endothelial repair or result in the maintenance of high numbers of inflammatory cells within the lung, representing possible factors that determine whether complete repair takes place or fibrogenesis is initiated. In the case of chronic exposures to mineral dusts, the resident alveolar macrophages are thought to be kept at a high level of activation, resulting in the sustained release of proinflammatory cytokines that promote neutrophil, monocyte-macrophage, and lymphocyte accumulations. Sustained levels of these cells within the interstitium in an activated state have the potential to enhance the production of fibronectin, and the release of factors such as insulin-like growth factors (IGF), platelet-derived growth factor (PDGF), TGFs that have been identified with promoting fibroblast migration and proliferation. Inhalation exposures to antigenic material results in a complex series of inflammatory and immunological events whereby alveolar macrophages and T-lymphocytes release proinflammatory cytokines (including IL-1, interferon-γ (IFNγ), and colony stimulating factors) that promote lymphocyte and monocyte recruitment and proliferation that can lead to a mononuclear cell infiltrate/granuloma (Rose and Newman, 1993; Semenzato and Agostini, 1993). Under certain circumstances the cells of these lesions release chemoattractants for monocytes and neutrophils, and fibronectin and growth factors that cause fibroblast proliferation and enhancement of collagen deposition associated with the fibrogenesis found in sarcoidosis (Semenzato and Agostini, 1993) and delayed hypersensitivity disorders (Rose and Newman, 1993).

Importance of Epithelial Cell Repair

The role of lung capillary endothelium in controlling inflammatory events has been described above. The airway epithelium also has a controlling role, since it is often involved in the initial injury, inflammatory reactions and the subsequent repair processes. Failure or delay in the repair of ciliated airway epithelium, damaged by either infection or oxidant injury, can decrease lung clearance of airway-deposited and macrophage-engulfed fibrogenic agents, increasing the risk of sustained injury and inflammation, and so facilitating fibrogenesis. Loss of epithelial and endothelial integrity in acute lung injuries can result in fibronectin release into the airways and alveolar accumulation of blood clotting components that can lead to fibrin deposition (Idell, 1995). Fibrin formation from fibrinogen has been shown to be influenced by both alveolar epithelium and macrophages (McGee and Rothberger, 1985; Gross et al., 1991). Under conditions when fibrin and other exudative components are not effectively removed by inflammatory phagocytes, fibroblasts can migrate into the fibrin matrix and deposit interstitial collagens, leading to scar formation and/or progressive fibrosis.

Regeneration of damaged alveolar surface epithelium (type I) involves proliferation of the surfactant-generating type II cell and their subsequent transformation to new type I cells (Adamson and Bowden, 1974, Evans et. al., 1974, 1976). Oxidant inhibition of this proliferative repair process following butylhydroxytoluene (BHT)-induced damage has been implicated in causing uncontrolled fibroblast proliferation, collagen deposition, and fibrogenesis in mice (Haschek and Witschi, 1979; Haschek et al., 1983).

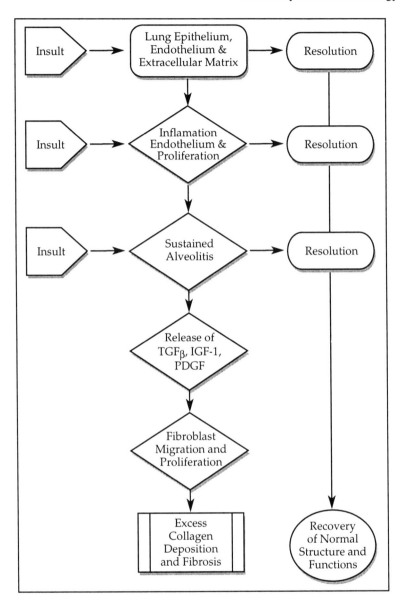

Figure 2. Sequence of cellular events leading to fibrosis.

Understanding the control of epithelial repair and the potential importance of an intact array of basement membrane components for such activity is also an important area of research interest (Kuhn et al., 1989; Sannes, 1991). Alveolar endothelial and epithelial cells have both been shown to synthesize extracellular matrix components that make up the basement membrane, that include type IV collagen, fibronectin, laminin, and proteoglycans. Severe damage and disorganization of this complex series of components that make up the basement membrane separating endothelium from epithelium in the alveolar septa, can lead to failure in repair, alveolar collapse, and proliferation of resident fibroblasts. Fibroblast migration and proliferation might also be influenced by altered epithelial synthesis of fibronectin, prostaglandins, cytokines, and other growth factors. Transgenic mice with respiratory epithelium expressing human TGFα have been shown to contain fibrotic lesions (Korfhagen et al., 1994), demonstrating a major role of the epithelium in controlling extracellular matrix metabolism.

Involvement of Neutrophils

As noted above, the margination and infiltration of neutrophils represents an early inflammatory event in acute lung injuries. These cells have been implicated as being responsible for causing or amplifying the destruction of tissue elements as a result of their ability to generate reactive oxygen and nitrogen intermediates and to release various proteinases into the interstitial environment (Henson and Johnston, 1987). Alveolar macrophages have similar potential but are thought to have more of a controlling role in subsequent events rather than being a major destructive influence. Neutrophil accumulation alone does not necessarily result in lung damage, since macrophage-mediated recruitment of neutrophils in response to airway exposure to chemoattractants has demonstrated no associated extracellular matrix or permeability damage (Staub et al., 1985; Rheinhart et al., 1998). However, an association between neutrophil infiltration and amplification of acute lung injury have been extensively demonstrated in other types of inflammatory model (Repine, 1985; Bassett et al., 1989; Patterson et al., 1989; Sibille and Reynolds, 1990; Bhalla et al., 1992; Kleeberger and Hudak, 1992; Pino et al., 1992; Ghio and Hatch, 1996).

Stimulated neutrophils have the capacity to generate superoxide anion, hydrogen peroxide, nitric oxide, hyroxyl radicals, and hypochlorous acid as well as release such proteinases as elastase, cathepsin G, collagenase, and gelatinase (Weiss, 1989). These proteinases target specific protein elements of the extracellular matrix. Other potentially damaging components include cationic proteins, peptides, and polyamines. It is thought that under normal conditions of neutrophil chemotaxis and phagocytosis, there are sufficient antioxidant and antiproteinase activities within the neutrophil itself, potential target cells, the alveolar lining fluid, and the interstitial environment. However, excess activity and imbalances between oxidant and antioxidant mechanisms, and between proteinases and antiproteinases activities would be expected to adversely affect a neutrophil-enriched lung. For example, oxidant exposures have been associated with inactivation of the methionine residue of α1-antiproteinase. Neutrophils have a relatively short half-life within the lung airways of less than 24 - 48 hr, during which time they undergo programmed cell death (apoptosis) and are then removed by macrophages (Savill et al., 1989; Savill, 1992). It has been proposed that interference with these processes that include failure of macrophages to recognize and remove apoptotic neutrophils, could result in release of excess proteinases and myeloperoxidase from necrotic neutrophils into the lung interstitial environment.

The conditions under which neutrophils contribute to the permeability damage observed in the early stages of acute oxidant injuries remain controversial and under investigation (Repine, 1985; Bassett et al., 1989; Bhalla et al., 1992; Kleeberger and Hudak, 1992; Pino et al., 1992; Ghio and Hatch, 1996). Their role at different stages of fibrogenesis is also not fully understood. Collagen levels during the first 7 d following bleomycin administration have been found to be enhanced in treated rats that have been neutrophil-depleted, suggesting a protective effect (Thrall et al. 1981; Clark and Kuhn, 1982; Thrall and Scalise, 1995). In contrast, sustained levels of neutrophils and hyroxyproline-enriched collagen degradation products in the lavage fluid recoveries from patients with IPF and from experimental animals undergoing progressive fibrosis, suggests a role in later events. Neutrophil recoveries in BAL have been correlated with the severity of the disease. It is interesting to note that in the case of granulomatous disease, the possibility that fibrosis might subsequently develop is indicated by the presence of a greater ratio of lavage recovered T-helper type 1 lymphocytes (T_H1) to type 2 cells (T_H2) have been discussed (Rose and Freeman, 1993; Semenzato and Agostini, 1993; Ward and Hunninghake, 1998). T_H1 cells are a source of IL-2 and IFNγ that would be expected to promote granuloma formation, while T_H2 cells release the anti-inflammatory cytokines IL-4 and IL-10 that could reduce neutrophil accumulations by their ability to influence macrophages to decrease TNFα generation. In addition to a potential role in sustaining tissue injury during fibrogenesis, neutrophil myeloperoxidase production of hypochlorous acid has been shown to activate collagenase and gelatinase enzymes (Brieland and Fantone, 1995). Neutrophil elastase has been shown *in vitro* to activate TGFβ, the main growth factor implicated in controlling fibroblast proliferation and metabolic function (Murphy and Docherty, 1992).

Macrophages as a Source of Fibrogenic Mediators

As previously noted, the determinants and mechanisms of fibrogenesis are far from clear, but the enhanced presence of fibroblasts in the progressive stages of the disease is a common feature. The macrophage has a central role in generating growth factors that promote fibroblast proliferation, migration, and collagen deposition include TGFβs, PDGF, IGFs, and granulocyte-macrophage colony stimulating factor (GM-CSF). However, it should be noted that fibroblasts also represent an important source of these factors under normal and inflamed conditions (Tremblay et al., 1995).

TGFβ1 is considered to be a major regulatory cytokine as it is generated as an inactive precursor that requires enzymatic modification for its activation. TGFβ1 has a wide spectrum of activity that includes being a chemoattractant for fibroblasts and a main stimulating factor in their synthesis of collagens and other extracellular matrix components (Khalil and OíConner, 1995). TGFβs have also been implicated in increasing matrix proteins by their inhibition of proteinase production with a concomitant stimulation of proteinase inhibitor generation (Kelley et al., 1991, 1992). In addition to alveolar and interstitial and alveolar macrophages, PDGF is also generated in the lung by endothelial and epithelial cells and the mesenchymal fibroblasts and smooth muscle cells. It has the potential to cause proliferation of fibroblasts and act as a chemoattractant, stimulating their migration into the affected region of the lung together with monocytes and neutrophils. IGFs have been associated with cell proliferation and differentiation. GM-CSF is usually associated with stimulation of myeloid stem cell differentiation to granulocytes and mono-cytes-macrophages. However, in fibrogenesis, GM-CSF is thought to have a direct role by its potential to induce macrophage production of TGFβ, resulting in enhanced accumulation of myofibroblasts and fibroblasts (Xing et al., 1997).

Fibroblasts and Collagen Metabolism

The disorganization and remodeling of the interstitium associated with fibrogenesis is accompanied by excessive production of structural collagens. Although epithelial and endothelial cells can produce basement membrane collagens, the major collagen in the lung is the fibrillar type I collagen with some type III. These structural proteins, together with elastin and fibronectin, are mainly produced by interstitial fibroblasts. Early studies suggested that collagen turnover in the adult lung was extremely slow. However, use of radiotracers that cannot be reutilized has demonstrated very high rates of collagen turnover that in experimental animals indicate that as much as a tenth of all lung collagen can be replaced every day. This finding has lead to the recognition that alterations in either collagen protein synthesis or collagen degradation can result in a net increase in lung collagens and fibrosis within the lung. The biochemical processes by which collagens are synthe-sized has been extensively investigated (Berg, 1986; Goldstein, 1991; McAnulty and Laurent, 1995; Bienkowski and Gotkin, 1995). In summary, procollagen poly-peptide α chains are generated and subjected to post-translational modifications that include hydroxylation of proline residues and hydroxylation and some glycosylation of lysine residues. Triple helix formation takes place with some disulphide bond formation. Following cell excretion and protease-facilitated removal of propeptide chains, collagen molecules form into fibrils, stabilized by crosslinks involving the action of lysyl oxidase. Hydroxyproline measurements have been extensively used as a measure of lung collagen content and as an indicator of its degradation when recovered in BAL fluid, blood, and urine. Enhanced lysyl oxidase activity has been used as a marker of increased collagen deposition in animal models of fibrogenesis.

It should be noted that studies with rats (Mays et al., 1989) and isolated fibroblasts (Rennard et al., 1982) have demonstrated that a very high percentage of newly formed procollagen is most likely degraded by intracellular proteinases before it is secreted from the cells (Berg, 1986; Murphy and Docherty, 1992). Extracellular collagen and collagen fibers can be degraded by cathepsins, collagenases and gelatinases generated by neutrophils, macrophages, and fibroblasts (Murphy and Docherty, 1992). Control of the activity of these destructive enzymes is provided by a need for their activation, and by several serum and extracellularly located inhibitors such as α2-macroglobulin and tissue inhibitor of metalloproteinases (TIMP).

Heterogeneity in fibroblast populations exists between different tissues and within the same tissues when inflamed and undergoing fibrogenesis. Their metabolic activities appear to be influenced by age, hormones, mechanical stresses, and the presence of growth factors. Comparisons of fibroblasts from normal and inflamed tissues have demonstrated enhanced ability to proliferate *in vitro*, in the presence of stimulating cytokines and other growth factors such as serum proteins. Fibroblasts have also been shown to be effector cells in the fibrogenic process by their autocrine release and response to a similar series of growth factors as those observed to be released by macrophages from fibrotic lungs (Tremblay et al., 1995). Myofibroblasts, that are identified by the presence of the contractile element smooth muscle cell actin, have also been observed in chronically-inflamed lungs, in bleomycin-induced lung fibrogenesis, and in lung biopsies from patients with IPF. The presence of these cells might represent a permanent change in phenotype, or a transient stage in the development of mature fibroblasts (Leslie et al. 1992).

SUMMARY

Lung fibrosis can result from a wide range of different types of insults that include acute exposures to reactive substances and chronic long term exposures to mineral and antigenic dusts and particulates. In the case of idiopathic fibrosis, the etiological origins often remain unknown, although in some cases the involvement of both the humoral and cellular immune systems and the formation of damaging immune complexes have been implicated in the early stages of fibrogenesis. In the case of mineral dust exposures, focal fibrotic lesions are often observed within the lung parencyhma in association with regions of normal tissue architecture. However, following an acute exposure to an oxidant, more diffuse lesions might be observed that could progress to cover a wide area of the lung parencyhma leading to a more rapid decline in pulmonary function. In spite of the diversity of pathogenesis, there are many common features that have linked the development of fibrosis with a persistent lung inflammatory state, characterized by a sustained aveolitis or mononuclear cell-enriched granuloma.

The inflammatory process involving the activation of the cellular immune system, infiltration of phagocytic cells into the lung and alterations in lung permeability, can be considered representative of protective mechanisms that: a) reduce exposure of lung cells to harmful substances; and b) orchestrate and/or initiate repair processes. However, under certain circumstances of either extreme acute exposure or sustained toxic exposure over prolonged periods of time, the inflammatory process can alternatively cause or contribute to a sustained level of lung damage or interfere with normal endothelial, epithelial, or extracellular matrix proliferative repair processes. Infiltrating inflammatory cells can cause damage by excess release of oxidants/proteinases. The neutrophil is recognized as having one of the greatest potentials to cause damage. Its continued presence within an inflamed lung has been associated with the development of fibrosis. Under a range of different pathological conditions, resident and infiltrating inflammatory cells as well as endothelial, epithelial, and mesenchymal cells of the lung release a complex series of lipid and peptide mediators and growth factors that can result in sustained tissue damage and inflammation, and subsequent migration and proliferation of fibroblasts. The associated derangement of the extracellular matrix with excess interstitial collagen synthesis and deposition represents a common pathological feature of lung fibrosis.

Detailed mechanisms associated with the inflammatory process have been examined extensively using modern molecular and cell biological approaches. The results of these studies have lead to the experimental evaluation of several different types of intervention that have reduced fibrogenesis. Such investigations have used antibodies against key inflammatory mediator receptors, and treatments with agents that affect the function and availability of inflammatory cells. Although such studies have delineated many of the inflammatory determinants of fibrogenesis, fewer intervention studies have been conducted to probe the complex series of events associated with fibrogenic progression.

In order to improve the diagnosis, prevention and clinical management of pulmonary fibrosis, continued research is required to obtain a greater understanding of: a) the biochemical mechanisms that determine the balances between oxidant and antioxidant, and between proteinase and antiproteinase activities within the interstium of the lung; b) the relationships between extracellular matrix components that include proteoglycans, collagens, and elastin, and the epithelium and endothelium during proliferative repair activities; c) factors that influence whether inflammatory cells release pro- or anti-inflammatory cytokines; and d) the mechanisms that control growth factor generation, fibroblast migration, proliferation and collagen deposition.

REFERENCES

Adamson IYR, Bowden DH. The pathogenesis on bleomycin-induced pulmonary fibrosis in mice. Am. J. Pathol., 1974;77:185-198.

Albelda SM. Endothelial and epithelial cell adhesion molecules. Am. J. Respir. Cell Mol. Biol., 1991;4:195-203.

Argenbright LW, Barton RW. Interactions of leukocyte integrins with intercellular adhesion molecule-1 in the production of inflammatory vascular injury *in vivo*. J. Clin. Invest., 1992;89:259-272.

Arnaout MA. Structure and function of the leukocyte adhesion molecules CD11/CD18. Blood, 1990;75:1037-1050.

Bassett DJP, Elbon CL, Reichenbaugh SS, Boswell GA, Stevens TM, McGowan MC, Kerr JS. Pretreatment with EDU decreases rat lung cellular responses to ozone. Toxicol. Appl. Pharmacol., 1989;100:32-40.

Berg RA. "Intracellular Turnover of Collagen." In *Regulation of Matrix Accumulation*, RP Mecham, ed. New York: Academic Press, 1986, pp 29-52.

Bhalla DK, Daniels DS, Luu NT. Attenuation of ozone-induced airway permeability in rats by pretreatment with cyclophosphamide, FPL 55712, and indomethacin. Am. J. Respir. Cell Mol. Biol., 1992;7:73-80.

Bhalla DK. Alteration of alveolar macrophage chemotaxis, cell adhesion, and cell adhesion molecules following ozone exposure of rats. J. Cell. Physiol., 1996;169:429-438.

Bienkowski RS, Gotkin MG. Control of collagen deposition in mammalian lung. Proc. Soc. Exp. Biol. Med., 1995;209:118-140.

Bigby TD, Nadel JA. *Asthma Inflammation: Basic Principles and Clinical Correlates.* JI Gallin, IM Goldstein, R Snyderman, eds. New York: Raven Press, 1988;36:679.

Begin R, Martel M, Desmarais Y, Drapeau G, Boileau R, Rola-Pleszczynski M, Masse S. Fibronectin and procollagen-3 levels in bronchoalveolar lavage of asbestos-exposed human subjects and sheep. Chest, 1986;89:237-243.

Bittleman DB, Casale TB. Interleukin-8 mediates interleukin-1α-induced neutrophil transcellular migration. Am. J. Respir. Cell Mol. Biol., 1995;13:323-329.

Bochner BS, Klunk DA, Sterbinsky SA, Coffman RL, Schleimer RP. Interleukin-13 selectively induces vascular cell adhesion molecule-1 (VCAM-1) expression in human endothelial cells. J. Immunol., 1995;154:799-803.

Brieland JK, Fantone JC. "Neutrophils and Pulmonary Fibrosis." In *Pulmonary Fibrosis*, S Phan, R Thrall, eds. New York: Marcel Dekker, 1995, pp. 383-404

Broekelmann TJ, Limper AH, Colby TV, McDonald JA. Transforming growth factor-β1 is present at sites of extracellular matrix gene expression in human pulmonary fibrosis. Proc. Natl. Acad. Sci. USA, 1991;88:6642-6646.

Burns AB, Takei F, Doerschuk CM. ICAM-1 expression in mouse lung during pneumonia. J. Immunol., 1994a;153:3189-3198.

Burns AB, Doerschuk CM. L-selectin and CD18 expression on rabbit neutrophils during CD18-independent and CD18-dependent emigration in the lung. J. Immunol., 1994b;153:3177-3188.

Burridge K, Fath K, Kelly T, Nuckolls G, Turner C. Focal adhesions: Transmembrane junctions between the extracellular matrix and the cytoskeleton. Ann. Rev. Cell Biol., 1988;4:487-525.

Carre PC, Mortenson RL, King TE, Noble PW, Stable CL, Riches DWH. Increased expression of the interleukin-8 gene by alveolar macrophages in idiopathic pulmonary fibrosis. J. Clin. Invest. 1991;88:1802-1811.

Choi AM, Elbon CL, Bruce SA, Bassett DJP. Messenger RNA levels of lung extracellular matrix proteins during ozone exposure. Lung, 1994;172:15-30.

Clark JG, Kuhn C. III. Bleomycin-induced pulmonary fibrosis in hamsters: The effect of neutrophil depletion on collagen synthesis. Am. Rev. Respir. Dis., 1982;126:737-739.

Dean DC. Expression of the fibronectin gene. Am. J. Respir. Cell Mol. Biol., 1989;1:5-10.

Denholm EM, Wolber FM, Pan SH. Secretion of monocyte chemotactic activity by alveolar macrophages. Am. J. Pathol., 1989;135:571-580.

Deno DC, Saba TM, Lewis EP. Kinetics of endogenously labeled plasma fibronectin: Incorporation into tissues. Am. J. Physiol., 1983;245:R564-R575.

Doerschuk CM, Downey GP, Doherty DE, English D, Gie RP, Ohgami M, Worthen GS, Henson PM, Hogg JC. Leukocyte and platelet margination within microvasculature of rabbit lungs. J. Appl. Physiol., 1990;68:1956-1961.

Doherty DE, Worthen GC, Henson PM."Inflammation in Interstitial Disease." In *Interstitial Lung Disease*, M Schwartz, T King, Jr., eds. St. Louis: Mosby - Year Book, 1993, pp. 23-43.

Driscoll KE, Hassewnbein DG, Carter J, Poynter J, Asquith TN, Grant RA, Whitten J, Purdon MP, Takigiku R. Macrophage inflammatory proteins 1 and 2: Expression by rat alveolar macrophages, fibroblasts, and epithelial cells and in rat lung after mineral dust exposure. Am. J. Respir. Mol. Biol., 1993a;8:311-318.

Driscoll KE, Simpson L, Carter J, Hassenbein D, Leikauf GD. Ozone inhalation stimulates expression of a neutrophil chemotactic protein, macrophage inflammatory protein-2. Toxicol. Appl. Pharmacol., 1993b;119:306-309.

Dustin ML, Rothlein R, Bhan AK, Dinarello CA, Springer TA. Induction by IL 1 and interferon-γ: Tissue distribution, biochemistry, and function of a natural adherence molecule (ICAM-1). J. Immunol., 1986;137:245-252.

Evans MJ, Johnson LV, Stephens RJ, Freeman G. Cell renewal in the lungs of rats exposed to low levels of ozone. Exp. Mol. Pathol., 1976;24:70-83.

Everitt EA, Malik AB, Hendey B. Fibronectin enhances the migration rate of human neutrophil *in vitro*. J. Leukocyte Biol., 1996;60:199-206.

Fullbrigge RC, Chaplin DD, Kiely JM, Unanue ER. Regulation of interleukin-1 gene expression by adherence and lipopolysaccharide. J. Immunol., 1987;138:3799-3802.

Gee JBL, Mossman BT. "Basic Mechanisms in Occupational Lung Diseases Including Lung Cancer and Mesothelioma." In *Occupational Lung Diseases*, W Morgan, A Seaton, eds. Philadelphia: W.B. Saunders Company, 1995, pp. 191-221.

Georgilis K, Schaefer C, Dinarello CA, Klempner MS. Human recombinant interleukin-1β has no effect on intracellular calcium or on functional responses of human neutrophils. J. Immunol., 1987;138:3403-3407.

Ghio AJ, Hatch GE. Tolerance to phosgene is associated with a neutrophilic influx into the rat lung. Am. J. Respir. Crit. Care Med., 1996;153:1064-1071.

Goldstein RH. Control of type I collagen formation in the lung. Am. J. Pathol., 1991;261:L29-L40.

Graves KL, Roman J. Fibronectin modulates expression of interleukin-1β and its receptor antagonist in human mononuclear cells. Am. J. Physiol., 1996;271:L61-L69.

Gross TJ, Simon RH, Sitrin RG. Expression of urokinase-type plasminogen activator by rat pulmonary alveolar epithelial cells. Am. J. Respir. Cell Mol. Biol., 1991;3:449-456.

Gupta S, Feng L, Yoshimura T, Redick J, Fu SM, Rose CE. Intra-alveolar macrophage inflammatory peptide-2 induces rapid neutrophil localization in the lung. Am. J. Respir. Cell Mol. Biol., 1996;15:656-663

Gupta SK, Reinhart PG, Bhalla DK. Enhancement of fibronectin expression in rat lung by ozone and an inflammatory stimulus. Am. J. Physiol., 1998;275:L330-L335.

Gupta SK, Bhalla DK. Airway changes in rats exposed to ozone: Elevation of bronchoalveolar lavage protein, albumin, inflammatory cells, fibronectin and cell adhesion molecules. Toxicol. Sci., 1998;42(1S):70 (abstract).

Haschek W, Witschi H. Pulmonary fibrosis - a possible mechanism. Toxicol. Appl. Pharmacol., 1979;51:475-487.

Haschek W, Reiser KM, Klein-Szanto AJ, Kehrer JP, Smith LH, Last JA, Witschi HP. Potentiation of butylhydroxytoluene-induced acute lung damage by oxygen. Cell kinetics and collagen metabolism. Am. Rev. Respir. Dis., 1983;127:28-34.

Haskill S, Johnson C, Eierman D, Becker S, Warren K. Adherence induces selective mRNA expression of monocyte mediators and proto-oncogenes. J. Immunol., 1988;140:1690-1694.

Henson PM, Johnson Jr. RB. Tissue injury in inflammation: Oxidants, proteinases, and cationic proteins. J. Clin. Invest., 1987;79:669-674.

Hogg N. The leukocyte integrins. Immunol. Today, 1989;10:111-114.

Huang S, Paulauskis JD, Godleski JJ, Kobzik L. Expression of macrophage inflammatory protein-2 and KC mRNA in pulmonary inflammation. Am. J. Pathol., 1992;141:981-988.

Huber AR, Kunkel SL, Todd RF, III, Weiss SJ. Regulation of transendothelial neutrophil migration by endogenous interleukin-8. Science, 1991;254:99-102.

Hunninghake GW. Release of interleukin-1 by alveolar macrophages of patients with active pulmonary sarcoidosis. Am. Rev. Respir. Dis., 1984;129:569-572.

Hynes RO. "Introduction and Historical Perspective." In *Fibronectin*, A Rich, ed. New York: Springer-Verlag, 1990a, pp. 1-6.

Hynes RO. "Wound Healing, Inflammation and Fibrosis." In *Fibronectin*, A Rich, ed. New York: Springer-Verlag, 1990b, pp. 349-364.

Hynes RO. Integrins; versality, modulation, and signaling in cell adhesion. Cell, 1992;69:11-25.

Idell S."Coagulation, Fibrinolysis, and Fibrin Deposition in Lung Injury and Repair." In *Pulmonary Fibrosis*, S Phan, R Thrall, eds. New York: Marcel Dekker, 1995, pp. 743-776.

Kang BH, Crapo JD, Wegner CD, Letts LG, Chang LY. Intercellular adhesion molecule-1 expression on the alveolar epithelium and its modification by hyperoxia. Am. J. Respir. Cell Mol. Biol., 1993;9:350-355.

Kelley J, Farbisiak JP, Hawes K, Absher M. Cytokine signaling in the lung: Transforming growth factor-β secretion by lung fibroblasts. Am. J. Physiol., 1991;260:L123-L128.

Kelley J. "Transforming Growth Factor-β." In *Cytokines of the Lung*, J Kelley, ed. New York: Marcel Dekker, 1992, pp. 101-137.

Khalil N, Greenberg AH. The role of TGF-β in pulmonary fibrosis. Ciba Found. Symp., 1991;157:194-207.

Khalil N, O'Connor RN, Unruh HW, Warren PW, Flanders KC, Kemp A, Bereznay OH, Greenberg AH. Increased production and immunochemical localization of transforming growth factor-β in idiopathic pulmonary fibrosis. Am. J. Respir. Cell Mol. Biol., 1991;5:155-162.

Khalil N, O'Conner RN. "Cytokine Regulation in Pulmonary Fibrosis: Transforming Growth Factor-β." In *Pulmonary Fibrosis*, S Phan, R Thrall, eds. New York: Marcel Dekker, 1995, pp. 627-645.

Kleeberger SR, Hudak BB. Acute ozone induced change in airway permeability: Role of infiltrating leukocytes. J. Appl. Physiol. 1992;72:670-676.

Koto H, Salmon M, Haddad E, Huang T, Zagorski J, Chung KF. Role of cytokine-induced neutrophil chemoattractant (CINC) in ozone-induced airway inflammation and hyperresponsiveness. Am. J. Respir. Cell Mol. Biol., 1997;156:234-239.

Korfhagen TR, Swantz RJ, Wert SE, McCarty JM, Kerlakian CB, Glasser SW, Whitsett JA. Respiratory epithelial expression of human transforming growth factor-alpha induces lung fibrosis in transgenic mice. J. Clin. Invest., 1994;93:1691-1699.

Kovacs EJ. Fibrogenic cytokines: The role of immune mediators in the development of scar tissue. Immunol. Today, 1991;12:17-23.

Kovacs EJ, Brock B, Silber IE, Neuman JE. Production of fibrogenic cytokines by interleukin-2-treated peripheral blood leukocytes: Expression of transforming growth factor-β and platelet-derived growth factor B chain genes. Obs. Gyn., 1993;82:29-36.

Kovacs EJ, DiPietro LA. Fibrogenic cytokines and connective tissue production. FASEB J., 1994;8:854-861.

Kucik DF, Dustin ML, Miller JM, Brown EJ. Adhesion-activating phorbol ester increases the mobility of leukocyte integrin LFA-1 in cultured lymphocytes. J.Clin. Invest., 1996;97:2139-2144.

Kuhn C, III, Boldt J, King Jr. TE, Crouch E, Vartio T, McDonald JA. An immunohistochemical study of architectural remodeling and connective tissue synthesis in pulmonary fibrosis. Am. Rev. Respir. Dis., 1989;140:1693-1703.

Kuhn C, III. "Pathology." In *Pulmonary Fibrosis*, S Phan, R Thrall, eds. New York: Marcel Dekker, 1995, pp. 59-83.

Kuhnle GEH, Kuebler WM, Groh J, Goetz AE. Effect of blood flow on the leukocyte-endothelium interaction in pulmonary microvessels. Am. J. Respir. Crit. Care Med., 1995;152:1221-1228.

Kuwahara M, Kuwahara M, Verma K, Ando T, Hemenway DR, Kagan E. Asbestos exposure stimulates pleural mesothelial cells to secrete the fibroblast chemoattractant, fibronectin. Am. J. Respir. Cell Mol. Biol., 1994;10:167-176

La Fleur M, Beaulieu AD, Kreis C, Poubelle P. Fibronectin gene expression in polymorphonuclear leukocytes. Accumulation of mRNA in inflammatory cells. J. Biol. Chem., 1987;262:2111-2115.

Last JA, Gelzleicher TR, Pinkerton KE, Walker RM, Witschi H. A new model of progressive fibrosis in rats. Am. Rev. Respir. Dis., 1993;148:487-494.

Leff JA, Wilke CP, Hybertson BM, Shanley PF, Beehler CJ, Repine JE. Postinsult treatment with N-acetyl-L-cysteine decreases IL-1-induced neutrophil influx and lung leak in rats. Am. J. Physiol., 1993;265:L501-L506.

LeMaire I. "Silica- and Asbestos-Induced Pulmonary Fibrosis." In *Pulmonary Fibrosis*, S Phan, R Thrall, eds. New York: Marcel Dekker, 1995, pp. 319-362.

Leslie KO, Mitchell J, Low R. Lung myofibroblasts. Cell Motil. Cytoskel., 1992;22:92-98.

Li W, Kumar RK, OíGrady R, Velan GM. Role of lymphocytes in silicosis: Regulation of secretion of macrophage-derived mitogenic activity for fibroblasts. Int. J. Exp. Pathol., 1992;73:793-800.

Limper AH, Roman J. Fibronectin. A versatile matrix protein with roles in thoracic development, repair and infection. Chest, 1992;101:1663-1673.

Liu L, Mul FPJ, Lutter R, Roos D, Knol EF. Transmigration of human neutrophils across airway epithelial cell monolayers is preferentially in the physiologic basolateral-to-apical direction. Am. J. Respir. Cell Mol. Biol., 1996;15:771-780.

Lollo BA, Chan KW, Hanson EM, Moy VT, Brian AA. Direct evidence for two affinity states for lymphocyte function-associated antigen-1 on activated T-cells (published erratum in J. Biol. Chem. 1994;269:10184). J. Biol. Chem., 1993;268:21693-21700.

Lub M, Van Kooyk Y, Van Vliet SJ, Figdor CG. Dual role of the action cytoskeleton in regulating cell adhesion mediated by the integrin lymphocyte function-associated molecule-1. Mol. Biol., 1997;8:341-351.

Lugano EM, Dauber JH, Elias JA. The regulation of lung fibroblast proliferation by alveolar macrophages in experimental silicosis. Am. Rev. Respir. Dis., 1984;1239:767-771.

Marshall RP, McAnulty RJ, Laurent GJ. The pathogenesis of pulmonary fibrosis: Is there a fibrosis gene? Int. J. Biochem. Cell Biol., 1997;29:107-120.

Malizia G, Calabrese A, Cottone M, Raimondo M, Treidosiewicz LK, Smart CJ. Expression of leukocyte adhesion molecules by mucosal mononuclear phagocytes in inflammatory bowel disease: Immunohistological evidence for enhanced antigen presenting capacity. Gastroenterology, 1991;100:150-159.

Mays PK, McAnulty RJ, Laurent GJ. Age-related changes in collagen metabolism: A role for degradation in regulating lung collagen production. Am. Rev. Respir. Dis., 1989;140:410-416.

McAnulty RJ, Laurent GJ. "Collagen and its Regulation in Pulmonary Fibrosis." In *Pulmonary Fibrosis*, S Phan, R Thrall, eds. New York: Marcel Dekker, 1995, pp. 135-171

McElvaney NG, Nakamura H, Birrer P, Hebert CA, Wong WL, Alphonso M, Baker JB, Catalano MA, Crystal RG. Modulation of airway inflammation in cystic fibrosis. J. Clin. Invest., 1992;90:1296-1301.

McGee MP, Rothberger H. Tissue factor in bronchoalveolar lavage fluids. Evidence for an alveolar macrophage source. Am. Rev. Respir. Dis., 1985;131:331-336.

McKeown-Longo PJ. Fibronectin-cell surface interactions. Rev. Infect. Dis., 1987;9:S322-S334.

Merrill WW, Naegel GP, Matthay RA, Reynolds HY. Alveolar macrophage-derived chemotactic factor: Kinetics of *in vitro* production and partial characterization. J. Clin. Invest., 1980;65:268-276.

Miller EJ, Cohen AB, Nagao S, Griffith D, Maunder RF, Martin TR, Weiner-Kronish JP, Sticherling M, Christophers E, Matthay MA. Elevated levels of NAP-1/interleukin-8 are present in the airspaces of patients with the adult respiratory distress syndrome and are associated with increased mortality. Am. Rev. Respir. Dis., 1992;146:427-432.

Muller WA, Ratti CM, McDonnell SL, Cohn ZA. A human endothelial cell-restricted, externally disposed plasmalemmal protein enriched in intercellular junctions. J. Exp. Med., 1989;170:399-414.

Mulligan MS, Varani J, Warren JS, Till GO, Smith CW, Anderson DC, Todd RF, Ward PA. Roles of β2 integrins of rat neutrophils in complement- and oxygen radical-mediated acute inflammatory injury. J. Immunol., 1992;148:1847-1857.

Mulligan MS, Watson SR, Fennie C, Ward PA. Protective effect of selectin chimeras in neutrophil-mediated lung injury. J. Immunol., 1993a;151:6410-6417.

Mulligan MS, Wilson GP, Todd RF, Smith CW, Anderson DC, Varani J, Issekutz TB, Myasaka M, Tamatani T, Rusche R, Vaporciyan AA, Ward PA. Role of β1, β2 integrins and ICAM-1 in lung injury after deposition of IgG and IgA immune complexes. J. Immunol., 1993b;150:2407-2417.

Mulligan MS, Vaporciyan AA, Myasaka M, Tamatani T, Ward PA. Tumor necrosis factor-α regulates *in vivo* intrapulmonary expression of ICAM-1. Am. J. Pathol., 1993c;142:1739-1749.

Mulligan MS, Jones ML, Bolanowski MA, Baganoff MP, Deppeler CL, Meyers DM, Ryan US, Ward PA. Inhibition of lung inflammatory reactions in rats by an anti-human IL-8 antibody. J. Immunol., 1993d;150:5585-5595.

Mulligan M S, Lentsch AB, Shanley TP, Miyasaka M, Johnson KJ, Ward PA. Cytokine and adhesion molecule requirements for lung injury induced by anti-glomerular basement membrane antibody. Inflammation, 1998;22:403-417.

Murphy G, Docherty JP. The matrix metalloproteinases and their inhibitors. Am. J. Respir. Cell Mol. Biol., 1992;7:120-125.

Newman PJ, Berndt MC, Gorski J, White GC, Lyman S, Paddock C, Muller WA. PECAM-1 (CD31) cloning and relation to adhesion molecules of the immunoglobulin gene superfamily. Science, 1990;247:1219-1222.

Newman SL, Tucci MA. Regulation of human monocyte macrophage function by extracellular matrix. Adherence of monocytes to collagen matrices enhances phagocytosis of opsonized bacteria by activation of complement receptors and enhancement of F_c receptor function. J. Clin. Invest., 1990;86:703-714.

Patel JA, Kunimoto M, Sim TC, Garofalo R, Eliott T, Baron S, Ruuskanen O, Chonmaitree T, Ogra PL, Schmalstieg F. Interleukin-1α mediates the enhanced expression of intercellular adhesion molecule-1 in pulmonary epithelial cells infected with respiratory syncytial virus. Am. J. Respir. Cell Mol. Biol., 1995;13:602-609.

Patterson CE, Barnard JW, Lafuze JE, Hull MT, Baldwin SJ, Rhoades RA. The role of activation of neutrophils and microvascular pressure in acute pulmonary edema. Am. Rev. Respir. Dis., 1989;140:1052-1062.

Pearson AC, Bhalla DK. Effects of ozone on macrophage adhesion *in vitro* and epithelial and inflammatory responses *in vivo*: The role of cytokines. J. Toxicol. Environ. Health, 1997;50:143-157.

Pendino KJ, Shuler RL, Laskin JD, Laskin DL. Enhanced production of interleukin-1, tumor necrosis factor-α, and fibronectin by rat lung phagocytes following inhalation of a pulmonary irritant. Am. J. Respir. Cell Mol. Biol., 1994;11:279-286.

Piedboeuf B, Frenett J, Petrov P, Welty SE, Kazzaz JA, Horowitz S. *In vivo* expression of intercellular adhesion molecule-1 in type II pneumocytes during hyperoxia. Am. J. Respir. Cell Mol. Biol., 1996;15:71-77.

Pierce LM, Alessandrini F, Godleski JJ, Paulauskis JD. Vanadium-induced chemokine mRNA expression and pulmonary inflammation. Toxicol. Appl. Pharmacol., 1996;138:1-11.

Pickrell JA, Abdel-Mageed AB. "Radiation-Induced Pulmonary Fibrosis." In *Pulmonary Disease*, S Phan, R Thrall, eds. New York: Marcel Dekker, 1995, pp. 363-381.

Piguet PF, Vesin C. Treatment by human recombinant soluble TNF receptor of pulmonary fibrosis induced by bleomycin or silica in mice. Eur. Respir. J., 1994;7:515-518.

Piguet PF, Vesin C, Grau GE, Thompson RC. Interleukin 1 receptor antagonist (IL-1ra) prevents or cures pulmonary fibrosis elicited in mice by bleomycin or silica. Cytokine, 1993c;5:57-61.

Piguet PF, Rosen H, Vesin C, Grau GE. Effective treatment of pulmonary fibrosis elicited in mice by bleomycin or silica with anti-CD11 antibodies. Am. Rev. Respir. Dis., 1993a;147:435-441.

Piguet PF, Rebaux C, Karpuz V, Grau GE, Kapanci Y. Expression and localization of tumor necrosis factor-alpha and its mRNA in idiopathic pulmonary fibrosis. Am. J. Pathol., 1993b;143:651-655.

Piguet PF, Collart MA, Grau GE, Sappino AP, Vassalli P. Requirements of tumor necrosis factor for development of silica-induced pulmonary fibrosis. Nature, 1990;344:245-247.

Pino MV, Stovall MY, Levin JR, Devlin RB, Koren HS, Hyde DM. Acute ozone-induced lung injury in neutrophil depleted rats. Toxicol. Appl. Pharmacol., 1992;114:268-276.

Reinhart PG, Bassett DJP, Bhalla DK. The influence of polymorphonuclear leukocytes on altered pulmonary epithelial permeability during ozone exposure. Toxicology, 1998;127:17-28.

Rennard SI, Hunninghake GW, Bitterman PB, Crystal RG. Production of fibronectin by the human alveolar macrophage: Mechanism for the recruitment of fibroblasts to sites of tissue injury in interstitial lung diseases. Proc. Natl. Acad. Sci. USA, 1981;78:7147-7151.

Rennard SI, Crystal RG. Fibronectin in human bronchopulmonary lavage fluid. Elevation in patients with interstitial lung disease. J. Clin. Invest., 1981;69:113-122.

Rennard SI, Stier LE, Crystal RG. Intracellular degradation of newly synthesized collagen. J. Invest. Dermatol., 1982;79:77s-82s.

Repine JE. "Neutrophils, Oxygen Radicals, and Adult Respiratory Distress Syndrome." In *The Pulmonary Circulation and Acute Lung Injury*, S Said, ed. Mount Kisco, NY: Futura, 1985, pp. 248-281.

Riley DJ, Poiani GJ. "Hyperoxia and Oxidants in Pulmonary Fibrosis." In *Pulmonary Fibrosis*, S Phan, R Thrall, eds. New York: Marcel Dekker, 1995, pp. 293-317.

Rochester C, Elias JA. Cytokines and cytokine networking in the pathogenesis of interstitial and fibrotic lung disorders. Semin Respir Med 1993;14:389-416.

Roman J, Limper AH, McDonald JA. Lung extracellular matrix: Physiology and pathophysiology. Hosp. Pract., 1990;25:125-140.

Rose CS, Newman LS. "Hypersensitivity Pneumonitis and Chronic Beryllium Disease." In *Interstitial Lung Disease*, M Schwartz, T King, Jr., eds. St. Louis: Mosby - Year Book, 1993, pp. 231-253

Ruoslahti E. Fibronectin and its receptors. Ann. Rev. Biochem., 1988;57:375-413.

Sannes PL. Differences in basement membrane associated microdomains of type I and type II pneumocytes in the rat and rabbit lung. J. Histochem. Cytochem., 1984;32:827-833.

Sannes PL. The structural and functional relationship between type II pneumocytes and components of extracellular matrices. Exp. Lung Res., 1991;17:639-659.

Savill JS, Wyllie AH, Henson JE, Walport MJ, Henson PM, Haslett C. Macrophage phagocytosis of aging neutrophils in inflammation. J. Clin. Invest., 1989;83:865-875.

Savill J. Macrophage recognition of senescent neutrophils. Clin. Sci., 1992;83:649-655.

Schleimer RP, Sterbinsky SA, Kaiser J, Bickel CA, Klunk DA, Tomioka K, Newman W, Luscinskas FW, Gimbrone MA Jr, McIntyre BW, Bochner BS. Interleukin-4 induces adherence of human eosinophils and basophils but not neutrophils to endothelium: Association with expression of VCAM-1. J. Immunol., 1992;148:1086-1092.

Schroeder JM, Mrowietz U, Moreta E, Christophers E. Purification and partial biochemical characterization of a human monocyte-derived, neutrophil-activation peptide that lacks interleukin-1 activity. J. Immunol., 1987;139:3474-3483.

Semenzato G, Agostini C. "Immunology of Sarcoidosis." In *Interstitial Lung Disease*, M Schwartz, T King, Jr., eds. St. Louis: Mosby - Year Book, 1993, pp. 127-158.

Sheterline P, Rickard JE, Richards RC. F_c receptor-directed phagocytic stimuli induce transient actin assembly at an early stage of phagocytosis in neutrophil leukocytes. Eur. J. Cell Biol., 1984;34:80-87.

Sheterline P, Rickard JE, Boothroyd B, Richards RC. Phorbol ester induces rapid actin assembly in neutrophil leukocytes independently of changes in $[Ca^{2+}]$ and pH. J. Muscle Res. Cell Motility, 1986;7:405-412.

Sheterline P, Rickard JE. "The Cortical Actin Filament Network of Neutrophil Leukocytes During Phagocytosis and Chemotaxis." In *Neutrophil: Cellular Biochemistry*, MB Hallet, ed. Boca Raton, FL: CRC Press, 1989, pp. 141-165.

Shijubo N, Imai K, Shigehara K, Honda Y, Koba H, Tsujisaki M, Hinoda Y, Yachi A, Ohmichi M, Hiraga Y. Soluble intercellular adhesion molecule-1 (ICAM-1) in sera and bronchoalveolar lavage fluid of patients with idiopathic pulmonary fibrosis and pulmonary sarcoidosis. Clin. Exp. Immunol., 1994;95:156-161.

Shock A, Laurent GJ. Production of a factor by human neutrophils capable of modulating fibroblast replication. Thorax, 1989;44:336P (abstract).

Shock A, Laurent GJ. Adhesive interactions between fibroblasts and polymorphonuclear neutrophils in vitro. Eur. J. Cell Biol., 1991;54:211-216.

Sibille Y, Rynolds HY. Macrophages and polymorphonuclear neutrophils in lung defense and injury. Am. Rev. Respir. Dis., 1990;141:471-501.

Sjostrand M, Absher PM, Hemenway DR. Comparison of lung alveolar and tissue cells in silica-induced inflammation. Am. Rev. Respir. Dis., 1991;143:147.

Smart SJ, Casale TB. Pulmonary epithelial cells facilitate TNF-α-induced neutrophil chemotaxis. A role for cytokine networking. J. Immunol., 1994;152:4087-4094.

Springer TA. Traffic signals for lymphocyte recirculation and leukocyte emigration: The multistep paradigm. Cell, 1994;76:301-314.

Standiford TJ, Kunkel SL, Kasahara K, Milia MJ, Rolfe MW, Strieter RM. Interleukin-8 gene expression from human alveolar macrophages: The role of adherence. Am. J. Respir. Cell Mol. Biol., 1991;5:579-585.

Stanislawski L, Huu TP, Perianin A. Priming effect of fibronectin on respiratory burst of human neutrophils induced by formyl peptides and platelet-activating factor. Inflammation, 1990;14:523-530.

Strieter RM, Kunkel SL, Showell HJ, Remick DG, Phan SH, Ward PA, Marks RM. Endothelial cell gene expression of a neutrophil chemotactic factor by TNFα, LPS and IL-1β. Science, 1989;243:1467-1469.

Stringer B, Imrich A, Kobzic L. Lung epithelial cell (A549) interaction with unopsonized environmental particulates: Quantitation of particle-specific binding and IL-8 production. Exp. Lung Res., 1996;22:495-508.

Thrall RS, Phan SH, McCormick JR, Ward PA. The development of bleomycin-induced pulmonary fibrosis in neutrophil-depleted and complement-depleted rats. Am. J. Pathol., 1981;105:76-81.

Thrall RS, Scalise PJ. "Bleomycin." In Pulmonary Fibrosis, S Phan, R Thrall, eds. New York: Marcel Dekker, 1995; pp. 231-292.

Tremblay GM, Jordana M, Gauldie J, Sørnstrand B. "Fibroblasts as Effector Cells in Fibrosis." In Pulmonary Fibrosis, S Phan, R Thrall, eds. New York: Marcel Dekker, 1995, pp. 541-577.

Torikata C, Villiger D, Kuhn C, McDonald JA. Ultrastructural distribution of fibronectin in normal and fibrotic human lung. Lab. Invest., 1985;52:399-408.

Tosi MF, Stark JM, Smith CW, Hamedani A, Gruenert DC, Infeld MD. Induction of ICAM-1 expression on human airway epithelial cells by inflammatory cyokines: Effects on neutrophil epithelial cell adhesion. Am. J. Respir. Cell. Mol. Biol., 1992;7:214-221.

van Kooyk Y, Weder P, Heije K, Figdor CG. Extracellular Ca^{2+} modulates leukocyte function-associated antigen-1 cell surface distribution on T lymphocytes and consequently affects cell adhesion. J. Cell Biol., 1994;124:1061-1070.

Vaporciyan AA, DeLisser HM, Yan HC, Mendiguren II, Thom SR, Jones ML, Ward PA, Albelda SM. Involvement of platelet-endothelial cell adhesion molecule-1 in neutrophil recruitment in vivo. Science, 1993;262:1580-1582.

Vignola AM, Chanez P, Chiappara G, Merendino A, Zinnanti E, Bousquet J, Bellia V, Bonsignore G. Release of transforming growth factor-β (TGF-β) and fibronectin by alveolar macrophages in airway diseases. Clin. Exp. Immunol., 1996;106:114-119.

Villiger B, Broekelmann T, Kelley D, Heymach GJ 3d, McDonald JA. Bronchoalveolar fibronectin in smokers and nonsmokers. Am. Rev. Respir. Dis., 1981;124:652-654.

von-Andrian UH, Chambers JD, McEvoy LM, Bargatze RF, Arfors KE, Butcher EC. Two-step model of leukocyte-endothelial cell interaction in inflammation: Distinct roles for LECAM-1 and the leukocyte β-2 integrins *in vivo*. Proc. Natl. Acad. Sci. USA, 1991;88:7538-7542.

Ward PA, Hunninghake GW. Lung inflammation and fibrosis. Am. J. Respir. Crit. Care Med., 1998;157: S123-S129.

Watanabe K, Koizumi F, Kurashige Y, Tsurufuji S, Nakagawa H. Rat CINC, a member of the interleukin-8 family, is a neutrophil-specific chemoattractant *in vivo*. Exp. Mol. Pathol., 1991;55:30-37.

Wegner CD, Gundel RH, Reilly P, Haynes N, Letts LG, Rothlein R. Intercellular adhesion molecule-1 (ICAM-1) in the pathogenesis of asthma. Science, 1990;247:456-459.

Wegner CD. "Role of ICAM-1 in Airway and Parenchymal Inflammation and Dysfunction." In *Adhesion Molecules and the Lung*, PA Ward, JC Fantone, eds. New York: Marcel Dekker, 1996, pp. 253-257.

Weiss SJ. Tissue destruction by neutrophils. N. Engl. J. Med., 1989;320:365-376.

Westergren-Thorsson G, Hernnas J, Sarnstrand B, Oldberg A, Heinegard D, Malmstrom A. Altered expression of small proteoglycans, collagen, and transforming growth factor-β1 in developing bleomycin-induced pulmonary fibrosis in rats. J. Clin. Invest., 1993;92:632-637.

Xing Z, Tremblay GM, Sime PJ, Gauldie J. Overexpression of granulocyte-macrophage colony-stimulating factor induces pulmonary granulation tissue formation and fibrosis by induction of transforming growth factor-β1 and myofibroblast accumulation. Am. J. Pathol., 1997;150:59-66.

Yamada KM. "Fibronectin Domains and Receptors." In *Fibronectin*, DF Mosher, ed. San Diego: Academic Press, 1989, pp. 48-121.

Yang KD, Augustine NH, Shaio M, Bohnsack JF, Hill HR. Effects of fibronectin on actin organization and respiratory burst activity in neutrophils, monocytes, and macrophages. J. Cell. Physiol., 1994;158:347-353.

Yuen IS, Hartsky MA, Snajdr SI, Warheit DB. Time course of chemotactic factor generation and neutrophil recruitment in the lungs of dust-exposed rats. Am. J. Respir. Cell Mol. Biol., 1996;15:268-274.

6. PULMONARY AUTOIMMUNITY AND INFLAMMATION

Thomas M. Jeitner, Ph.D. and David Lawrence, Ph.D.
Laboratory of Environmental & Clinical Immunology
Wadsworth Center
New York State Department of Health
Empire State Plaza
Albany, NY 12201-509

INTRODUCTION

Of the epithelial surfaces exposed to environmental chemicals and pathogens, the lungs are unique, in that, they have the largest surface area in the body and one that is constantly being exposed to airborne xenobiotics due to respiration. Thus, there must be means to regulate tissue entry and elimination of the foreign constituents in a controlled fashion so that immune hyperresponsiveness is minimized. The ciliary and mucous-producing cells of the respiratory system are aided in this process by leukocytes (white blood cells). Leukocytes are derived from hematopoietic multipotential stem cells located mainly within fetal liver and adult bone marrow. Leukocytes provide all aspects of immune defense which includes: 1) innate (non-antigen specific) reactivity by such cell types as polymorphonuclear cells (PMNs or neutrophils), monocytes/macrophages, and natural killer (NK) cells, which are large granular lymphocytes (not B- or T-lymphocytes), and 2) acquired, antigen-specific immunity involving bone marrow-derived B-lymphocytes and thymus-derived T-lymphocytes. Other cells of the hematopoietic system such as mast cells and eosinophils also play significant roles in immune processes within the respiratory system as well as other systems. The biological and chemical reactivities of some of these cell types will be discussed with regard to the delicate homeostatic balance that must be maintained to provide protection against airborne constituents and concomitant limitation of reactions to self.

As with immune defense of all organ systems, there are different types, rates and degrees of immune reactions. The initial immune response is provided by innate defenses. Macrophages, which develop from blood monocytes, and neutrophils are the major phagocytes to capture and slow the invasiveness of respiratory bacteria and viruses. The more pathogenic microbes cannot be readily phagocytosed or killed once phagocytosed without products from antigen-specific lymphocytes (T- and B-lymphocytes). Antibodies from B-lymphocytes and cytokines from T-lymphocytes aid these processes. The usual levels of immune cells in respiratory tissue is low, so that many of these cells need to be recruited to mount appropriate defenses, which requires increased accumulation in the lung from blood and lymphatics by production of chemotactic factors from the inflammatory sites.

Low immune reactivity would potentially allow invasion and dissemination of respiratory pathogens, whereas exaggerated immunity could culminate in destruction of self tissue as can occur with the response initiated by *Mycobacterium tuberculosis* which can lead to excessive cell-mediated immunity (type-1 immunity) and tuberculoid lesions (necrosis of lung tissue). Pathological outcomes are not restricted to inhalation of pathogens; inhaled exogenous environmental non-viable factors (particles and/or non-particle bound chemicals) also could induce respiratory problems by fostering detrimental immune responses to themself or to self-constituents damaged by these factors and/or released from cryptic sites, and/or damaged by immune responses initially directed toward the exogenous factor.

Autoimmune diseases arise when the body generates an immune response to a self-constituent and the response progresses to the point of causing pathophysiological effects. This is distinct from autoimmune responses where immune reactions occur to self-constituents to clear the body of senescent cells, self debris, etc. Autoimmune diseases are classified according to whether the immune reactivities are restricted to an organ or are systemic. The lung is a particularly relevant site for autoimmunity because it represents the largest vascular bed in the body and is exposed to all manner of airborne toxicants and pathogens. Indeed, a number of systemic autoimmune diseases originate in the lung. The etiology of most autoimmune disease is still a matter of conjecture. In this Chapter, autoimmunity is discussed in the context of responses to airborne initiators of inflammation in the lung and the potential for reduced oxygen species to cause tissue damage and generate self-antigens which promote autoimmune problems. Inflammatory products posited to be implicated in this process are hypohalous acids.

Non-Organ-Specific Pulmonary Autoimmunity: Wegener's Granulomatosis

Wegener's granulomatosis is a disease of ambiguous etiology that presents with necrotizing granulomas of the upper and lower respiratory tract and attendant glomerulonephritis and vasculitis (Sneller and Fauci, 1997). This disease is often classed as a vasculitis even though it is preceded by pulmonary granulomatosis (Niles, 1996). Wegener's granulomatosis can also be restricted to the respiratory tract with no systemic vasculitis (DeRemee, 1995), however, more often that not, vasculitis is an attending complication. Two forms of anti-neutrophil cytoplasmic antibodies (ANCA) are diagnostic for Wegener's granulomatosis: cytoplasmic (C-ANCA) which recognize proteinase 3 (PR3) and perinuclear (P-ANCA) which react with myeloperoxidase (MPO) (Niles, 1996). Both of these antigens are contained within and released from the primary granules of neutrophils (Borregaard and Cowland, 1997). Of these antibodies, C-ANCA are predominantly associated with

Wegener's granulomatosis and very rarely do both P-ANCA and C-ANCA appear in the same individual (Niles, 1996).

ANCA are thought to be important in the pathogenesis of Wegener's granulomatosis for the following reasons. First, since patients have either P-ANCA or C-ANCA, the generation of the autoantibodies are considered a precipitating event in the etiology of the disease (Niles, 1996), because if antibody production was a secondary event in the pathogenesis, autoantibodies to both PR3 and MPO would be expected, since they both originate from the primary granules of neutrophils (Borregaard and Cowland, 1997). Second, Cohen-Tervaert et al. (1989) have reported that ANCA titers increase with the onset of disease flares although this is not a universal observation (DeRemee, 1995). Third, lowering the level of circulating ANCA with either plasma exchange (Frasca et al., 1992) or pooled immunoglobulin (Jayne et al., 1991) ameliorates the disease symptoms. Pooled immunoglobulins contain antibodies against the antigenic determinants of ANCA (autoidiotypic antibodies) which bind ANCA and facilitate their removal (Jayne et al., 1991). Fourth, immunization of mice with human ANCA results in the production of idiotypic mouse ANCA and symptoms characteristic of Wegener's syndrome, including vasculitis, focal mononuclear infiltration in the lungs and immunoglobulin deposits in the kidneys (Shoenfeld et al., 1995). Finally, ANCA have direct effects on leukocytes and endothelial cells indicative of their involvement in virtually every aspect of the inflammatory response.

Deposition of Immune Complexes

Exacerbation of vasculitis is usually initiated by a bacterial infection of the upper respiratory tract which results in the release of cytokines from infected endothelial cells and macrophages, as well as from T-lymphocytes responding to antigens presented on the surface of the infected cells (Hoffman et al., 1992). The inflammatory cytokines tumor necrosis factor (TNF)-α and interferon (IFN)-γ cause endothelial cells to express both PR3 (Sibelius et al., 1998) and $F_{c\gamma}$ receptors (Pan et al., 1998) on their cell surface, allowing anti-PR3 antibodies to bind these cells through their F_{ab} and F_c fragments, thereby forming an immune complex. This sequence of events also suggests how immune complexes of PR3 and anti-PR3 antibodies may deposit specifically in the vascular bed. Sibelius et al. (1998) demonstrated that the formation of this complex is sufficient to perturb endothelial barrier function *in vitro*. Another suggested mechanism for immune complex formation relies on the cationic nature of both PR3 and MPO which favors the binding of these molecules to the endothelium and the subsequent binding of the relevant ANCA to these antigens (Savage et al., 1993; Varagunam et al., 1993).

Neutrophil Activation by ANCA

Immune complex deposition generates an inflammatory response because the F_c portion of the ANCA bind $F_{c\gamma}RIIa$ receptors on neutrophils causing them to release reduced oxygen species ($\cdot O_2^-$, H_2O_2, HOCl, etc.) and lytic enzymes (Porges et al., 1994). This release of toxic metabolites by neutrophils requires degranulation: the fusion of intracellular granules containing the antigens for which ACNA are specific (MPO, PR3, lactoferritin, and elastase), with the plasmalemma, exposing these antigens exofacially. Indeed, cell surface expression of MPO (Falk ett al., 1990; Mulder et al., 1994), PR3 (Mayet et al., 1993; Csernok et al., 1994; Mulder et al., 1994), lactoferritin (Mulder et al., 1994), and elastase (Csernok et al., 1994) can be induced by inflammatory cytokines such as TNFα, interleukin (IL)-1, IL-8 or IFNγ *in vitro* and PR3 has been detected on neutrophils from patients with active ANCA-associated vasculitis (Csernok et al., 1994). *In vitro* studies have shown that ANCA bind these antigens and $F_{c\gamma}$RIIa receptors to stimulate neutrophils to (i) degranulate (Falk ett al., 1990; Charles et al., 1991; Keogan et al., 1992; Mulder et al., 1994), (ii) bind endothelial cells (Mayet and Meyer zum Buschenfelde, 1993) and (iii) produce reduced oxygen species (Falk et al., 1990; Charles et al., 1991; Mulder et al., 1994; Prges et al., 1994), which can ultimately injure endothelial cells and compromise the endothelial permeability barrier (Ewert et al., 1992; Savage et al., 1992). It has been suggested that activation of neutrophils by ANCA is mediated via binding to both the antigen and the $F_{c\gamma}$RIIa receptor, because saturating concentrations of anti-$F_{c\gamma}$RIIa antibodies do not completely block the effects of ANCA on neutrophils (Porges et al., 1994) and $F_{(ab')2}$ fragments of ANCA are able to stimulate neutrophil degranulation and reduced oxygen species production (Falk et al., 1990; Keogan et al., 1992). Unfortunately, the effects of ANCA $F_{(ab')2}$ have not been confirmed by other studies and remain controversial (Mulder et al., 1994).

Role of ANCA in Neutrophil Recruitment to Inflammatory Sites

In addition to promoting immune complex formation and neutrophil activation, anti-PR3 also stimulates neutrophil binding to the endothelium (Mayet and Meyer zum Buschenfelde, 1993). The binding of neutrophils to endothelium involves selectins, integrins and intercellular adhesion molecules, and ANCA elevate the cell surface expression of endothelial-leukocyte adhesion molecule-1 (ELAM-1) and intercellular adhesion molecule-1 (ICAM-1) on endothelial cells *in vitro* (Mayet and Meyer zum Buschenfelde, 1993). ICAM-1 (CD54) also promotes the binding of T-lymphocytes to endothelial cells, as does VCAM-1, the expression of which is enhanced by anti-PR3 antibodies (Mayet et al., 1996). Increased ICAM-1 and VCAM expression have been observed in renal biopsy tissue and the soluble form of these molecules is elevated in the serum of patients with Wegener's granulomatosis (Stegeman et al., 1994; Mrowka and Sieberth, 1995a and b; Boehme et al., 1996; Pall et al., 1996). ELAM-1 binds sialylated Lewis X glycoproteins constitutively-present on the surface of neutrophils, whereas the expression of a co-receptor for ICAM-1, CD11b/CD18 (Mac-1), requires neutrophil degranulation (Borregaard and Cowland, 1997). Although it has not been shown that either anti-PR3 or anti-MPO antibodies specifically stimulate the expression of integrins on neutrophils necessary for adhesion to the endothelium, this can be assumed because the release of another constituent of CD11b-bearing granules, lactoferritin, is enhanced by ANCA (Mulder et al., 1994). The process of neutrophil binding to the endothelium is greatly enhanced by chemoattractants, such as IL-8 and IL-1β, which are secreted from monocytes and neutrophils in response to ANCA (Brooks et al., 1996; Ralston et al., 1997). ANCA also augment neutrophil chemotaxis in response to formylated methione-leucine-phenylalanine (Keogan et al., 1992).

Once attached to endothelium, the neutrophils are able to either damage or cause the endothelial cells to retract from each other due to the release of granular products like MPO and proteases and the production of $\cdot O_2^-$, H_2O_2, and hypohalous acids. Elastase is the one of the proteases released during neutrophil degranulation and its plasma levels are elevated in patients with active ANCA-associated vasculitis (Haubitz et al., 1997). This is pertinent because *in vitro* studies have established that elastase is able to damage endothelial monolayers (Harlan et al., 1981; Schmid et al., 1985; Smedly et al., 1986; Sugahara et al., 1986). Consquently, endothelial cells no longer present a barrier to the proteinaceous components of blood and proteins and fluids accumulate in the interstitial space. If the fluid accumulation exceeds the capacity of lymphatics to drain it, the lungs become edematous and damaged. Another danger is the exposure of subendothelial matrix to further damage by phagocytes and recognition by autoantibodies against its components. Indeed, a significant proportion of patients with ANCA also have antiglomerular basement membrane antibodies which can deposit in the lungs or kidneys (Niles, 1996).

Involvement of T-Lymphocytes in Wegener's Granulomatosis

The currently favored hypothesis for the etiology of Wegener's granulomatosis is that T-lymphocytes, responding to an unknown antigen, direct a T-helper-1 (T_H1)-type immune response against components of the vascular wall (Lúdvíksson et al., 1998). This hypothesis is based on the following observations. Attempts to isolate the putative infectious agents from Wegener's granulomatosis lesions have been unsuccessful (Hoffman et al., 1992), but these lesions have significant infiltration of CD4[+] T-lymphocytes and macrophages (ten Berg et al., 1985; Rasmussen and Petersen, 1993). In addition, patients with active symptoms have increased numbers of circulating HLA-DR[+] and CD25[+] T-lymphocytes (Schlesier et al., 1995) and elevated plasma levels of CD30 and the soluble IL-2 receptor (Nassonov et al., 1997; Wang et al., 1997), indicating that these cells have been activated *in vivo*. The HLA-DR[+] T-lymphocytes from patients with active Wegener's granulomatosis can be induced to proliferate in response to mitogens and show a T_H1-type response, in that, the proliferating cells preferentially secrete IFNγ and TNFα rather than IL-4, -5, and -10 (Lúdvíksson et al., 1998). This skewing of the T-lymphocyte response is usually attributed to secretion of IL-12 by monocytes. Indeed, mitogenically-stimulated peripheral blood mononuclear cells from Wegener's patients do produce IL-12; this production can be blocked with exogenous IL-10 and this cytokine also prevents generation of the T_H1 phenotype by mitogenically-stimulated T-lymphocytes isolated from these patients (Lúdvíksson et al., 1998).

Based on the above observations, Lúdvíksson et al. (1998) proposed that Wegener's granulomatosis begins with presentation of an antigen to cells which, in conjunction with excess IL-12 production, generates a T_H1-specific response. This is an attractive proposal because IFNγ and TNFα promote the recruitment of T-lymphocytes, neutrophils, and monocytes to sites of antigen presentation, thereby stimulating the subsequent granulomatous inflammation (Sneller and Fauci, 1987). These cytokines are also important in priming neutrophils and endothelial cells to respond to ANCA, as discussed above. What remains unanswered is the identity of antigen that causes the initial T-lymphocyte activation. Fauci and his colleagues (Lúdvíksson et al., 1998) have argued that ANCA are not the disease-causing antigens based principally on the lack of correlation of ANCA titers with disease activity (Sneller and Fauci, 1997). While this may be the case, it ignores observations that T-lymphocytes from vasculitis patients do respond to both PR3 and MPO, and do so irrespective of disease status of the patients (Griffith et al., 1996; King et al., 1998). Thus, elevations in ANCA would be expected to stimulate T-lymphocyte proliferation *in vivo*. Examination of sequences encoding the variable regions of ANCA suggests that antigenic determinants arise through specific somatic mutations to match the antigens (Finnern et al., 1997; Sibilia et al., 1997), a process that is usually driven by a T_H2-type response. It may be that different T-lymphocyte responses predominate at various stages of the disease, so that a T_H1-type response may mark the early stages of disease exacerbation and a T_H2-specific response, later stages. The balance of these responses could determine ANCA titers and may explain why titers do not always correspond with disease status. Elucidating the role of $CD4^+$ T-lymphocyte subsets at all stages of the disease is necessary to explain both the cell-mediated and humoral immune responses in vasculitis.

Organ-Specific Pulmonary Autoimmune Disease: Idiopathic Pulmonary Fibrosis

There are no clear examples of autoimmune disease with antibodies directed against unique lung antigens, although autoimmunity to collagen is considered to be relevant in the pathogenesis of idiopathic pulmonary fibrosis. This disease is characterized by chronic inflammation and fibrosis consisting principally of collagen (Reynolds et al., 1977; Turner-Warick et al., 1980; Madi and Furthmayr, 1998). The aberrant accumulation and deposition of collagen correlates with both the stage of disease and its prognosis. Consequently, it has been hypothesized that collagen acts as an autoantigen in the etiology of idiopathic pulmonary fibrosis (Kravis et al., 1976; Nakos et al., 1993). This is supported by observations that patients with idiopathic pulmonary fibrosis have autoantibodies (IgG and IgM) specific for collagen (Nakos et al., 1993) and T-lymphocytes that respond to collagen by producing migration inhibitory factor (Kravis et al., 1976). In addition, IgG is markedly elevated in the bronchoalveolar lavage fluid of idiopathic pulmonary fibrosis patients, although the antigen for the IgG has not been identified (Reynolds et al., 1977). Taken together, these observations suggest that autoantibodies develop against collagen during idiopathic pulmonary fibrosis resulting in an inflammatory response against the collagen-anticollagen complexes that perpetrates the disease by damaging the vasculature and exposing more collagenous antigenic sites.

One interesting feature of idiopathic pulmonary fibrosis is that pathologies typically do not develop at sites other than the lung (Turner-Warick et al., 1980). This is noteworthy for two reasons. First, idiopathic pulmonary fibrosis patients have antibodies to a variety of forms of collagen (Types I, II, III, and IV) (Nakos et al., 1993) and their T-lymphocytes respond to collagens from different species and tissues (Kravis et al., 1976). Second, the basement membrane is exposed in tissues where the endothelium is fenestrated, like the choroid plexus, liver and kidney, but the anticollagen antibodies do not react at these sites. This suggest that the autoimmunity of idiopathic pulmonary fibrosis is specific to this organ. It also suggests a number of possibilities for this specificity including that antibodies are recognizing collagen epitopes that were modified as a result of some pathological event in the lung, such as inflammation or exposure to some inhaled xenobiotic. During inflammation, collagen could be proteolytically or oxidatively altered by the products of neutrophils, eosinophils and macrophages, as will be discussed later in this Chapter. Another possibility is that collagen remains in its native state and the local immune response becomes aberrant (Ha and Waksman, 1973). Antibodies might also be raised against collagen-like molecules from pathogens, which precipitate an self-directed immune response (antigenic mimicry).

Inflammation and Hypohalous Acids

In the examples given above, inflammation plays a crucial role in the etiology of pulmonary autoimmune disease by damaging tissue and possibly generating or exposing novel autoantigens. The cells most likely to perpetrate tissue damage or chemical modifications are neutrophils and macrophages because they release a number of damaging species including lytic enzymes and reduced oxygen species. Of the reduced oxygen species, hypochlorous acid (HOCl) is one of the most reactive and prevalent species produced by neutrophils and will be discussed further in the context of endothelial damage and chemical modification of cellular constituents.

Hypochlorous acid is the major bactericidal oxidant produced by neutrophils and is generated through the combined actions of NADPH oxidase and MPO. The destruction of bacteria by neutrophils begins with engulfment of the pathogen into the phagocytic vacuole. As bacteria are ingested, NADPH oxidase is assembled within unique rod-shaped, alkaline phosphatase-containing granules which presumably fuse with the portion of the plasmalemma that becomes the phagocytic vacuole (Kobayashi et al., 1998). Once the oxidase is assembled, it begins to oxidize cytoplasmic NADPH, vectorially shuttling the solvated electrons into the lumen of the phagocytic vacuole (DeLeo and Quinn, 1996). As oxygen is present in bodily fluids at between 2 and 18% (30 to 140 torr), it is the most likely receptor for the electron and so is reduced to superoxide anion ($\cdot O_2^-$).

$$2 O_2 + NADPH \rightarrow 2 \cdot O_2^- + NADP^+ + H^+ \qquad (1)$$

Superoxide, having one unpaired electron, is highly reactive and reacts rapidly with neighboring $\cdot O_2^-$ radicals at a rate of ~5 x 10^5 $M^{-1} \cdot sec^{-1}$ (at neutral pH; Halliwell and Gutteridge, 1989) to form hydrogen peroxide (H_2O_2) via dismutation.

$$2 \cdot O_2^- + 2 H^+ \rightarrow 2 H_2O_2 \qquad (2)$$

The spontaneous rate of $\cdot O_2^-$ dismutation is faster in the phagocytic vacuole because its internal pH is between 6.0 - 6.5 (Segal et al., 1981) and protonation of $\cdot O_2^-$ overcomes the restriction imposed by two species of equal charge reacting with each other (Halliwell and Gutteridge, 1989). This reaction could also be accelerated catalytically by a superoxide dismutase (SOD), which associates with the exofacial plasmalemma by binding heparan sulphate polyglycans (Adachi et al. 1995, 1996). It is not known if SOD is bound to the surface of neutrophils.

The peroxide derived from dismutation of $\cdot O_2^-$ becomes a substrate for MPO released from primary granules subsequent to their fusion with the phagocytic vacuole. Myeloperoxidase catalyzes the oxidation of halides and the pseudohalide thiocyanate, to their corresponding hypohalous acids.

$$H_2O_2 + X^- + H^+ \rightarrow HOX + H_2O \qquad\qquad (3)$$
$$(X^- = Cl^-, Br^-, I^- \text{ or } SCN^-)$$

Of these species, Cl^- and thiocyanate (SCN^-) are the major substrates for MPO. The rates of the reactions for these species with MPO are $k \sim 4.5 \times 10^2$ and 3.3×10^5 $M^{-1}\cdot sec^{-1}$, for Cl^- and SCN^-, respectively (van Dalen et al., 1997). Chloride is a major substrate for MPO because its concentration in biological fluids commonly exceeds that of the other halides by ≈ 1000-fold; consequently, HOCl is a major oxidation product (Weiss et al., 1982). One exception to this generalization is saliva which contains 10^3 M SCN (Pettigrew and Fell, 1972). Unlike HOCl, hypothiocyanous acid is not harmful to mammalian cells and is not considered a pulmonary toxicant (Stungaard and Mahoney, 1991). Myeloperoxidase is not alone in mediating the oxidation of halides. Eosinophil peroxidase also shares this ability and may account for some of the adverse effects of the eosinophilia which occurs in asthma (Agosti et al., 1987; Stungaard and Mahoney, 1991).

Hypochlorous Acid as a Bactericidal Agent

While $\cdot O_2^-$ and H_2O_2 have microcidal activity in their own right, they do not contribute significantly to the microcidal activity of neutrophils for the following reasons. The catalytic rate of MPO is $\approx 4.5 \times 10^2$ $M^{-1}\cdot sec^{-1}$ at pH 7.4 (van Dalen et al., 1997) and ensures that $\approx 40\%$ of the NADPH oxidase-derived H_2O_2 is converted to HOCl (Weiss et al., 1982). Since only a small amount of H_2O_2 is detectable outside of neutrophils during phagocytosis, the remaining H_2O_2 is probably catabolized by glutathione peroxidase (GSHPX) and catalase (Test and Weiss, 1984). The toxicity of H_2O_2 depends to a large extent on the presence of reduced iron to facilitate production of the hydroxyl ion ($\cdot OH$); however, neutrophils and some bacteria release lactoferritin and sideropores, respectively, which sequester iron and prevent $\cdot OH$ production (Halliwell and Gutteridge, 1989). Finally, H_2O_2 is 500 - 1000 times less effective than HOCl at killing *Escherichia coli* (Chesney et al., 1996).

The production of $\cdot OH$ may not always be generated in iron-dependent manner. Candeias et al. (1993) have suggested that $\cdot OH$ could be formed by the reaction of HOCl and O_2^-, as shown below.

$$HOCl + \cdot O_2^- \rightarrow \cdot OH + Cl^- + O_2 \qquad\qquad (4)$$

This reaction is analogous to the classic Haber-Weiss reaction (which requires iron) (Halliwell and Gutteridge, 1989) but is at least six orders of magnitude faster (Candeias et al., 1993).

The pK_a of HOCl is 7.42 (Izatt and Christensen, 1976) so that at physiological pH $\approx$50% is present as hypochlorite ($^-$OCl). For the sake of brevity, the effects of HOCl will be discussed without distinguishing between those effects that are or may be due to either the acidic or anionic forms of the molecule. Hypochlorous acid reacts with biological molecules with the following rank order: iron-sulfur proteins > β-carotene > nucleotides > porphyrins > heme proteins, indicating that components of the bacterial respiratory chain would be most sensitive to the bleaching activity of HOCl (Albrich et al., 1981; Rosen and Klebanoff, 1985; Winterbourn, 1985; Prutz, 1996). Exposure of bacteria to HOCl results in the loss of respiration concomitant with irreversible oxidation of cytochromes, adenine nucleotides and carotene pigments, all within a few minutes (Albrich et al., 1981).

Hypochlorous Acid as an Endothelial Toxicant

While the reactions described above occur inside the phagocytic vacuole, there is some leakage of HOCl because the plasmalemma does not represent an efficient barrier to this molecule (Vissers and Winterbourn, 1995; Carr and Winterbourn, 1997). This leakage of HOCl has the potential to damage the surrounding host tissues if it exceeds the extracellular and cellular defenses. The main scavengers of HOCl in bodily fluids appear to be the thiol-based antioxidants: glutathione (GSH), cysteine, and albumin (Wasil et al., 1987; Aruoma et al., 1988). In addition, both MPO and H_2O_2 can be released extracellularly in response to phorbol ester tumor promoters (Qian and Eaton, 1994) and this is thought to be crucial in certain carcinogenic processes (Robertson et al., 1994; Vasunia et al., 1994). Myeloperoxidase may also be expelled as part of the inflammatory response and thereby contribute to extracellular HOCl production (Albrecht and Jungi, 1993). Neutrophil attachment to the endothelium is accompanied by the fusion of primary granules with the plasmalemma to release among other proteins, the serine protease elastase and MPO (Owen et al., 1995). In this context, it is thought that elastase acts as an antagonist for the CD11b/CD18- and ICAM-dependent adhesion of neutrophils to endothelial cells, allowing the detachment of the leading edge of neutrophils facilitating the movement of these cells into tissues and bronchoalveolar fluids (diapedesis) (Cai and Wright, 1996).

The evidence that HOCl is cytotoxic to the pulmonary endothelium comes from the demonstration that the instillation of MPO and H_2O_2 into the lungs causes severe pulmonary damage (Johnson et al., 1981; Tatsumi and Fliss, 1994) and that MPO is commonly recovered in the bronchial lavage fluid of patients and experimental animals with pulmonary injury (Weiland et al., 1986; Watts et al., 1990). Similarly, perfusion of rat kidneys with MPO-H_2O_2-halide results in endothelial lysis, proteinuria, and halogenation of glomerular basement proteins (Johnson et al., 1987a and b). As halogenation was only observed in the presence of

MPO, peroxide, and (^{125}I]-iodine, this was thought to reflect iodination by hypo-iodinous acid (Johnson et al., 1987a). In these studies, MPO was also found deposited along glomerular capillary walls.

The Reaction of Hypochlorous Acid and Thiol Groups

Pulmonary damage probably arises partly from the oxidation of thiol moieties associated with GSH and cellular enzymes (Albrich et al., 1981; Vissers and Winterbourn, 1995). Although there are chemical species other than thiols that react with HOCl, thiols are ~100 times more reactive than amines or heme proteins (Winterbourn, 1985; Folkes et al., 1995). A number of studies have indicated that cellular GSH is the major sacrificial defense against HOCl for erythrocytes (Vissers and Winterbourn, 1995), neutrophils (Carr and Winterbourn, 1997), bacteria (Chesney et al., 1996), and endothelial cells (Tatsumi and Fliss, 1994). This is not surprising given that the rate constant for the reaction between GSH and HOCl is $\geq$ 10^7 $M^{-1} \cdot sec^{-1}$ (Folkes et al., 1995) and that the intracellular concentration of GSH, (typically in the 10^{-2} M range) vastly exceeds the 10^{-5} M concentrations of HOCl generated by neutrophils. Consequently, up to 75% of the GSH content of neutro-phils and erythrocytes is lost within 5 min of exposure to 10^{-5} M HOCl (Vissers and Winterbourn, 1995; Carr and Winterbourn, 1997). This decrement may not be responsible for the toxic effects of HOCl on the endothelium because some cells, such as lymphocytes but not hepatocytes (Fariss et al., 1985), are able to sustain losses of $\approx$90% of their GSH, and remain viable and responsive to mitogenic stimuli (Messina and Lawrence, 1989; Jeitner and Lawrence, unpublished observations). Winterbourn and Brennan (1997) have proposed that under physiological conditions (i.e., [GSH] >> [HOCl]), GSH reacts with HOCl to produce stable adducts including GSH disulfide, GSH sulphonic acid, GSH thiolsulphonate and a unique internal sulphonamide of GSH (Winterbourn and Brennan, 1997). Of these, GSH disulfide (Vissers and Winterbourn, 1995; Carr and Winterbourn, 1997) and the intermolecular GSH sulphonamide (Carr and Winterbourn, 1997) have been identified in HOCl-treated cells. Unfortunately, the structure and possible toxicity of internal GSH sulphonamide is not known. Glutathione disulfide is unlikely to be toxic to the pulmonary endothelium because it is either reduced to GSH by GSH reductase (GSHRX) or rapidly expelled from cells (Akerboom and Sies, 1994); however, GSH disulfide could also participate in thiol:disulfide interchange with proteins and thereby inactivate modified proteins. Protein thiols are also lost following treatment of cells with HOCl (Schraufstatter et al., 1990; Tatsumi and Fliss, 1994; Vissers and Winterbourn, 1995; Carr and Winterbourn, 1997), per-oxides (Brodie and Reed, 1987; Hyslop et al., 1988; Baker et al., 1989), and, oddly enough, with GSH-depleting agents (Messina and Lawrence, 1989). It is assumed that the loss of these protein thiols contributes to the toxicity of, at least, the peroxides (Brodie and Reed, 1987; Hyslop et al., 1988; Baker et al., 1989). The full extent of protein thiol modifications perpetrated by reduced oxygen species has not been elucidated, but one of the more studied alterations is the formation of mixed disulfides between low molecular weight, cellular thiols (usually GSH) and protein thiols, known as "S-thiolation" (Cotgreave and Gerdes, 1994; Thomas et al., 1995).

A number of significant proteins such as carbonic anhydrase III (Rokutan et al., 1989; Chai et al., 1991), creatine kinase, glycogen phosphorylase b (Collison and Thomas, 1987; Park and Thomas, 1988; Miller et al., 1990), glyceraldehyde-3-phosphate dehydrogenase (Ravichandran et al., 1994; Schuppe-Koistinen et al., 1994; Lind et al., 1998), and actin (Chai et al., 1994; Rokutan et al., 1994; Schuppe-Koistinen et al., 1995) have been shown to be S-thiolated by oxidizing species, both *in vivo* and *in vitro*. The S-thiolation of actin is particularly relevant to the effects of oxidants on the endothelium because it has been associated with the loss of normal stress fibers and a concomitant increase in actin polymerization in conjunction with the contraction of gastric mucosal cells (Rokutan et al., 1994). Physiologically-attainable levels of HOCl (i.e., 10^{-5} M) (Weiss et al., 1982; Qian and Eaton, 1994) cause confluent endothelial cells to contract, in association with actin microfilament shortening, cell retraction, decreases in electrical resistance and increases in protein permeability, all within minutes of the exposure (Tatsumi and Fliss, 1994; Ochoa et al., 1997). It is not known if HOCl oxidizes thiol groups in actin although protein thiols, in general, disappear rapidly in endothelial (Tatsumi and Fliss, 1994) and other cell types (Vissers and Winterbourn, 1995; Carr and Winterbourn, 1997) treated with this acid. Taken together, these observations suggest that HOCl could oxidize cysteinyl residues in actin to promote filament contraction and cellular retraction. This oxidation could occur directly, as has been described for GSH (Winterbourn and Brennan, 1997) or cysteine (Pereira et al., 1973; Winterbourn, 1985), or indirectly through the generation of ·OH (Reaction 4) and, consequently, glutathione thiyl (GS·) radical or GSH disulfide. Thomas and his colleagues (1995) have argued that thiol:disulfide interchange is not a significant mechanism for S-thiolation; however others have suggested that GSH disulfide might contribute to S-thiolation by acting as a substrate for glutaredoxin (Lind et al., 1998). The possibility exists that higher oxidation states of the sulfur in the protein thiols and perhaps intermolecular sulfonamides (as has been described for GSH; Winterbourn and Brennan, 1997), might also occur.

Hypochlorous Acid as an Inducer of Cell Shape Changes

Hypocholorous acid causes endothelial cells to retract from each other and it has been proposed that this is a physiological mechanism to allow diapedesis (Tatsumi and Fliss, 1994). This is reasonable in that the retraction occurs within minutes, is reversible, and is enacted by concentrations of HOCl that can be formed by neutrophils *in vivo* (Tatsumi and Fliss, 1994; Ochoa et al., 1997). However, the factors that attenuate HOCl-dependent cell retraction still require elucidation. Some possibilities include: internalization of the chemotaxic receptor; the distance between the neutrophil and endothelial cells; and, the scavenging of the HOCl.

The restoration of cellular shape is an important consideration in determining the deleterious effects of HOCl. Physiological levels of HOCl are not toxic to endothelial cells (Weiss, 1989; Tatsumi and Fliss). In fact, HOCl is much less effective than H_2O_2 in promoting endothelial cell lysis despite its greater reactivity with cellular thiols and amines (Weiss, 1981; Harlan et al., 1984; Ward, 1991). Yet, HOCl is far more effective at inducing endothelial cell retraction and albumin permeability than H_2O_2 (Ochoa et al., 1997) and these changes rapidly become irreversible after 2 min of HOCl exposure (Tatsumi and Fliss, 1994). Increases in endothelial permeability to proteins and the development of pulmonary edema are etiologic factors in the lung pathology associated with inflammation, embolism, sepsis and trauma. Thus, the pulmonary toxicity of HOCl may be due to the loss of endothelial barrier function rather than a frank loss of cell viability. Moreover, concentrations of HOCl that are not toxic to endothelial cells cause pronounced edema in isolated lungs within 10 min (Tatsumi and Fliss, 1994)

Reactions of the MPO-H_2O_2-Cl⁻ System and Amino Acids

Cysteine contributes the thiol moieties of virtually all proteins that bear them and the thiol of GSH. So, the reactions of thiols with hypohalous acid and MPO (Compound I) described above, pertain to cysteine (with some caveats and exceptions [see Winterbourn and Brennan, 1997]) and will not be discussed further. While chemical studies have established that HOCl reacts more rapidly with thiols than with amino groups (Winterbourn, 1985; Folkes et al., 1995), Heinecke's group noted that oxidation of amino groups by HOCl occurs at a rate sufficient to produce highly-reactive aldehydes that are detectable at inflammatory sites *in vivo* (Hazen et al., 1997a). All α-amino acids, except sulfur-bearing amino acids and the imino acid proline, can form reactive aldehydes upon exposure either to HOCl, a mixture of H_2O_2, MPO and Cl⁻, or activated neutrophils (Anderson et al., 1997; Hazen et al., 1996, 1998a and b). In contrast, β- and ε-amino acids react with HOCl and the combination of H_2O_2, MPO and Cl⁻ to form stable chloramines (Hazen et al., 1998a). Reaction of HOCl with relevant α-amino acids results in the formation of unstable α-monochloramines which subsequently decompose to yield aldehydes (Hazen et al., 1998a). Among the aldehydes formed are glycoaldehyde and 2-hydroxypropanal from L-serine and L-threonine, respectively (Anderson et al., 1997). The 2-hydroxypropanal undergoes further dehydration to form acrolein which, together with glycoaldehyde and 2-hydroxypropanal, have been observed as products of the combination of either HOCl, H_2O_2, MPO, and Cl⁻, or activated neutrophils with L-serine and L-threonine (Anderson et al., 1997). Glycoaldehyde and acrolein are highly reactive and have been implicated in a number of disease states (Witz, 1989), however, the relevance of these species in inflammatory lung diseases remains to be determined.

Kinetic studies have shown that aldehydes constitute a significant fraction of the HOCl consumed in the reactions with non-sulfur containing α-amino acids and that thiols are able to successfully compete with this reaction (Hazen et al., 1998b). Nevertheless, Hazen et al. (1997a) have shown that HOCl oxidizes amino acids to aldehydes *in vivo*. The product of HOCl and L-tyrosine, *p*-hydoxyphenylacetaldehyde, reacts with lysine to form Schiff base adducts. The generation of these adducts is attenuated in the presence of serum but is still detectable in samples taken from sites of inflammation (Hazen et al., 1997a). As amino acids are present at mM levels in serum (Linder, 1992) while those of thiols are in the μM range (Stamler and Slivka, 1996), this concentration difference probably accounts for the generation of amino acid-derived aldehydic products despite the greater reactivity of the thiols for HOCl. Investigations into the toxicity of aldehydes are still in their infancy.

Chlorination of tyrosine by H_2O_2, MPO, and Cl^- also warrants comment as this is the most likely pathway for this modification in humans and so could be a potential indicator of neutrophil activation *in vivo* (Kettle, 1996; Hazen et al., 1997b and c). This reaction involves initial chlorination of an amine by HOCl followed by an intramolecular reaction that chlorinates the tyrosine aromatic ring (Domigan et al., 1995). The formation of 3-chlorotyrosine in peptides and proteins is catalyzed by activated neutrophils, the H_2O_2, MPO, and Cl^- system and HOCl (Domigan et al., 1995; Kettle, 1996; Hazen et al., 1997b and c), and has been detected in atherosclerotic lesions known to contain phagocytes (Hazen et al., 1997b and c), active MPO (Daugherty et al., 1994) and HOCl-modified proteins (Hazell et al., 1998). The ability of 3-chlorotyrosine to reflect *in vivo* inflammation depends on the relative efficiency of this reaction compared to the MPO-mediated oxidations of amino and thiol groups, and the extent to which these reactions will predominate at various sites of inflammation, all of which are not known.

It is important to note that while a number of the above modifications have been identified at sites of inflammation, this does not imply they are causally involved in the disease process (Hazen et al., 1997a,b, and c). The fact that antibodies can be raised against HOCl-modified proteins (Malle et al., 1995) suggests that novel epitopes may contribute to autoimmunity, but this requires further investigation. It is also instructive to consider that products produced by the interaction of HOCl and any amino acid may differ for the same amino acid in the context of a protein, where neighbouring groups can contribute to the reactions. A case in point, is the complete oxidation of cysteine to cystine by HOCl, as compared to the oxidation of this amino acid in GSH by HOCl which produces GSH-disulfide, -sulphonic acid, -thiolsulphonate, and -sulphonamide (Winterbourn and Brennan, 1997). These observations indicate that a great number of protein modifications that can be effected by the H_2O_2, MPO, and Cl^- system that have the potential to modify protein function and antigenicity and contribute to pulmonary disease. Finally, although much of the above is speculative, it is known that gaseous chloramines do cause pulmonary disease (Gapany-Gapanavicius et al., 1982).

Toxicant Generation from Xenobiotics by MPO-H_2O_2-Cl^- System

The MPO-H_2O_2-Cl^- system is able to metabolize a variety of xenobiotics into toxicants by either reaction with hypohalous acid or MPO in its Compound I form. Some examples of this metabolism include oxidation of sulfadiazine to hydroxylamine; acetaminophen to benzoquinone, *N*-acetyl-*p*-benzoquinoneimine, chloroacetaminophen, and dichloroacetaminophen; hydralazine to phthalazine, phthalazinone and *N*-chlorophthalazinone; and propylthiouracil (PTU) to PTU disulfide, PTU-2-sulfinate and PTU-2-sulfonate (Hofstra and Uetrecht, 1993). These reactions are significant because they have profound ramifications for the activity of neutrophils at sites of inflammation and may contribute to disease. For example, hydralazine (Nassberger et al., 1991) and PTU (Dolman et al., 1993) treatments can exacerbate ANCA-associated vasculitis by unknown mechanisms. It might be possible that the oxidation of these drugs by activated neutrophils (Hofstra and Uetrecht, 1993) produces metabolites (listed above) that contributes to the vasculitis.

Neutrophils can oxidize xenobiotics through the generation of an oxidized form of MPO (Compound I) or hypohalous acids. Complete reduction of H_2O_2 to H_2O requires oxidation of ferric (Fe^{3+}) ion in MPO to the ferryl (Fe^{5+}) form. This MPO (Compound I) form is a strong oxidant capable of catalyzing the 2-electron oxidation of chloride ion to HOCl and so regenerate the enzyme. One of the difficulties in assessing the contribution of Compound I-mediated oxidations to the generation of toxicants is distinguishing these from the 2-electron oxidations by HOCl. Nonetheless, direct oxidation of a wide range of xenobiotics by Compound I has been established in experiments performed in the absence of halides (Duescher and Elfarra, 1992; Kettle et al., 1992; Hofstra and Uetrecht, 1993). It is assumed that these oxidations will take place *in vivo* because a plethora of available HOCl scavengers implies that a proportion of xenobiotics will also be oxidized by MPO. Furthermore, some xenobiotics that act as substrates for MPO inhibit the chlorinating activity of the enzyme (Kettle et al., 1992; Hofstra and Uetrecht, 1993; Frimat et al., 1997), thus favoring Compound I-dependent drug oxidation rather than one mediated by HOCl. Finally, HOCl also reacts with a variety of xenobiotic compounds to alter their biological functions (Kozumbo et al., 1992; Hofstra and Uetrecht, 1993; Miyamoto et al., 1993; Parrish et al., 1997; Prutz, 1998).

The metabolism of xenobiotics also inhibits neutrophil function by reacting with, and so removing, HOCl or inhibiting the chlorinating reactions of MPO. Some products of these reactions are also cytotoxic for neutrophils and blunt the inflammatory response (Hofstra and Uetrecht, 1993). The loss of neutrophil function or number would protract the infection and increase the possibility of aberrant immune responses. Other possibilities for pathologies include generation of novel metabolites that are immunologically reactive. So far, much of the attention on MPO-mediated xenobiotic metabolism has focused upon pharmaceutical agents. It would be pertinent to also consider inhaled pollutants and antigens as substrates for MPO, given the wide range of compounds that can be oxidized by MPO Compound I and HOCl and the potential immune or chemical reactivity of the products. A candidate for such studies would be benzene-containing compounds in various smokes. Neutrophils promote ring chlorination of substituted aryl compounds (Foote et al., 1983; She and Davis, 1991) and benzene metabolites (Hofstra and Uetrecht, 1993).

The above reactions are potentially important for pulmonary pathology, as MPO has been found in the bronchial lavage fluid of diseased lungs (Weiland et al., 1986; Watts et al., 1990). In addition, MPO is highly cationic (pI > 9) (Agner, 1941) and readily binds to the surface of cells (Johnson et al., 1987a). Thus, MPO bound to respiratory tissue has the potential to affect or damage these cells by generating hypohalous acids or converting xenobiotics into toxicants. MPO might also expose and bind the subendothelial matrix because a number of highly cationic proteins, such as elastase (Peterson, et al., 1987, 1995) and gelatinase (Peterson, 1989), cause the endothelial cells to contract to lose barrier function. The loss of endothelial barrier function has been not demonstrated for MPO alone, but is an interesting possibility because MPO constitutes 5% of the dry weight of neutrophils; of this amount, ~10% is released upon cell stimulation (Jeitner and Eaton, unpublished studies).

Hypohalous Acids, Chlorinated Amines, and Protease Activation in Pulmonary Injury

Hypochlorous acid reacts with a critical methionine residue of the major circulating inhibitor of elastase, α1-antiprotease, abolishing its ability to bind to elastase (Johnson and Travis, 1979). Thus, HOCl produced during neutrophil migration may act to allow elastase to bind CD11b/CD18 and cause the leading edge of the cells to dissociate from the endothelium to facilitate further movement. Excess HOCl production also could contribute to pulmonary toxicity by accentuating the net elastolytic activity of neutrophils. Both elastase and oxidatively-inactivated α1-antiproteinase have been found in lungs of patients with pulmonary disease (Lee et al., 1981; Wewers et al., 1988). α1-Antiproteinase also binds and neutralizes the proteolytic activity of PR3; consequently, excess HOCl may expose PR3 as an antigen in C-ANCA-associated autoimmunity, such as in Wegener's granulomatosis.

In addition to elastase, neutrophils also release collagenase metalloproteinase. This enzyme, which has been implicated in lung injury, is released from granules in a latent form requiring either gelatinase-dependent proteolysis or oxidation by HOCl and chloramines for activation (Claesson et al., 1996). The half-life of HOCl in buffer is relatively short; however, chloramines are long-lived species with the oxidizing equivalents of HOCl and may be more relevant in collagenase activation. These observations suggest that part of the toxicity of HOCl is due to generation of chloramines and an increase in the activities of elastase and collagenase through inactivation of α1-antiproteinase or enzyme activation, respectively.

It has been argued that neutrophils damage endothelial tissue through the combined actions of HOCl, elastase, collagenase, and gelatinase (Weiss, 1989). Elastase has been found in the lavage fluid of patients with adult respiratory distress syndrome (ARDS) (Lee et al., 1981; Wewers et al., 1988). Addition of elastase alone increases permeability of endothelial cell monolayers *in vitro* (Harlan et al., 1981; Schmid et al., 1985; Smedly et al., 1986; Sugahara et al., 1986) and infusion of the protease into mini-pigs is sufficient to induce a compromised lung function characteristic of ARDS (Burchardi and Stokke, 1988). Also, administration of elastase inhibitors prevents both sequestration of neutrophils and pulmonary manifestations of endotoxin-induced respiratory distress (Redens et al., 1988; Birrer et al., 1992; Kubo et al., 1994; Fugimoto et al., 1995).

Elastase participates in the migration of neutrophils on endothelial cells (Owen et al., 1995; Cai and Wright, 1996). As such, this protease also could potentially be involved in diapedesis because, as a cationic species it causes epithelial and, presumably, endothelial cell, to contract and so facilitate movement of neutrophils between the cells within a monolayer (Peterson et al., 1995). This is an important consideration as the substrates for neutrophil elastase (i.e., elastin, collagen, gelatin, laminin, and fibronectin) (Roughley, 1977; McDonald et al., 1979; Mainardi et al., 1980; McDonald and Kelley, 1980; Gadek et al., 1980; Daudi et al., 1991) exist in the subendothelial matrix. Proteolytic cleavage of these proteins is thought to disrupt critical integrin- and selectin-dependent linkages between the extracellular matrix and endothelial cells resulting in cell contraction and passage of fluid through the endothelial layer. This loss of barrier function is thought to be a principle damaging event in many forms of pulmonary injury.

HOCl and the Immune Response

Though the role of the innate immune response in protection against aerosolic pathogens and xenobiotics, as well as damage to host tissue, has been emphasized, T-lymphocytes are principally responsible for overall coordination of the immune response to pathogens and resolution of disease they cause. Thus, any modulation of T-lymphocyte function by neutrophil and macrophage products has profound ramifications for pulmonary immunity. In this regard, generation of hypohalous acid and their products have the potential to alter antigen-specific and non-specific immune reactivities of T- and, thus, B-lymphocytes. The ability of lymphocytes to proliferate depends on their GSH content (Kavanagh et al., 1990) and can be influenced by reduced oxygen species in a biphasic manner. Modest concentrations of reduced oxygen species promote lymphocyte proliferation while higher concentrations inhibit cell division. Low level oxidative stress is thought to enhance T-lymphocyte proliferation, in part, by stimulating NF-κB-dependent transcription (Schreck et al., 1991; Anderson et al., 1994; Goldstone et al., 1995); HOCl stimulates NF-κB activation in a T-lymphocyte line (Schoonbroodt et al., 1997). Unfortunately, in the latter study, HOCl was added to cells in the presence of amino acids, so reported effects could be due to generation of reactive aldehydic products and chlorinated amines, as well as the HOCl. Aldehydes capable of forming Schiff bases are co-stimulatory for CD4$^+$ T-lymphocyte proliferation and promote a T_H1-type response (Rhodes et al., 1995). This response depends on the nature of the aldehydes and their cellular products because studies have revealed that acrolein preferentially inhibits proliferation of human peripheral blood mononuclear cells stimulated with tetanus toxin rather than staphylococcal enterotoxin B or PHA (Barlin and Lawrence, unpublished observation). Acrolein is a potential product of HOCl reaction with L-threonine (Anderson et al., 1997), however, many other potential products could be generated to perturb immune responses (Witz, 1989; Hazen et al., 1996, 1997a, 1998a and b; Anderson et al., 1997).

Previously, we have suggested that reduced oxygen species act on lymphocytes to retard their development and their ability to produce cytokines (Krieger et al., 1998). The overall consequences of these modifications were postulated to include a lessened immune potential against pathogens, coupled with an overzealous reactivity to self. The respiratory system is constantly being challenged by potential initiators of immune responses, in an environment favoring the generation of reduced oxygen species, namely, high oxygen tension. Thus, the potential for aberrance is relatively high and may account for the prevalence of asthma and respiratory infections. Compounding these factors is the influence of aging which is associated with lower cellular thiol concentrations, a greater incidence of respiratory infections and more frequent requirement for vaccinations to augment the immune response. All of these facets of the immune response presumably impinge on each other.

The effect of reduced oxygen species is reflected in loss of cellular thiols, particularly GSH and can be modeled by depleting GSH by inhibiting its synthesis with buthionine sulfoxamine (Griffith and Meister, 1979). Depletion of GSH in lymphocytes prior and during the G_1 phase of the cell cycle results in a profound inhibition of their proliferation and decrement in their cell surface and intracellular thiol pools (Messina and Lawrence, 1989; Suthanthiran et al., 1990). Thiol groups are a major target for neutrophil-derived HOCl and aldehydes (Witz, 1989) and so these inflammatory products have the potential to modify lymphocyte proliferation. Thiol oxidation may affect specific lymphocyte subsets because they can be distinguished on the basis on their GSH and cell surface thiol content (Lawrence, 1981; Lawrence et al., 1996). Different oxidants produce different patterns of thiol oxidation (Thomas et al., 1995; Cotgreave and Gerdes, 1994) and cell surface thiols appear to be especially susceptible to oxidation by HOCl (Schraufstatter et al., 1990). Taken together, these observations suggest mechanisms for modulation of lymphocyte reactivity by HOCl and its products and may provide some explanation for the loss of immune control after respiratory exposure to irritant and pathogens.

SUMMARY

The lung is affected by both specific and non-specific autoimmune disease processes. Inflammation plays an important role in the etiologies in these diseases and this Chapter has examined how the major reduced oxygen species produced by neutrophils, hypohalous acid, might contribute to pulmonary damage and autoimmunity. Hypohalous acids react vigorously with amine and thiol moieties, and so have the potential to modify the antigenicity and function of virtually every protein at sites of inflammation. There appear to be few, if any, incidences of pulmonary autoimmune disease with antibodies directed against specific lung antigens. Thus, the development of pulmonary autoimmunity probably reflects the exposure of this organ to airborne toxicants and pathogens and inappropriate immune responses, including excessive inflammation. Consideration of the reactions of hypohalous acids and their products may aid in understanding the sensitivity of the lung tissues and the immune system to airborne pathogens and toxicants, and thereby the means to protect against disease.

ACKNOWLEDGEMENTS

The Authors would like to thank Fiona Jeitner for her careful reading of the manuscript. Ulrich Rudofsky and Fred Minnear are also acknowledged for their helpful discussions of this manuscript.

REFERENCES

Adachi T, Yamada H, Futenma A, Kato K, Hirano K. Heparin-induced release of extracellular-superoxide dismutase form (V) to plasma. J. Biochem., 1995;117:586-590.

Adachi T, Yamada H, Yamada Y, Morihara N, Yamazaki N, Murakami T, Futenma A, Kato K, Hirano K. Substitution of glycine for arginine-213 in extracellular dismutase impairs affinity for heparin and endothelial cell surface. Biochem. J., 1996;313:235-239.

Agner K. Verdoperoxidase. A ferment isolated from leucocytes. Acta Physiol. Scand., 1941;2:1-62.

Agosti JM, Altman LC, Ayars GH, Loegering DA, Gleich GJ, Klebanoff SJ. The injurious effects of eosinophil peroxidase, hydrogen peroxide, and halides on pneumocytes *in vitro*. J. Allergy Clin. Immunol., 1987;79:496-504.

Akerboom TP, Sies H. Transport of glutathione disulfide and glutathione S-conjugates in hepatocyte plasma membrane vesicles. Meth Enzymol., 1994;233:416-425.

Albrecht D, Jungi TW. Luminol-enhanced chemiluminescence induced in peripheral blood-derived human phagocytes: Obligatory requirement of myeloperoxidase exocytosis by monocytes. J. Leukocyte Biol., 1993;54:300-306.

Albrich JM, McCarthy CA, Hurst JK. Biological reactivity of hypochlorous acid: Implications for microbicidal mechanism of leukocyte myeloperoxidase. Proc. Natl. Acad. Sci. USA., 1981;78:210-214.

Anderson MM, Hazen LS, Hsu FF, Heinecke JW. Human neutrophils employ the myeloperoxidase-hydrogen peroxide-chloride system to convert hydroxy-amino acids into glycoaldehyde, 2-hydroxypropanal, and acrolein. J. Clin. Invest., 1997;99:424-432.

Anderson MT, Staal JT, Gilter C, Herzenberg LA, Herzenberg LA. Separation of oxidant-initiated and redox-regulated in the NF-κB signal transduction pathway. Proc. Natl. Acad. Sci. USA, 1994;91:11527-11531.

Aruoma OI, Halliwell B, Hoey BM, Butler J. The antioxidant action of taurine, hypotaurine and their metabolic precursors. Biochem. J., 1988;256:251-255.

Baker MS, Feigan J, Lowther DA. The mechanism of chondrocyte hydrogen peroxide damage. Depletion of intracellular ATP due to suppression of glycolysis caused by oxidation of glyceraldehyde-3-dehydrogenase. J. Rheumatol., 1989;16:7-14.

Birrer P, McElvaney NG, Gillissen A, Hoyt RF, Bloedow DC, Hubbard RC, Crystal RG. Intravenous recombinant secretory leukoprotease inhibitor augments anti-neutrophil elastase defense. J. Appl. Physiol., 1992;73:317-323.

Boehme MW, Schmitt WH, Youinou P, Stremmel WR, Gross WL. Clinical relevance of elevated serum thrombomodulin and soluble E-selectin in patients with Wegener's granulomatosis and other systemic vasculitides. Am. J. Med., 1996;101:387-394.

Borregaard N, Cowland JB. Granules of the human neutrophilic polymorphonuclear leukocyte. Blood, 1997;89:3503-3521.

Brodie AE, Reed DJ. Reversible oxidation of glyceraldehyde 3-phosphate dehydrogenase thiols in human lung carcinoma cells by hydrogen peroxide. Biochem. Biophys. Res. Commun., 1987;148:120-125.

Brooks CJ, King WJ, Radford DJ, Adu D, McGrath M, Savage CO. IL-1β producton by human polymorphonuclear leukocytes stimulated by anti-neutrophil cytoplasmic autoantibodies: Relevance to systemic vasculitis. Clin. Exp. Immunol., 1996;106:273-279.

Burchardi H, Stokke T. Experimental studies on the adult respiratory syndrome: Elastase infusion in normal and agranulocytic mini-pigs. Adv. Exp. Med. Biol., 1988;240:149-157.

Cai TQ, Wright SD. Human leukocyte elastase is an endogenous ligand for the integrin CR3 (CD11b/CD18, Mac-1, $\alpha_M\beta_2$) and modulates polymorphonuclear leukocyte adhesion. J. Exp. Med., 1996;184:1213-1223.

Candeias LP, Patel KB, Stratford MR, Wardman P. Free hydroxyl radicals are formed on reaction between the neutrophil-derived species superoxide anion and hyperchlorous acid. FEBS Lett., 1993;333:151-153.

Carr AC, Winterbourn CC. Oxidation of neutrophil glutathione and protein thiols by myeloperoxidase-derived hypochlorous acid. Biochem. J., 1997;327:275-281.

Chai YC, Ashraf SS, Rokutan K, Johnson RB, Thomas JA. S-Thiolation of individual human neutrophil proteins including actin by stimulation of the respiratory burst: Evidence against a role for glutathione disulfide. Arch. Biochem. Biophys., 1994;310:273-281.

Chai YC, Jung CH, Lii CK, Ashraf SS, Hendrich S, Wolf B, Sies H, Thomas JA. Identification of an abundant S-thiolated rat liver protein as carbonic anhydrase III: Characterization of S-thiolation and dethiolation reactions. Arch. Biochem. Biophys., 1991;284:270-278.

Charles LA, Caldas ML, Falk RJ, Terrel RS, Jennette JC. Antibodies against granule proteins activate neutrophils in vitro. J. Leukocyte Biol., 1991;50:539-546.

Chesney JA, Eaton JW, Mahoney JR. Bacterial glutathione: A sacrificial defense against chloride compounds. J. Bacteriol., 1996;178:2131-2135.

Claesson R, Karlsson M, Zhang Y, Carlsson J. Relative role of chloramines, hypochlorous acid, and protease in the activation of human polymorphonuclear leukocyte collagenase. J. Leukocyte Biol., 1996;60:598-602.

Cohen Tervaert JW, van der Woude FJ, Fauci AS, Ambrus JL, Velosa J, Keane WF, Meijer S, van der G, van der Hem GM, The TH, van der Hem GK, Kallenberg CG. Association between active Wegener's granulomatosis and anti-neutrophil antibodies. Arch. Int. Med., 1989;149:2461-2465.

Collison MW, Thomas JA. S-Thiolation of cytoplasmic cardiac creatine kinase in heart cells treated with diamide. Biochim. Biophys. Acta, 1987;928:121-129.

Cotgreave IA, Gerdes RG. Recent trends in glutathione biochemistry-glutathione-protein interactions: A molecular link between oxidative stress and cell proliferation. Biochem. Biophys. Res. Commun., 1994;242:1-9.

Csernok E, Ernst M, Schmitt W, Bainton DF, Gross WL. Activated neutrophils express proteinase-3 on their plasma membrane in vitro and in vivo. Clin. Exp. Immunol., 1994;95:244-250.

Daudi I, Gudewicz PW, Saba TM, Cho E, Vincent P. Proteolysis of gelatin-bound fibronectin by activated leukocytes: A role for leukocyte elastase. J. Leukocyte Biol., 1991;50:331-340.

Daugherty A, Dunn JL, Rateri DL, Heinecke JW. Myeloperoxidase, a catalyst for lipoprotein oxidation, is expressed in human atherosclerotic lesions. J. Clin. Invest., 1994;94:437-444.

DeLeo FR, Quinn MT. Assembly of the phagocyte NADPH oxidase: Molecular interaction of the oxidase proteins. J. Leukocyte Biol., 1996;60:677-691.

DeRemee RA. The treatment of Wegener's granulomatosis. Clin. Exp. Immunol., 1995;101(Suppl. 1):23-26, 1995.

Dolman KM, Gans RO, Veraat TJ, Zevenbergen G, Maingay D, Nikkels RE, Donker AJ, von der Borne AE, Goldscheding R. Vasculitis and anti-neutrophil cytoplasmic autoantibodies associated with propylthiouracil therapy. Lancet, 1993;342:651-652.

Domigan NM, Charlton TS, Duncan MW, Winterbourn CC, Kettle AJ. Chlorination of tyrosyl residues in peptides by myeloperoxidase and human neutrophils. J. Biol. Chem., 1995;270:16542-16548.

Duescher RJ, Elfarra AA. 1,3-Butadiene oxidation by human myeloperoxidase. J. Biol. Chem., 1992;267:19859-19865.

Ewert BH, Jennette JC and Falk RJ, Anti-myeloperoxidase antibodies stimulate primed neutrophils to damage human endothelial cells. Kidney Intl., 1992;41:375-383.

Falk RJ, Terrell RS, Charles LA, Jennette JC. Anti-neutrophil cytoplasmic autoantibodies induce neutrophils to degranulate and produce oxygen radicals in vitro. Proc. Natl. Acad. Sci. USA, 1990;87:4115-4119.

Fariss MW, Brown MK, Schmitz JA, Reed DJ. Mechanism of chemical-induced toxicity. I. Use of a rapid centrifugation technique for the separation of viable and nonviable hepatocytes. Toxicol. Appl. Phamacol., 1985;79:283-295.

Finnern R, Pedrollo E, Fisch I, Weislander J, Marks JD, Lockwood CM, Ouwehand WH. Human autoimmune anti-protease 3 scFv from a phage display library. Clin. Exp. Immunol., 1997;107:269-281.

Folkes LK, Candeias LP, Wardman P. Kinetics and mechanisms of hypochlorous acid reactions. Arch. Biochem. Biophys., 1995;323:120-126.

Foote CF, Goyne TE, Leher RI. Assement of chlorination by human neutrophils. Nature, 1983;301:715-716.

Frasca GM, Zoumparidis NG, Borgnino LC, Neri L, Vangelista A, Bonomini V. Plasma exchange treatment in rapidly-progressive glomerulonephritis associated with anti-neutrophil antibodies. Intl. J. Artificial Organs, 1992;15:181-184.

Frimat B, Gressier B, Odou P, Brunet C, Dine T, Lukcky M, Cazin M, Cazin JC. Metabolism of clozapine by human neutrophils: Evidence for a specific oxidation of clozapine by the myeloperoxidase system with inhibition of enzymatic chlorination cycle. Fundam. Clin. Pharmacol., 1997;11:267-274.

Fugimoto K, Kubo K, Shinozaki S, Okada K, Matsuzawa Y, Kobayashi T, Sugane K. Neutrophil elastase inhibitor reduces asthmatic response in allergic sheep. Respir. Physiol., 1995;100:91-100.

Gadek JE, Fells GA, Wright DG, Crystal RG. Human neutrophil elastase functions as a type III collagen "collagenase". Biochem. Biophys. Res. Commun., 1980;95:1815-1822.

Gapany-Gapanavicius M, Molho M, Tirosh M. Chloramine-induced pneumonitis from mixing household cleaning agents. Br. Med.J. Clin. Res. Ed., 1982;285:1086.

Goldstone SD, Fragonas JC, Jeitner TM, Hunt NH. Transcription factors as targets for oxidative signaling during lymphocyte activation. Biochim. Biophys. Acta, 1995;1263:114-122.

Griffith ME, Coulthart A, Pusey CD. T-cell responses to myeloperoxidase (MPO) and proteinase-3 (PR3) in patients with systemic vasculitis. Clin. Exp. Immunol., 1996;103:253-258.

Griffith OW, Meister A. Potent and specific inhibitor of glutathione synthesis by buthionine sulfoximine (S-n-butyl homocysteine sulfoximine). J. Biol. Chem., 1979;254:7558-7560.

Ha TY, Waksman BH. Role of the thymus in tolerance. X. "Suppressor" activity of antigen-stimulated rat thymocytes transferred to normal recipients. J. Immunol., 1973;110:1290-1299.

Halliwell B, Gutteridge JM. (eds.) Free Radicals in Biology and Medicine. Oxford: Clarendon Press, 1989.

Harlan JM, Killen PD, Harker LA, Striker GE, Wright DG. Neutrophil-mediated endothelial injury in vitro mechanisms of cell detachment. J. Clin. Invest., 1981;68:1394-1403.

Harlan JM, Levine JD, Callahan KS, Schwartz BR, Harker LA. Glutathione redox cycle protects cultured endothelial cells against lysis by extracellularly-generated hydrogen peroxide. J. Clin. Invest., 1984;73:706-713.

Haubitz M, Schulzeck P, Schellong S, Schulze M, Koch KM, Brunkhorst R. Complexed plasma elastase as in vivo marker for leukocyte activation in anti-neutrophil cytoplasmic antibody-associated vasculitis. Arthritis Rheum., 1997;40:1680-1684.

Hazell LJ, Arnold L, Flowers D, Waeg G, Malle E, Stocker R. Presence of hypochlorite-modified proteins in human atherosclerotic lesions. J. Clin. Invest., 1998;97:1535-1544.

Hazen SL, d'Avignon A, Anderson MM, Hsu FF, Heinecke JW. Human neutrophils employ the myeloperoxidase-hydrogen peroxide-chlorine system to oxidize α-amino aicds to a family of reactive aldehydes. J. Biol. Chem., 1998a;273:4997-5005.

Hazen SL, Gaut JP, Hsu FF, Growley JR, d'Avignon A, Heinecke JW. p-Hydroxyphenylacetaldehyde, the major product of L-tyrosine oxidation by the myeloperoxidase-H_2O_2-chlorine system of phagocytes, covalently modifies ϵ-amino groups of protein lysine residues. J. Biol. Chem., 1997a;272:16990-16998.

Hazen SL, Growley JR, Mueller DM, Heinecke JW. 3-Chlorotyrosine, a specific marker of myeloperoxidase-catalyzed oxidation, is markedly elevated in low density lipoprotein isolated from human atherosclerotic intima. J. Clin. Invest., 1997c;99:2075-2081.

Hazen SL, Growley JR, Mueller DM, Heinecke JW. Mass spectrometric quantification of 3-chlorotyrosine in human tissues with attomole sensitivity: A sensitive and speicific marker for myeloperoxidase-catalyzed chlorination at sites of inflammation. Free Rad. Biol. Med., 1997b;23:909-916.

Hazen SL, Hsu FF, d'Avignon A, Heinecke JW. Human neutrophils employ the myeloperoxidase to convert α-amino acids to a battery of reactive aldehydes: A pathway for aldehyde generation at sites of inflammation. Biochemistry, 1998b;37:6864-6873.

Hazen SL, Hsu FF, Heinecke JW. p-Hydroxyphenylacetaldehyde is the major product of L-tyrosine by activated human phagocytes: A chloride-dependent mechanism for the conversion of free amino acids into reactive aldehydes by myeloperoxidase. J. Biol. Chem., 1996;271:1861-1867.

Hoffman GS, Kerr GS, Leavitt RY, Hallahan CW, Lebovics RS, Travis WD, Rottem M, Fauci AS. Wegener's granulomatosis: An analysis of 158 patients. Ann. Int. Med., 1992;116:488-498.

Hofstra AH, Uetrecht JP. Myeloperoxidase-mediated activation of xenobiotics by human leukocytes. Toxicology, 1993;82:221-242.

Hyslop PA, Hinshaw DB, Halsey WA, Schraufstatter IU, Sauerheber RD, Spragg RG, Jackson JH, Cochrane CG. Mechanisms of oxidant-mediated cell injury: The glycolytic and mitochondrial pathways of ADP phosphorylation are major intracellular targets inactivated by hydrogen peroxide. J. Biol. Chem., 1988;263:1665-1675.

Izatt RM, Christensen JJ. (eds.) "Heats of Proton Ionization, pK_a, and Related Thermodynamic Quantities." In *Handbook of Biochemistry and Molecular Biology: Physical and Chemical Data*. Boca Raton, FL: CRC Press, 1976.

Jayne DR, Davies MJ, Fox CJ, Black CM, Lookwood CM. Treament of systemic vasculitis with pooled intravenous immunoglobulin. Lancet, 1991;337:1137-1139.

Johnson D, Travis J. The oxidative inactivation of human α1-proteinase inhibitor. J. Biol. Chem., 1979;254:4022-4026.

Johnson KJ, Fantone JC, Kaplan J, Ward PA. *In vivo* damage of rat lungs by oxygen metabolites. J. Clin. Invest., 1981;67:983-993.

Johnson RJ, Couser WG, Chi EY, Adler S, Klebanoff SJ. New mechanism for glomerular injury: Myeloperoxidase-hydrogen peroxide system. J. Clin. Invest., 1987;79:1379-1387.

Johnson RJ, Klebanoff SJ, Ochi RF, Adler S, Baker P, Sparks L. Couser WG. Participation of the myeloperoxidase-H_2O_2-halide system in the immune complex nephritis. Kidney Intl., 1987;32:342-349.

Kavanagh TJ, Grossmann A, Jaecks EP, Jinneman JC, Eaton DL, Martin GM, Rabinovitch PS. Proliferative capacity of human peripheral blood lymphocytes sorted on the basis of glutathione content. J. Cell. Physiol., 1990;145:472-480.

Keogan MT, Esnault VL, Green AJ, Lockwood CM, Brown DL. Activation of normal neutrophils by anti-neutrophil cytoplasm antibodies. Clin. Exp. Immunol., 1992;90:228-234.

Kettle AJ, Robertson IG, Palmer BD, Anderson RF, Patel KB, Winter CC. Oxidative metabolism of amsacrine by the neutrophil enzyme myeloperoxidase. Biochem. Pharmacol., 1992;44:1731-1738, 1992.

Kettle AJ. Neutrophils convert tyrosyl residues in albumin to chlorotyrosine. FEBS Lett., 1996;379:103-1066.

King WJ, Brooks CJ, Holder R, Hughes P, Adu D, Savage CO. T-Lymphocyte responses to anti-neutrophil cytoplasmic autoantibodies (ANCA) antigens are present in patients with ANCA-associated vasculitis and persit during disease remission. Clin. Exp. Immunol., 1998;112:539-546.

Kobayashi T, Robinson JM, Seguchi H. Identification of intracellular sites of superoxide production in stimulated neutrophils. J. Cell Sci., 1998;111:81-91.

Kozumbo WJ, Agarwal S, Koren HS. Breakage and binding of DNA by reaction products of hypochlorous acid with aniline, 1-naphthylamine, or 1-naphthol. Toxicol. Appl. Pharmacol., 1992;115:107-115.

Kravis TC, Ahmed A, Brown TE, Fulmer JD, Crystal RG. Pathogenic mechanisms in pulmonary fibrosis: Collagen-induced migration inhibitory factor production and cytotoxicity mediated by lymphocytes. J. Clin. Invest., 1976;58:1223-1232.

Krieger JA, Heo Y, Lawrence DA. "Oxidative Stress and Heavy Metal Stress Modification of T-Lymphocyte Subsets." In *T-Lymphocyte Subpopulations in Immunotoxicology*, I Kimber, MJK Selgrade, eds. Chichester, NY: John Wiley and Sons, 1998, pp. 103-120.

Kubo K, Kobayashi T, Hayano T, Koizumi T, Honda T, Sekiguchi M, Sakai A. Effects of ONO-5046, a specific neutrophil elastase inhibitor, on endotoxin-induced lung injury in sheep. J. Appl. Physiol., 1994;77:1333-1340.

Lawrence DA, Colinas RJ, Walsh AC. Influence of oxygen partial pressure on human and mouse myeloid cell line characteristics. Fundam. Appl. Pharmacol., 1996;29:287-293.

Lawrence DA. "Antigen Activation of T-Cells." In *Tumor Antigens: Structure and Function*, H Waters, ed. New York: Garland STPM, 1981, pp. 257-330.

Lee CT, Fein AM, Lippmann M, Holtzman H, Kimbel P, Weinbaum G. Elastolytic activity in pulmonary lavage fluid from patients with adult respiratory-distress syndrome. N. Engl. J. Med., 1981;304:192-196.

Lind C, Gerdes R, Schuppe-Koistinen I, Cotgreave IA. Studies on the mechanism of oxidative modification of human glyceraldehyde-3-phosphate dehydrogenase by glutathione: Catalysis by glutaredoxin. Biochem. Biophys. Res. Commun., 1998;247:481-486

Linder M. (ed.) *Nutritional Biochemistry and Metabolism*. New York: Elsevier Science Publishing Company, 1992.

Lúdvíksson BR, Sneller MC, Chua KS, Talar-Willians C, Langford CA, Ehrhardt RO, Fauci AS, Strober W. Active Wegener's granulomatosis is associated with HLA-DR$^+$ CD4$^+$ T-cells exhibiting an unbalanced T$_H$1-type T-cell cytokine pattern: Reversal with IL-10. J. Immunol., 1998;160:3602-3609.

Madri JA, Furthmayr H. Collagen polymorphism in the lung: An immunochemical study of pulmonary fibrosis. Hum. Pathol., 1998;11:353-366.

Mainardi CL, Dixit SN, Kang AH. Degradation of type IV (basement membrane) collagen by a proteinase isolated from human polymorphonuclear leukocyte granules. J. Biol. Chem., 1980;255:5435-5441.

Malle E, Hazell LJ, Stocker R, Sattler W, Arnold L, Esterbauer H, Waeg G. Immunologic detection and measurement of hypochlorite-modifed LDL with specific monoclonal antibodies. Arterioscler., Thromb. Vascul. Biol., 1995;15:982-989.

Mayet WJ, Csernok E, Szymkowiak C, Gross WL, Meyer zum Buschenfelde KH. Human endothelial cells express proteinase-3, the target antigen of anti-cytoplasmic antibodies in Wegener's granulomatosis. Blood, 1993;82:1221-1229.

Mayet WJ, Meyer zum Buschenfelde KH. Antibodies to proteinase-3 increase adhesion of neutrophils to human endothelial cells. Clin.Exp. Immunol., 1993;94:440-446.

Mayet WJ, Schwarting A, Orth T, Duchmann R, Meyer zum Buschenfelde KH. Antibodies to proteinase-3 mediate expression of vascular cell adhesion molecule-1 (VCAM-1). Clin. Exp. Immunol., 1996;103:259-267.

McDonald JA, Baum BJ, Rosenberg DM, Kelman JA, Brin SC, Cystal RG. Destruction of a major extracellular adhesive glycoprotein (fibronectin) on human fibrobalsts by neutral proteases from polymorphonuclear leukocyte granules. Lab. Invest., 1979;40:350-357.

McDonald JA, Kelley DG. Degradation of fibronectin by human leukocyte elastase. Release of biologically-active fragments. J. Biol. Chem., 1980;255:8848-8858.

Messina JP, Lawrence DA. Cell cycle progression of glutathione-depleted human peripheral blood mononuclear cells is inhibited at S phase. J. Immunol., 1989;143:1974-1981.

Miller RMH, Sies H, Park EM, Thomas JA. Phosphorylase and creatine kinase modification by thiol-disulfide exchange and by xanthine oxidase-initiated S-thiolation. Arch. Biochem. Biophys., 1990;276:355-363.

Miyamoto G, Zahid N, Uetrecht JP. Oxidation of diclofenac to reactive intermediates by neutrophils, myeloperoxidase, and hypochlorous acid. Chem. Res. Toxicol., 1993;10:414-419.

Mrowka C, Sieberth HG. Circulating adhesion molecules ICAM-1, VCAM-1, and E-selectin in systemic vasculitis: Marked differences between Wegener's granulomatosis and systemic lupus erythematosus. Clin. Invest., 1995b;72:762-768.

Mrowka C, Sieberth HG. Detection of circulating adhesion molecules ICAM-1, VCAM-1, and E-selectin in Wegener's granulomatosis, systemic lupus erythematosus, and chronic renal failure. Clin. Nephrol., 1995a;43:288-296.

Mulder AH, Heeringa P, Brouwer E, Limburg PC, Kallenberg CG. Activation of granulocytes by anti-neutrophil-cytoplasmic antibodies (ANCA): An $F_{c\gamma}RII$-dependent process. Clin. Exp. Immunol., 1994;98:270-278.

Nakos G, Adams A, Andriopoulos N. Antibodies to collagen in patients with idiopathic pulmonary fibrosis. Chest, 1993;103:1051-1058.

Nassberger L, Johansson AC, Bjork S, Sjoholm AG. Antibodies to neutrophil granulocyte myeloperoxidase and elastase: Autoimmune responses in glomerulonephritis due to hydralazine treatment. J. Intern. Med., 1991;229:261-265.

Nassonov EL, Samsonov MY, Tilz GP, Beketova TV, Semenkova EN, Baranov A, Wachter H, Fuchs D. Serum concentrations of neopterin, soluble interleukin-2 receptor, and soluble tumor necrosis factor receptor in Wegener's granulomatosis. J. Rheumatol., 1997;24:666-670.

Niles JL. Anti-neutrophil cytoplasmic antibodies in the classification of vasculitis. Ann. Rev. Med., 1996;47:303-313.

Ochoa L, Waypa G, Mahoney JR, Rodriguez L, Minnear FL. Contrasting effects of hypochlorous acid and hydrogen peroxide on endothelial permeability. Am. J. Crit. Care Med., 1997;156:1247-1255.

Owen CA, Campbell MA, Sannes PL, Boukedes SS, Campbell EJ. Cell surface-bound elastase and cathepsin G on human neutrophils: A novel, non-oxidative mechanism by which neutrophils focus and preserve catalytic activity of serine proteinases. J. Cell Biol., 1995;131:775-789.

Pall AA, Howie AJ, Adu D, Richards GM, Inward DV, Milford DV, Richards NT, Michael J, Taylor CM. Glomerular vascular cell adhesion molecule-1 in renal vasculitis. J. Clin. Pathol., 1996;49:238-242.

Pan LP, Kreisle RA, Shi YD. Detection of $F_{c\gamma}$ receptors on human endothelial cells stimulated with cytokines tumor necrosis factor-alpha (TNF-α) and interferon-gamma (IFN-γ). Clin. Exp. Immunol., 1998;112:533-538.

Park EM, Thomas JA. S-Thiolation of creatine kinase and glycogen phosphorylase b initiated by partially-reduced oxygen species. Biochim. Biophys. Acta, 1988;964:151-160.

Parrish DD, Schlosser MJ, Kapeghian JC, Traina VM. Activation of CGS 12094 (prinimide metabolite) to 1,4-benzoquinone by myeloperoxidase: Implication for human idiosyncratic agranulocytosis. Fundam. Appl. Toxicol., 1997;35:197-204.

Pereira WE, Hoyano Y, Summons RE, Bacon VA, Duffield AM. Chlorination studies II. The reaction of aqueous hypochlorous acid with α-amino acids and dipeptides. Biochim. Biophys. Acta, 1973;313:170-180.

Peterson MW, Stone P, Shasby DM. Cationic neutrophil proteins increase transendothelial albumin movement. J. Appl. Physiol., 1987;62:1521-1530

Peterson MW, Walter ME, Nygaard SD. Effect of neutrophil mediators on epithelial permeability. Am. J. Respir. Cell Mol. Biol., 1995;13:719-727.

Peterson MW. Neutrophil cathepsin G increases transendothelial albumin flux. J. Lab. Clin. Med., 1989;113:297-308.

Pettigrew AR, Fell GS. Simplified colorimetric determination of thiocyanate in biological fluids, and its application to investigation to the toxic amblyopias. Clin.Chem., 1972;18:996-1000.

Porges AJ, Redecha PB, Kimberly WT, Csernok E, Gross WL, Kimberly RP. Anti-neutrophil cytoplasmic antibodies engage and activate neutrophils via $F_{c\gamma}RIIa$. J. Immunol. 1994;153:1271-1280.

PrÜtz WA. Hypochlorous acid interactions with thiols, nucleotides, DNA, and biological substrates. Arch. Biochem. Biophys., 1996;332:110-120.

PrÜtz WA. Reaction of hypochlorous acid with biological substrates are activated catalytically by tertiary amine. Arch. Biochem. Biophys., 1998;357:265-273.

Qian M, Eaton JW. Free fatty acids enhance hypochlorous acid production by activated neutrophils. J. Lab. Clin. Med., 1994;124:86-95.

Ralston DR, Marsh CB, Lowe MP, Wewers MD. Anti-neutrophil cytoplasmic antibodies induce monocyte IL-8 release. J. Clin. Invest., 1997;100:1416-1424.

Rasmussen N, Petersen J. Cellular immune responses and pathogenesis in c-ANCA-positive vasculitides. J. Autoimmunol., 1993;6:227-236.

Ravichandran V, Seres T, Moriguchi T, Thomas JA, Johnson RB. S-Thiolation of glyceraldehyde-3-phosphate dehydrogenase induced by the phagocytosis-associated respiratory burst in blood monocytes. J. Biol. Chem., 1994;269:25010-25015.

Redens TB, Leach WJ, Bogdanoff DA, Emerson TE. Synergistic protection from lung damage by combining antithrombin-III and α1-proteinase inhibitor in the E. coli endotoxemic sheep pulmonary dysfunction model. Circ. Shock, 1988;26:15-26.

Reynolds HY, Fulmer JD, Kazierowski JA, Roberts WC, Frank MM, Crystal RG. Analysis of cellular and protein content of bronchoalveolar lavage fluid from patients with idiopathic pulmonary fibrosis and chronic hypersensitivity pneumonitis. J. Clin. Invest., 1977;59:165-175.

Rhodes J, Chen H, Hall SR, Beesley JE, Jenkins DC, Collins P, Zheng B. Therapeutic potentiation of the immune system by costimulatory Shiff-base-forming drugs. Nature, 1995;377:71-75.

Robertson FM, Bijur GN, Obersyzyn AS, Pellegrini AE, Boros LG, Sabourin LK, Oberyszyn TM. Granulocyte-macrophage colony stimulating factor gene expression and function during tumor promotion. Carcinogenesis, 1994;15:1017-1029.

Rokutan K, Johnston RB, Kawai K. Oxidative stress induces S-thiolation of specific proteins in cultured gastric mucosal cells. Am. J. Physiol., 1994;266:G247-G254.

Rokutan K, Thomas JA, Sies H. Specific S-thiolation of a 30-kDa cytosolic protein from rat liver under oxidative stress. Eur. J. Biochem., 1989;179:233-239.

Rosen H, Klebanoff SJ. Oxidation of microbial iron-sulfur centers by the myeloperoxidase-H_2O_2-halide antimicrobial system. Infect. Immmun., 1985;47:613-618.

Roughley PJ. The degradation of proteoglycans by leukocyte elastase. Biochem. Soc. Trans., 1977;5:443-445.

Savage CO, Gaskin G, Pusey CD, Pearson JD. Anti-neutrophil cytoplasm antibodies can recognize vascular endothelial cell-bound anti-neutrophil cytoplasm antibody-associated autoantigens. Exp. Nephrol., 1993;1:190-195.

Savage CO, Pottinger BE, Gaskin G, Pusey CD and Pearson JD, Autoantibodies developing to myeloperoxidase and proteinase-3 in systemic vasculitis stimulate neutrophil cytotoxicity toward cultured endothelial cells. Am. J. Pathol. 1992;141:335-342.

Schlesier M, Kaspar T, Gutfleisch J, Wolff-Vorbeck G, Peter HH. Activated CD4+ and CD8+ T-cell subsets in Wegener's granulomatosis. Rheumatol. Intl., 1995;14:213-219.

Schmid RP, Wangensteen S, Hoidal J, Gosnell B, Niewoehner D. Effects of elastase and cigarette smoke on alveolar epithelial permeability. J. Appl. Physiol. 1985;58:96-100.

Schoonbroodt S, Legrand-Poels S, Best-Belpomme M, Piette J. Activation of the NF-κB transcription factor in a T-lymphocytic cell line by hypochlorous acid. Biochem. J., 1997;321:777-785.

Schraufstatter IU, Browne K, Harris A, Hyslop PA, Jackson JH, Quehenberger O, Cochrane CG. Mechanisms of hypochlorite injury of target cells. J. Clin. Invest., 1990;85:554-562.

Schreck R, Rieber P, Baeuerle PA. Reactive oxygen intermediates as apparently widely-used messengers in the activation of the NF-κB transcription factor and HIV-1. EMBO J., 1991;10:2247-2258.

Schuppe-Koistinen I, Moldeus P, Bergman T, Cotgreave IA. Reversible S-thiolation of human endothelial cell actin accompanies a structural reorganization of the cytoskeleton. Endothelium, 1995;3:301-308.

Schuppe-Koistinen I, Moldeus P, Bergman T, Cotgreave IA. S-Thiolation of human endothelial cell glyceraldehyde-3-phosphate dehydrogenase after hydrogen peroxide treatment. J. Eur. Biochem., 1994;221:1033-1037.

Segal AW, Geisow M, Garcia R, Harper A, Miller R. The respiratory burst of phagocytic cells is associated with a rise in vacuolar pH. Nature, 1981;290:406-409.

She ZW, Davis WB. Use of p-aminosalicylic acid as a trapping agent of hypochlorous acid. Am. Rev. Respir. Dis., 1991;143:A742.

Shoenfeld Y, Tomer Y, Blank M. A new experimental model for Wegener's granulomatosis. Isr. J. Med. Sci., 1995;31:13-16.

Sibelius U, Hattar K, Schenkel A, Noll T, Csernok E, Gross L, Mayet W-J, Piper HP, Seeger W, Grimminger F. Wegener's granulomatosis: Anti-proteinase-3 antibodies are potent inducers of human endothelial cell signaling and leakage response. J. Exp. Med., 1998;187:497-503.

Sibilia J, Benlagha K, Vanhille P, Ronco P, Brouet JC, Mariette X. Structural analysis of human antibodies to proteinase-3 from patients with Wegener's granulomatosis. J. Immunol., 1997;159:712-719.

Slungaard A, Mahoney JR. Thiocyanate is the major substrate for eosinophil peroxidase in physiologic fluids. J. Biol. Chem., 1991;266:4903-4910.

Smedly LA, Tonnesen MG, Sandhaus RA, Haslett C, Guthrie LA, Johnson RB, Henson PM, Worthen GS. Neutrophil-mediated injury to endothelial cells: Enhancement by endotoxin and essential role for neutrophil elastase. J. Clin. Invest., 1986;77:1233-1243.

Sneller MC, Fauci AS. Pathology of vasculitis syndromes. Med. Clin. N. Amer., 1997;81:221-242.

Stamler JS, Slivka A. Biological chemistry of thiols in the vasculature and in vascular-related disease. Nutr. Rev., 1996;54:1-30.

Stegeman CA, Tervaert JW, Huitema MG, de Jong P, Kallenberg CG. Serum levels of soluble adhesion molecules intercellular adhesion molecule-1, vascular adhesion molecule-1, and E-selectin in patients with Wegener's granulomatosis. Relationship to disease activity and relevance during follow-up. Arthritis Rheum., 1994;37:1228-1235.

Sugahara K, Cott GR, Parsons PE, Mason RJ, Sandhaus RA, Henson PM. Epithelial permeability produced by phagocytosing neutrophils *in vitro*. Am. Rev. Respir. Dis., 1986;133:875-881.

Suthanthiran M, Anderson ME, Shama VK, Meister A. Glutathione regulates activation-dependent DNA synthesis in highly-purified normal human T-lymphocytes stimulated via the CD2 and CD3 antigens. Proc. Natl. Acad. Sci. USA, 1990;87:3343-3347.

Tatsumi T, Fliss H. Hypochlorous acid and chloramines increase endothelial permeability: Possible involvement of cellular zinc. Am. J. Physiol., 1994;267:H1597-H1607.

ten Berg I, Wilmink JM, Meyer CJ, Surachno, ten Veen K, Balk TG, Schellenkens PT. Clinical and immunological follow-up of patients with severe renal disease in Wegener's granulomatosis. Am. J. Nephrol., 1985;5:21-29.

Test ST, Weiss SJ. Quantitative and temporal characterization of the extracellular H_2O_2 pool generated by human neutrophils. J.Biol. Chem., 1984;259:399-405.

Thomas JA, Poland B, Honzatko R. Protein sulphydryls and their role in the antioxidant function of protein S-thiolation. Arch. Biochem. Biophys., 1995;319:1-9.

Turner-Warick M, Burrows B, Johnson A. Cryptogenic fibrosing alveolitis: Clinical features and their influence on survival. Thorax, 1980;35:171-180.

van Dalen CJ, Whitehouse MW, Winterbourn CC, Kettle AJ. Thiocyanate and chloride as competing substrates for myeloperoxidase. Biochem. J., 1997;327:487-492.

Varagunam M, Adu D, Taylor CM, Michael J, Richards N, Neuberger J, Thompson RA. Endothelium, myeloperoxidase, anti-myeloperoxidase interaction in vasculitis. Adv. Exp. Med. Biol., 1993;336:129-132.

Vasunia KB, Miler ML, Puga A, Baxter CS. Granulocyte-macrophage colony-stimulating factor (GM-CSF) is expressed in mouse skin response to tumor-promoting agents and modulates dermal inflammation and epidermal dark cell numbers. Carcinogenesis, 1994;15:653-660.

Vissers MC, Winterbourn CC. Oxidation of intracellular glutathione after exposure of human red blood cells to hypochlorous acid. Biochem. J., 1995;307:57-62.

Wang G, Hansen H, Tatsis E, Csernok E, Lemke H, Gross WL. High plasma levels of the soluble form of CD30 activation molecule reflect disease activity in patients with Wegener's granulomatosis. Am. J. Med., 1997;102:517-523.

Ward PA. Mechanism of endothelial killing by H_2O_2 or products of activated neutrophils. Am. J. Med., 1991;91(Suppl. 3C):89S-94S.

Wasil M, Halliwell B, Hutchison DC, Baum H. The antioxidant action of human extracellular fluids. Effect of human serum and its protein components on the inactivation of α1-antiprotease by hypochlorous acid and by hydrogen peroxide. Biochem. J., 1987;243:219-223.

Watts FL, Oliver BL, Johnson GM, Thrall RS. Superoxide production by rat neutrophils in the oleic acid model of lung injury. Free Rad. Biol. Med., 1990;9:327-332.

Weiland JE, Davis WB, Holter JF, Mohammed JR, Dorinsky PM, Gadek JE. Lung neutrophils in the adult respiratory distress syndrome. Am. Rev. Respir. Dis., 1986;133:218-225.

Weiss JW, Young J, LoBuglio AF, Slivka A, Nimeh NF. Role of hydrogen peroxide in neutrophil-mediated destruction of cultured endothelial cells. J. Clin. Invest., 1981;68:714-721.

Weiss SJ, Klein R, Slivka A, Wei M. Chlorination of taurine by human neutrophils. Evidence for hypochlorous acid generation. J. Clin.Invest., 1982;70:598-607.

Weiss SJ. Tissue destruction by neutrophils. N. Engl. J. Med., 1989;320:365-376.

Wewers MD, Herzyk DJ, Gadek JE. Alveolar fluid neutrophil elastase activity in the adult respiratory distress syndrome is complexed to α2-macroglobulin. J. Clin. Invest., 1988;82:1260-1267.

Winterbourn CC, Brennan SO. Characterization of the oxidation products of the reaction between glutathione and hypochlorous acid. Biochem. J., 1997;326:87-92, 1997.

Winterbourn CC, Comparative reactivities of various biological compounds with myeloperoxidase-hydrogen peroxidase-chloride, and similarity of oxidant to hypochlorite. Biochem. Biophys. Acta, 1985;840:204-210.

Witz G. Biological interaction of α,β-unsaturated aldehydes. Free Rad. Biol. Med., 1989;7:333-349.

7. BIOLOGICAL AGENTS

Robert L. Sherwood, Ph.D.
Manager of Research
Microbiology & Immunology Division
IIT Research Institute
10 West 35th Street
Chicago, IL 60616

Pulmonary immunotoxicity from microbial exposure is increasing in many aspects of modern society and presents new challenges to health professionals. The ability of microorganisms to cause disease through non-infectious mechanisms is well known and is the focus of this Chapter. The ability of microorganisms to cause pulmonary disease through infectious mechanisms is also well known, but will not be extensively reviewed in this treatise.

The pulmonary immune system is exquisitely designed to maintain a sterile oxygen exchange system. The interior surface of the lung represents the largest surface area of the body that is directly exposed to the external environment. The respiratory system uses ciliated cells, mucosal cells, mucus, antibodies, and alveolar macrophages in an attempt to protect this tremendous surface area (Quie, 1986). Despite this intricate system, inhaled microbial agents are occasionally able to circumvent the host's defenses and cause illness through infectious processes. Microbial agents are also able to cause illness through non-infectious means such as immune hypersensitivity reactions.

Microbial deposition in the lung may have different health outcomes based upon the type of microbe. For example, infectious agents, such as rhinoviruses that cause the common cold, are able to survive and replicate in susceptible hosts despite the body's attempt to prevent infection. Other agents are rapidly inactivated with no apparent effect on health. In simplistic terms, microbes are classified as pathogens, opportunists or non-pathogens (Salyers and Whitt, 1997). Generally, this classification is based upon the ability of a microbe to cause disease, not via an immunologic mechanism, but through an infectious process.

Microbial pathogens are able to circumvent the host's normal pulmonary defenses, multiply, and cause disease. Illness associated with inhaled microbes is generally attributed to pathogenic microbes such as *Mycobacterium tuberculosis*, the causative agent of tuberculosis. Opportunistic pathogens are microbes that are unable to cause disease in healthy immunocompetent people, but are able to cause disease in people with impaired immune defenses. Microbial opportunists may be able to multiply in a host, but they are only able to cause disease in immuno-compromised hosts or when they are present in numbers sufficient to overwhelm normal defenses. An example of an opportunistic pathogen is *Pseudomonas aeruginosa* which is the predominant cause of mortality in persons suffering from burns. *P. aeruginosa* also causes significant morbidity and mortality in cystic fibrosis patients. In both cases, the immunocompromised host is unable to kill this ubiquitous microbe that multiplies and eventually causes disease. Sometimes, as in the case of the pathogen *M. tuberculosis*, disease is a result of damage induced by the body's own host defense mechanisms. Opportunists are able to replicate in the host and with assistance, evade the host's defenses. The difference between opportunists and non-pathogens was shown in a study where an opportunistic strain of *Pseudomonas* (*P. aeruginosa* AC869) survived at least 14 d in mice challenged via intranasal inoculation, while a larger dose of a non-pathogenic pseudomonad (*B. cepacia* AC1100) resulted in a shorter colonization period (only 7 d) (George et al., 1991). Non-pathogens are not able to cause infectious disease through replication or infection. Most microbes fall into this category. However, pathogens, opportunists and non-pathogens all may cause illness through adverse immunologic reactions.

Biologic agents have been associated with various immunologic diseases. Inhalation of thermophilic actinomycetes (Reijula, 1993; Nakagawa-Yoshida et al., 1997) has been associated with Farmer's Lung (Mundt et al., 1996). Similar lung diseases resulting in hypersensitivity pneumonitis are caused by inhalation of bacteria-associated products, such as lipopolysaccharide (LPS) (Burrel and Ye, 1990; Clapp et al., 1993; Zhiping et al., 1996). Incidence of asthma has also been associated with inhalation of microbes (Michel et al., 1992). Inhalation of certain fungi by sensitive individuals has also been associated with an illness termed sick building syndrome (Rautiala et al., 1996).

Man has always been exposed to low levels of aerosolized bacteria. Microbes are present whenever the wind entrains dirt particles from the soil (Kwaasi et al., 1998). For reasons primarily associated with ecologic concerns, technologies such as biodegradation, biologic pesticides and genetically recombinant organisms are presenting added opportunities for microbial challenges of the pulmonary immune system. Grain elevators, indoor air conditioning, sprayed biopesticides, composting and recycling are some of the technologies leading to increased pulmonary exposure to microbes. Microbes have been recovered down wind of various types of treatment facilities, including wastewater treatment (Goff et al., 1973; Hickey and Reist, 1975; Fannin et al., 1976; Teltsch and Katzenelson, 1978), and composting (Brown et al., 1995). Aerosolized microbes are occasionally found in significant numbers in buildings with poorly maintained air conditioning systems. Legionnaire's disease was first recognized following pulmonary exposure of a large number of veterans in just such an environment (Weisse, 1992). Other examples of human exposure from newer technologies include contaminated cutting oils (Smith, 1970; Kreiss and Cox-Ganser, 1997; Robins et al., 1997), bioreactors of many designs (Feng et al., 1997; Cesario et al., 1998; Zhukov et al., 1998), microbial pesticides sprayed on crops, ultrasonic humidifiers (Alvarez-Fernandez et al., 1998), ice nucleation sprays used in enhancing snowfall or preventing frost damage to crops (Wolber, 1993), and bioremediation microbes sprayed on chemical spills (Brubaker and Exner, 1988).

A good example of increasing human exposure to microbes is the growing use of biologic pesticides (biopesticides). Biopesticides are microbial agents that have adverse effects on specific products such as bacteria, fungi, insects or plants. Biopesticides are an important addition to the world's anti-pest arsenal because they are generally much less toxic than the currently available chemical pesticides. Examples include insecticidal agents such as *Bacillus thuringiensis* var. *israelensis* (Bti), *B. thuringiensis* var. *kurstaki* (Btk), *B. cereus, Beauveria bassiana, Metarrhizium anisopliae, Paecilomyces* sp., and nuclear polyhedrosis viruses. Examples of fungicidal agents include *Trichoderma harzianum, Burkholderia cepacia, Pseudomonas fluorescens,* and *P. antimicrobica* (Table 1). More biopesticides, biofertilizers, and bioremediation microbes are discovered and developed each year.

Immunologic reactions in the lung may be classified as primary, secondary or tertiary responses (Bellanti, 1971). Primary immunologic responses involve immediate inflammatory reactions without extensive cellular processing and antigen recognition. Inflammation may consist of cellular and/or mucosal or lymphatic infiltration to the affected site. Microbes that cause physical damage may produce this type of response. For example, pulmonary instillation of large numbers of *B. thuringiensis* var. *kurstaki* (Btk) spores in mice causes mortality in less than 24 hr (Sherwood, unpublished observations). This response is not a result of an infectious process as Btk spores are incapable of replication in a mammalian host. Lung weights of dosed mice are significantly increased compared to controls indicating the presence of a significant inflammatory response. The crystal polypeptides produced by most strains of *B. thuringiensis* may cause the pulmonary inflammation seen in animal studies. Purified 28K crystal polypeptide and delta endotoxin of *B. thuringiensis* var. *israelensis* have been shown to be hemolytic and toxic in mice and rats (Armstrong et al., 1985; Mayes et al., 1989). Similar effects have been observed following intranasal instillation of more than 1×10^7 colony forming units (cfu) of *P. aeruginosa* in mice (George et al., 1991). In the case of *P. aeruginosa*, it is felt that the LPS of the Gram-negative *Pseudomonas* causes an overwhelming inflammatory response resulting in rapid mortality (less than 24 hr).

Studies performed in our laboratories with various strains of Bt have indicated that toxicity is strain dependent. Mice challenged once intranasally with vegetative or spore preparations of Btk, a crystal-minus Btk mutant, *B. cereus*, or *B. subtilis* had differing responses (Table 2). Both vegetative and spore preparations of *B. cereus* and crystal-minus Btk killed mice. However, heat inactivation of the preparations inactivated toxic factors in *B. cereus*, but not those of crystal-minus Btk. Vegetative, but not spore, preparations of Btk killed mice; no effects were caused by vegetative or spore preparations of *B. subtilis*. These data suggest that specific factors in microbes may result in adverse effects and that these factors are not always possessed by all members of the genus or species or indeed by all strains within the species.

Other examples of the specific reactions caused by various inhaled microbes are shown in studies performed by Onofrio and colleagues (1981, 1983). Studies performed with *Staphylococcus aureus* demonstrated that the host response changed as the pulmonary dose increased. Low numbers of bacteria were rapidly cleared without neutrophil influx while higher numbers took longer to clear and initiated a significant influx of neutrophils (Onofrio et al., 1983). Additional studies were done

TABLE 1. MICROBES USED IN ENVIRONMENTAL APPLICATIONS

Microbe	Use	Reference
B. thuringiensis *var.* kurstaki	Insecticide	Damgaard et al., 1996
Bacillus cereus	Fungicide	Silo Suh et al., 1994
Bacillus thuringiensis var. *israelensis*	Insecticide	Armstrong et al., 1985
Beauveria bassiana	Insecticide	Wraight et al., 1998
Burkholderia cepacia	Bioremediation, Fungicide	Govan et al., 1996
Colletotrichum gloeosporioides	Herbicide	Rikkerink et al., 1994
Comamonas testosteroni	Bioremediation	Bae et al., 1996
Entomophaga grylli	Insecticide	Bidochka et al., 1996
Metarrhizium anisopliae	Insecticide	Burgner et al., 1998; De Garcia et al., 1997
Nuclear polyhedrosis viruses	Insecticides	Castro et al., 1997; Andrews et al., 1980; McClintock et al., 1991
Paecilomyces sp.	Insecticide	Wraight et al., 1998
Pediococcus pentosaceus	Hay preservative	Duchaine et al., 1996
Pseudomonas syringae	Ice-nucleation	Goodnow et al., 1990
Pseudomonas antimicrobica	Fungicide	Walker et al., 1996
Pseudomonas fluorescens	Fungicide	Laville et al., 1998
Pseudomonas putida	Bioremediation	Ronchel et al., 1995; Guerin and Boud, 1995
Rhizobia	Bioremediation	Damaj and Ahmad, 1996
Rhizobium meliloti	Fertilizer	Dammann et al., 1996
Trichoderma harzianum	Fungicide	Grondona et al., 1997; Flores et al., 1997
White rot fungi	Bioremediation	Barr and Aust, 1994

TABLE 2. EFFECT OF SINGLE INTRANASAL CHALLENGE WITH VARIOUS BACILLUS FORMULATIONS ON MORTALITY OF FEMALE CD-1 MICE

Strain	% Spore	CFU Administered Per Mouse (Log10)	Dead/Total	MTD[a]
B. thuringiensis var. kurstaki HD73	0	5.96	1/10	-
		6.96	7/10	1
		7.96	9/10	2
		(7.96)[b]	0/10	-
	100	5.97	0/10	-
		6.97	0/10	-
		7.97	0/10	-
		(7.97)	0/10	-
B. thuringiensis var. kurstaki HD31 (crystal -)	0	6.36	7/10	1
		7.36	10/10	1
		8.36	10/10	0
		(8.36)	1/10	1
	100	5.61	0/10	-
		6.61	2/10	1
		7.61	8/10	1
		(7.61)	4/10	2
B. cereus 14579	0	4.98	0/10	-
		5.98	1/10	1
		6.98	10/10	1
		7.98	10/10	0
		(7.98)	0/10	-
	100	3.88	0/9	-
		4.88	0/10	-
		5.88	6/10	1
		6.88	9/10	1
		(6.88)	0/10	-
B. subtilis 6051	0	6.40	0/10	-
		7.40	0/10	-
		8.40	0/10	-
		(8.40)	0/10	-
	100	6.49	0/10	-
		7.49	0/10	-
		8.49	0/10	-
		(8.49)	0/10	-

[a] MTD = mean time to death (days)
[b] ( ) – number of heat-killed bacteria

with *Streptococcus sanguis, S. salivarius* and *Neisseria catarrhalis,* bacteria that are commonly found in the human pharynx. These studies showed that *S. sanguis* was cleared more easily from the lung than *S. salivarius,* with *N. catarrhalis* taking the longest. Further, the predominant pulmonary cell responsible for bacterial clearance was the alveolar macrophage for *S. sanguis,* while *S. salivarius* and *N. catarrhalis* caused a significant influx of neutrophils (Onofrio et al., 1981).

Microbial opportunists vary in their ability to colonize or cause disease. Colonization capacity is not common to all members of a genus and indeed varies between strains. Mice challenged via intranasal instillation with strains of *P. aeruginosa* (AC869) and *B. cepacia* (AC1100) had varying pulmonary clearance rates (George et al., 1991). Clearance of microbes from the lung is dependent upon the host response and the microbe. Strains of *B. thuringiensis* act as particulates and are cleared in a biphasic manner by the pulmonary defense system. Cells are first phagocytized and then removed via the lymphatics or via the mucociliary escalator to the stomach. While most non-pathogenic microbes are killed in less than 7 d, *B. thuringiensis* strains often remain viable in the lung for at least 30 d and sometimes over 3 mo (Sherwood, unpublished observations; Siegel et al., 1987). Other microbes, such as *S. aureus,* cause virtually no inflammation at low doses and are easily cleared by the alveolar macrophage, which is the basic pulmonary cellular defense mechanism (Goldstein et al., 1974). Higher initial pulmonary doses overwhelm the ability of the alveolar macrophage to respond and cause a polymorphonuclear (PMN) cellular infiltration (Onofrio et al., 1983). Still other microbes, such as *Streptococcus zooepidemicus,* are rapidly cleared from the lung upon initial challenge, but small numbers escape detection and create foci of infection. If the host immune response is unable to combat these foci that become filled with PMN cells and fluid, the infection escapes to the bloodstream and the animals succumb to septicemic infection (Sherwood et al., 1981, 1988). In yet another mechanism of evading the host defenses, *P. aeruginosa* strains produce a polysaccharide glycocalyx that inhibits phagocytosis. The cellular rhamnolipids of *P. aeruginosa* have also been shown to inhibit phagocytosis (McClure and Schiller, 1996).

Secondary immune responses result in stimulation of humoral and cell-mediated cells and production of antibodies and activated T-lymphocytes. This, of course, is the body's normal response to foreign materials and our bodies regularly respond to inhaled biologic materials with no adverse response. Only the rare exposure that results in illness is recognized.

Tertiary immune responses may be produced whenever antigen excess drives the secondary immune response to hypersensitivity. Four types of immunologically -mediated hypersensitivity disease occur: Types I, II, III, and IV. Three of the these types (i.e., Types I, III and IV) have been associated with microbial exposure.

Type I disease, also known as immediate hypersensitivity, is caused by antigenic elicitation of IgE antibody in humans. IgE binds to basophils causing release of preformed mediators such as histamine. Based upon the location and severity of the response, the immediate hypersensitivity disease is classified as allergic rhinitis, asthma, urticaria, or angioedema (Bellanti, 1971). Several types of microbes have been implicated in causation of type I hypersensitivity disease, including several types of fungi (*Fusarium vasinfectum, Aspergillus fumigatus, Cladosporium* sp, and *Altemaria* sp.) and LPS associated with Gram-negative bacteria (Table 3).

TABLE 3. MICROORGANISMS IMPLICATED IN CAUSATION OF PULMONARY TYPE I HYPERSENSITIVITY DISEASE

Microbe	Disease	Reference
Aspergillus fumigatus	Asthma (Allergic bronchopulmonary aspergillosis)	Zhaoming and Lockey, 1996; Greenberger, 1997; Chauhan et al., 1997
Fusarium vasinfectum	Asthma	Saini et al., 1998
Gram-negative bacteria	Organic dust toxic syndrome	Zhiping et al., 1996
Gram-negative bacteria	Asthma	Michel et al., 1992
Molds (*Cladosporium, Alternaria, Aspergillus*)	Asthma	Cross, 1997

Pulmonary diseases associated with microorganisms and type III hypersensitivity include farmer's lung, malt worker's lung, wheat weevil disease (Bellanti, 1971), machine operator's lung (Bernstein et al., 1995), organic dust toxic syndrome (Zhiping et al., 1996). New diseases are also included such as composter's lung (Brown et al., 1995), humidifier lung, air-conditioner disease (Ando and Suga, 1997), and sump bay fever (Anderson et al., 1996). Adult respiratory distress syndrome (ARDS) may be triggered by pulmonary challenge with significant amounts of biologic products such as LPS (Simpson and Casey, 1989; Burrell and Ye, 1990). Exposure to microbes associated with cutting oils has also been associated with illness (Bernstein et al., 1995; Kriebel et al., 1997). Similar illnesses have been reported in personnel working in waste sorting (Poulson et al., 1995a and b). Type III hypersensitivity disease is characterized by production of antibody-antigen complexes and subsequent localization of those complexes. The complexes bind complement causing vascular permeability with resultant edema and influx of neutrophils. Release of cellular enzymes from the activated cells causes tissue damage (Bellanti, 1971). A wide variety of microbes have been implicated in causation of this disease including Gram-positive bacteria (*Bacillus pumilus, Staphylococcus capitis, Rhodococcus*, and *Pediococcus*), Gram-negative bacteria (*Pseudomonas fluorescens*), actinomycetes (*Thermoactinomyces*), and fungi (*Penicillium, Aspergillus*, and *Saccharopolyspora*) (Table 4).

Type IV hypersensitivity caused by microbes is classically associated with species of *Mycobacteria* and specifically *M. tuberculosis*. Because the manifestation of type IV hypersensitivity generally occurs 24 - 48 hr after induction, type IV hypersensitivity is also known as delayed-type or cellular hypersensitivity. This immune disease is characterized by activation of T-lymphocytes that recognize specific proteins or protein-hapten conjugates. After antigen recognition, many T-lymphocyte factors are released such as transfer factor, macrophage activating factor, skin reaction factors, chemotactic factors, mitogenic factors, lymphotoxin, and inter-

TABLE 4. MICROORGANISMS IMPLICATED IN CAUSATION OF PULMONARY TYPE III HYPERSENSITIVITY DISEASE

Microbe	Disease	Reference
Aspergillus fumigatus	Hypersensitivity pneumonitis	Hinojosa et al., 1996; Madan et al., 1997
Aspergillus niger	Machine operator's lung	Bernstein et al., 1995
Aspergillus spp.	Hypersensitivity pneumonitis	Moreno-Ancillo et al., 1997
Bacillus pumilus	Machine operator's lung	Bernstein et al., 1995
Epicoccum nigrum	Hypersensitivity pneumonitis	Hogan et al., 1996
Micropolyspora faeni	Hypersensitivity pneumonitis	Hinojosa et al., 1996
Pediococcus pentosaceus	Farmer's lung	Duchaine et al., 1996
Penicillium brevicompactum	Farmer's lung	Nakagawa-Yoshida et al., 1997
Penicillium olivicolor	Farmer's lung	Nakagawa-Yoshida et al., 1997
Pseudomonas fluorescens	Machine operator's lung	Bernstein et al., 1995
Rhodococcus sp.	Machine operator's lung	Bernstein et al., 1995
Rhodotorula spp.	Hypersensitivity pneumonitis	Alvarez-Fernandez et al., 1998
Saccharopolyspora rectivirgula	Farmer's lung	Mundt et al., 1996; Gudmundsson et al., 1998; Reijula, 1993
Staphylococcus capitis	Machine operator's lung	Bernstein et al., 1995
Thermoactinomyces vulgaris	Farmer's lung	Reijula, 1993
Thermoactinomyces vulgaris	Hypersensitivity pneumonitis	Hinojosa et al., 1996

feron. Release of these factors results in an intense cellular infiltrate consisting of activated mononuclear cells whose byproducts cause swelling, pain and tissue destruction. Although *M. tuberculosis* has typically been implicated in causation of tuberculosis, many species of *Mycobacteria* are now recognized as causative agents of tuberculosis, particularly in immunocompromised individuals. Examples include, *M. avium* complex, *M. kansasii*, *M. chelonei*, and *M. flavescens* (Table 5).

TABLE 5. MICROORGANISMS IMPLICATED IN CAUSATION OF PULMONARY TYPE IV HYPERSENSITIVITY

Microbe	Disease	Reference
Mycobacterium avium complex	Tuberculosis	Martinez Moragon et al., 1996
Mycobacterium chelonei	Tuberculosis	Martinez Moragon et al., 1996
Mycobacterium flavescens	Tuberculosis	Martinez Moragon et al., 1996
Mycobacterium kansasii	Tuberculosis	Martinez Moragon et al., 1996
Mycobacterium tuberculosis	Tuberculosis	Dannenberg, 1989

Mycobacteria cause disease almost completely through host immunologic reactions. *M. tuberculosis* and related microbes evade host inactivation because of their unique ability to evade the macrophage phagolysosome. They then replicate within the cytoplasm of the host macrophage. In order to prevent the spread of infection, the host forms thick granulomatous structures called tubercles that wall up the microbes and prevent spread. At the same time the host develops a strong cellular immune response to the microbe. A complex lipopolysaccharide called wax D is the mycobacterial component contained by these bacteria that is often implicated in activation of cellular immunity (Hamamoto et al., 1981). Tubercles occasionally rupture, thus exposing cellular antigens to the host's highly active immune system. The subsequent inflammatory response causes significant tissue destruction (Dannenberg, 1989). Release of tumor necrosis factor from interferon-activated macrophages is one factor that has been associated with tissue destruction in tuberculosis (Rook et al., 1987). Although this disease is not currently of great concern in most developed countries, it still is the leading cause of mortality from infectious disease in the world with over 2 - 3 million deaths per year (WHO, 1992; Murray and Lopez, 1997).

Virtually all microbes are nonpathogenic. However, exposure to nonpathogenic microbes may not be as innocuous as sometimes thought. In a study performed in our laboratories we found that daily low level pulmonary challenge (≈ 1 x 10^3 viable bacteria) with nonpathogenic microbes (once per day for 5 d) enhanced susceptibility to concurrent infection with the pathogenic microbe, *Streptococcus zooepidemicus* (Tables 6 and 7). Daily pulmonary challenge with higher levels of nonpathogenic microbes (≈ 1 x 10^6 viable bacteria) was protective to concurrent *Streptococcus* infection even when challenged with a higher infectious dose (68% mortality vs. 15% in controls) (Table 8).

TABLE 6. EFFECT OF INTRANASAL CHALLENGE OF MICROBES ON *STREPTOCOCCUS* MORTALITY IN CD-1 MICE

Group[a]	Daily Dose[b]	# of Mice	Percent Mortality	MST[c]	MTD[d]
Saline	0	20	15	13.3	10.0
Vegetative *B. thuringiensis* var. kurstaki HD31	1.60×10^3	19	32	11.5	7.0*
	1.60×10^6	20	0	14.0	-
B. thuringiensis var. *kurstaki* HD31 Spores	2.31×10^3	20	75*	7.9*	6.9*
	2.33×10^6	5	0	14.0	-

[a]Mice challenged i.n. daily on Days –2 to 2 relative to infectious challenge on Day 0 with *Streptococcus*
[b]Viable colony forming units per animal
[c]Mean survival time (days)
[d]Mean time to death (days)
*p<0.05 vs saline control; chi-square (% mortality) and Dunnett's test (MST and MTD)

TABLE 7. EFFECT OF INTRANASAL CHALLENGE OF MICROBES ON *STREPTOCOCCUS* MORTALITY IN CD-1 MICE

Group[a]	Daily Dose[b]	# of Mice	Percent Mortality	MST[c]	MTD[d]
Saline	0	21	5	14.0	14.0
Pseudomonas aeruginosa AC869	1.40×10^2	17	41*	12.1	9.7
	1.40×10^5	16	56*	7.3*	3.0*
Burkholderia cepacia AC1100	2.23×10^3	20	55*	10.2	8.0*
	2.21×10^6	20	0	14.0	-

[a]Mice challenged i.n. daily on Days –2 to 2 relative to infectious challenge on Day 0 with *Streptococcus*
[b]Viable colony forming units per animal
[c]Mean survival time (days)
[d]Mean time to death (days)
*p<0.05 vs. saline control; chi-square (% mortality) and Dunnett's test (MST and MTD)

TABLE 8. EFFECT OF INTRANASAL CHALLENGE OF MICROBES ON *STREPTOCOCCUS* MORTALITY IN CD-1 MICE

Group[a]	Daily Dose[b]	# of Mice	Percent Mortality	MST[c]	MTD[d]
Saline	0	19	68	9.4	8.0
Vegetative *B. thuringiensis* var.	1.40×10^2	18	44	10.5	7.0
kurstaki HD31	1.40×10^5	16	0 *	14.0*	-*
Burkholderia cepacia AC1100	2.23×10^3	20	55	9.4	6.5
	2.21×10^6	20	0*	14.0*	-*

[a] Mice challenged i.n. daily on Days −2 to 2 relative to infectious challenge on Day 0 with *Streptococcus*

[b] Viable colony forming units per animal

[c] Mean survival time (days)

[d] Mean time to death (days)

*p<0.05 vs. saline control; chi-square (% mortality) and Dunnett's test (MST and MTD)

Daily pulmonary exposure to low levels of nonpathogenic microbes caused no change in numbers or types of cells isolated by bronchopulmonary lavage (data not shown). However, daily exposure to higher levels of nonpathogenic microbes caused significant increases in numbers of neutrophils. Similar effects have been reported with *Hartmanella vermiformis* enhancing disease caused by *Legionella pneumophila* (Brieland et al., 1996). This effect does not appear to be consistent for all infectious microbes or even for all routes of exposure as daily intraperitoneal challenge with nonpathogenic microbes had no effect on concurrent intraperitoneal infectious challenge with *Listeria monocytogenes* (Table 9). These data suggest that repetitive pulmonary challenge with levels of microbes that fail to elicit an immune response may distract the normal functioning of the pulmonary immune system and render the host more susceptible to infectious disease. Thus, the increasing use of microbes may have a detrimental effect on health if the low-level effect on the immune response is not considered.

Pulmonary exposure to microbes can cause disease through infectious and immunologic mechanisms. Modern technology is creating many new opportunities for significant human pulmonary exposure to microbes and microbial products. Because of the possible adverse consequences in some people, efforts need to be increased to better understand immunologic disease mechanisms and methods for alleviating immunologic disease caused by inadvertent exposure to microbes or microbial products.

TABLE 9. EFFECT OF INTRAPERITONEAL CHALLENGE OF MICROBES ON *LISTERIA* MORTALITY IN $B_6C_3F_1$ MICE

Group[a]	Daily Dose[b]	# of Mice	Percent Mortality	MST[c]	MTD d
Saline	0	20	10	9.6	7.0
Vegetative *Bacillus thuringiensis* var. kurstaki HD31	1.60×10^3	20	0	10.0	-
	1.60×10^6	20	10	9.6	6.5
Bacillus thuringiensis var. *kurstaki* HD31 spores	2.31×10^3	20	5	9.8	7.0
	2.33×10^6	20	75*	5.4*	4.9*
Pseudomonas aeruginosa AC869	1.40×10^2	20	0	10.0	-
	1.40×10^5	20	10	9.7	7.5
Burkholderia cepacia AC1100	2.23×10^3	20	0	10.0	-
	2.21×10^6	20	0	10.0	-
Beauveria bassiana 48023	2.75×10^3	20	0	10.0	-
	2.75×10^3	20	0	10.0	-

[a] Mice challenged i.n. daily on Days –2 to 2 relative to infectious challenge on Day 0 with *Listeria*
[b] Viable colony forming units per animal
[c] Mean survival time (days)
[d] Mean time to death (days)
*p<0.05 vs. saline control; chi-square (% mortality) and Dunnett's test (MST and MTD)

REFERENCES

Alvarez-Fernandez JA, Quirce S, Calleja JL, Cuevas M, Losada E. Hypersensitivity pneumonitis due to an ultrasonic humidifier. Allergy, 1998;53:210-212.

Anderson K, McSharry CP, Clark C, Clark CJ, Barclay GR, Morris GP. Sump bay fever: Inhalational fever associated with a biologically-contaminated water aerosol. Occup. Environ. Med., 1996;53:106-111.

Ando M, Suga M. Hypersensitivity pneumonitis. Curr. Opinion Pulm. Med., 1997;3:391-395.

Andrews, Jr. R.E., Spence KD, Miller LK. Virulence of cloned variants of *Autographica californica* nuclear polyhedrosis virus. Appl. Environ. Microbiol., 1980;39:932-933.

Armstrong JL, Rohrmann GF, Beaudreau GS. Delta endotoxin of *Bacillus thuringiensis* subsp. *israelensis*. J. Bacteriol., 1985;161:39-46.

Bae HS, Lee JM, Kim YB, Lee ST. Biodegradation of the mixtures of 4-chlorophenol and phenol by *Comamonas testosteroni* CPW301. Biodegradation, 1996;7:463-469, 1996.

Barr DP, Aust SD. Pollutant degradation by white rot fungi. Rev. Environ. Contam. Toxicol., 1994;138:49-72.

Bellanti JA. "Chapter 15. Immunologically-Mediated Diseases." In *Immunology*, NJ Zvaifler, ed. Philadelphia: W.B. Saunders Company, 1971, pp. 341-431.

Bernstein DI, Lummus ZL, Santilli G, Siskosky J, Bernstein IL. Machine operator's lung. A hypersensitivity pneumonitis disorder associated with exposure to metalworking fluid aerosols. Chest, 1995;108:636-641.

Bidochka MJ, Walsh SR, Ramos ME, St. Leger RJ, Silver JC, Robert DW. Fate of biological control introductions: Monitoring an Australian fungal pathogen of grasshoppers in North America. Proc. Natl. Acad. Sci., USA, 1996;93:918-921.

Brieland J, McClain M, Heath L, Chrisp C, Huffnagle G, LeGendre M, Hurley M, Fantone J, Engleberg C. Coinoculation with *Hartmannella vermiformis* enhances replicative *Legionella pneumophila* lung infection in a murine model of Legionnaire's disease. Infect. Immun., 1996;64:2449-2456.

Brown JE, Masoud D, Couser JI, Patterson R. Hypersensitivity pneumonitis from residential composting: Residential composter's lung. Ann. Allergy Asthma Immunol., 1995;74:45-47.

Brubaker GR, Exner JH. Bioremediation of chemical spills. Basic Life Sci., 1988;45:163-171.

Burgner D, Eagles G, Burgess M, Procopis P, Rogers M, Muir D, Pritchard R, Hocking A, Priest M. Disseminated invasive infection due to *Metarrhizium anisopliae* in an immunocompromised child. J. Clin. Micro., 1998;36:1146-1150.

Burrell R, Ye SH. Toxic risks from inhalation of bacterial endotoxin. Br. J. Ind. Med., 1990;47:688-691.

Castro ME, Souza ML, Aruajo S, Bilimoria SL. Replication of *Anticarsia gemmatilis* nuclear polyhedrosis virus in four lepidopteran cell lines. J. Invert. Pathol., 1997;69:40-45.

Cesario MT, Brandsma JB, Boon MA, Tramper J, Beeftink HH. Ethene removal from gas by recycling a water-immiscible solvent through a packed absorber and a bioreactor. J. Biotechnol., 1998;62:105-18.

Chauhan B, Santiago L, Kirschmann DA, Hauptfeld V, Knutsen AP, P.S. Hutcheson PS, S.L. Woulfe SL, Slavin RG, Schwartz HJ, Bellone CJ. The association of HLA-DR alleles and T-cell activation with allergic bronchopulmonary aspergillosis. J. Immunol., 1997;159:4072-4076.

Clapp WD, Thorne PS, Frees KL, Zhang X, Lux CR, Schwartz DA. The effects of inhalation of grain dust extract and endotoxin on upper and lower airways. Chest, 1993;104:825-830.

Cross S. Mould spores: The unusual suspects in hay fever. Community Nurse, 1997;3:25-26.

Damaj M, Ahmad D. Biodegradation of polychlorinated biphenyls by rhizobia: A novel finding. Biochem. Biophys. Res. Commun., 1996;218:908-915.

Damgaard PH, Larsen HD, Hansen BM, Bresciani J, Jorgensen K. Enterotoxin-producing strains of *Bacillus thuringiensis* isolated from food. Lett. Appl. Microbiol., 1996;23:146-150.

Dammann Kalinowski T, Niemann S, Keller M, Selbitschka W, Tebbe CC, Puhler A. Characterization of two bioluminescent *Rhizobium melitoti* strains constructed for field releases. Appl. Microbiol. Biotechnol., 1996;45:509-512.

Dannenberg, AM Jr. Immune mechanisms in the pathogenesis of pulmonary tuberculosis. Rev. Infect. Dis., 1989; Suppl 2:S369-S378.

de Garcia MC, Arboleda ML, Barraquer F, Grose E. Fungal keratitis caused by *Metarhizium anisopliae* var. *anisopliae*. J. Med. Vet. Mycol., 1997;35:361-363.

Duchaine C, Israel Assayag E, Fournier M, Cormie Y. Proinflammatory effect of *Pediococcus pentosaceus*, a bacterium used as hay preservative. Eur. Respir. J., 1996;9:2508-2512.

Fannin KF, Spendlove JC, Cochran KW, Gannon JJ. Airborne coliphages from wastewater treatment facilities. Appl. Environ. Microbiol., 1976;31:705-710.

Feng Y, Racke KD, Bollag JM. Use of immobilized bacteria to treat industrial wastewater containing a chlorinated pyridinol. Appl. Microbiol. Biotechnol., 1997;47:73-77.

Flores A, Chet I, Herrara Estrella A. Improved biocontrol activity of *Trichoderma harzianum* by over-expression of the proteinase-encoding gene prb1. Curr. Genet., 1997;31:30-37.

George SE, Kohan MJ, Whitehouse DA, Creason JP, Kawanishi CY, Sherwood RL, Claxton LD. Distribution, clearance, and mortality of environmental pseudomonads in mice upon intranasal exposure. Appl. Environ. Microbiol., 1991;57:2420-2425.

Goff GD, Spendlove JC, Adams AP, Nicholes PS. Emission of microbial aerosols from sewage treatment plants that use trickling filters. Health Serv. Rep., 1973;88:640-652.

Goldstein E, Lippert W, Warshauer D. Pulmonary alveolar macrophage. Defender against bacterial infection of the lung. J. Clin. Invest., 1974;54:519-528.

Goodnow RA, Katz G, Haines DC, Terrill JB. Subacute inhalation toxicity study of an ice-nucleation-active *Pseudomonas syringae* administered as a respirable aerosol to rats. Toxicol. Lett., 1990;54:157-167.

Govan JR, Hughes JE, Vandamme P. *Burkholderia cepacia*: Medical, taxonomic and ecological issues. J. Med. Microbiol., 1996;45:395-407.

Greenberger PA. Immunologic aspects of lung diseases and cystic fibrosis. J. Am. Med. Assoc., 1997;10:1924-1930.

Grondona I, Hermosa R, Tejada M, Gomis MD, Mateos PF, Bridge PD, Monte E, Garcia Acha I. Physiological and biochemical characterization of *Trichoderma harzianum*, a biological control agent against soilborne fungal plant pathogens. Appl. Environ. Microbiol., 1997;63:3189-3198.

Gudmundsson G, Monick MM, Hunninghake GW. IL-12 modulates expression of hypersensitivity pneumonitis. J. Immunol., 1998;161:991-999.

Guerin WF, Boyd SA. Maintenance and induction of naphthalene degradation activity in *Pseudomonas putida* and an *Alcaligenes* sp. under different culture conditions. Appl. Environ. Microbiol., 1995;61:4061-4068.

Hamammoto, Y., Kobara Y, Kojima A, Kumazawa Y, Yahuhira K. Experimental production of pulmonary granulomas. I. Immune granulomas induced by chemically modified cell walls and their constituents. Br. J. Exp. Pathol., 1981;62:259-269.

Hickey JL, Reist PC. Health significance of airborne microorganisms from wastewater treatment processes. Part I: Summary of investigations. J. Water Pollut. Control Fed., 1975;47:2741-2757.

Hinojosa M, Fraj J, de la Hoz B, Alcazar R, Sueiro A. Hypersensitivity pneumonitis in workers exposed to esparto grass (*Stipa tenacissima*) fibers. J. Allergy Clin. Immunol., 1996;98:985-991.

Hogan MB, Patterson R, Pore RS, Corder WT, Wilson NW. Basement shower hypersensitivity pneumonitis secondary to *Epicoccum nigrum*. Chest, 1996;110:854-856.

Kreiss K, Cox-Ganser J. Metalworking fluid-associated hypersensitivity pneumonitis: A workshop summary. Am. J. Ind. Med., 1997;32:423-432.

Kriebel D, Sama SR, Woskie S, Christiani DC, Eisen EA, Hammond SK, Milton DK, Smith M, Virji MA. A field investigation of the acute respiratory effects of metal working fluids. I. Effects of aerosol exposures. Am. J. Ind. Med., 1997;31:756-766.

Kwaasi AA, Parhar RS, al-Mohanna FA, Harfi HA, Collison KS, al-Sedairy ST. Aeroallergens and viable microbes in sandstorm dust. Potential triggers of allergic and non-allergic respiratory ailments. Allergy, 1998;53:255-265.

Laville J, Blumer C, von Schroetter C, Gaia V, Defaga G, Keel C, Haas D. Characterization of the hcnABC gene cluster encoding hydrogen cyanide synthase and anaerobic regulation by ANR in the strictly aerobic biocontrol agent *Pseudomonas fluorescens* CHA0. J. Bacteriol., 1998;180:3187-3196, 1998.

Madan T, Kishore U, Shah A, Eggleton P, Strong P, Wang JY, Aggrawal SS, Sarma PU, Reid KB. Lung surfactant proteins A and D can inhibit specific IgE binding to the allergens of *Aspergillus fumigatus* and block allergen-induced histamine release from human basophils. Clin. Exp. Immunol., 1997;110:241-249.

Martinez Moragon E, Menendez R, Santos M, Lorente R, Marco V. Lung diseases due to opportunistic environmental *Mycobacteria* in patients uninfected with human immunodeficiency virus. Risk factors, clinical and diagnostic aspects and course. Arch. Bronchoneumol., 1996;32:170-175.

Mayes ME, Held GA, Lau C, Seely JC, Roe RM, Dauterman WC, Kawanishi CY. Characterization of the mammalian toxicity of the crystal polypeptides of *Bacillus thuringiensis* subsp. *israelensis*. Fundam. Appl. Toxicol., 1989;13:310-322.

McClintock JT, Guzo D, Guthrie KP, Dougherty EM. DNA-Binding proteins of baculovirus-infected cells during permissive and semipermissive replication. Virus Res., 1991;20:133-145.

McClure CD, Schiller NL. Inhibition of macrophage phagocytosis by *Pseudomonas aeruginosa* rhamnolipids *in vitro* and *in vivo*. Curr. Microbiol., 1996;33:109-117.

Michel O, Ginanni R, Le Bon B, Content J, Duchateau J, Sergysels R. Inflammatory response to acute inhalation of endotoxin in asthmatic patients. Am. Rev. Respir. Dis., 1992;146:352-357.

Moreno-Ancillo A, Padial MA, Lopez-Serrano MC, Granado S. Hypersensitivity pneumonitis due to inhalation of fungi-contaminated esparto dust in a plaster worker. Allergy Asthma Proc., 1997;18:355-357.

Mundt C, Becker WM, Schlaak M. Farmer's lung: Patients IgG$_2$ antibodies specifically recognize *Saccharopolyspora rectivirgula* proteins and carbohydrate structures. J. Allergy Clin. Immunol., 1996;98:441-450.

Murray CJ, Lopez AD. Mortality by cause for eight regions of the world: Global burden of disease study. Lancet, 1997;349:1269-1276.

Nakagawa-Yoshida K, Ando M, Etches RI, and Dosman JA. Fatal cases of farmer's lung in a Canadian family. Probable new antigens, *Penicillium brevicompactum* and *P. olivicolor*. Chest, 1997;111:245-248.

Onofrio JM, Shulkin AN, Heidbrink PJ, Toews GB, Pierce AK. Pulmonary clearance and phagocytic cell response to normal pharyngeal flora. Am. Rev. Respir. Dis., 1981;123:222-225.

Onofrio JM, Toews GB, Lipscomb MF, Pierce AK. Granulocyte-alveolar-macrophage interaction in the pulmonary clearance of *Staphylococcus aureus*. Am. Rev. Respir. Dis., 1983;127:335-341.

Poulsen OM, Breum NO, Ebbeh JN, Hansen AM, Ivens UI, van Lelieveld D, Malmros P, Matthiasen L, Nielsen BH, Nielsen EM. Sorting and recycling of domestic waste. Review of occupational health problems and their possible causes. Sci. Total Environ., 1995a:168:33-56.

Poulsen OM, Breum NO, Ebbeh JN, Hansen AM, Ivens UI, van Lelieveld D, Malmros P, Matthiasen L, Nielsen BH, Nielsen EM. Collection of domestic waste. Review of occupational health problems and their possible causes. Sci. Total Environ., 1995b:170:1-19.

Quie PG. Lung defense against infection. J. Pediatr., 1986;108:813-816.

Rautiala S, Reponen T, Hyvarinen A, Nevalainen A, Husman T, Vehvilainen A, Kalliokoski P. Exposure to airborne microbes during the repair of moldy buildings. Am. Ind. Hyg. Assoc. J., 1996;57:279-284.

Reijula KE. Two bacteria causing farmer's lung: Fine structure of *Thermoactinomyces vulgaris* and *Saccharopolyspora rectivirgula*. Mycopathologia, 1993;121:143-147.

Rikkerink EH, Solon SL, Crowhurst RN, Templeton MD. Integration of *vectors by homologous recombination in the plant pathogen Glomerella* cingulata. Curr. Genet., 1994;25:202-208.

Robins T, Seixas N, Franzblau A, Abrams L, Minick S, Burge H, Schork MA. Acute respiratory effects on workers exposed to metalworking fluid aerosols in an automotive transmission plant. Am. J. Ind. Med., 1997;31:510-524.

Ronchel MC, Ramos C, Jensen LB, Molin S, Ramos JL. Construction and behavior of biologically-contained bacteria for environmental applications in bioremediation. Appl. Environ. Microbiol., 1995;61:2990-2994.

Rook GA, Taverne J, Leveton C, Steele J. The role of gamma-interferon, vitamin D_3 metabolites, and tumor necrosis factor in the pathogenesis of tuberculosis. Immunology, 1987;62:229-234.

Saini SK, Boas SR, Jerath A, Roberts M, Greenberger PA. Allergic bronchopulmonary mycosis to *Fusarium vasinfectum* in a child. Ann. Allergy Asthma Immunol., 1998;80:377-380.

Salyers AA, Whitt DD. "Chapter 3. Virulence Factors that Promote Colonization." In *Bacterial Pathogeneis, A Molecular Approach*, DD Whitt, ed. Washington, DC: ASM Press, 1994, pp. 30-46.

Sherwood RL, Tarkington B, Lippert WE, Goldstein E. Effects of ferrous sulfate aerosols and nitrogen dioxide on murine pulmonary defense. Arch. Environ. Health, 1981;36:130-135.

Sherwood RL, Thomas PT, Kawanishi CY, Fenters JD. Comparison of *Streptococcus zooepidemicus* and influenza virus pathogenicity in mice by three pulmonary exposure routes. Appl. Environ. Microbiol., 1988;54:1744-1751.

Siegel JP, Shadduck JA, Szabo J. Safety of the entomopathogen *Bacillus thuringiensis* var. *israelensis* for mammals. J. Econ. Entomol., 1987;80:717-723.

Silo Suh LA, Lethbridge BJ, Raffel SJ, He H, Clardy J, Handelsman J. Biological activities of two fungistatic antibiotics produced by *Bacillus cereus* UW85. Appl. Environ. Microbiol., 1994;60:2023-2030.

Simpson SQ, Casey LC. Role of tumor necrosis factor in sepsis and acute lung injury. Crit. Care Clin., 1989;5:27-47.

Smith TH. Toxicological and microbiological aspects of cutting fluid preservatives. IMS Ind. Med. Surg., 1970;39:56-64.

Teltsch B, Katzenelson E. Airborne enteric bacteria and viruses from spray irrigation with wastewater. Appl. Environ. Microbiol., 1978;35:290-296.

Walker R, Emslie KA, Allan EJ. Bioassay methods for the detection of antifungal activity by *Pseudomonas antimicrobica* against the grey mould pathogen *Botrytis cinerea*. J. Appl. Bacteriol., 1996;81:531-537.

WHO: World Health Organization. Tuberculosis control and research strategies for the 1990s: Memorandum from a WHO meeting. Bull. World Health Org., 1992;70:17-21.

Wolber PK. Bacterial ice nucleation. Adv. Microb. Physiol., 1993;34:203-207.

Weisse AB. A plague in Philadelphia. The story of Legionnaire's disease. Hosp. Prac., 1992;27:151-4, 157, 161-168.

Wraight SP, Carruthers RI, Bradley CA, Jaronski ST, Lacey LA, Wood P, Galaini Wraight S. Pathogenicity of the entomopathogenic fungi *Paecilomyces* sp. and *Beauveria bassiana* against the silverleaf whitefly, *Bemisia argentifolii*. J. Invert. Pathol., 1998;71:217-226.

Zhaoming W, Lockey RF. A review of allergic bronchopulmonary aspergillosis. J. Invest. Allergy Clin. Immunol., 1996;6:144-151.

Zhiping W, Malmberg P, Larsson BM, Larsson K, Larsson L, Saraf A. Exposure to bacteria in swine-house dust and acute inflammatory reactions in humans. Am. J. Respir. Crit. Care Med., 1996;154:1261-1266.

Zhukov VG, Rogozhin IS, Ushakova NA, Zaguitina NA, Popov VO, Bezborodov AM. Development of microbiological technology of air deodoration in laboratory-industrial conditions using a pilot plant. Prikl. Biokhim Mikrobiol., 1998;34:370-6.

8. PESTICIDES, SOLVENTS, AND POLYCYCLIC AGENTS

Kathleen E. Rodgers, Ph.D.
University of Southern California
Livingston Research Institute
1321 North Mission Road
Los Angeles, CA 90033

INTRODUCTION

Compared with the literature that has assessed the immunotoxic potential of a variety of pesticides, solvents and polyaromatic hydrocarbons by oral, intraperitoneal, and dermal routes, very little has been done to assess the effect of the majority of substances from these chemical classes on the immune system after exposure via inhalation. With the exception of benzene, only one citation was found that discussed the immunotoxicity of the chemical under consideration by inhalation. Further, the effect of the chemical on the pulmonary immune defense system has not been assessed with any of these molecules based upon published literature. This is surprising given that occupational (and therefore the highest long-term exposures) exposures to these chemicals would probably occur via inhalation. Further, from a pharmacokinetic standpoint, the blood supply, surface area, membrane thickness of the lung and lack of first pass metabolism by the liver would maximize systemic exposure due to rapid uptake. For example, Hwang and Schanker (1974) found that the pulmonary absorption of carbaryl was 2.5 times faster than intestinal absorption. With oral administration, however, the liver is initially exposed to a larger dose of carbaryl (i.e., the first-pass effect) compared with inhalation administration.

In one study discussed below where inhalation, oral and dermal routes were compared (Ladics et al., 1996), an effect on the immune system was observed only after exposure via inhalation. This emphasizes the need for assessment of this route of exposure as part of the determination of the immunotoxic potential of a compound where human exposure by this route is likely (e.g. solvents that would have a high vapor pressure under normal conditions).

PESTICIDES

The major route of exposure of the general population to pesticides is through residues in water and food (in particular fruits). For example, the carbamate pesticides carbofuran and aldicarb have been detected in groundwater throughout the United States at 1 - 50 ppb (Cohen et al., 1986). Exposure to pesticides, however, can also occur in a wide variety of occupational settings, ranging from manufacturing plants and formulation facilities to agricultural settings. Pesticides present a hazard to agricultural workers in three forms: as concentrates, as diluted sprays or dusts, or as residues in the environment (Popendorf and Leffingwell, 1982; Turnball, 1985; Hayes, 1991). Although the dermal route represents the major source of exposure, exposure to pesticide dusts and sprays via inhalation also represents a potential route of exposure. Seiber et al. (1980) found that the respiratory exposure of sprayers could be up to milligrams per cubic centimeter concentrations which can diminish to airborne concentrations in the micrograms per cubic centimeter range above treated fields or in adjacent areas. The potential exposures of carbaryl formulators and applicators has been calculated to be 74 and 49 mg/hr, respectively, by the dermal route and 1 mg/hr by inhalation (Comer et al., 1975).

These compounds are widely used occupationally and domestically via spray making exposure highly likely via inhalation of droplets, if not vapors, of pesticides. These compounds are, by nature and intent, biocidal with the safety of non-target organisms relied upon due to sensitivities in the target population (metabolism differences, uptake) that is not present in other species. Substantial work has been done on a few pesticides after exposure via oral and intraperitoneal routes. However, only two literature references could be uncovered which examined these compounds after inhalation exposure; these will be discussed in detail below.

Carbaryl

Carbaryl is a carbamate pesticide that is widely used occupationally for control of pests. In a recent study, the effect of carbaryl on the murine immune system was assessed after exposure via inhalation, ingestion and dermal contact (Ladics et al., 1994). In the inhalation study, the animals received 36, 137 or 335 mg/m^3 for 6 hr/d, 5 d/wk for 2 wk. Immune function was assessed by measuring the response to intravenous injection with sheep red blood cells (SRBC), a T-lymphocyte-dependent antigen, given 4 d prior to necropsy. At the time of necropsy, several parameters were assessed. These included splenic weight and cell number, thymic weight, red blood cell number, white blood cell number, the number of antibody forming units to SRBC in the spleen and the antibody titer (IgM) to SRBC in the sera of the animals. After inhalation, there was a dose-dependent decrease in thymic weight, splenic cell number, the number of antibody-forming cells per spleen and per 10^6 splenic cells, and serum IgM to SRBC. This suppression of the humoral immune response was observed only after inhalation of carbaryl and not after dermal or oral administration. This report emphasizes the need to assess the effect exposure via this route in animal studies. Although studies have not been conducted to assess the mechanism of this systemic immune suppression after pulmonary exposure, carbaryl is biocidal due to its ability to inhibit acetyl cholinesterase (a serine esterase). Therefore, it is possible that carbaryl inhibits immune function by inhibition of an esterase important in the generation of an immune response. Other studies based upon oral administration have shown that administration of pesticides of the class of anticholinesterases (as carbaryl is one), immune suppression has been most often observed after stress due to neurotoxicity. However, as emphasized above, sufficient study has not been done via this route of exposure to generalize that the stress of neurotoxicity contributes to immune suppression in this model.

Aminocarb

The only other pesticide assessed after exposure via inhalation was aminocarb (Bernier et al., 1995). Again, in this study, the effect of aminocarb on the immune system after exposure via several routes (oral, dermal, intraperitoneal and inhalation) was compared. However, only the effect on the generation of a humoral immune response was assessed. After oral or dermal exposure to very low concentrations (1/256 of the LD_{50}), the ability of the animal to generate a humoral immune response was elevated. This is consistent with other observations where oral administration of a very low concentration of an anticholinesterase pesticide, malathion, elevated the ability of splenocytes from exposed animals to generate a humoral immune response to SRBC (Rodgers et al., 1986, 1996; Rodgers and Ellefson, 1992). In contrast, intraperitoneal administration of aminocarb suppressed the generation of a humoral immune response. Finally, in contrast to the data presented above for carbaryl, inhalation of aminocarb, another carbamate insecticide that acts to inhibit acetyl cholinesterase, had no effect on the generation of a humoral immune response. The reason for the observed difference between these two structurally related compounds is not evident at this time. However, there may be a difference between the dosing schedule or level of exposure that would explain this dichotomy. Further, these studies emphasize the need to study the effect of compounds on the immune system via routes relevant to human exposure. That is, the only risk of immune suppression observed, in this study, was after exposure via intraperitoneal injection which is irrelevant to human exposure.

POLYAROMATIC HYDROCARBONS

A multitude of studies have examined the effect of polyaromatic hydrocarbons given via oral gavage or intraperitoneal injection. These have been extensively reviewed elsewhere and are beyond the scope of this chapter. For the most part, these studies have shown what exposure to polyaromatic hydrocarbons can lead to immunosuppression, reduced host resistance and alterations in lymphoid subpopulations. As with pesticides, there has been a paucity of studies of the effect of inhalation of polyaromatic hydrocarbons on immune function.

Organic matter can be found at the core or layered on the surface of urban particulate matter including smog and diesel exhaust particles (DEP). Estimates of the carbonaceous content of the particles vary considerably but are nominally considered to be about 50 - 60% of the total mass of the fine particulate material. The sources vary and include natural smoke (e.g., forest fires), stationary-source combustion products (e.g., fly ash), and DEP matter. The diesel particle itself is largely carbon with small amount of various combustion-derived complex nitroaromatics, but is not particularly toxic when administered acutely, even by intratracheal instillation. The limited studies with coal fly ash, which typically has a considerable organic fraction, have shown that it can induce some inflammation, but much of the response appears to be related to the metal content of the dust. The same appears to be true for samples of urban dust, which when instilled into the lungs of rodents suppress host defenses against bacterial challenge (Hatch et al., 1985).

There has been a recent emphasis on the possibility that inhalation of DEP may alter immune responses. Initial studies showed that immunosuppression could be achieved after inhalation of DEP (Vostal, 1983). However, it was felt that the polyaromatic hydrocarbons coating the DEP were instrumental in achieving that immunosuppression as organic solvent stripping of the DEP rendered the particles non-immunosuppressive at high concentrations and that immunosuppression could be reinstated by inclusion of the organic solvent extracts by intratracheal administration in the test system.

Similar studies, which confirm this, have been carried out in human populations (Diaz-Sanchez et al., 1997; Tsien et al., 1997; Winker et al., 1997). In the study by Winker and colleagues (1997), the effect of polyaromatic hydrocarbon inhalation with particles was assessed in chronic, occupational exposure of coke oven workers. Two groups were compared in this study. The differentiating factor between the two groups was the level of emissions to which the individuals were exposed due to differences in the ovens used in the work place. In the workers exposed to the higher levels of emissions and thereby the higher levels of polyaromatic hydrocarbons with particulates, the mitogenic responses to phytohemagglutinin, the level of interleukin-2 (IL-2) receptor expression, antibody production after exposure to *Staphylococcus aureus in vitro* and the respiratory burst activity of peripheral blood mononuclear cells in response to *in vitro* exposure to *Escherichia coli* generated were suppressed. In contrast to this, work by Tsien and by Diaz-Sanchez et al. (1997) suggested that exposure to DEP may be an adjuvant in the express of a respiratory allergic response. In these studies, individuals allergic to ragweed allergen were exposed to either ragweed allergen alone or in combination with DEP. In this study, several parameters that indicate increased responsivity to the ragweed allergen was increased in persons exposed to the antigen in combination with DEP. These parameters include: increased antigen-specific (but not total) IgE, increased antigen-specific and total IgG4, decreased mRNA to T_H1 cytokines (γ-interferon and IL-2) and increased mRNA to T_H2 cytokines (IL-4, -5, -6, -10 and -13). Further, the contribution of the polyaromatic hydrocarbon portion of the DEP contributing to the observed increases in allergic potential was confirmed by Tsien et al. (1997). In this study, DEP were extracted with solvent and found to have phenanthrene on their surface. Exposure of allergic individuals to ragweed allergen in combination with solvent extract of DEP or phenanthrene increased IgE production *in vitro*. These studies suggest an immunotoxic effect, both suppressive and immunoenhancing (in conjunction with a known allergen) effect by polyaromatic hydrocarbons administered via inhalation in the presence of particulate matter.

SOLVENTS

Formaldehyde

Formaldehyde (HC[O]H) is a ubiquitous chemical to which exposure can occur through multiple sources including cigarette smoke, automotive exhaust, photochemical smog, incinerators, industrial processes, biomedical research, tissue processing in a histopathology laboratory, and degassing of urea-formaldehyde resinous products. Cigarette smoke contains as much as 40 ppm HC[O]H by volume (Kensler and Battista, 1963). Formaldehyde is also a normal metabolite in mammalian systems, and in small quantities is rapidly metabolized to formate through an initial reaction with glutathione to form through a hemiacetal CO_2 and H_2O (Akabane, 1970; Goodman and Tephly, 1971).

Concern about the safety of formaldehyde after inhalation exposure came from studies which showed 50% (103/206) incidence of squamous cell carcinoma in the nasal cavities of F-344 rats (14.3 ppm, 6 hr/d for 2 yrs), but only a 3.3% (2/60) incidence in $B_6C_3F_1$ mice (Swenberg et al., 1980; Kerns et al., 1983). Differences in nasal carcinogenicity appear to be due to reflex apnea in the mice thereby reducing the delivered formaldehyde dose per nasal surface area in the mouse (Chang et al., 1983). The carcinogenicity of formaldehyde was confirmed by the induction of squamous cell carcinomas in Sprague-Dawley rats exposed to 14.1 ppm HC[O]H (Albert et al., 1982). Swenberg et al. (1983) observed severe inflammation and acute cytotoxicity in the nasal turbinates of rats and mice exposed for only 6 hr to 15 ppm HC[O]H. The predominant response to this local irritation was acute inflammation with accompanying mononuclear cell infiltration followed by restorative cell proliferation and hyperplasia.

Only one citation thoroughly assessed the immunotoxicologic potential of formaldehyde administered via inhalation (Dean et al., 1984). In this study, 15 ppm (the maximum tolerated dose for mice) HC[O]H was administered for 6 hr/d for 21 d. Several immunological parameters were assessed including delayed type hypersensitivity, antibody-forming cells to T-dependent and -independent antigens, the weight of lymphoid organs, hematologic parameters (white blood cell number, differential counts, hemoglobin, hematocrit, etc.), the number of nucleated cells in the bone marrow, the number of colony-forming progenitors in the bone marrow, the lymphocyte subpopulations in lymphoid organs, proliferative responses to mitogen stimulation, macrophage activity, and host resistance to *Listeria monocytogenes* and tumor challenge. Although this extensive battery of parameters were assessed at the maximum tolerated dose, the only changes observed were the respiratory burst activity of resident macrophages, the bactericidal activity of macrophages and, not unexpectedly, the host resistance to challenge with *Listeria*. The acute inflammation and restorative process present in the nasal mucosa of rodents exposed to formaldehyde may provide the general stimulus for activation of the mononuclear phagocytes and the accompanying increased resistance to *Listeria* challenge. A second possibility is that formaldehyde produced a direct systemic effect on the mononuclear phagocyte system. These data suggest that the immunologic hazard associated with exposure to formaldehyde via inhalation is minimal in the absence of chemical irritation contributing to bronchospasm and asthma.

The only immunotoxic effects previously associated with formaldehyde exposure have been allergic contact dermatitis and skin sensitization (Pirila and Kilpio, 1949; Glass, 1961; Epstein and Maibach, 1966; Maibach, 1983 Arts et al., 1997) and occupational asthma (Hendrick and Lane, 1977; Hendrick et al., 1982). This is consistent with studies that showed that exposure to formalin vapors can increase bronchial spastic reaction and bronchial hypersensitivity in response to exposure to a respiratory allergy (Pindel et al., 1995). Allergic asthma was reported in a limited number of individuals exposed for prolonged periods to formaldehyde (Hendrick and Lane, 1977; Hendrick et al., 1982). The incidence of this syndrome appears rare and the hypersensitivity is partially reversible once the formaldehyde exposure is removed. There was a case report of urea-formaldehyde foam insulation (UFFI) dust producing severe asthma in one individual (Frigas et al., 1981). This was confirmed by bronchial challenge with UFFI, but dust challenged with formaldehyde had no effect. Recent studies by Dearman et al (1996) showed that dermal administration of formaldehyde as a contact sensitizer increased gamma interferon and IL-2 mRNA levels while exposure to toluene diamine, a respiratory sensitizer, increased the levels of IL-4 and IL-10 mRNA expression. These data suggest that respiratory and dermal sensitizers can be distinguished based upon the cytokines and induced by exposure to them.

Methyl Isocyanate

Methyl isocyanate (MIC) is recognized as an extremely hazardous chemical due to its volatility, flammability, and reactivity. The studies of Kimmerele and Eben (1964) indicated the irritating nature of MIC to skin and mucous membranes and emphasized the lung as a primary target organ with the LC_{50} in rats being 5 ppm following a single 4 hr exposure. That the respiratory system is the primary target for acute MIC exposure in rodents has been confirmed by several investigators (Nemery et al., 1985; Salmon et al., 1985; Ferguson et al., 1986). Irritation of the respiratory system, which results in coughing and nasal secretion has been reported in MIC-exposed workers; a second type of response is allergic in nature, occurs only after repeated exposure and can be manifested as mild hayfever or an asthmatic attack (Rye, 1973).

Additional toxicological evaluation of MIC which focused on its immunotoxic potential was prompted by a disaster in the mid-1980s in Bhopal, India, in which many people were exposed to MIC at concentrations up to and including lethal doses (Kamat et al., 1985). In this study, mice were exposed to 1, 3 or 6 ppm MIC for 6 hr/d for 4 d (Luster et al., 1986) and a battery of immunologic parameters were then evaluated. These included: the humoral immune response to SRBC; natural killer activity; mitogenic responses to concanavalin A, phytohemagglutinin and lipopolysaccharide; mixed lymphocyte reaction; and, host resistance to *L. monocytogenes*, malaria, influenza virus, or tumor. No affect was observed on a majority of these parameters; only a small decrease in proliferative responses (mitogenic and mixed lymphocyte reactions) was observed, but this was felt to be due to secondary to general toxicity due to MIC exposure.

Benzene

The development of the steel industry and its need for coke led to a readily available source of benzene as a byproduct in the petroleum industry to produce fuels resulted in an abundance of benzene, as both a fuel component and byproduct in the twentieth century. Benzene has been used as a solvent for rubber, inks and other materials and as a starting material in chemical synthesis. Today it remains one of the largest volume chemicals produced and used in the world.

The heavy industrial production and use of benzene has resulted in a plethora of reports relating to its toxic effects in humans and to studies of its toxicity in animals and other non-human biological systems (Snyder and Kocsis, 1975; Snyder et al., 1993). Benzene toxicity in the workplace is most frequently thought to be the result of the inhalation of benzene vapors with some undefined contribution from skin absorption. Current environmental concerns also include the potential effects of benzene found in samples of well water used for drinking.

Exposure to benzene, a widely used intermediate in the production of a variety of chemicals, has long been associated with hematopoietic disorders in animals and humans. Exposure induces a variety of blood dyscrasias, the most common form of hematotoxicity associated with benzene exposure is pancytopenia, with an associated reduction in bone marrow stem cells, as well as granulocyte/monocyte and erythroid precursor cells (Snyder and Kocsis, 1975; Goldstein, 1977; Green et al., 1981; Baarson et al., 1984). There is firm epidemiological evidence for an association between human exposure to benzene and leukemia (Laskin and Goldstein, 1977; Boyland, 1981; Sun 1982). Human occupational exposure to benzene is now regulated at a threshold limit value (TLV) of 10 ppm. However, the safety of this current limit is in dispute and attempts have been made to lower it (Snyder, 1984; Luken and Miller, 1981). It is generally believed that the agents ultimately responsible for benzene toxicity are the semiquinone and quinone metabolities of benzene which act on cycling hematopoietic precursor cells, interfering with normal differentiation and/or proliferation (Rushmore, et al., 1984).

Benzene can damage stem cells in bone marrow (Uyeki et al., 1977; Tunek et al., 1981) which ultimately induces leukopenia (Snyder and Kocsis, 1975; Goldstein, 1977). Studies in CD-1 mice exposed to 1.1 - 4862 ppm benzene for 6 hr/day for 5 d/wk, showed a depression in the number of nucleated cells in the bone marrow at 100 ppm; however, marrow GM-CFU were not depressed until the exposure level used was 302 ppm (Green et al., 1981). Exposure of mouse pups to benzene via inhalation by the dams during days 6 to 15 of gestation can also affect hematopoiesis (Keller and Snyder, 1986). If hematopoiesis was assessed in the fetus, exposure of the dam to 20 ppm benzene lead to an increase in CFU-E and BFU-E, indicating an effect on erythroid development. If an effect of *in utero* exposure to benzene (inhalation of 20 ppm benzene by the dam on day 6 to 15 of gestation) on the neonate was assessed, an increase in both CFU-E and GM-CFC was observed indicating an effect on both erythroid and myeloid lineages. From the immunological aspect of benzene toxicity, changes in immune reactions have been observed in animals exposed to relatively high doses of the vapor (Browning, 1952; Snyder and Kocsis, 1975). A more severe depression of leukocytes was found after a 7 d exposure at 200 ppm. These findings are similar to the observations by Green et al

(1981) that leukopenia and detectable changes in bone marrow are found in mice exposed to ≥103 ppm benzene 6 hr/day for 5 d. Therefore, it is estimated that hematopoietic disorders in bone marrow were induced after at least a 14 d exposure to 50 ppm benzene.

Numerous studies have also suggested that lymphocytes are particularly sensitive to benzene toxicity (Snyder et al., 1980; Pfeifer and Irons, 1982). The effects of benzene inhalation on T-lymphocytes are well documented, including induction of thymic involution (Baarson et al., 1982) and depression of the lympho-proliferative responses to a T-lymphocyte mitogen (Rozen et al., 1984). More recent studies have shown that this decrease in femoral, splenic and thymic lymphocyte number may be due to an increase in apoptosis (Farris et al., 1997). In this study, exposure of mice to 100 - 200 ppm benzene up to 8 wk both decreased lymphocyte number and increased apoptosis 6 - 15-fold in lymphocyte populations. This was associated with an increase in lymphoid replication in the bone marrow, which may explain the association of benzene with leukemia and blood dyscrasia. Microscopically, atrophy of white pulp and loss of small lymphocytes in lymphoid nodules are found in spleen after inhalation of benzene. Reduction or depletion of thymocytes from the cortex and medulla was found in thymus (Aoyama et al., 1982).

Previous studies on the mitogenic blastogenesis of lymphocytes exposed to benzene have demonstrated a significant difference between the mitogenic responses induced by phytohemagglutinin (PHA) and by Con A at 50 ppm benzene, and no difference between the two responses induced at 200 ppm (Aoyama and Matsushita, 1984). Further, exposure to 50 or 200 ppm benzene for 5 hr/d for 14 d suppressed the generation of a humoral immune response while having no effect or increasing the response to contact sensitizer (Aoyama, 1986). Several additional studies have been conducted on the effect of long term exposure to benzene via inhalation on immune function. Mice were exposed to 300 ppm benzene for 6 hr/d, 5 d/wk up to 23 wk (Rozen and Snyder, 1985). As with other studies, the number of B-lymphocytes in the bone marrow and spleen and T-lymphocytes in the thymus and spleen was decreased at all timepoints tested. However, the effect on immune function changed over time. There was a progressive decrease in the mitogenic response of lymphocytes in the bone marrow and the spleen observed in this study. Further, exposure of mice to lower concentrations (10 - 100 ppm) of benzene for 6 hr/d, 5 d/wk for 20 wk suppressed mixed-lymphocyte reactions but did not induce suppressor lymphocyte activity or change the percentage of B-, T-, or T-lymphocyte subsets in the spleen (Rosenthal and Snyder, 1987). Exposure to 100 ppm benzene for 20 wk did suppress host resistance to tumor challenge (PYB6 tumor) and the generation of a cytotoxic T-lymphocyte response. The effect of exposure to benzene on host resistance to bacterial infection has also been assessed (Rosenthal and Snyder, 1985). In this study, mice were exposed to benzene (10, 20,100, or 300 ppm, 6hr/day) either prior (prior) to or prior to and during (continuous) infectious challenge with *L. monocytogenes*. Exposure to 300 ppm benzene prior to infection increased the number of splenic bacteria 4 d after exposure. However, continuous exposure to benzene increased this parameter at concentrations of 30 ppm or greater. As before, exposure to 30 ppm or greater benzene decreased the total number of splenic lymphocytes, B-lymphocytes, and T-lymphocytes on days 1, 4 and 7 after infection with *Listeria*.

After human exposure to benzene, some studies have shown (in addition to the blood dyscrasias and leukemias described above) adverse effects immune function, including alterations of serum immunoglobulin and complement levels and development of autoantibodies (Cohen et al., 1978). Studies have also been conducted to assess immune function in workers exposed to benzene, toluene, and xylene (Moszczynski and Lisiewicz, 1984). The only effect noted in this study was decrease in the number of circulating T-lymphocytes. No effect was noted on the immune functions tested including mitogenic response to phytohemagglutinin, and delayed-type hypersensitivity responses to tuberculin (ppd) and distreptase. These data suggest that benzene has a high potential to modulate immune function after various routes of exposure, including inhalation.

Miscellaneous Solvents

Although the potential for inhalation of solvent vapors, given the vapor pressure of many of these organic molecules, is high, the vast majority of them have not received attention as agents that are potentially immunotoxic via inhalation. A variety of studies have been reported that suggest some immunotoxic potential for solvents of different classes.

Benzidine and β-Naphthylamine

In 1990, a study was reported with compared the natural killer activity of workers occupationally exposed to benzidine and β-naphthylamine. This study was conducted because of the concern due to increased bladder cancer incidence in workers. Natural killer activity was examined because of the theory that these cells contribute to tumor survellience and therefore inhibition of the activity of these cells may contribute to the progression of cancer. In these workers, bulk natural killer activity was not affected by exposure to these two solvents. However, an increase in the number of Leu 11+ cells was observed; therefore, the Authors suggested that the natural killer activity that each cell was expressing was suppressed as a result of exposure. Further, when workers diagnosed with bladder cancer were separated out from the group with exposure, but no bladder cancer, no difference was observed with these two populations. Given these caveats, the significance of the observations with regard to bladder cancer in these persons is unclear.

Ethylbenzene and 1,3,5-Trimethyl Benzene

Recently, ethylbenzene has become one of the chemical substances attracting an increased attention. The reason lies evidently in its large-scale synthesis as a raw material for manufacture of styrene. This polymer is suspected by some as being carcinogenic, through its intermediate phenylethylene oxide formed in the organism. Ethylbenzene is also present in some motor fuels and industrial solvents containing aromatic compounds. Its carcinogenic potential became a matter of concern not long ago in that epidemiologic studies suggested an association between occupational exposure to ethylbenzene and cancer. A study was undertaken which summarized the results of a 20 year biomonitoring effort to assess the magnitude of the effect of this production intermediate on the health of a group of ethylbenzene production workers in Czechoslovakia (Bardodej and Cirek, 1985). The level of exposure to ethylbenzene was followed by assessing the urinary levels of madelic acid, a metabolite, which did not exceed 3.25 mM at any time during the test period. In this study, no effect on the hematopoiesis of white blood cell number was observed. On the other hand, one study showed that exposure to high doses of 1,3,5-trimethylbenzene can cause leukopenia which may be secondary to other toxicaties.

2-Methoxyethanol and 2-Ethoxyethanol

Recent studies have shown that exposure to greater than 9213 mg/m^3 2-methoxyethanol or 1450 mg/m^3 2-ethoxyethanol for 90 d reduced the number of white blood cells and thymic lymphocytes (WHO, 1996). This was confirmed in another study in which exposure to 93 to 930 mg/m^3 2-methoxyethanol reduced the number of white blood cells in the rats and the thymic size in the rabbit.

Trichloroethane

A recent case report was published which suggested that exposure to trichloroethane, a solvent widely used occupationally as a degreaser, led to multiple symptoms, including rash, itching and an increase in atypical lymphocytes (Bond, 1996). These symptoms disappeared after cessation of exposure and recurred with diffuse rash and itching upon re-exposure. Although this is consistent with a few other case reports, the rarity of this reaction in the face of wide spread use makes an association difficult to discern.

Methyl Pyrrilidinone

A recent study showed that exposure to 1 mg/L methyl pyrrilidinone resulted in focal pneumonia, bone marrow hypoplasia, atrophy of the spleen and thymus (Nordic, 1992). This overall leukopenia was accompanied by an increase in polymorphonuclear neutrophil and a decrease in lymphocytes.

SUMMARY

The available studies assessing the effect of inhalation of pesticides, polyaromatic hydrocarbons and solvents on immune function are summarized. Numerous chemicals of these classes have been shown to modulate systemic immune function (i.e., cause immunosuppression or act as an adjuvant to enhance allergic responses). However, the paucity of experiments which assessed their immunotoxic potential after exposure via inhalation for many of these chemicals indicate that this field of experimentation is wide open for future studies.

REFERENCES

Akabane J. "Aldehydes and Related Compounds." In *International Encyclopedia of Pharmacological Therapy, Vol. 2*, Oxford, England: Pergamon Press, 1970, pp. 523-560.

Albert RE, Sellakumar AR, Laskin S, Kuschner M, Nelson N, Snyder CA. Gaseous formaldehyde and hydrogen chloride induction of nasal cancer in the rat. J. Natl. Cancer Inst., 1982;68:597-603.

Aoyama K, Matsushita T. Organic solvents and immune response: Benzene. Japan. J. Ind. Health, 1984;27:256-268.

Aoyama K, Yoshimi K, Matsushita T. The effect of benzene exposure on the immune response in mice. Japan. J. Ind. Health, 1982;24:769-776.

Baarson KA, Snyder CA, Green JD, Sellakumar A, Goldstein BD Albert RE. The hematotoxic effects of inhaled benzene on peripheral blood, bone marrow and spleen cells are increased by ingested ethanol. Toxicol. Appl. Pharmacol., 1982;64:393-440.

Baarson KA, Snyder, CA, and Albert RE. Repeated exposure of C57Bl mice to inhaled benzene at 10 ppm markedly depressed erythropoietic colony formation. Toxicol. Lett., 1984;20:337-342.

Bond GR. Hepatitis, rash and eosinophilia following trichloroethylene exposure: A case report and speculation on mechanistic similarity to halothane-induced hepatitis. J. Toxicol. Clin. Toxicol., 1996;34:461-466.

Boyland E. Role of benzene in carcinogenesis. IRCS Med. Sci. Libr. Compend., 1981;9:560-562.

Browning E. (ed.) *Toxicity of Industrial Organic Solvents. Revised Edition.* London: Her Majesty's Stationery Office, 1952, pp. 3-46.

Chang JC, Gross EA, Swenberg JA, Barrow CS. Nasal cavity deposition, histopathology and cell proliferation after single or repeated formaldehyde exposures in $B_6C_3F_1$ mice and F-344 rats. Toxicol. Appl. Pharmacol., 1983;68:161-167

Cohen HS, Freedman ML, Goldstein BD. Invited review: The problem of benzene in our environment: Clinical and molecular considerations. Am. J. Med. Sci., 1978;275:124-136.

Cohen SZ, Eiden C, Lorber MN. "Monitoring Groundwater for Pesticides." In *Evaluation of Pesticides in Groundwater*, WY Garner, RC Honeycutt, HN Nigg, eds. Washington, DC: American Chemical Society, 1986, pp. 170-196.

Comer WW, Staiff DC, Armstrong JF, Wolfe HR. Exposure of workers to carbaryl. Bull. Environ. Contam. Toxicol., 1975;13:285-391.

Cranmer MF. Carbaryl – A toxicological review and risk analysis. Neurotoxicology, 1986;10:247-332.

Dean JH, Lauer LD, House RV, Murray MJ, Stillman WS, Irons RD, Steinhagen WH, Phelps MC, Adams DO. Studies of immune function and host resistance in $B_6C_3F_1$ mice exposed to formaldehyde. Toxicol. Appl. Pharmacol., 1984;72:519-529.

Dearman RG, Moussavi A, Kemeny DM, Kimber I. Contribution of CD4+ and CD8+ T-lymphocyte subsets to the cytokine secretion patterns induced in mice during sensitization to contact and respiratory chemical allergens. Immunology, 1996;89:502-510.

Diaz-Sanchez D, Tsien A, Fleming J, Saxon A. Combined diesel exhaust particulate and ragweed allergen challenge markedly enhances human *in vivo* nasal ragweed-specific IgE and skews cytokine production to a T-helper cell 2-type pattern. J. Immunol., 1997;158:2406-2413.

Epstein X, Maibach H. Formaldehyde allergy: Incidence and patch test problems. Arch. Dermatol., 1966;94:186-190.

Ferguson JS, Schaper M, Stock MF, Weyel DA, Alare Y. Sensory and pulmonary irritation with exposure to methyl isocyanate. Toxicol. Appl. Pharmacol., 1986;82:329-335.

Frigas E, Filley WV, Reed CE. Asthma induced by dust from urea-formaldehyde foam insulating material. Chest, 1981;79:706-707.

Gatrell MJ, Craun JC, Podrebarac DS, Gunderson EL. Pesticides, selected elements, and other chemicals in adult total diet samples, October 1980-March 1982. J. Assoc. Off. Anal. Chem., 1986;69:146-161.

Glass WI. An outbreak of formaldehyde dermatitis. N. Zeal. Med. J., 1961;60:423-427.

Goldstein BD. *Benzene Toxicity, A Critical Evaluation*, S. Laskin, BD Goldstein, eds. New York: McGraw-Hill, 1977, pp. 69-105.

Goldstein BD. Hematotoxicity in humans. J. Toxicol. Environ. Health (Supplement) 1977;2:69-106.

Goodman JI, Tephly TR. A comparison of rat and human liver formaldehyde dehydrogenase. Biochim. Biophys. Acta, 1971;252:489-505.

Green JD, Snyder CA, Lobue J, Goldstein BD, Albert RE. Acute and chronic dose/response effects of benzene inhalation on the peripheral blood, bone marrow, and spleen cells of CD-1 male mice. Toxicol. Appl. Pharmacol., 1981;59:204-214.

Hayes WJ, Jr. "Production and Use of Pesticides." In *Handbook of Pesticide Toxicology*, WJ Hayes, Jr., ER Laws, Jr., eds. New York: Academic Press, 1991, pp. 22-45.

Hendrick DJ, Lane DJ. Occupational formalin asthma. Br. J. Ind. Med., 1977;34:11-19.

Hendrick DJ, Rando RJ, Lane DJ, Morris MJ. Formaldehyde asthma: Challenge exposure levels and fate after five years. J. Occup. Med., 1982;24:893-897.

Hwang SW, Schanker LS. Absorption of carbaryl from the lung and small intestine of the rat. Environ. Res., 1974;7:206-211.

Kamat SR, McHashur AA, Tiwari AK, Potdar PV, Gaur M, Kolhatkar VP, Vaidya P, Parmar D, Rupwate R, Chatterjee TS, Jain K, Kellar MD, Kinare SG. Early observation on pulmonary changes and clinical morbidity due to isocyanate gas leak at Bhopal. J. Postgrad. Med., 1985;31:63-92.

Kensler CJ, Battista SP. Components of cigarette smoke with ciliary-depressant activity: Their selective removal by filters containing activated charcoal granules. N. Engl. J. Med., 1963;269:1161-1166.

Kerns WD, Pavkov KL, Donofrio DJ, Gralla EJ, Swenger JA. Carcinogenicity of formaldehyde in rats and mice after long-term inhalation exposure. Cancer Res., 1983;43:4389-4392.

Kimmerele G, Eben A. Toxicity of methyl isocyanate and how to determine its quantity in air. Arch. Toxicol., 1964;20:235-241.

Laskin S, Goldstein BD, eds. Benzene toxicity: A critical evaluation. J. Toxicol. Environ. Health, 1977;Suppl. 2.

Luken RH, Miller SG. The benefits and costs of regulating benzene. J. Air Pollut. Control Assoc., 1981;31:1254-1259.

Maibach H. "Formaldehyde: Effects on Animal and Human Skin." In *Formaldehyde Toxicity*, JE Gibson, ed. New York: Hemisphere Publishing, 1983, pp 166-174.

Moszcynski P, Lisiewicz J. Occupational exposure to benzene, toluene, and xylene and the T-lymphocyte functions. Haematologica (Budap.), 1984;17:449-453.

Nemery B, Dinsdale D, Sparrow S, Ray DE. Effects of methyl isocyanate on the respiratory tract of rats. Br. J. Ind. Med., 1985;42:799-805.

Pfeifer RW, Irons RD. Effects of benzene metabolites on phytohemagglutinin-stimulated lymphopoiesis in rat bone marrow. J. Reticuloendothel. Soc., 1982;31:155-170.

Pindel B, Jarzab J, Sach K, Rogala E. Influence of bacterial infection and formalin vapors on the course of ovalbumin induced experimental bronchospastic reaction in guinea pigs. Pneumonol. Alergol. Pol. 1995;62:253-258.

Pirila V, Kilpio O. On dermatitis caused by formaldehyde and its compounds. Ann. Med. Intern. Fenn., 1949;38:38-51.

Popendorf WJ, Leffingwell JT. Regulating OP pesticide residues for farmworker protection. Res. Rev., 1982;82:1-200.

Rodgers KE, Ellefson DD. Mechanism of the modulation of murine peritoneal cell function and mast cell degranulation by low doses of malathion. Agents Action, 1992;35:57-63.

Rodgers KE, Leung N, Devens B, Ware CF, Imamura T. Lack of immunosuppressive effects of acute and subacute administration of malathion on murine cellular and humoral immune responses. Pesticide Biochem. Physiol., 1986;25:358-365.

Rodgers KE, St Amand K, Xiong S. Effects of malathion on the humoral immunity and macrophage function In mast cell-deficient mice. Fundam. Appl. Toxicol., 1996;31:252-258.

Rozen MG, Snyder CA, Albert RE. Depression in B-and T-lymphocyte mitogen induced blastogenesis in mice exposed to low concentrations of benzene. Toxicol. Lett., 1984;20:343-349.

Rushmore T, Snyder R, Kalf G. Covalent binding of benzene and its metabolities to DNA in rabbit bone marrow mitochondria *in vitro*. Chem.-Biol. Interact., 1984;49:133-154.

Rye WA. Human responses to isocyanate exposure. J. Occup. Med., 1973;15:306-307.

Salmon AG, Kerrmuir M, Andersson A. Acute toxicity of methylisocyante: A preliminary study of the dose response for eye and other effects. Br. J. Ind. Med., 1985;42:795-798.

Seiber JN, Ferreira GA, Hermann B, Woodrow JE. Analysis of pesticidal residues in the air near agricultural treatment sites. ACS Symp. Series, 1980;136:177-207.

Snyder CA, Goldstein BD, Sellakumar AR, Bromberg I, Laskin S, Albert RE. The inhalation toxicology of benzene. Incidence of hematopoietic neoplasms and hematotoxicity in AKR/J and C57B1/6 mice. Toxicol. Appl. Pharmacol., 1980;54:323-331.

Snyder R, Kocsis JJ. Current concepts of chronic benzene toxicity. CRC Crit. Rev. Toxicol., 1975;3:265-288.

Snyder R. The benzene problem in historical perspective. Fundam. Appl. Toxicol., 1984;4:692-699.

Sun M. Risk estimate vanishes from benzene report. Science, 1982;217:914-915

Swenberg JA, Kerns WD, Mitchell RI, Gralla EJ, Pavkov KL. Induction of squamous cell carcinomas of the rat nasal cavity by inhalation exposure to formaldehyde vapor. Cancer Res., 1980;40:3398-3402.

Tsien A, Diaz-Sanchez D, Ma J, Saxon A. The organic component of diesel exhaust particles and phenanthrene, a major polyaromatic hydrocarbon constituent, enhances IgE production by IgE-secreting EBV-transformed human B cells *in vitro*. Toxicol. Appl. Pharmacol., 1997;143:256-263.

Tunek A, Olofsson T, Berlin M. Toxic effects of benzene and benzene metabolites on granulopoietic stem cells and bone marrow cellularity in mice. Toxicol. Appl. Pharmacol., 1981;59:149-156.

Turnbull GJ, Sanderson DM, Crome SJ. "Exposure to Pesticides During Application." In *Occupational Hazards of Pesticide Use*, GJ Turnbull, ed. London: Taylor and Francis, 1985, pp 35-49.

Uyeki EM, Ashkar AE, Shoeman DW, Bisel TU. Acute toxicity of benzene inhalation to hemopoietic precursor cells. Toxicol. Appl. Pharamacol., 1977;40:49-57.

Winker N, Tushl H, Kovac R, Weber E. Immunological investigations in a group of workers exposed to various levels of polycyclic aromatic hydrocarbons. J. Appl. Toxicol., 1997;17:23-29.

9. BERYLLIUM

Gregory L. Finch, Ph.D.
Inhalation Toxicology Laboratory
Lovelace Respiratory Research Institute
PO Box 5890
Albuquerque, NM 87185

INTRODUCTION

Although first discovered in the late 1700s, health effects resulting from beryllium (Be) exposure are diseases of the 20th Century. This Chapter reviews current information regarding the health effects of Be, with emphasis on chronic beryllium disease (CBD), the principal adverse health risk from inhaled Be today. This disease is immunologically mediated through a T-lymphocyte-dependent response to inhaled Be particles, resulting in a persistent granulomatous lung disease with pulmonary debilitation.

The first section of this Chapter provides summaries of the sources and uses of Be and briefly describes human exposures that have occurred. The section also discusses the evolution and current status of occupational and environmental standards for airborne Be. The second section focuses on CBD, including its characteristics, the potential for progression from sensitization to disease, and its epidemiology. Also reviewed are acute beryllium disease, a non-specific chemical pneumonitis-type response rarely if ever encountered today, and lung cancer. The section also discusses animal models of Be-induced disease, with emphasis on models having features of human CBD. The third section reviews mechanistic studies into the nature of CBD. An overview of the specific nature of responding lymphocytes, the cascade of cytokines released in the disease process, and the understanding of genetic markers for susceptible individuals is given. The concluding section describes some unresolved issues surrounding induction of Be-specific sensitization and progression to CBD.

SOURCES, EXPOSURES, AND STANDARDS

The toxicity of Be became apparent in the 1930s and has been generally recognized since the 1940s. This section describes the important sources and uses of Be, and current environmental and occupational exposure standards. The knowledge to date regarding the toxicity of Be to humans and experimental animals is briefly reviewed, and the section concludes with a discussion of several current issues. Where appropriate, selected key references are cited. This section does not attempt to thoroughly review the topic of Be toxicology and biokinetics; for more in-depth coverage of these topics, those interested may read numerous recent reviews presented in the published scientific literature (Kriebel et al., 1988; Rossman et al., 1991; Newman, 1992; WHO, 1993) or in the reports of national and international governmental agencies (USEPA, 1987, 1998; WHO, 1990, 1993; ATSDR, 1993).

Sources and Uses

Beryllium is a naturally-occurring element found in numerous minerals. The most commercially-significant minerals are beryl and bertrandite. Extraction and refining of Be yields the metal that can also be used as an oxide or alloyed with other metals such as aluminum (Al) or copper (Cu). Beryllium is a silver-gray metal with a low density, relatively high melting point, high stiffness, and excellent neutron reflection properties. Beryllium oxide, when formed into desired shapes and fired at temperatures of 1350 - 1500°C, forms a ceramic having high thermal conductivity and electrical resistivity, and good dielectric properties. Beryllium is commonly alloyed with other metals such as Cu, Al, or nickel. The addition of 2% or less Be to Cu forms an alloy with high strength and hardness; the other alloys are less important commercially. In contrast to these relatively insoluble forms, Be is also prepared as soluble salts, including beryllium sulfate ($BeSO_4$), fluoride (BeF_2), and chloride ($BeCl_2$) (Powers, 1991; Stonehouse and Zenczak, 1991).

Human Exposures

The principal setting for human exposure to Be is in the workplace, and the primary routes of significant human exposure are inhalation and dermal contact. Oral intake from food/water is not of major concern because Be is poorly absorbed from the gastrointestinal tract (Reeves, 1991b). Exposures of humans in the ambient atmosphere are also not of significant concern, although the greatest anthropogenic release of Be to the atmosphere is from coal-fired power plants (USEPA, 1987).

Industries where potential inhalation exposures exist include the basic production industry, beryllium alloy casting operations, ceramic beryllium oxide (BeO) parts manufacturing, and metallic Be and Be alloy fabrication and working (Preuss, 1991). The worker population potentially exposed ranges from 30,000 (NIOSH, 1972) to 800,000 (NIOSH 1978), although this latter estimate may be an overestimate (Preuss, 1991). To provide for review and ongoing evaluation of acute and chronic forms of Be-induced disease, a Beryllium Case Registry (BCR) was established at the Massachusetts Institute of Technology in 1952. The BCR was subsequently relocated to the Massachusetts General Hospital in the 1960s, then to the National Institute for Occupational Safety and Health in 1978; however, it has not been actively or uniformly updated in recent decades.

Health effects resulting from Be exposure were reported as early as the 1930s in Europe and the 1940s in the United States (Eisenbud, 1982). Workers in Be extraction plants and in manufacturing Be-containing phosphors for fluorescent lamps exhibited a severe, acute chemical pneumonitis now called acute beryllium disease; this disease was first reported in the United States by van Ordstrand et al. (1943). Individuals in Be extraction and fluorescent lamp manufacturing industries were also exposed (reviewed by Eisenbud, 1982), and it is within this population that CBD was first described (Hardy and Tabershaw, 1946).

A curious feature of the early literature regarding CBD was the so-called "neighborhood cases", several tens of cases in people residing in the vicinity of Be plants. It was not always possible to estimate Be concentrations in the air near these plants, but it was clear that air concentrations could not have been as large as in the plants. Investigations headed by Merril Eisenbud led to the conclusion that insufficient personnel hygiene practices of plant workers, largely in the form of the laundering contaminated work clothing in the home, were responsible in some cases (Eisenbud, 1982, 1990). These investigations also led to the adoption in 1950 of the $0.01 \ \mu g/m^3$ concentration standard for air in the vicinity of beryllium plants.

The United States Department of Energy (DOE) uses a substantial amount of Be and compounds in nuclear weapons and reactors, and the adverse health effects on DOE workers is a continuing concern. Efforts are underway to test current and/or former Be workers for Be-specific sensitization using the blood lymphocyte proliferation test as described below. These efforts are being undertaken largely at two DOE sites, the Rocky Flats Plant near Golden, CO (Stange et al., 1996) where several thousand workers have been tested, and the Y-12 plant at Oak Ridge, TN, where additional workers are being tested. In addition, the United States DOE is currently formalizing a Chronic Beryllium Disease Prevention Program for its facilities and subcontractors. Workers in the private sector are also being monitored.

Environmental and Occupational Standards

Origins of occupational and environmental standards for allowable exposures to Be date from the late 1940s (Eisenbud, 1982, 1998). Sampling done in conjunction with the neighborhood CBD cases described above was the basis for the allowable Be concentration in ambient air. Environmental sampling in the vicinity of Be extraction plants indicated that the lowest limit of Be that caused disease was probably between 0.01 and 0.1 $\mu g/m^3$ (Eisenbud et al., 1949; Eisenbud, 1982). Therefore, the recommended maximum Be concentration was set at 0.01 $\mu g/m^3$, a limit which was widely accepted and subsequently adopted by the United States Environmental Protection Agency (USEPA). This limit still stands.

The issue of allowable occupational threshold limit values (TLVs) was also examined in conjunction with the issue of an ambient standard. This matter was troublesome because epidemiological and industrial hygiene studies could not identify a dose-response relationship between levels of exposure and disease. Because there was no firm basis to establish a standard, Eisenbud and Machle (see Eisenbud, 1990) recommended that the standard be based on the assumption that Be was as toxic on a per-atom basis as some of the heavy metals, and that an exposure limit could be calculated from existing standards based on the ratios of atomic weights. An additional safety factor led to the recommendation in 1948 of a TLV of 2 $\mu g/m^3$ averaged over an 8-hr workday (Eisenbud, 1982, 1998). This level was adopted by the Atomic Energy Commission (AEC, precursor to the DOE) in 1949, and was also adopted by the Occupational Safety and Health Administration (OSHA) as the Permissible Exposure Level (PEL) and by the American Conference of Governmental Industrial Hygienists (ACGIH) as the TLV. OSHA also lists a short-term exposure limit of 5 $\mu g/m^3$ for 30 min and a ceiling of 25 $\mu g/m^3$. The 2 $\mu g/m^3$ 8-hr standard still stands, although there is considerable discussion about the extent to which the standard is protective for the most sensitive individuals.

It should be noted that the ACGIH TLV committee has announced in its notification of intended changes for 1999 that the TLV for Be may be changed from 2 $\mu g/m^3$ airborne Be to 0.2 $\mu g/m^3$ inhalable fraction of airborne Be (ACGIH 1998a). Inhalable particulate mass TLVs are applicable to those materials that are hazardous when deposited anywhere in the respiratory tract. For the purpose of applying this TLV, the inhalable fraction is determined from the fraction of particulates passing a size-selector with the characteristics defined in Appendix D, paragraph A of the ACGIH TLV book (ACGIH, 1998b). Some of the notable deposition efficiencies in this definition are 50% inhalability for particles with 100 μm aerodynamic diameter, 77% inhalability for particles with 10 μm aerodynamic diameter, and 100% inhalability for particles with 1 μm aerodynamic diameter. It remains to be seen whether the proposed change will be accepted, but it focuses the debate on lowering the limit for particles that can enter the respiratory tract.

HEALTH/IMMUNOMODULATORY EFFECTS OF BERYLLIUM

The adverse health effects of Be have historically been classified into acute beryllium disease, CBD, and lung cancer. Of principal concern for human health effects today is the immunologically-mediated CBD. This disease is generally recognized as being a T-lymphocyte-dependent, cell-mediated response with little if any humoral components. Acute disease and lung cancer will also be mentioned briefly for completeness, as well as animal models of Be-induced disease, especially those models having features of human CBD.

Chronic Beryllium Disease

The clinical features, pathology, diagnosis, and treatment of CBD have been reviewed (Freiman and Hardy, 1970; Hardy, 1980; Kriebel et al., 1988; Rossman and Jones-Williams, 1991; Newman, 1992). The disease, also known as berylliosis, results from inhaling airborne Be, most often in relatively insoluble forms (Eisenbud and Lisson, 1983). Hardy and Tabershaw (1946) first described CBD as a chronic, delayed-type chemical pneumonitis. The disease is characterized clinically by pulmonary symptoms that include dyspnea, nonproductive cough, and detriments in lung function, although symptoms can include progressive weakness and fatigue, pain, and anorexia. Histologically, features of CBD include the presence of progressive, non-caseating granulomas, mononuclear cell infiltrates, and calcific inclusions (Freiman and Hardy, 1970). The latency period for CBD ranges from 1 - 40 yrs, with an incidence in exposed populations of from 1 - 10%. Features of the disease led Sterner and Eisenbud (1951) to suggest an immunologically-mediated basis for the disease. Hardy (1980) suggested that other host factors such as pregnancy may influence the induction of CBD. The hypothesis that genetic susceptibilities to the induction of CBD may exist has rested on the observation of CBD in identical twins (McConnochie et al., 1988) and the difference in Be-induced immunogenicity in different strains of guinea pigs (Barna et al., 1984a and b). Dermal effects ranging from contact dermatitis to dermal granulomas may develop in response to skin contact with soluble salts or Be slivers.

Humoral Immunity

Although principally a cell-mediated disease, some reports suggest humoral immune responses may play a role in or at least be associated with CBD. Increased serum immunoglobulin levels in CBD patients have been recognized for some time (see, for example, Deodhar et al., 1973), but Be specificity of these antibodies had not been demonstrated for decades. More recently, Clarke (1991) measured elevated Be-specific antibodies in the plasma of one CBD patient and one Be-sensitized individual, compared with pooled controls. This antibody assay was subsequently proposed for use as a risk assessment and health surveillance tool for CBD monitoring (Clarke et al., 1995); however, the assay has not been widely used. In work with experimental animals, Newman and Campbell (1987) demonstrated that $BeSO_4$ was mitogenic to murine B-lymphocytes *in vitro*. Further, Nikula et al. (1997), in characterizing a Be metal-induced murine model producing many features of human CBD, demonstrated a pronounced component of B-lymphocytes in interstitial lymphocyte aggregates, and a lesser component of B-lymphocytes in microgranulomas.

Cell-Mediated Immunity

The primary diagnostic tool to confirm CBD in patients presenting with pulmonary symptoms is the Be-specific lymphocyte proliferation test (BeLPT), in which a patient's blood or lung lymphocytes are placed in cell culture with a soluble form of Be, and cell proliferation is measured (Figure 1). The BeLPT is also the principal screening tool being used in current epidemiologic screening studies. Possible alternative tests based on flow cytometry are being developed (B. Marrone, unpublished).

In humans, CBD is associated with the *in vitro* clonal expansion of T-helper lymphocytes in response to Be. This proliferation is presumed to occur within the lung, although this has not been demonstrated directly in humans. Significant proliferation in response to the Be exposure indicates the presence of Be-specific sensitization. In asymptomatic individuals, a positive BeLPT is usually repeated to confirm that the individual is a positive responder. Two repeat positives signal that medical follow-up is indicated. This typically consists of a bronchoalveolar lavage

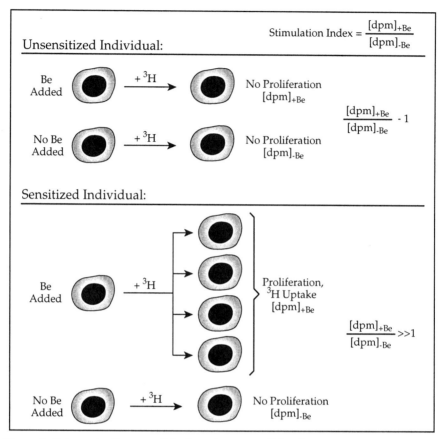

Figure 1. Schematic diagram of the lymphocyte proliferation test.

to obtain lung lymphocytes for the BeLPT test and collection of lung tissue by biopsy for histologic confirmation of the presence of granulomas. Chest radiographs may be performed although they are considered to be insensitive in detecting disease (Newman, 1996); thin-section computer tomography scans may offer an improvement over conventional radiographs. Pulmonary function tests, with focus on diffusing capacity (DL_{CO}), and exercise testing may be used to detect clinical disease, although these too are considered relatively insensitive (Newman, 1996).

Since the advent of the BeLPT, new diagnostic criteria have been proposed that include a diagnosis of subclinical disease (Table 1) (Newman et al., 1989; Newman, 1996). Beryllium sensitization, subclinical CBD, and CBD are defined under these criteria. The subclinical disease is defined as Be-specific BeLPT tests, granulomas upon biopsy, but with no clinical sign of disease. Once confirmed, clinical CBD can be treated with corticosteroids, although the disease is variably responsive to this treatment which often has complicating side effects. There are no clear guidelines regarding the extent to which the subclinical form of CBD should be treated. Methotrexate may be a potential treatment tool in some patients, but the steroid-sparing effect is inconsistent and its side effects may be limiting (Daniloff et al., 1998).

TABLE 1. DIAGNOSTIC CRITERIA FOR Be SENSITIZATION, SUBCLINICAL DISEASE, AND CHRONIC Be DISEASE (adapted from Newman et al 1989; Newman 1992)

Criteria for Sensitization	Criteria for Subclinical Disease	Criteria for CBD	Feature
X	X	X	History of exposure to beryllium
X	X	X	Beryllium hypersensitivity as demonstrated by positive blood and/or lung BeLPT response
	X	X	Lung biopsy histopathology consistent with granulomas or mononuclear cell infiltrates
		X	Pulmonary clinical abnormalities including one or more of: - respiratory symptoms or adverse signs - reticulonodular infiltrates on radio-graphy - altered pulmonary physiology

The specificity of the proliferative lymphocytic responses to Be has been shown by demonstrating a lack of proliferative response to other stimuli. Saltini et al. (1989) showed that cell lines derived from T-lymphocytes from CBD patients proliferated in response to Be but not to other metals or to common recall antigens. Haley et al. (1997) reported similar results in canine T-lymphocyte lines. In addition, Ojo-Amaize et al. (1994) showed the specificity to Be of lymphocytes from naive individuals stimulated *in vitro*; no proliferative responses of the Be-stimulated cells were observed when cultures were exposed to various other metals.

It is of great interest to determine the extent to which Be-sensitized individuals without disease might progress to CBD. Unfortunately, many efforts have been focused on initial worker population screening; where there has been follow-up screening of individuals, the duration of follow-up has been short in terms of the clinical course of CBD. Newman et al. (1992b) have reported that sensitization can progress to CBD. In a population of six sensitized workers who were followed over a 5-yr period, three of the individuals progressed to CBD. Newman and co-workers have subsequently attempted to quantify the rate of progression from sensitization to disease. In a recent abstract (Newman et al., 1998), the Investigators updated their ongoing epidemiological studies in which they are following at ≈2-yr intervals 28 sensitized individuals without evidence of CBD. Of the 16 evaluable patients, seven progressed from sensitization to CBD during an average 2.3 yr follow-up period, giving an annualized rate of 19%. Conservatively, assuming none of the other 12 progressed, the researchers reported an annual conversion rate of 9 - 19%.

Given the number of Be-sensitized individuals recently identified in large-scale screening studies (see below), understanding both the rate of progression of sensitized, asymptomatic individuals to CBD and the potential for early treatment to intervene in the disease process is of great importance. Ongoing studies following sensitized, asymptomatic individuals will provide important information on the relationship between Be-specific sensitization and the development of clinical CBD.

Human Epidemiology

The primary focus of this section is to review recent and ongoing human epidemiology studies, principally over the last 20 years. Table 2 contains selected reports relevant to this topic.

As noted above, Hardy and Tabershaw (1946) reported the first CBD cases in fluorescent lamp manufacturing workers. At that time, Be was a constituent of the phosphors used in fluorescent lamps. Significant efforts were expended during the mid- to late-1940s in investigations into both clinical and industrial hygiene aspects of CBD (as reviewed by Hardy, 1980; Eisenbud, 1982, 1990, 1998; Eisenbud and Lisson, 1983). Cases of both acute and chronic beryllium disease were identified in other populations including extraction workers and at AEC laboratories. Industrial processes were often poorly controlled and exposures were often very high (over 100 $\mu g/m^3$). Largely under the purview of the AEC Health and Safety Laboratory in New York, investigators led by Merril Eisenbud studied the airborne Be concentrations and health hazards in and around two plants in Ohio. These investigations led to recommendation and adoption of the current TLV, and a better understanding of the relationship between exposures and the induction of the acute and chronic forms of the disease. Interestingly, an Eisenbud community exposure study (Eisenbud et al., 1949) currently serves as one of the two principal studies supporting the USEPA reference concentration for chronic inhalation of Be and non-cancer health effects (USEPA, 1998).

Shortly thereafter, Sterner and Eisenbud (1951) proposed an immunologic basis for response to Be in CBD. Epidemiological investigations into CBD continued throughout the 1950s, 1960s, and 1970s, and Kriebel et al. (1988) reviewed several of these studies. During this period, records of some 900 cases of acute and CBD were accumulated in the BCR. However, not all cases were reported, and the BCR has remained static since the 1980s.

In 1983, Eisenbud and Lisson reviewed the current status of human CBD, noting that the acute disease had not been observed since 1968, and that the number of CBD cases had decreased markedly since the adoption of the workplace exposure standard. A decrease in the latency of CBD was reported, and it was reported that no cases had been observed to date in individuals first exposed after 1972. The Authors further noted that CBD rates decreased in conjunction with a significant increase in the usage of Be in the U.S.

The suggestion that CBD was largely a disease of the past was dispelled with identification of new CBD cases in 1983 and 1984. Johnson (1983) reported three new cases in a spacecraft manufacturing plant and further reported that the BeLPT was used to prove Be-specific sensitization. Analysis of workplace records indicated

Table 2. Selected Reports of Human CBD in the 1980s and 1990s

Authors	Population	Findings	Comments
Eisenbud and Lisson (1983)	Review	Reviewed history of CBD from 1940s to 1980	Concluded that exposure standards have effectively controlled acute and chronic beryllium disease
Cotes et al (1983)	Private plant	Four of 146 workers from 1950s developed CBD	Cast doubt on adequacy of exposure standards
Johnson (1983)	Aerospace plant	Report of a new CBD case	Began era of renewed scrutiny for CBD cases
US DOE (1985)	Rocky Flats	Report of a new CBD case	Began renewed scrutiny for CBD in DOE facilities
Cullen et al (1987)	Scrap metal recovery	Attributed two CBD cases to exposures $< 2 \ \mu g/m^3$	Cast doubt on adequacy of exposure standards
Newman et al (1989)	Aerospace and ceramics plants	Suggested re-evaluation of definition of CBD	Increased role for BeLPT; called attention to questions about relationship between sensitization and CBD
Kriess et al (1989)	Private plant	Identified sensitized, asymptomatic workers, and an unexpectedly high rate of CBD	Epidemiology study; increased scrutiny about relationship between sensitization and CBD
Levin (1991)	Private plant machinist	CBD case attributed to exposure to beryllium-copper alloy	Indicated Be-alloys might also present a CBD risk

Table 2. Selected Reports of Human CBD in the 1980s and 1990s (cont'd)

Authors	Population	Findings	Comments
Newman and Kriess (1992)	Spouse of a beryllium worker	CBD diagnosed in an individual not occupationally exposed	Non-occupational CBD case
Kotloff et al (1993)	Dental technician	CBD case reported in a dental technician	CBD case from unexpected occupational setting
Kriess et al (1993b)	Be ceramics plant	Epidemiology study identified new cases; disease rates related to work tasks	Reported CBD rates from 2.9 to 15.8% depending on work task
Kriess et al (1993a)	Rocky Flats	Screened workers using BeLPT; described new cases; disease rates related to work tasks	12 of 18 sensitized workers had CBD; CBD rates higher for certain tasks and for those overexposed
Stange et al (1996)	Rocky Flats	Initial report of results from DOE's Beryllium Health Surveillance Program; identified 29 cases of CBD and 76 sensitized workers	Report of DOE's large-scale screening program; identified cases with apparently low exposure
Kriess et al (1996)	Private ceramics plant	136 workers studied; six of eight sensitized workers had CBD; machinists had highest sensitization rate	Study combined screening and confirmation of CBD with workplace monitoring; concluded many CBD cases had median exposures < 2 $\mu g/m^3$
Kriess et al (1997)	Private mixed-operation plant	59 of 627 individuals were sensitized; 24 new CBD cases were identified	Highest disease rates were associated with ceramics and production of beryllium metal

that many of the air samples taken from 1963-1973 exceeded the permissible TLV. A separate case was reported at the DOE Rocky Flats plant in Colorado (USDOE 1985) in a machinist working there from 1970-1984. An investigation reported a number of industrial hygiene deficiencies leading to overexposures.

Studies published in the 1980s questioned the adequacy of the existing 2 $\mu g/m^3$ exposure standard. Cotes et al. (1983) described a 30-yr follow-up study of workers at a plant manufacturing Be products. Four workers developed CBD out of 146 who had worked in the plant from 1952-1963, and monitoring data from 1952-1960 suggested that the estimated daily concentration did not exceed 2 $\mu g/m^3$ in the plant. Cullen et al. (1987) identified additional CBD cases in scrap metal recoverers. These Investigators found five cases of granulomatous lung disease, with four of the individuals having BeLPT-demonstrated hypersensitivity. Beryllium air concentrations in the plant were monitored after the cases were identified (no earlier monitoring results were available), and results suggested that exposures in the furnace area, the work site for four of the cases, might have been consistently less than 2 $\mu g/m^3$. Although the relationship between the sampling and exposures that caused the disease and the potential for unmeasured spikes in exposures could be questioned, these studies raised concerns that existing workplace PELs might not be protective.

Two epidemiology studies published by National Jewish Center researchers relied heavily on BeLPT for screening worker populations. Kriess et al. (1989) screened nuclear workers using an initial questionnaire followed by the BeLPT. Six of 51 individuals showed signs of hypersensitivity to Be. Four out of five of these individuals who underwent further clinical evaluation were found to have CBD, as determined by lung granulomas upon biopsy and positive lung lymphocyte BeLPT results. The Authors concluded that the BeLPT can detect non-diseased individuals and those with minimally symptomatic CBD. In a similar vein, Newman et al. (1989) reported results from a screening of workers from two different plants in Colorado. By using BeLPT screening to identify hypersensitive individuals, followed by a more rigorous clinical workup including lung biopsy, the Investigators identified 12 new CBD cases. Interestingly, only five of the 12 individuals had previously sought medical attention because of respiratory symptoms. The Investigators concluded that BeLPT screening can identify sensitized individuals and those with subclinical disease (granulomas upon biopsy but no clinical symptoms).

Although most "new" CBD cases were (and continue to be) discovered in plants where Be metal or oxides are used, cases with exposures presumably outside these industries have been recently identified. Levin (1991) reported a CBD case in a machinist whose only Be exposure appeared to be due to working with Be-Cu alloy. Balkissoon and Newman (1999) reported two additional CBD cases in individuals working with Be-Cu alloy. Newman and Kriess (1992) reported a CBD case in the spouse of a Be worker whose exposure was probably minimal; this constituted the first non-occupational case found using the BeLPT. In yet another worker population, Kotloff et al. (1993) reported a CBD case in a dental technician.

A series of epidemiology studies published in the 1990s were generally of sufficient size to begin associating disease rates with work tasks in a statistically meaningful fashion. Some studies, principally at private Be-working plants where Be is the primary business, screened virtually the entire workforce, whereas others, principally at the DOE Rocky Flats plant and Oak Ridge facilities, used questionnaires to distinguish Be workers from the larger overall workforce. These studies used blood BeLPT as an initial screening tool to first identify sensitized individuals, then to invite individuals having two positive BeLPT responses to participate in a more aggressive clinical workup including the lung lymphocyte BeLPT and collecting lung tissue via biopsy for histological evaluation. It is important to note that as research, these studies can only invite subject participation. As with all human research, appropriate informed consent must be obtained from each participant before either the initial screening or more invasive clinical workup. Individuals are occasionally lost to follow-up or decline participation in the clinical workups following a positive BeLPT response. In addition, medical counseling of identified individuals is an important part of these programs, and removal from Be-related work is considered prudent, although this is problematic for some individuals in the private sector where working with Be may be the sole business activity. Several issues related to the ethics of screening workers with the BeLPT have recently been discussed (Marshall, 1999).

Kriess et al. (1993a) evaluated the epidemiology of CBD among 895 workers at the Rocky Flats plant. Some 18 new cases of Be sensitization were identified, and 12 had disease. Three of the remaining six developed lung granulomas over a succeeding 2-yr period. The six sensitized cases without initial disease had a greater pack-year cigarette smoking history than the CBD cases. Sensitization occurred in individuals with only inadvertent or bystander exposures. The greatest risks for sensitization, in the range of 3.4 - 7.4%, occurred in groups of individuals with exposure beginning before 1970, consistent exposure, reported overexposures, or work activities involving Be metal sawing, bandsawing, or machining.

Stange et al. (1996) reported results from the large-scale, DOE-funded Beryllium Health Surveillance Program at Rocky Flats. Through March 31, 1995, 4397 current and former workers completed screening; 28 new CBD cases and 69 sensitized, non-diseased individuals were identified. Five hundred eighteen workers were retested within 1 - 3 yrs, and one new CBD and nine new sensitized cases were identified. Overall, a 2.43% sensitization rate (sensitization with or without disease) was reported. The Authors concluded that the sensitization rate was similar to that observed in the earlier Kriess et al. (1993a) study, and that BeLPT retesting after 1 - 3 yrs is useful in detecting new cases of sensitization and/or disease. Results from a companion industrial hygiene study (Barnard et al., 1996) indicated that many exposure excursions occurred in the plant, particularly prior to 1984.

Kriess and colleagues published results from BeLPT screening studies coupled with industrial hygiene monitoring in two private Be facilities. Eight Be-sensitized employees were identified from a population of 136 screened in a ceramics plant in operation since 1980; of these, six had granulomatous lung lesions (Kriess et al., 1996). Machinists had a 14.3% sensitization rate compared to 1.2% among non-machining employees; machining also had higher median Be air concentrations as well as excursions beyond the standard. The Authors concluded that lowering exposures may be important to lowering CBD risk. This study serves as the second of two principal studies for the USEPA reference concentration for chronically-inhaled Be (USEPA, 1998).

Kriess et al. (1997) examined workers in a plant producing pure metal, oxide, alloys, and ceramics; 59 of 627 workers tested positive in the blood BeLPT, and 24 of 47 employees accepting clinical followup were found to have CBD. Including five known cases, the disease prevalence rate was 4.6% (29/632). The greatest risk factor was working in the ceramics or pebble plant areas. There was a significantly reduced cigarette smoking history for sensitized cases vs non-sensitized employees, but not for CBD cases vs non-cases. This finding is consistent with their earlier report (Kriess et al., 1993a) for sensitized individuals. Interestingly, however, the Kriess et al. (1996, 1997) reports did not find a significant smoking history differences for the CBD cases vs non-diseased employees. Thus, in two separate studies an association between a lesser cigarette smoking history and sensitization but not disease has been suggested. The Authors concluded that CBD remains a contemporary problem, that exposure and disease are related in some fashion, and that further characterizations of exposure-response relationships for CBD are warranted.

General characteristics of the current epidemiology of CBD can be summarized as follows. First, given the improvements of industrial hygiene controls for airborne Be over the last 50 yrs, it is surprising that the rate of CBD or sensitization in current studies ranges from ≈1 - 15%. These rates are similar to reported rates in early studies when exposures were probably much higher. On the other hand, certain jobs (such as machining) and rates of disease and/or sensitization are clearly associated. Thus, some degree of dose-response relationship may be present. An additional complication to this picture is that the role of the physicochemical properties of exposure aerosols to disease is not understood, and additional research is needed. Certain genetic factors also appear to be associated with CBD rates. The possible genetic screening of potential Be workers has been mentioned (Eisenbud, 1998), but no markers are completely predictive of risk, and pre-employment screening requires additional social and ethical debate.

The relationship between sensitization and progression to CBD is possibly the most pressing issue today, and certainly the most important to asymptomatic blood BeLPT-positive individuals. Given the poor understanding of Be-specific antigenic stimulus, the resulting cascade of immunological responses leading to disease, and the lack of an animal model that could be used to test this relationship, epidemiologic followup of currently monitored worker populations will be needed to further understand the progression from sensitization to disease.

Acute Beryllium Disease

Acute Be-induced pulmonary disease results from exposure to relatively high levels of airborne Be, typically in the form of soluble salts and/or low-fired oxides (reviewed by Kreibel et al., 1988; Ridenour and Preuss, 1991; Newman, 1992). The disease was characterized by acute nasopharyngitis, tracheobronchitis, and chemical pneumonitis rather than an immunologically-mediated syndrome. The acute response occurred in virtually all workers exposed to concentrations above 1,000 µg Be/m^3, but in none exposed to less than 100 µg/m^3. With removal from exposure, the subject improves gradually from the acute symptoms over a period of several weeks to a few months. It has been generally accepted that this form of the disease virtually ceased to occur after adoption of exposure standards in 1950. A few instances of acute lung disease from inhaled Be have been reported in post-1950 literature, but the true etiology of these reports has been questioned (Ridenour and Preuss, 1991). The most likely cause of any future acute Be disease cases would be accidental exposure to extremely high concentrations of airborne Be .

Lung Cancer

The carcinogenicity of a variety of Be compounds in various tissues of several species of laboratory animals has been recently reviewed (EPA, 1987, 1998; WHO, 1990, 1993; Reeves, 1991a and c; ATSDR, 1993). Most notable have been bone cancer induced by various routes of exposure, and more pertinent to humans, lung cancer induced by inhalation or intratracheal instillation of several different Be-containing materials. Finch et al. (1996) reported marked species differences between rats (relatively susceptible) and mice (relatively resistant) in pulmonary carcinogenic responses to inhaled Be metal. Experimental data are sufficient to permit both the USEPA (1987, 1998) and International Agency for Research on Cancer (WHO, 1993) to conclude that Be and its compounds are carcinogenic in animals.

The extent to which Be and its compounds should be regarded as a human lung carcinogen has been extensively debated. Several groups have undertaken epidemiological research on Be workers from two general populations: all identifiable Be workers and those included within the BCR. Throughout the 1970s, the former group of workers was the basis of reports by Wagoner et al. (1980) and Mancuso (1980), whereas the latter was examined by Infante et al. (1980). These studies have been extensively reviewed and critiqued (USEPA, 1987; ATSDR, 1993; WHO, 1993), and their general tone is that serious methodological problems with the studies prevented definitive conclusions.

More recently, these same two cohorts have been reanalyzed. Using the BCR population, Steenland and Ward (1991) described a small yet statistically-significant increase in lung cancer vs the United States population; interestingly, excess mortality for lung cancer was greater in individuals experiencing acute Be-induced lung disease (possibly having greater Be exposure) compared with CBD cases. Ward et al. (1992) examined some 9225 Be workers from seven Be plants and found a small yet significant increase in excess mortality for lung cancer in workers from some - but not all - of the plants. On the basis of these studies, IARC (WHO, 1993) classified Be as a demonstrated lung carcinogen in humans. In the United States, however, Be is classified as a confirmed carcinogen by the ACGIH (ACGIH, 1998b) and as a probable human carcinogen by the EPA (USEPA, 1998).

Animal Models

Information is limited exists regarding the biokinetics of excretion of Be in humans (see, for example, WHO, 1993). Most of our knowledge of lung clearance, translocation to other tissues, and excretion comes from experimental studies in laboratory animals. Available data have been recently summarized (USEPA, 1987;

Reeves 1991a, b, and c; ATSDR, 1993; WHO, 1993; Finch et al., 1996). There is low systemic absorption of Be through the skin after dermal exposure or through the gastrointestinal tract after oral ingestion. Upon inhalation, retention in the lung, distribution of Be to other tissues and organs, and Be excretion are largely dictated by the physicochemical form of the inhaled Be. More soluble Be is cleared relatively quickly from the lungs, but a longer-term retained component does exist, and is presumably due to precipitation into relatively insoluble forms. Forms having lower solubility, such as the metal or oxide, can be cleared (1) following conventional lung kinetics of clearance of relatively insoluble particles, although toxicity induces delayed clearance (Finch et al., 1996), or (2) by dissolution. Systemically-distributed Be is retained predominantly in bone; however, retention in other compartments has been noted. Excretion in feces is largely due to unabsorbed Be passing from the respiratory to gastrointestinal tract, and urinary elimination is more important for systemically distributed Be (Reeves, 1991b).

A number of animal models of Be-induced toxicity were investigated beginning in the 1940s; much of this early work was published as the proceedings of the 1947 Saranac Symposium (Vorward, 1950). These studies demonstrated the acute to chronic toxicity of Be, and in conjunction with concurrent findings in humans, helped to overturn the early United States Public Health Service view that Be was not the toxic component of the exposures (Hyslop et al., 1943). Subsequent work over the next several decades has been summarized (Reeves, 1991a; WHO, 1993). Findings included production of a pneumonitis-like response similar to that seen in acutely-exposed humans. Osteosarcomas and lung cancer were produced in various animal models. More pertinent to this Chapter were investigations into immunological responses of animals to Be, and to the production of immune granulomas.

Barna and colleagues (1981, 1984a and b) exposed guinea pigs to BeO via intratracheal (IT) instillation and found that Strain 2 but not Strain 13 animals develop lung granulomas. The Investigators demonstrated that compared to Strain 13 animals, the Strain 2 guinea pigs have decreased chemiluminescence and bactericidal activity of bronchoalveolar lavage cells, and increased *in vitro* uptake of tritiated thymidine in blood, lymph node, and spleen cells. These differences in response were attributed to the difference in the major histocompatibility complex (MHC) Ia region between these two strains.

Many components of human CBD have been demonstrated in dogs or monkeys. Haley et al. (1989) exposed dogs by acute inhalation to BeO prepared at either 500° or 1000°C, then examined blood and lavage cell BeLPT responses over time. Positive blood BeLPT responses were observed sporadically after exposure, but in dogs exposed to the 500°C oxide, lymphocytes constituted over 50% of the lavage cells recovered, and these cells responded with a positive *in vitro* BeLPT most notably some 6 - 8 mo after exposure. In a companion group of dogs sacrificed periodically after exposure, lung lesions included granulomas, macrophage hyperplasia, lymphocyte infiltrates, fibrosis, and proliferation of the alveolar epithelium (Finch et al., 1990); these responses were most evident ≈2 mo after dogs inhaled the 500°C oxide. In a follow-on study, groups of dogs were re-exposed to the 500°C oxide some 30 mo after the first exposure (Haley et al., 1992). Responses observed were similar to the first exposure, and Haley et al. (1997) described the immunologic specificity to Be in these cells. The Authors suggested that at least under this exposure regime in this species, any potential immune memory from the first exposure does not amplify the response after the second exposure.

Haley et al. (1994) also exposed cynomolgus monkeys to a single bolus of BeO and metal via IT instillation, then periodically bled and lavaged the animals. Numbers of lung lymphocytes were elevated, but only metal-exposed animals had positive lung BeLPT responses. Monkeys were sacrificed after 6 mo, and lesions consistent with human CBD were observed; some had immune granulomas having tightly-organized lymphocytic cuffs surrounding epithelioid macrophage aggregates. The Authors concluded that the metal produces a greater response than the oxide at presumably comparable levels of Be ion bioavailability, and that both monkey and dog models constitute good, if expensive, models of many features of human CBD.

Numerous investigators have attempted to produce a rodent model of human CBD. Votto et al. (1987) described granulomatous lung lesions in rats produced by $BeSO_4$ and albumin, and delivered via IT instillation to animals previously-immunized subcutaneously with $BeSO_4$. In contrast, Haley (1991) described in the rat a foreign body-type response to Be without a significant component of immuno-logic recognition.

Huang et al. (1992) described a granulomatous lung disease model in Strain A/J but not C57Bl6/J or BALB/c mice. The Investigators immunized mice with $BeSO_4$, then administered BeO by IT instillation. Treated mice exhibited transient increases in lavageable lymphocytes, microgranulomas, and fibrosis. A mild degree of Be-specific stimulation in lavage cells was demonstrated using the BeLPT.

Nikula et al. (1977) described another mouse model having many features of human CBD. The Investigators exposed A/J and C3H mice via acute inhalation to Be metal and sacrificed animals after 6 mo. Exposed lungs had interstitial aggregates of lymphocytes and a granulomatous pneumonia. $CD4^+$ lymphocytes constituted the bulk of lymphocytes in the interstitial infiltrates and microgranulomas, and distribution of the $CD4^+$ cells correlated with the location of proliferating cells. However, Be-specific sensitization using BeLPT was not demonstrated. The Authors concluded that this model mimics several features of human CBD and is useful for investigations into the roles of Be dose, exposure pattern, and physicochemical form in CBD. These studies were extended to evaluate the time course of pulmonary histological responses in C3H mice (Finch et al 1998), and to demonstrate the role of MHC-Ia antigenic characteristics in response to Be (Benson et al., 1998).

MECHANISMS OF RESPONSES TO BERYLLIUM

Despite numerous investigations, specific mechanisms of host immunologic responses leading to the induction of CBD remain unknown. Kriebel et al. (1988) suggested that there is no convincing evidence that humoral immune responses play an important role in the induction of CBD, and although there is evidence for B-lymphocyte mitogenesis and presence within granulomatous lesions, this view is commonly accepted. Clearly, the principal driving factor in induction of CBD is the $CD4^+$ lymphocyte, cell-mediated, delayed-type hypersensitivity response.

Several recent publications have examined the roles of factors such as various T-lymphocyte subsets with differing receptors, cytokines, and host genetic factors in the induction of CBD. The marker HLA-DPB1Glu69 (discussed below) is the principal susceptibility gene alteration in CBD identified to date; however, this marker does not completely account for susceptibility to CBD induction upon Be exposure. Kriess et al. (1994) suggested that this predisposing gene alteration is complemented by other genetic factors such as various T-lymphocyte repertoires, suppressor circuits, and immune response modifiers such as cytokines. Alternatively, they propose that exposure conditions may be the principal factor in CBD. It is likely that both predisposing host genetic characteristics and exposure factors are involved in disease induction, probably in a complicated fashion. At any rate, additional immunogenetic studies linked with worker epidemiology studies are needed to better understand the mechanistic interplay of factors causing CBD.

Cell-Mediated Immunity

Responding Cell Types

As noted above, the principal cell type involved in the lymphocytic response seen in Be-induced sensitization and disease is the $CD4^+$ lymphocyte (Rossman et al., 1988; Saltini et al., 1989). However, despite significant effort, neither the specific nature of the antigenic form of Be nor the mechanism of antigenic recognition is currently known. Based on patterns of cytokine secretion and mRNA message levels, Maier et al (1998) suggested that the principal lymphocytic response during proliferation of $CD4^+$ cells from CBD patients is characteristic of a T-helper 1 (T_H1)-type response.

Research has been focused on identifying specific T-lymphocyte receptors involved in the response to Be. Saltini et al. (1989) demonstrated that different T-lymphocyte receptors are present in lines of T-lymphocyte clones from different CBD patients. Similarly, Fontenot et al. (1998) found a marked heterogeneity in the β and α variable regions of the T-lymphocyte receptor in CBD patients, and that responses are similar between lymphocytes from lung and blood. Eleven of 28 CBD patients demonstrated 16 different T-lymphocyte subset expansions in lung lymphocytes, but only one expansion in cells from 10 controls. Furthermore, additional subsets of cells expanded upon *in vitro* challenge with $BeSO_4$.

Involvement of Cytokines and Other Mediators

Lung lymphocytes from CBD patients produce a variety of cytokines. Tinkle et al. (1997) obtained lung lymphocytes by bronchoalveolar lavage and examined the cytokines produced by stimulation with $BeSO_4$ *in vitro*. The Investigators observed a brief, rapid increase in interleukin (IL)-2 protein and mRNA, and a more protracted increase of interferon (IFN)-γ protein and message. Neutralization of IL-2 only partially blocked Be-specific T-lymphocyte proliferation, and investigators hypothesized that the α subunit of the soluble IL-2 receptor may be a biomarker of disease severity. These Investigators also found Be-induced increases in levels of tumor necrosis factor (TNF)-α and IL-6, but not of IL-4, in similarly-treated lymphocytes from CBD patients vs controls (Tinkle et al., 1996). This finding suggests the participation of T_H1 lymphocytes in the Be disease process. Tinkle et al. (1997) subsequently found that the α subunit of the soluble IL-2 receptor was elevated in serum and bronchoalveolar lavage fluid (BALF) of CBD patients, and that levels of this subunit correlated with clinical measures of disease severity. Similar findings were also observed for soluble $TNF\alpha$ and IL-6 receptors (Tinkle and Newman, 1997). The correlation demonstrated between levels of these soluble receptors and clinical stage of CBD suggests their potential use for markers of disease severity. These Investigators subsequently described a partial, IL-10 induced inhibition of the production of various cytokines in Be-stimulated BAL cells from CBD patients, but no effect of excess IL-10 on T-lymphocyte proliferation (Tinkle et al., 1999).

Other investigators have examined the expression of macrophage/monocyte-derived cytokines. Bost et al. (1994) found that BALF levels of $TNF\alpha$ but not IL-1 or IL-6 was increased in CBD patients vs normal subjects. They further examined macrophage mRNA and found elevated mRNA expression for $TNF\alpha$ and IL-6 in CBD patients vs controls. Galbraith et al (1996) found elevated levels of $TNF\alpha$ and IL-1β mRNA in a human monocytic leukemia cell line exposed *in vitro* to Be in the presence of fluoride, compared with cells not stimulated by Be. Thomassen et al. (1998) found increased secretion of $IFN\gamma$ following $BeSO_4$ stimulation in cultured BAL cells from CBD patients compared with exposed, non-diseased individuals. Attempts have been made to modulate the *in vitro* expression of cytokines produced by lymphocytes from CBD patients. Maier et al. (1998), noting the elevated levels in CBD patients of the T_H1 lymphocyte-derived cytokines $IFN\gamma$ and IL-2, and macrophage-derived $TNF\alpha$, attempted to block secretion of these cytokines with IL-4 pretreatment of CBD patient-derived BAL cells. No effect was noted, and the Authors concluded that IL-4 cannot modulate the T_H1-type response generated during the *in vitro* proliferation of lymphocytes derived from CBD patients.

Other mediators have been associated with CBD. Harris et al. (1997) found elevated serum levels of nepoterin in CBD patients compared with either sensitized, asymptomatic or control individuals, and concluded that this marker has a high positive predictive value for distinguishing CBD cases from non-diseased, sensitized individuals. Finally, Comhair et al. (1999) demonstrated elevated levels of gluta-thione and glutathione peroxidase in BALF from CBD patients compared with non-diseased controls.

Thus, a variety of cytokines appear to be involved in CBD-related pulmonary responses. No specific cytokine has been identified as key in inducing the granulomatous response. Several cytokines show promise for use in helping to assess the severity of CBD and for possibly distinguishing between sensitized and diseased individuals.

Genetic Markers

An important aspect of current research into responses to Be is the issue of genetic susceptibility to induction of CBD. As noted above, the work of Barna and colleagues (1981, 1984a and b), which demonstrated differential responses of two guinea pig strains, suggested that strain differences in response can be attributed to differences in the MHC Ia region. The work of Benson et al. (1998) supported these findings by comparing responses of C_3H/HeJ mice to inhaled Be with those of selected wild-type, congenic, and knockout mice. C_3H mice developed lung granulomas suggestive of a $CD4^+$-dependent, cell-mediated immune response. Other mice of various MHC-II IA region haplotypes had lesser granulomatous-type responses, suggesting the importance of the k/k haplotype at the H-2 IA/IE allele. In a similar vein, Huang et al. (1992) IT-exposed mice to Be salts or to BeO, and examined lung lesions and properties of lung cells obtained by lavage. The greatest Be-induced responses were observed in Strain A/J mice; neither BALB/c nor C57Bl6/J mice produced minimal responses. The Authors suggested that strain differences at the MHC H-2 gene complex may have accounted for the strain differences observed.

In humans, Richeldi et al. (1993) examined MHC Class II genes involved in CBD and identified a marker in the HLA-DPB1 region of the human MHC-II complex associated with CBD cases. These MHC genes code for cell-surface glycoproteins that may participate in antigen recognition. Compared with 44 Be-exposed individuals without CBD, 33 CBD patients examined had a significantly-increased frequency of the DPB1*0201 allele, and a significantly decreased frequency of the DPB1*0401 allele. The most notable amino acid substitution attributed to the DPB1*0201 allele is the substitution of Glu for Lys at position 69; thus, the Investigators termed the marker Glu69. It was found that 97% of the CBD patients examined expressed the Glu69 marker in the HLA-DPB1 region, whereas only 30% of the Be-exposed controls had the marker. Whereas the Glu69 marker had a significant association with CBD, it did not completely explain the occurrence of CBD in this population.

Richeldi et al. (1997) extended their study of this marker into another Be-exposed population of workers in a private ceramics plant. These workers had been enrolled in a Be disease surveillance program that included interviews, job history characterization, BeLPT, and bronchoalveolar lavage with lung BeLPT and biopsy sampling for individuals positive in the blood BeLPT. HLA-DPB1Glu69 marker was present in 36 of 119 unsensitized individuals (30%), in five of the six CBD-afflicted individuals (83%), and in neither of the two Be-sensitized, asymptomatic individuals examined. Among machinists, CBD was found in four of the 16 individuals positive for Glu69, whereas CBD occurred in only one of 29 Glu69-negative individuals. A logistic regression model demonstrated an at least additive effect between Glu69 marker and machining in production of CBD cases in this workforce.

A significant increase in frequency of HLA-DPB1Glu69-positive alleles was also observed in sarcoidosis patients compared with controls (Lympany et al., 1996). Like CBD, sarcoidosis is a chronic granulomatous disorder characterized by accumulation of $CD4^+$ lymphocytes. The association between the Glu69 marker in both sarcoidosis and CBD indicates that this alteration may be important in antigen presentation/recognition in chronic granulomatous diseases in general.

Further work continues into the identification of other markers associated with CBD continues. Richeldi et al. (1993) hypothesized that other MHC genes may be associated with CBD or may amplify the effect of HLA-DPB1Glu69. They examined the DRB4*0101 and TNFB*1 alleles (associated with rheumatoid arthritis and autoimmunity, respectively), but found no association with CBD. Stubbs and co-workers (1996) screened individuals for marker alleles in the HLA-DP and HLA-

DR regions. HLA-DPB1Glu69 was positive in 25 of 29 individuals with demonstrated Be-specific hypersensitivity, but only 42 of 86 Be-exposed, non-hypersensitive individuals were positive for Glu69. Several HLA-DRB1 alleles were preferentially-associated with hypersensitive vs exposed, non-hypersensitive individuals, and other HLA-DRB1 alleles were preferentially-associated with exposure vs hypersensitivity. In addition, an anti-HLA-DR antibody was found to block the proliferative response to Be. The Investigators concluded that HLA-DR is involved in presenting Be to lymphocytes, but that no specific HLA-DR allele is strongly associated with hypersensitivity.

More recent work has demonstrated highly significant differences in Glu69-containing alleles between CBD patients and Be-exposed, non-diseased controls (Wang et al., 1999). Glu69 carriers in the control group tended to have the Glu69 substitution at the DPB1*0201 allele, whereas in CBD patients, the substitution tended to be at other alleles. Copy number (homozygous or heterozygous status) was also associated with CBD susceptibility.

Other genetic markers are being investigated. Maier et al. (1999) reported an association between angiotensin-1 converting enzyme (ACE) polymorphisms, serum ACE activities, and CBD status, and suggested that ACE genotype might be important in the progression to CBD.

Taken together, these data suggest that individuals having certain genetic characteristics are particularly susceptible to the induction of CBD following Be exposure. Despite the association between genetic makeup and CBD, however, circumstances of exposure appear to contribute to the induction of disease as well. Although potentially important, These genetic susceptibility findings, although potentially important, are not yet sufficiently well characterized such that they can be effectively used for disease monitoring or early disease detection. Social and ethical issues must be carefully considered before the establishment of any potential genetic screening program. Population-based studies are needed to determine the extent to which various markers will be predictive of individuals at risk for CBD.

CONCLUSIONS AND UNRESOLVED ISSUES

Beryllium and its compounds have found unique and increased industrial use since early this century despite having significant toxicity. The purpose of this Chapter was to acquaint the reader with characteristics of Be-related diseases, and in particular, the immunologic basis for pulmonary responses that lead to the induction of CBD. A related purpose was to describe current issues and controversies. Despite investigations over several decades into the effects of exposure to Be in both humans and in experimental systems, significant questions remain unanswered.

Toxic effects of inhaled Be range from acute effects at high levels to chronic granulomatous lung disease to lung cancer, although the latter is controversial in humans. While acute disease is principally historical in interest, new cases of CBD continue to arise. The specific nature of exposures to Be and host-related genetic factors leading to induction of CBD remains in question. Neither the natural history of Be sensitization nor relationships between Be-specific sensitization, asymptomatic CBD, and full clinical CBD (i.e., does sensitization or subclinical disease necessarily progress to disease?) are understood. Furthermore, the extent to which markers such as genetic characteristics might predict individuals at increased risk for CBD is not known. Lack of an animal model having all features of human CBD hampers further understanding of these issues.

The relationship between Be exposure and CBD has been recognized for some 50 years; however, industrial hygiene assessments have not definitively identified relationships between specific forms of Be, the frequency of exposure (single acute, chronic, or multiple episodic exposure), and the disease. Exposure to less soluble forms of Be, such as Be metal or BeO, is apparently more likely to be associated with CBD cases compared to the more soluble forms. Epidemiological studies generally show that machinists working with the Be metal or oxide have the greatest sensitization and/or disease incidence. A few cases are attributable to exposures to Be-containing alloys, although these are apparently more rare. CBD cases are also attributable to a single or only a few exposures. Regarding the relative importance of low-level, chronic exposure versus periodic "spikes" of exposure, neither industrial hygiene data on human exposures nor data from experimentation on animals permit this question to be adequately answered. The ACGIH (1998a) proposal to base the workplace Be TLV on inhalable fraction should help focus discussion on the role of Be physicochemical form on health effects.

Although not a focus of this Chapter, the classification of Be and its compounds as a demonstrated human lung carcinogen remains inconsistent throughout the world. The IARC (WHO, 1993) and the ACGIH (1998b) classified Be as a demonstrated human carcinogen principally on the basis of cohort epidemiology studies (Ward et al 1992; Steenland and Ward 1991). However, interpretation of the findings of these studies is controversial, and factors other than Be exposure might explain the results (MacMahon, 1994). In this vein, the EPA still classifies Be as a suspect human carcinogen (USEPA, 1998).

Finally, the adequacy of the existing workplace PEL to protect against CBD is in serious question. Although the incidence of CBD cases was markedly reduced by adoption of the 2 $\mu g/m^3$ 8-hr exposure limit 50 years ago, cases of sensitization and CBD are still being identified in recent and ongoing epidemiology studies. It is difficult to retrospectively determine actual worker exposures; however, sensitization and CBD are increasingly being attributed to median exposures less than 2 $\mu g/m^3$.

ACKNOWLEDGEMENTS

The Author acknowledges the stimulating collaborations with colleagues and the support of the technical staff into Be-related health effects studies conducted at the Lovelace Respiratory Research Institute (LRRI), formerly the Inhalation Toxicology Research Institute. These studies investigated the physicochemical properties of Be aerosols related to health effects and were largely conducted in experimental animals. I also thank a larger group of colleagues from government, private concerns, and other research organizations for intriguing discussions related to the nature of Be-induced adverse health effects. Funding for the preparation of this Chapter and for our studies at LRRI was provided by the U.S Department of Energy, Office of Defense Programs and Office of Biological and Environmental Research, under

Cooperative Agreement No. DE-FC04-96AL76406, in facilities fully accredited by the Association for the Accreditation of Laboratory Animal Care International.

REFERENCES

ATSDR: Agency for Toxic Substances and Disease Registry, Public Health Service, U.S. DHHS. *Toxicological Profile for Beryllium.* ATSDR/TP-92/04. Washington, DC: Department of Health and Human Services, 1993.

ACGIH: American Conference of Governmental Industrial Hygienists. *Annual Reports of the Threshold Limit Value and Biological Exposure Indices Committees.* Cincinnati, OH, 1998a.

ACGIH: American Conference of Governmental Industrial Hygienists. *Threshold Limit Values for Chemical Substances and Physical Agents and Biological Exposure Indices.* Cincinnati, OH, 1998b.

Balkissoon R, Newman LS. Beryllium copper alloy (2%) causes chronic beryllium disease. J. Occup. Environ. Med., 1999;41:304-308.

Barna BP, Chiang T, Pillarisetti SG, Deodhar SD. Immunologic studies of experimental beryllium lung disease in the guinea pig. Clin. Immunol. Immunopathol., 1981;20:402-411.

Barna BP, Deodhar SD, Chiang T, Gautam S, Edinger M. Experimental beryllium-induced lung disease. I. Differences in immunologic responses to beryllium compounds in Strain 2 and 13 guinea pigs. Intl. Arch. Allergy Appl. Immunol., 1984a;73:42-48.

Barna BP, Deodhar SD, Gautam S, Edinger M, Chiang T, McMahon JT. Experimental beryllium-induced lung disease. II. Analyses of bronchial lavage cells in Strains 2 and 13 guinea pigs. Intl. Arch. Allergy Appl. Immunol., 1984b;73:49-55.

Barnard AE, Torma-Krajewski J, Viet SM. Retrospective beryllium exposure assessment at the Rocky Flats Environmental Technology site. Am. Ind. Hyg. Assoc. J., 1996;57:804-808.

Benson JM, Barr EB, Nikula KJ. Role of the Major Histocompatibility II complex in mediating beryllium-induced pulmonary lesions in mice. The Toxicologist, 1998;42:350.

Bost TW, Riches DW, Schumacher B, Carre PC, Khan TZ, Martinez JA, Newman LS. Alveolar Macrophages from patients with beryllium disease and sarcoidosis express increased levels of mRNA for tumor necrosis factor-alpha and interleukin-6 but not interleukin-1 beta. Am. J. Respir. Cell Mol. Biol., 1994;10:506-513.

Clarke SM. A novel enzyme-linked immunosorbent assay (ELISA) for the detection of beryllium antibodies. J. Immunol. Meth., 1991;137:65-72.

Clarke SM, Thurlow SM, Hilmas DE. Application of beryllium antibodies in risk assessment and health surveillance: Two case studies. Toxicol. Ind. Health, 1995;11:399-411.

Comhair SAA, Lewis MJ, Bhathena PR, Hammel JP, Erzurum SC. Increased glutathione and glutathione peroxidase in lungs of individuals with chronic beryllium disease. Am. J. Respir. Crit. Care Med., 1999;159:1824-1829.

Cotes JE, Gilson JC, McKerrow CB, Oldham PD. A long-term follow-up of workers exposed to beryllium. Br. J. Ind. Med., 1983;40:13-21.

Cullen MR, Kominsky JR, Rossman MD, Cherniack MG, Rankin JA, Balmes JR, Kern JA, Daniele RP, Palmer L, Naegel GP, McManus K, Cruz R. Chronic beryllium disease in a precious metal refinery: Clinical epidemiologic and immunologic evidence for continuing risk from exposure to low level beryllium fumes. Am. Rev. Respir. Dis., 1987;135:201-208.

Daniloff E, Barnard J, Barker E, Solida M, Newman LS. Methotrexate treatment in chronic beryllium disease. Am. J. Respir. Crit. Care Med., 1998;157:A146.

Deodhar SD, Barna B, Van Ordstrand HS. A study of the immunological aspects of chronic berylliosis. Chest, 1973;63:309-313.

Eisenbud M, Wanta RC, Dustan C, Steadman LT, Harris WB, Wolf BS. Non-occupational berylliosis. J. Ind. Hyg. Assoc., 1949;31:282-294.

Eisenbud M. Origins of the standards for control of beryllium disease (1947-1949). Environ. Res., 1982;27:79-88.

Eisenbud M. *An Environmental Odyssey: People, Pollution, Politics in the Life of a Practical Scientist.* University of Washington Press, Seattle, WA, 1990.

Eisenbud M. The standard for control of chronic beryllium disease. Appl. Occup. Environ. Hyg., 1998;13:25-31.

Eisenbud M, Lisson J. Epidemiological aspects of beryllium-induced nonmalignant lung disease: A 30-year update. J. Occup. Med., 1983;25:196-202.

Finch GL, Mewhinney JA, Hoover MD, Eidson AF, Haley PJ. Clearance, translocation, and excretion of beryllium following inhalation of beryllium oxide by beagle dogs. Fundam. Appl. Toxico.,l 1990;15:231-241.

Finch GL, Hoover MD, Hahn FF, Nikula KJ, Belinsky SA, Haley PJ, Griffith WC. Animal models of beryllium-induced lung disease. Environ. Health Perspect., 1996;104(Suppl 5):973-979.

Finch GL, Nikula KJ, Hoover MD. Dose-response relationships between inhaled beryllium metal and lung toxicity in C_3H mice. Toxicol. Sci., 1998;42:36-48.

Fontenot AP, Kotzin BL, Comment CE, Newman LS. Expansions of T-cell subsets expressing particular T-cell receptor variable regions in chronic beryllium disease. Am. J. Respir. Cell Mol. Biol., 1998;18:581-589.

Freiman DG, Hardy HL. Beryllium Disease. The relation of pulmonary pathology to clinical course and prognosis based on a study of 130 cases from the U.S. Beryllium Case Registry. Human Pathol., 1970;1:25-44.

Galbraith GM, Pandey JP, Schmidt MG, Arnaud P, Goust JM. Tumor necrosis factor-alpha gene expression in human monocytic THP-1 cells exposed to beryllium. Arch. Environ. Health, 1996;51:29-33.

Haley PJ, Finch GL, Mewhinney JA, Harmsen AG, Hahn FF, Hoover MD, Bice DE. A canine model of beryllium-induced granulomatous lung disease. Lab. Invest., 1989;61:219-227.

Haley PJ. Mechanisms of granulomatous lung disease from inhaled beryllium: The role of antigenicity in granuloma formation. Toxicol. Pathol., 1991;19:514-525.

Haley PJ, Finch FL, Hoover MD, Mewhinney JA, Bice DE, Muggenburg BA. Beryllium-induced lung disease in the dog following repeated BeO exposure. Environ. Res., 1992;59:400-415.

Haley PJ, Pavia KF, Swafford DS, Davila DR, Hoover MD, Finch GL. The comparative pulmonary toxicity of beryllium metal and beryllium oxide in cynomolgus monkeys. Immunopharmacol. Immunotoxicol., 1994;16:627-644.

Haley PJ, Swafford DS, Finch GL, Hoover MD, Muggenburg BA, Johnson NF. Immunologic specificity of lymphocyte cell lines from dogs exposed to beryllium oxide. Immunopharmacol. Immunotoxicol., 1997;19:459-471.

Hardy HL, Tabershaw IR. Delayed chemical pneumonitis occurring in workers exposed to beryllium compounds. J. Ind. Hyg. Toxicol., 1946;28:197-211.

Hardy H. Beryllium disease: A clinical perspective. Environ. Res., 1980;21:1-9.

Harris J, Bartelson BB, Barker E, Balkissoon R, Kriess K, Newman LS. Serum neopterin in chronic beryllium disease. Am. J. Ind. Med., 1997;32:21-26.

Huang H, Meyer KC, Kubai L, Auerbach R. An immune model of beryllium-induced pulmonary granulomata in mice: Histopathology, immune reactivity, and flow-cytometric analysis of broncho-alveolar lavage-derived cells. Lab. Invest., 1992;67:138-146.

Hyslop F, Palmes ED, Alford WC, Monaco AR, Fairhall LT. *The Toxicology of Beryllium.* U.S. Public Health Service, Washington, DC, National Institute of Health - Bulletin No. 181, 1943.

Infante PF, Wagoner JK, Sprince NL. Mortality patterns from lung cancer and non-neoplastic respiratory disease among white males in the Beryllium Case Registry. Environ. Res., 1980;21:35-43.

Johnson NR. Beryllium disease among workers in a spacecraft-manufacturing plant - California, MMWR, 1983;32:419-420,425.

Kotloff RM, Richman PS, Greenacre JK, Rossman MD. Chronic beryllium disease in a dental laboratory technician. Am. Rev. Respir. Dis., 1993;147:205-207.

Kriebel D, Brain JD, Sprince NL, Kazemi H. The pulmonary toxicity of beryllium. Am. Rev. Respir. Dis., 1988;137:464-473.

Kriess K, Newman LS, Mroz MM, Campbell PA. Screening blood test identifies subclinical beryllium disease. J. Occup. Med., 1989;31:603-608.

Kriess K, Mroz MM, Zhen B, Martyny JW, Newman LS. Epidemiology of beryllium sensitization and disease in nuclear workers. Am. Rev. Respir. Dis., 1993a;148:985-991.

Kriess K, Wasserman S, Mroz MM, Newman LS. Beryllium disease screening in the ceramics industry. J. Occup. Med., 1993b;35:267-274.

Kriess K, Miller F, Newman LS, Ojo-Amaize EA, Rossman MD, Saltini C. Chronic beryllium disease: From the workplace to cellular immunology, molecular immunogenetics, and back. Clin. Immunol. Immunopathol., 1994;71:123-129.

Kriess K, Mroz MM, Newman LS, Martyny J, Zhen B. Machining risk of beryllium disease and sensitization with median exposures below 2 μg/m^3. Am. J. Ind. Med., 1996;30:16-25.

Kriess K, Mroz MM, Zhen B, Wiedemann H, Barna B. Risks of beryllium disease related to work processes at a metal, alloy and oxide production plant. Occup. Environ. Med., 1997;54:605-612.

Levin L. Letter to the Editor. Appl. Occup. Environ. Hyg., 1991;6:567.

Lympany PA, Petrek M, Southcott AM, Newman-Taylor AJ, Welsh KI, du Bois RM. HLA-DPB polymorphisms: Glu69 association with sarcoidosis. Eur. J. Immunogenet., 1996;23:353-359.

MacMahon B. The epidemiological evidence on the carcinogenicity of beryllium in humans. J. Occup. Med., 1994;36:15-24.

McConnochie K, Williams WR, Kilpatrick GS, Jones-Williams W. Chronic beryllium disease in identical twins. Br. J. Dis. Chest, 1988;82:431-435.

Maier LA, Kittle LA, Tinkle SS, Newman LS. IL-4 regulation of cytokine production and gene expression in chronic beryllium disease. Am. J. Respir. Crit. Care Med., 1998;157:A30.

Maier LA, Raynolds MV, Young DA, Barker EA, Newman LS. Angiotensin-1 converting enzyme polymorphisms in chronic beryllium disease. Am. J. Respir. Crit. Care Med., 1999;159:1342-1350.

Mancuso TF. Mortality study of beryllium industry worker's occupational lung cancer. Environ. Res., 1980;21:48-55.

Marshall, E. Beryllium screening raises ethical issues. Science, 1999;285:178-179.

Newman LS, Campbell PA. Mitogenic effect of beryllium sulfate on mouse B lymphocytes but not T lymphocytes *in vitro*. Intl. Arch. Allergy Appl. Immunol., 1987;84:223-227.

Newman LS, Kriess K, King TE, Seay S, Campbell PA. Pathologic and immunologic alterations in early stages of beryllium disease: Re-examination of disease definition and natural history. Am. Rev. Respir. Dis., 1989;139:1479-1486.

Newman LS, Kriess K. Nonoccupational beryllium disease masquerading as sarcoidosis: Identification by blood lymphocyte proliferation response to beryllium. Am. Rev. Respir. Dis., 1992a;145:1212-1214.

Newman LS, Mroz MM, Schumacher B, Daniloff E, Kreiss K. Beryllium sensitization precedes chronic beryllium disease. Am. Rev. Respir. Dis., (Suppl) 1992b;145:A324.

Newman LS. "Beryllium." Chapter 83. In *Hazardous Materials Toxicology*. 1992c;882-890.

Newman LS. The natural history of beryllium sensitization and chronic beryllium disease. Environ. Health Perspect., 1996;104(Suppl 5):937-943.

Newman LS, Balkissoon R, Daniloff E, Solida M, Mroz M. Rate of progression from beryllium sensitization to chronic beryllium disease is 9-19% per year. Am. J. Respir. Crit. Care Med., 1998;157:A145.

Nikula KJ, Swafford DS, Hoover MD, Tohulka MD, Finch GL. Chronic Granulomatous pneumonia and lymphocyte responses induced by inhaled beryllium metal in A/J and C_3H/HeJ mice. Toxicol. Pathol., 1997;25:2-12.

NIOSH:National Institute for Occupational Safety and Health [U.S. Department of Health, Education and Welfare] *Occupational Exposure to Beryllium*. Washington, DC: Health Services and Mental Health Administration- NIOSH, HSM 72-10268, 1972.

NIOSH: National Institute for Occupational Safety and Health [U.S. Department of Health, Education and Welfare] Health Hazard Evaluation Determination, *Report No. 78-28-480 Persolite Products Inc.: Florence, Colorado*. Cincinnati, OH: NIOSH, NIOSH-TR-HHE-78-028-480, 1978.

Ojo-Amaize EA, Agopian MS, Peter JB. Novel *in vitro* method for identification of individuals at risk for beryllium hypersensitivity. Clin. Diagn. Lab. Immunol., 1994;1:164-171.

Powers MB. "History of Beryllium." In *Beryllium: Biomedical and Environmental Aspects*, MD Rossman, OP Pruess, MB Powers, eds. Baltimore: Williams & Wilkins, 1991, pp. 9-24.

Preuss OP. "Assessment of Risk Potential." In *Beryllium: Biomedical and Enviromental Aspects*, MD Rossman, OP Pruess, MB Powers, eds. Baltimore: Williams & Wilkins, 1991, pp. 263-273.

Reeves AL. "Experimental Pathology." In *Beryllium: Biomedical and Environmental Aspects*, MD Rossman, OP Pruess, MB Powers, eds. Baltimore: Williams & Wilkins, 1991a, pp. 59-76.

Reeves AL. "Toxicokinetics." In *Beryllium: Biomedical and Environmental Aspects*, MD Rossman, OP Pruess, MB Powers, eds. Baltimore: Williams & Wilkins, 1991b, pp. 77-86.

Reeves AL. "Toxicodynamics." In *Beryllium: Biomedical and Environmental Aspects*, MD Rossman, OP Pruess, MB Powers, eds. Baltimore: Williams & Wilkins, 1991c, pp. 87-93.

Richeldi L, Sorrentino R, Saltini C. HLA-DPB 1 glutamate 69: A genetic marker of beryllium disease. Science, 1993;262:242-244.

Richeldi L, Kreiss K, Mroz MM, Zhen B, Tartoni P, Saltini C. Interaction of genetic and exposure factors in the prevalence of berylliosis. Am. J. Ind. Med., 1997;32:337-340.

Ridenour PK, Preuss OP. "Acute Pulmonary Beryllium Disease." In *Beryllium: Biomedical and Environmental Aspects*, MD Rossman, OP Pruess, MB Powers, eds. Baltimore: Williams & Wilkins, 1991, pp.. 103-112.

Rossman MD, Kern JA, Elias JA, Cullen MR, Epstein PE, Preuss OP, Markham TN, Daniele RP. Proliferative response of bronchoalveolar lymphocytes to beryllium. Ann. Intern. Med., 1988;108:687-693.

Rossman MD, Jones-Williams W. "Immunopathogenesis of Chronic Beryllium Disease." In *Beryllium: Biomedical and Environmental Aspects*, MD Rossman, OP Pruess, MB Powers, eds. Baltimore: Williams & Wilkins, 1991, pp. 121-132.

Saltini C, Winestock K, Kirby M, Pinkston P, Crystal RG. Maintenance of alveolitis in patients with chronic beryllium disease by beryllium-specific helper T-cells. N. Engl. J. Med., 1989;320:1103-1109.

Stange AW, Hilmas DE, Furman FJ. Possible health risks from low level exposure to beryllium. Toxicology, 1996;111:213-224.

Steenland K, Ward E. Lung cancer incidence among patients with beryllium disease: A cohort mortality study. J. Natl. Cancer Inst., 1991;83:1380-1385.

Sterner JH, Eisenbud M. Epidemiology of beryllium intoxication. Arch. Ind. Hyg. Occup. Med., 1951;4:123-151.

Stonehouse AJ, Zenczak S. "Properties, Production Processes, and Applications." In *Beryllium: Biomedical and Environmental Aspects*, MD Rossman, OP Pruess, MB Powers, eds. Baltimore: Williams & Wilkins, 1991, pp. 27-58.

Stubbs J, Argyris E, Lee CW, Monos D, Rossman M. Genetic markers in beryllium hypersensitivity. Chest, 1996;109(3 Suppl):45S.

Tabershaw IR. *The Toxicology of Beryllium*, Rockville, MD: U.S. Public Health Service, PHS-PUB-2173, 1972.

Thomassen MJ, Dweik RA, Buhrow LT, Jacobs BS, Saltini C, Deubner DC, Wiedemann HP, Barna BP. Beryllium-induced interferon-gamma production by bronchoalveolar lavage cells is diagnostic for berylliosis in beryllium-exposed workers. Am. J. Respir. Crit. Care Med., 1998;157:A145.

Tinkle SS, Schwitters PW, Newman LS. Cytokine production by bronchoalveolar lavage cells in chronic beryllium disease. Environ. Health Perspect., 1996;104(Suppl.5):969-971.

Tinkle SS, Little LA, Schumacher BA, Newman LS. Beryllium induces IL-2 and IFN-gamma in berylliosis. J. Immunol., 1997;158:518-526.

Tinkle SS, Newman LS. Beryllium-stimulated release of tumor necrosis factor-alpha, interleukin-6, and their soluble receptors in chronic beryllium disease. Am. J. Respir. Crit. Care Med., 1997;156:1884-1891.

Tinkle SS, Kittle LA, Newman LS. Partial IL-10 inhibition of the cell-mediated immune response in chronic beryllium disease. J. Immunol., 1999;163:2747-2753.

U.S. DOE: United States Department of Energy, Environment Health & Safety EH2. Case of berylliosis confirmed. Serious Accidents, 1985;9:1-3.

USEPA: United States Environmental Protection Agency. *Health Assessment Document for Beryllium*. Washington, DC: EPA, EPA/600/8-84/026F, 1987.

USEPA: United States Environmental Protection Agency. *Integrated Risk Information Systems: Beryllium and Compounds*. EPA, IRIS Database, April 3, 1998.

Van Ordstrand HS, Hughes R, Carmody MG. Chemical Pneumonitis in workers extracting beryllium oxide: Report of 3 cases. Cleveland Clin. Quart., 1943;10:10.

Vorwald AJ, ed. *Pneumoconiosis (Sixth Saranac Symposium). 1950.* New York: Paul B. Hoeber Publishing, 1950.

Votto JJ, Barton RW, Gionfriddo MA, Cole SR, McCormick JR, Thrall RS. A model of pulmonary granulomata induced by beryllium sulfate in the rat. Sarcoidosis, 1987;4:71-76.

Wagoner JK, Infante PF, Bayliss DL. Beryllium: An etiologic agent in the induction of lung cancer, non-neoplastic respiratory disease, and heart disease among industrially-exposed workers. Environ. Res., 1980;21:15-34.

Wang Z, White PS, Petrovic M, Tatum OL, Newman LS, Maier LA, Marrone BL. Differential susceptibilities to chronic beryllium disease contributed by different Glu69 HLA-DPB1 and –DPA1 alleles. J. Immunol., 1999;163:1647-1653.

Ward E, Okun A, Ruder A, Fingerhut M, Steenland K. A mortality study of workers at seven beryllium processing plants. Am J Ind Med 1992;22:885-904.

WHO: World Health Organization, International Programme on Chemical Safety, *Environmental Health Criteria 106: Beryllium*, Geneva: World Health Organization, 1990.

WHO: World Health Organization. *Beryllium, Cadmium, Mercury, and Exposures in the Glass Manufacturing Industry, IARC Monographs on the Evaluation of Carcinogenic Risk to Humans. Vol. 58,* Lyon, France: WHO, International Agency for Research on Cancer, 1993, pp. 41-117.

10. ARSENIC, CADMIUM, CHROMIUM, AND NICKEL

Leigh Ann Burns-Naas, Ph.D., D.A.B.T.
Toxicology and Product Safety
Health and Environmental Sciences
Dow Corning Corporation
Midland, MI 48611

INTRODUCTION

Of all the malignancies by which man can be afflicted, lung cancer is the most common in the United States, and is second only to bladder cancer in the relative proportion of cases thought to be a result of occupational exposures (Steenland et al., 1996). Although over half of these cases are a result of exposure to asbestos, excessive occupational exposures to arsenic (As), cadmium (Cd), chromium (Cr), and nickel (Ni) (among others) have been documented in a recent epidemiological review of occupational lung carcinogens to result in an increased risk for this malignancy (Steenland et al., 1996). Systemic exposure to these metals as a result of inhalation of particles or fumes may contribute to this increased risk. Additionally, metal-induced suppression of pulmonary immune function may either result in this increased risk or may exacerbate other more direct causes. Evidence is rapidly building that the immune system plays a role in tumor identification and rejection (immune surveillance). Xenobiotic-induced suppression of pulmonary immune function may also be a cause of increased pulmonary infection noted in workers exposed occupationally to some specific agents such as metals. Thus, it is important to understand and to continue to expand our knowledge base on how these materials can modulate immune function.

This chapter will explore the sources of and possible levels of exposure to As, Cd, Cr, and Ni, as well as what is currently known about the effects that pulmonary exposure has on lung immunity. Systemic immunomodulation by these metals will not be specifically addressed except where those studies may relate to possible mechanisms of action. The reader is referred to other recent reviews on systemic metal immunomodulation for more detailed information on that topic (Chang, 1996; Thomas and Zelikoff, 1998).

ARSENIC

Sources of Exposure and Encounterable Levels

Arsenic has held the attention of scientists, writers, and the general population for hundreds of years due to its use as both a medicinal agent and a poison. Its history can be traced back to the time of Hippocrates (460 - 377 BC) and is rather complex (reviewed by Dickerson, 1992). More recently, the use of arsenical compounds has been as toxic agents in pesticides, herbicides, fungicides, in metal smelters and mining, in glass production, in the pharmaceutical and microelectronics industries, and in chemical warfare (Hill et al., 1948; Watrous and McCaughey, 1977; Mabuchi et al., 1979; La Dou, 1986; Farmer and Johnson, 1990; Steenland, 1996). The recent focus in toxicology, and the "new frontier" has been the understanding of the toxicity of arsine (AsH_3). Together, these industries represent the majority of the current uses of both inorganic and organic arsenicals and, thus, the primary means of occupational exposures.

Although there are several potential routes of exposure to As-containing compounds, the primary means of occupational exposure is inhalation of dusts of As compounds or materials which contain or are contaminated with As (Landrigan, 1992). Because the particle size is often relatively large (~5 μm), these particles tend to settle in the upper airways. Smaller particles, capable of passing further into the alveolar space, represent a means of pulmonary absorption of As. The greatest occupational risk appears to be from inorganic as opposed to organic As.

The primary occupations exposed to significant inorganic As levels are metal smelting and manufacture of arsenical pesticides. Exposure also occurs in the blasting and mining of metallic ores, as As is naturally associated with several minerals, including cobalt, copper, gold, lead, and zinc (Maloney, 1996). In the early 1980's it was estimated that ≈58,000 U.S. workers were exposed to airborne As excluding the mining industry (NIOSH, 1990; Steenland, 1996). Other estimates place the potential for occupational exposure to inorganic As in the range of 1.5 million workers (including mining/smelting) with an annual consumption of As in the United States ranging between 25 - 35,000 tons annually (NIOSH, 1975; Landrigan, 1992). Emissions from smelters, geothermal and coal-burning power plants continue to be a source of environmental As contamination. Coal contains As in amounts that can be as great as 1.5 g As/kg; in the United States, average content is 5 mg As/kg (Dickerson, 1994). Recently, individuals employed in the semiconductor industry, producing gallium arsenide (GaAs) or indium arsenide (InAs), have also been determined to be at risk. Although manufacture of GaAs requires use of arsine gas, in the industrial setting the primary route of GaAs exposure is inhalation of GaAs particles generated in the sawing and polishing of GaAs wafers. It has been estimated that between 50 - 60% of the GaAs crystals are converted into GaAs dust in this processing step, and these particles are capable of penetrating deep into the lung (Briggs and Owen, 1980). In 1983, Willardson reported that manufacturing processes used 5 - 10 tons of As in GaAs devices and that production was expected to increase 3- to 10-fold by 1990. A study by SRI International projected that in 1995, over 4% of total integrated circuit sales would be GaAs (Harrison, 1986). During the 1950s it was observed that tobacco often contained As at concentrations as high as 40 mg

As/kg, possibly resulting in daily inhalation of up to 100 mg As (Dickerson, 1994). Although this is not necessarily related to occupational exposure, it is important to consider this information as this exposure may have contributed to the propensity of lung cancers noted in society today.

The Occupational Safety and Health Administration (OSHA) has a recommended time-weighted average (8 hr) permissible exposure limit (PEL) for inorganic As of 10 mg As/m^3 (OSHA, 1997). The National Institute for Occupational Safety and Health (NIOSH) recommends an exposure limit (REL) for As and all its compounds of 2 $\mu g/m^3$ as a ceiling in a 15 min air sampling (NIOSH, 1982). Finally, the threshold limit value (TLV) set by the American Conference of Government Industrial Hygienists (ACGIH) for As and its soluble compounds is 0.2 mg/m^3 (ACGIH, 1997). For arsine, the OSHA PEL and ACGIH TLV has been set at 0.05 ppm. Presently, GaAs is regulated on the basis of inorganic As toxicity data with a NIOSH REL of 2 $\mu g/m^3$ as a ceiling in a 15 minute air sampling (NIOSH, 1992). Although it has been demonstrated recently that GaAs dissociates into its component metals, gallium and As, following intratracheal (IT) instillation or oral exposure in rats (Webb et al., 1984) and IT exposure in mice (Burns et al., 1991), the actual levels of As exposure in GaAs-exposed workers remains unknown.

Pulmonary Immunomodulation

General Pulmonary Toxicity

Although there is considerable information available on the immunotoxicology of As and its compounds, the primary focus has been systemic immunity, a topic recently reviewed (Burns, 1998) and therefore not addressed here. This section will review only the immunological effects occurring in the lung as a result of exposure to arsenicals. In general, experimental pulmonary exposure to As-containing compounds has been shown to produce events in the lung which should be considered when attempting to draw conclusions from the existing database. IT-instillation of arsenic trioxide (As$_2$O$_3$; 13 mg/kg) or GaAs (1.5 - 52 mg/kg resulted in observations of marked irritation and hyperplasia in the lungs of rats and hamsters (Webb et al., 1986; Ohyama et al., 1988; Garcia-Vargas and Cebrian, 1996). In these studies, pulmonary toxicity was not observed to be specific to As exposure. Huaux et al. (1995) also observed pulmonary inflammation (macrophage and neutrophil influx) following IT instillation of 5 mg As$_2$O$_3$/kg in rats. Other studies in which hamsters were exposed to 7.5 mg As (as InAs) IT once per week for 15 wk evidenced several histopathological events in the lung including an increased rate of proteinosis-like lesions, alveolar or bronchiolar hyperplasia, pneumonia, and metaplasia (Tanaka et al., 1996).

Humoral Immunity

There are only limited studies on humoral immunity following pulmonary As exposure, and all examine systemic effects. Pulmonary exposure to GaAs (50 - 200 mg/kg) or sodium arsenite ($NaAsO_2$; 5.7 mg/kg) has been shown to suppress the ability of mice to mount systemic primary or secondary immune responses to the T-dependent sheep erythrocyte (SRBC) antigen (Sikorski et al., 1989). In an epidemiological study of workers in a thermal power plant burning coal containing 900 - 1500 g of As per dry ton weight, there was no effect of exposure on basal IgG, IgA or IgM levels in the sera of these workers when compared to control workers from a different power plant where the As content in the coal was 10-fold less. Apparent discrepancies between these studies may be related to exposure concentration and duration, basal vs antigen-specific responses, or other factors.

Cell-Mediated Immunity

No studies specifically examining pulmonary cellular immunity following As exposure could be found in the published literature. However, the existing database suggests that inorganic As (pulmonary or systemic exposure) can interfere with cellular immunity in both animals and humans and that overall, inorganic As interferes with immune function (Garcia-Vargas and Cebrian, 1996; Burns, 1998).

Innate/Non-Specific Immunity

The majority of studies associated with arsenical-induced pulmonary immuno-modulation have focused on innate immunity. Systemically, pulmonary exposure to arsenicals has been demonstrated to produce decreases in the levels of several complement proteins (5.7 mg $NaAsO_2$/kg; Sikorski et al., 1989) in mice and to alter that of several serum acute phase proteins (i.e., transferrin, orosomucoid, and ceruloplasmin) in humans (Bencko et al., 1988).

As mentioned previously, inhalation represents a primary route of exposure to As. Although a significant amount may be transported up the mucociliary escalator and subsequently swallowed or be detoxified in the lung itself, the alveolar macrophage plays a critical role in recognition and elimination of foreign materials within the respiratory tract. Phenylarsine oxide has been demonstrated to possess anti-inflammatory activity by inhibiting NADPH oxidase-dependent production of superoxide anion ($\cdot O_2^-$) in alveolar neutrophils (Roussin et al., 1997). Lantz et al. (1994) compared the *in vivo* (1 mg/kg, IT, 1 d) toxicity of both oxidation states of soluble As [sodium arsenite (III) or sodium arsenate (V)] upon rat alveolar macrophages, using the production of $\cdot O_2^-$, prostaglandin E_2 (PGE_2) and tumor necrosis factor-alpha ($TNF\alpha$) as endpoints. Alveolar macrophages from As(V)-exposed rats showed significant increases in $\cdot O_2^-$ production. Exposure to either form of As decreased baseline and LPS-induced $TNF\alpha$ production by alveolar macrophages and resulted in an inflammatory response in the lung (neutrophil influx). In similar *in vivo* studies using slightly soluble forms of As [arsenic (III) trisulfide or calcium arsenate (V)], Lantz et al. (1995) demonstrated rat alveolar macrophages exposed to slightly soluble As(V) had increased $\cdot O_2^-$ production and basal $TNF\alpha$ release, whereas responses from cells exposed to slightly soluble As(III) were not different from controls. The Authors concluded that alterations in alveolar macrophages may compromise host defense. These findings and conclusions are supported by observations by Aranyi et al. (1985) that inhalation of 0.94 mg As_2O_3/m^3 increased susceptibility of mice to infection with *Streptococcus pneumonia* and also decreased pulmonary bactericidal activity against radiolabeled *Klebsiella pneumoniae.*

A recent study evaluating the effect of IT-instilled coal fly ash or copper smelter dust (delivering 20 mg As/kg) showed a 90% reduction in LPS-induced $TNF\alpha$ release up to 30 d post-instillation (Broeckaert et al., 1997). The persistent suppression of $TNF\alpha$ production was suggested to be related to the ability of pulmonary phagocytes to clear dust particles and their metal content from the lung. This possible decreased clearance may be related to the loss of bactericidal activity in these cells observed by Aranyi et al. (1985). In contrast to the studies by Lantz et al. (1994, 1995) and Broeckaert et al. (1997), Huaux et al. (1995) found that IT instillation of 5 mg As_2O_3/kg stimulated the production of $TNF\alpha$ as well as interleukin (IL)-1 in rat phagocytes obtained by bronchoalveolar lavage. As_2O_3 was also found to stimulate production of fibronectin and cystatin-c. Although the composition of the culture wells from these exposed phagocytes contained similar numbers of alveolar macrophages, they also contained neutrophils, which may account for the differences observed between this study and others.

In vitro exposure of alveolar macrophages to trivalent As results in suppression of phagocytic ability as well as a diminution of $\cdot O_2^-$ production (Fisher et al., 1986; Labedzka et al., 1989). *In vitro* phenylarsine oxide exposure also inhibited the stimulus-induced generation of $\cdot O_2^-$ in both macrophages and neutrophils obtained from the BAL fluid of rats (Roussin et al., 1997); these studies are supported by the observations of Lantz and colleagues (1994, 1995). *In vitro* assessments (0.1 - 300 µg/ml) with slightly soluble forms of As - arsenic (III) trisulfide and calcium arsenate (V) - indicated both oxidation forms of the metal produced significant dose-related suppression of $\cdot O_2^-$ production and LPS-induced $TNF\alpha$ release by rat alveolar macrophages with no effect on PGE_2 production. In contrast, *in vitro* exposure to the soluble forms of As [sodium arsenite (III) or sodium arsenate (V)] showed that As(III) was 10-fold more potent than As(V) at inhibiting $\cdot O_2^-$ production and LPS-induced $TNF\alpha$ release. Curiously, As(V) produced suppression of LPS-induced PGE_2 release while As(III) had no effect. Other Investigators had previously been unable to demonstrate an effect of As(V) on $\cdot O_2^-$ production (Castranova et al., 1984). *In vitro* administration of As_2O_3 (50 ng/ml) to rat pulmonary phagocytes has been shown to have no effect on basal or stimulated $TNF\alpha$ or IL-1 release, or on fibronectin or cystatin-c production (Huaux et al., 1995), although suppression of stimulus-induced $\cdot O_2^-$ production by As_2O_3 (IC_{50} ~15 mM) has been reported (Geertz et al., 1994).

Mechanism of Action

What we have learned over these many years of experimentation with arsenicals is that there is no single mechanism associated with As toxicity (Squibb and Fowler, 1983; Dickerson, 1994; Goyer, 1996). Briefly, there are several possible broad mechanisms of toxicity of As. Arsenic may interact with critical enzyme thiol (-SH) groups or persulfide residues and inhibit function (trivalent). It may act by the inhibition of endogenous substrate binding through competitive interactions of structurally similar organic arsenicals (pentavalent). Finally, As may interact with secondary molecules or enzyme substrates to inhibit metabolic reactions, and As may disrupt energy systems or ion balances within cells.

In a broad sense, the mechanism of pulmonary immunotoxicity is likely to be related to functional deficits in the phagocytic and/or cytotoxic populations residing in or recruited to the lung. Decreased resistance to the B16F10 melanoma, which forms nodules in the lung, has been demonstrated following pulmonary exposure to GaAs (Sikorski et al., 1989). No pulmonary natural killer (NK) activity assessment was made to determine their function, although systemic NK activity was enhanced. As-Related decreases in host resistance to bacterial pathogens with concomitant decreases in *in vivo* bactericidal activity (Aranyi et al., 1985) support the hypothesis that alveolar macrophages are suppressed in their phagocytic capacity. Additionally, it appears that As exposure alters the release of several inflammatory mediators including $\cdot O_2^-$, TNFα, and IL-1. These changes, particularly the increased $\cdot O_2^-$ production, are likely related, at least in part, to the As-induced and histologically-diagnosed inflammatory responses reported by some Investigators after pulmonary exposure to arsenicals. The effects observed on these inflammatory mediators (enhancement or suppression) are typical of what is seen with systemic As immuno-modulation. That is, the effects observed appear to be dependent on the form of As (III or V), its solubility, concentration, and whether exposure occurred *in vivo* or *in vitro*. Although this muddles the mechanistic picture somewhat, it can be stated that exposure to As or As-containing compounds alters pulmonary phagocytic function, probably resulting in lower cytotoxicity and/or lower pathogen clearance from the lung. A biochemical or molecular immunologically-related mechanism is uncertain at this time. However, systemic alterations (apparently antigen-dependent) in phago-cytic ability have been shown following pulmonary GaAs exposure (Sikorski et al., 1989) and the expression of MHC Class II on splenocytes (macrophages and B-lymphocytes) appears to be decreased. Additionally, alterations in several adhesion molecules have been shown, suggesting the ability of macrophages to recognize and bind to or phagocytize non-self matter may be a targeted function.

As stated at the beginning of this section on As, lung cancer is the most common malignancy in the United States, and a substantial proportion of cases are thought to be a result of occupational exposures. The evidence for occupational or accidental exposure to many metals, including As, having a carcinogenic result is convincing and apparently dose-related (WHO, 1981; EPA, 1987; Snow, 1992). Trivalent As can interfere with replication of deoxyribonucleic acid (DNA), with alkyl transferase and excision repair (Petres et al., 1977; McCabe et al., 1983; Squibb and Fowler, 1983; Aposhian, 1989), while pentavalent forms may substitute for phosphate and form unstable esters with DNA (Squibb and Fowler, 1983; Aposhian, 1989). Arsenic has also been implicated in alterations of gene expression such as both induction and amplification (Squibb and Fowler, 1983; Aposhian, 1989). It has been demonstrated that pulmonary exposure to As_2O_3 increased the frequency of chromosomal aberrations in the peripheral lymphocytes of smelter workers (Beckman et al., 1977; Nordenson et al., 1978), and in general the data to date indicate that As is clastogenic (Garcia-Vargas and Cebrian, 1996). While As is not active in inducing specific gene mutations, it has been shown to cause cell transformation and an increased frequency of SA7 virus-induced transformations in Syrian hamster embryo cells (Casto et al., 1979; DiPaolo et al., 1979; Lee et al., 1985). Arsenic induces aberrant DNA synthesis and produces chromosomal aberrations in cultured cells (Sunderman, 1979; Lee et al., 1988). The International Agency for Research on Cancer (IARC) considers that there is sufficient evidence for human carcinogenicity associated with exposure to As, although until recently animal data has been lacking. Yamanaka et al. (1989, 1995) found that a metabolite of As, dimethylarsenic acid (DMA) was capable of inducing lung-specific DNA damage in mice and rats, findings which were confirmed in human pulmonary cells. A subsequent multi-organ promotional study (Yamamoto et al., 1995) demonstrated than DMA was capable of promoting carcinogenesis. This is the first demonstration of an animal model of As-induced promotion.

Can these observations regarding As and cancer be related to the observed immunomodulatory effects? Probably. In fact, several studies strongly suggest that the carcinogenic potential of As does not lie solely in the genetic activity of As, but rather in the alterations in immunocompetence induced by As exposure (reviewed in Burns, 1998). Suppression of cytotoxicity of pulmonary phagocytes and their ability to recruit other phagocytic cells may allow less soluble forms of As or As-contaminated dusts to remain in the lung for longer periods of time and thus exert or enhance carcinogenic events. Can systemic immunomodulatory effects of pulmonary As exposure be related to increased cancer risk? Probably, and for similar reasons. More work is needed to better understand the pulmonary immunomodulation produced by As exposure and how it is interrelated with systemic immunotoxicity and carcinogenic potential.

CADMIUM

Sources of Exposure and Encounterable Levels

Cadmium is a white-blue metallic element which arises mainly as a byproduct of the mining and refining of zinc ores, but may also be found complexed naturally with lead (Newman-Taylor, 1992). It has been estimated that the global annual Cd production capacity and consumption averages 25,700 metric tons in mining and 27,000 metric tons which are refined (Lauwerys, 1994), and NIOSH has estimated that approximately 250,000 U.S. workers were exposed occupationally to Cd in the early 1980's (Steenland, 1996). Cadmium is principally used in electroplating, in the production of alloys, brazing solders, and in Ni-Cd batteries, in the manufacture of compounds that serve as stabilizers for plastics, and as pigments (IARC, 1976; Shaller and Angerer, 1992; Steenland, 1996). Most occupational exposures occur via inhalation of Cd-containing fumes and dusts. Depending on the size of the particles, it has been estimated that between 10 - 50% of inhaled Cd may be deposited in the alveolar spaces of the lung. Although some is cleared through normal physiological mechanisms, absorption of the remaining metal may occur. Absorption depends on the chemical form and probably on its solubility (e.g., cadmium chloride ($CdCl_2$)>

cadmium sulfate (CdSO$_4$)> cadmium oxide (CdO)> cadmium sulfide) (Lauwerys, 1994; Koller, 1998).

Because of the relatively high volatility of the metal, inhalation hazards are particularly important for workers welding Cd-plated materials or using silver solder, which may contain up to 25% Cd (Lauwerys, 1994). In the past, Cd concentrations in workplaces were high; air levels of 1-10 mg Cd/m^3 were found in alkaline battery factories in the 1950s (Schaller and Angerer, 1992). The primary environmental source of inhaled Cd is tobacco smoke and it has been estimated that on average, each cigarette contains as much as 2 mg Cd, all of which may be inhaled (Newman-Taylor, 1992). Estimates of air levels of Cd in cities in the United State ranged from 5 -40 ng/m^3, most of which is in the relatively insoluble form of CdO (Koller, 1998). The IARC considers that there is sufficient evidence for human carcinogenicity associated with exposure to Cd and its compounds (IARC, 1993). OSHA has set a PEL of 0.1 mg/m^3 (8hr TWA) for Cd fumes (as oxides) and 0.2 mg/m^3 (8hr TWA) for Cd dust. The ACGIH TLVs are lower: 0.01 mg/m^3 for Cd in inhalable fractions and 0.002 mg/m^3 for Cd in respirable fractions/dust.

Pulmonary Immunomodulation

Humoral Immunity

The effects of Cd on the humoral immune system have been recently reviewed (Koller, 1998) and the majority of studies examine the systemic effects of Cd following exposures from routes other than inhalation or IT instillation. An acute (i.e., 2 hr) exposure of mice to 190 g CdCl$_2$/m^3 resulted in suppression of the primary (IgM) antibody-forming cell response (Graham et al., 1978). Similarly, mice exposed by inhalation to 880 mg CdCl$_2$/m^3 for 1 hr also had a decreased IgM AFC response which was accompanied by a marked decrease in splenocyte viability 1 wk following exposure (Krzystyniak et al., 1987). In that same study, mice displayed suppressed lymphoproliferative responses to LPS. *In vitro* studies with CdCl$_2$ have shown effects on LPS-induced proliferative responses ranging from suppression to enhancement (Henker et al., 1977; Gallager et al., 1979). Inhalation of 9 mg/m^3 (15 min) of the relatively insoluble CdO did not affect antibody titers in mice challenged with influenza virus (Chaumard et al., 1991). Koller and Brauner (1977) noted that mice exposed to 3, 30, or 300 ppm CdCl$_2$ for 10 wk displayed suppressed erythrocyte-antibody-complement rosette formation, and suggested that Cd may impair the ability of complement components to bind with receptors on B-lymphocytes, thereby inhibiting the effectiveness of antibody to eliminate or inactivate bacterial antigens. *In vitro* exposure to CdCl$_2$ has also been shown to inhibit RNA and DNA synthesis, and to inhibit IgG secretion (Daum et al., 1993). In general, B-lymphocyte mitogenicity appears to be enhanced while antibody synthesis and secretion is suppressed following exposure to Cd (principally CdCl$_2$). Interestingly, an epidemiological study of workers exposed to Cd in the zinc/cadmium smelting industry revealed no differences in basal IgM, IgG, or IgA serum levels when compared to unexposed controls (Karakaya et al., 1994).

Cell-Mediated Immunity

The effects of Cd on cell-mediated immunity have been recently reviewed (Koller, 1998) and the majority of studies examine the systemic effects of Cd following exposures from routes other than inhalation or IT instillation. In general, *in vivo* and *in vitro* studies suggest that Cd suppresses cell-mediated immunity, in particular the delayed hypersensitivity response, and may enhance allograft rejection. A single 1 hr inhalation exposure to 880 mg $CdCl_2/m^3$ significantly reduced T-lymphocyte proliferation to allogeneic antigens and mitogens (Krzystyniak et al., 1987); these data were correlated with alterations in splenic cell viability. Changes in lymphoproliferation (mitogen and MLR) in human lymphocytes exposed *in vitro* has also been reported (Kastelan et al., 1981). A shift in the lymphocyte populations (from small to large in size) was observed in both mice and rats exposed for 4 wk to 100 mg CdO/m^3 (Ohsawa and Kawai, 1981). Splenomegaly, anemia, neutropenia, and lymphopenia were also noted. Finally, like observations in B-lymphocytes, Gallagher and Gray (1982) observed that *in vitro* exposure of lymphocytes to Cd inhibited RNA formation. Whether this was primarily the B-lymphocyte population or both T- and B-lymphocytes is not clear. This does suggest, however, that in addition to the decrease in antibody synthesis and secretion, similar changes may occur for the cytokines in the T-lymphocyte population.

Innate/Non-Specific Immunity

Macrophage-mediated immune function is extremely sensitive to effects of Cd (Koller, 1998). While there is no definitive effect of Cd upon innate immunity, the tendency appears to be toward suppression. No published information on the effects of pulmonary exposure to Cd on NK cell activity was found, although exposure by other routes and examination of splenic NK function suggests that Cd suppresses NK cytotoxicity (Koller, 1998). This conclusion is supported by the demonstration that *in vitro* $CdCl_2$ exposure inhibited NK cytotoxicity and antibody-dependent cell cytotoxicity (ADCC) in human peripheral blood lymphocytes (PBLs; Cifone et al., 1990).

Reduced alveolar macrophage phagocytic activity has been observed in rabbits exposed to $CdCl_2$ by inhalation (Graham et al., 1975). Impaired rosette formation by alveolar macrophages has been demonstrated following oral exposure to Cd suggesting deficits (physical or functional) in the F_c receptor may exist (Hadley et al., 1977). These data are in agreement with other studies (reviewed by Koller, 1998) on the reticuloendothelial system following Cd exposure which indicated that Cd may alter either Fc or complement receptors on Kupffer cells.

The majority of studies examining the effects of Cd on pulmonary innate immunity have been conducted using *in vitro* exposures. Geertz and colleagues (1994) demonstrated that stimulus-induced $\cdot O_2^-$ production by rabbit alveolar macrophages was suppressed by $CdCl_2$ (IC_{50} ~7 mM). This study was in agreement with a study by other Investigators showing that the oxidative burst in rat alveolar macrophages cultured with $CdCl_2$ or cadmium acetate was suppressed (Castranova et al., 1980). These Investigators hypothesized that this effect inhibited the antibacterial activity of the alveolar macrophage. In contrast to these studies, no effect by $CdCl_2$ on guinea pig alveolar macrophage $\cdot O_2^-$ production was reported by Kramer and colleagues (1990). Similarly, Kramer reported that phagocytosis by guinea pig alveolar macrophages was not affected by $CdCl_2$ exposure. Other investigators, however, have shown that in vitro Cd is capable of inhibiting alveolar macrophage phagocytosis and microbicidal activity (Loose et al., 1977, 1978a and b). It should be considered that in several studies, inhibition of phagocytic activity occurred concomitantly with overt cytotoxicity by Cd. Nevertheless, the data are supported by the study of Graham et al. (1975) showing reduced phagocytosis in vivo. Additionally, the inhibition of macrophage motility, responsiveness to lymphokines, and cytotoxic activity has been demonstrated following in vitro cadmium exposure at concentrations presumably not overtly toxic. $CdCl_2$ has been shown to inhibit the production of IL-1 and TNFα from human peripheral blood mononuclear cells (PBMC; Theocharis et al., 1994). These observations were accompanied by suppressed mRNA levels indicating that Cd exposure can suppress production of these cytokines. The production of IL-6 and its mRNA have also been demonstrated to be reduced in a human monocyte cell line by in vitro exposure to $CdCl_2$ (Funkhouser et al., 1994), although production of IL-8 in human PBMC appears to be enhanced (Horiguchi et al., 1993).

Koller (1998) observed that, in general, mice treated with Cd (any form) and challenged with bacteria had decreased resistance to the pathogen, while Cd-exposed mice had increased resistance to viral pathogens. In support of this statement, animals exposed to aerosolized $CdCl_2$ and then challenged with streptococci had an increased mortality (Gardner et al., 1977). Likewise, exposure to CdO resulted in increased susceptibility to bacterial challenge with *Pasturella multocida* and a decreased susceptibility to orthomyxovirus influenza A (Chaumard et al., 1983).

Mechanism of Action

The mechanism of Cd toxicity in relation to pulmonary immunology has not been explored in great depth. The contribution of chemical form and solubility is an area which needs further study, as most studies have focused on soluble $CdCl_2$ and a few on less soluble CdO. Nevertheless, several possibilities exist as possible cellular mechanisms responsible for the immunomodulation observed. It appears that Cd exposure can alter alveolar macrophage phagocytic and cytotoxic functions. Exposure to Cd compounds also appears to affect receptors (for F_c or complement, on the surface of macrophages and other cells) responsible for effective binding and, ultimately, cytotoxicity. Also, production of $\cdot O_2^-$, IL-1, and TNFα are inhibited by Cd exposure, and data from lymphocyte studies suggest this may be at the transcriptional level. Together these data suggest that overall function of the alveolar macrophage is compromised and may lead to enhanced susceptibility to pathogens which require macrophages for clearance (e.g., bacterial pathogens). The fact that viral pathogens do not seem to present a significant threat to Cd-exposed animals may be reflective of the fact that although Cd does have effects on both humoral and cell-mediated immunity, generally high concentrations are required for suppression of these responses while lower concentrations tend to enhance these effects.

The necrosis of alveolar macrophages and ultimate release of enzymes has been proposed as a mediator of chronic pulmonary disease associated with long-term Cd exposure. Additionally, inhalation exposure to Cd has been demonstrated to cause an increased incidence in lung tumors in animals (reviewed in Goyer, 1996). Although a direct genotoxic effect of Cd on the cells in the lung may be responsible for the carcinogenic potential of Cd and its compounds, the contribution of Cd-induced pulmonary or systemic immune suppression to this observation cannot be ruled out.

CHROMIUM

Sources of Exposure and Encounterable Levels

Chromium is a Group VIB metal which can exist in any oxidation state from Cr^{2+} to Cr^{6+}. It is found most often in nature complexed with other minerals such as crocoite (the only natural source of hexavalent Cr) and iron (a principal source of trivalent Cr) and most natural Cr is found in oxides in chromite ore (Sawyer, 1994). Chromium has found significant use in the metallurgical industry for the manufacture of stainless steel, alloy cast irons, and other alloys, and is also used in high temperature furnaces in a magnesite-chrome brick form. Chemical industries employ Cr (both tri- and hexavalent forms) in manufacture of paints, dyes, and pigments, with the latter accounting for ≈33% of all Cr-containing chemicals used (Sawyer, 1994). Cr(III) has been used in leather tanning (≈25% of Cr-containing chemicals) while Cr(VI) is used today to preserve wood products (Wolmanizing) and as a corrosion control agent in a variety of products (≈25% of Cr-containing chemicals). Finally, small amounts of Cr find application in textiles, cement, magnetic tape, joint prostheses, and copying machine toners (reviewed by Arfsten et al., 1998).

Stainless steel accounted for 82% of all Cr consumption in the United States in 1987 (IARC, 1990). Worker exposures occur in the production of stainless steel, other chrome alloys, chrome-containing pigments, chrome plating, and welding (of stainless steel) and are most often by the dermal route. However, inhalation of particulates or fumes containing Cr may occur in welders, chromite miners, and workers involved in ore processing or purification. Estimates of the number of American workers exposed daily in the early 1980s to Cr(VI) range between 300,000 - 550,000 (Steenland, 1996; Arfsten et al., 1998). Values reported as typical airborne Cr(VI) concentrations in a variety of industries have been estimated to range between 5 - 25 mg Cr/m^3 (chromium plating) to as high as 400 - 600 mg Cr/m^3 (stainless steel welding/chromate production/chrome pigment) (Stern, 1982). Airborne Cr(III) concentrations have been estimated to be between 10 - 50 mg Cr/m^3 in industries using a 1-bath tanning process (Arfsten et al., 1998).

OSHA has established the following PELs (8 hr TWA) for exposure to several Cr compounds: 0.1 mg/m^3 for chromic acid and chromates (as CrO_3); 0.5 mg/m^3 for Cr(II) and Cr(III) compounds; and 1 mg/m^3 for Cr metal and insoluble salts. The ACGIH TLV (8 hr TWA) are: 0.05 mg/m^3 for chromate; 0.5 mg/m^3 for Cr metal and Cr(III) compounds; 0.05 mg/m^3 for water-soluble Cr(VI) compounds; and 0.01 mg/m^3 for insoluble Cr(VI) compounds. Historical exposures in chromate production and chromate plating were generally 10 times higher than the current

OSHA standard (IARC, 1990). Long-term exposure to Cr is associated with squamous cell carcinoma and adenocarcinoma in the lungs. IARC considers that there is sufficient evidence for human carcinogenicity associated with exposure to Cr. There is no significant evidence that inhaled Cr(III) at current exposure limits poses any occupational health risk (Sawyer, 1994). However, inhalation of Cr(VI) may be carcinogenic in sufficient concentrations.

Pulmonary Immunomodulation

As with other metals, chemical speciation and solubility are important in the ability of Cr to be absorbed and to exert toxic effects (reviewed in Arfsten et al., 1998). Although there are six possible valence states for Cr, only two have toxicological importance: Cr(III) and Cr(VI). Cr(VI) readily passes through the skin and biological membranes and is taken up rapidly by erythrocytes. Once inside cells, Cr(VI) is relatively unstable and rapidly converted to Cr(III) by cellular enzymes and other proteins. In contrast to Cr(VI), Cr(III) is a very reactive cation which binds in the stratum corneum and does not readily pass through the skin or cell membranes. Interestingly, it is Cr(III) within the cell (reduced from Cr(VI)) which most likely represents the toxic moiety.

Humoral Immunity

The effects of Cr on the humoral immune system have been recently reviewed (Arfsten et al., 1998); the majority of studies examined the effects of Cr following exposures from routes other than inhalation or IT instillation. In one inhalation study, humoral immunity to SRBC was increased in rats exposed for 25 or 90 d to concentrations of sodium dichromate ($Na_2Cr_2O_7$) up to 100 mg Cr/m^3, but suppressed at higher levels (Glaser et al., 1985). In a carcinogenicity study where rats were exposed to $Na_2Cr_2O_7$ or a pyrolyzed Cr(VI)/Cr(III) (3:2) oxide mixture and then allowed a 1-yr recovery, decreases in serum immunoglobulins (as well as elevated WBC and RBC counts) were observed at different stages of the study in the oxide group, possible related to Cr retention in the lung (Glaser et al., 1986). Interestingly, a clinical study by Boscolo and colleagues (1995) showed no differences in circulating serum IgM, IgG, or IgA levels in 15 dye workers despite reductions in total circulating lymphocytes. B-Lymphocyte proliferation has been shown to be inhibited by in vivo Cr exposure in animals (Snyder, 1994) and by in vitro exposure in human studies (Borella and Bargellini, 1993); in the latter, inhibited immunoglobulin production was also observed. These studies suggest that prolonged exposure to Cr compounds or exposure to high Cr concentrations has the potential to produce some immunosuppressive events.

Cell-Mediated Immunity

The effects of Cr on the cell-mediated immune response have been recently reviewed (Arfsten et al., 1998). The vast majority of studies examine the ability of

Cr to produce delayed hypersensitivity responses in both animals and humans. *In vitro* studies in human and animal lymphocytes indicate that Cr can inhibit T-lymphocyte proliferative responses (Borella et al., 1990; Snyder, 1994). In the human studies, proliferation was enhanced with low concentrations of Cr(VI) but suppressed at higher concentrations; Cr(III) was inactive at all concentrations tested.

Innate/Non-Specific Immunity

Few studies have been conducted examining the effects of Cr on the lung (reviewed by Arfsten et al., 1998). Glaser and colleagues (1985) observed that exposure of rats to 50 mg/m^3 Cr(VI) for 4 wk resulted in enhanced numbers of macrophages and enhanced phagocytic activity in the lung. When exposure was extended to 12 - 13 wk, a decreased number of alveolar macrophages was observed and, like other metals, Cr(VI) enhanced macrophages activity at a "lower" concentration (50 mg/m^3) and suppressed activity at the "high" concentration (200 mg/m^3). The Authors noted that the decreased numbers and activity correlated with a 4-fold reduction in pulmonary clearance of inhaled iron oxide. In a second study, rabbits exposed by inhalation to 900 mg Cr(VI)/m^3 for 4 - 6 wk showed an influx of macrophages into the lung, but no alteration in their functional activity (Johansson et al., 1986a). Cr(VI) exposure did cause morphological changes in the macrophages consisting primarily of enlarged lysosomes. In contrast, rabbits exposed to 600 mg Cr(III)/m^3 of had no increase in the number of macrophages in the lung, but did have decreased functional activities (metabolic and phagocytic) which correlated with the presence of enlarged lysosomes and intercellular inclusions containing high amounts of Cr. Further investigations confirmed the morphological changes in alveolar macrophages following Cr(VI) or Cr(III) exposure, indicating direct effects of Cr on these cells, and suggested that the effect in Cr(III)-exposed macrophages may be related to impaired catabolism of surfactant in the lung (Johansson et al., 1986b). Chromium exposure of the monocyte cell line U937 or isolated human monocytes/macrophages has been demonstrated to enhance release of IL-1 and TNFα while suppressing release of transforming growth factor-β (TGFβ) from these cells (Wang et al., 1996). In that study, Cr also induced proliferation of blood monocytes and macrophages. Taken together, these studies suggest that long-term exposure to higher concentrations of Cr may impair natural pulmonary clearance mechanisms (macrophages) and possibly lead to decreased resistance to opportunistic infection.

The involvement of oxidative stress in Cr toxicity has been proposed and investigated by several laboratories. Pulmonary exposure of rats to lower concentrations (2 mg) of either Cr(III) or Cr(VI) did not produce any effect on macrophage chemiluminescence or oxygen consumption. *In vitro* exposure to the same concentration, however, significantly reduced these parameters (Galvin and Oberg, 1984). Exposure of the macrophage cell line J774A.1 to either Cr(III) or Cr(VI) resulted in increases in nitric oxide (NO) and $\cdot O_2^-$ production (Hassoun and Stohs, 1995). Cr(III) also induced single strand DNA breaks, while Cr(VI) had no effect on this parameter. Although the concentrations used produced some overt toxicity to the cell line, the Authors concluded that Cr toxicity was at least in part, due to the induction of oxidative stress. Tian and Lawrence (1996) were not able to show an *in vitro* effect of Cr on NO production using splenic macrophages. Bagshi and colleagues (1997) also used the monocyte cell line J774A.1 to study the effects of Cr. In those studies, the effects of chromium (III) picolinate and nicotinate on reactive oxygen species, lipid peroxidation, and DNA fragmentation were assessed; both salts induced some degree of oxidative stress in this system.

Mechanism of Action

What is interesting in Cr toxicity is the prevalence of occupational asthma, subtypes of which (immediate and delayed) may have immunologic origins (reviewed in Arfsten et al., 1998). Immediate asthma, which occurs within minutes after exposure, is mediated by antigen binding to IgE-bound mast cells. The mast cells then degranulate releasing mediators of bronchoconstriction. Delayed asthma is dependent on proliferation of T-lymphocytes secreting lymphokines which promote chemotaxis, bronchoconstriction, and mucus secretion, and generally occurs hours after exposure. Both types of asthma have been reported in workers exposed to dichromates, ammonium bichromate, chromic acid, chromite ore, chromate pigments, and welding fumes. In some of these cases (but not all), hypersensitivity to Cr can be confirmed by diagnostic patch testing, suggesting both immunologic and non-immunologic origins. Although probably related to dermal hypersensitivity associated with Cr exposure, underlying mechanisms of these pulmonary reactions have yet to be explained. Similarly, descriptive studies and investigations into the mechanisms of Cr immunotoxicity, pulmonary or systemic, have yet to be fully explored. It would appear that the macrophage plays a role in the reduction of the putative carcinogenic Cr(VI) to the "less toxic" Cr(III) form (DeFlora et al., 1997). It remains to be determined whether reduction events occurring within the cell may lead to cell death or other immunotoxicological events such as decreased host resistance, as a result of possible genotoxic events occurring with intracellular Cr(III).

NICKEL

Sources of Exposure and Encounterable Levels

Nickel is a commonly-used, lustrous silvery-white malleable metal which occurs naturally in the Earth's crust (Snow and Costa, 1996). There are two general types of naturally-occurring Ni ores: nickel-iron sulfides and the oxide or silicate laterites (Smialowicz, 1998). Nickel can occur in a variety of valence states from No^0 to Ni^{4+}, but Ni^{2+} is the most common and toxicologically important. Annual world consumption of Ni (excluding the former USSR and Eastern bloc countries) has been estimated at over 670 kilotons (Morgan and Usher, 1994). The principal current uses of Ni and Ni salts are in the production of stainless steel, non-ferrous alloys, electroplating, high temperature and electrical resistance alloys, cast irons the manufacture of Ni-Cd batteries, and as a catalyst and pigment (Steenland, 1996). The volatile liquid nickel carbonyl ($Ni[CO]_4$) is an intermediate product of the Mond process used for refining of nickel sulfide ores.

Exposure to Ni occurs naturally in the atmosphere when Ni is released as a result of natural processes such as volcanic eruptions and wind-blown dust. Nickel is also released into the atmosphere as a result of the combustion of fossil fuels, Ni mining and refining, alloy production and waste incineration (ATSDR, 1993; Mastromatteo, 1996; Smialowicz, 1998). It is estimated than nearly 8.5 million kg of Ni are released naturally and possibly as much as 43 million kg are released from other sources noted previously. For the general population, the average inhalation intake of Ni ranges from 0.1 - 1.0 mg Ni/d. Individuals who smoke cigarettes may have higher exposures, as Ni is found in tobacco at concentrations as high as 1 - 3 mg Ni/cigarette (ATSDR, 1987; Mastromatteo, 1996). The primary means of occupational exposure is through dermal contact or inhalation; NIOSH (1990) has estimated that that as many as 730,000 workers in the United States were potentially exposed to metallic Ni and Ni compounds in the early 1980s. OSHA has established the following PELs (8 hr TWA) for exposure to Ni at 1 mg/m^3 for Ni metal and both soluble and insoluble compounds and 0.007 mg Ni/m^3 for Ni(CO)$_4$. The 1997 ACGIH TLVs (8 hr TWA) are the following (proposed changes noted in parentheses): 1 mg/m^3 (1.5 mg/m^3) for Ni metal; 1 mg/m^3 (0.2 mg/m^3) for insoluble Ni compounds; 0.1 mg/m^3 for soluble Ni compounds; and 0.12 mg Ni/m^3 for Ni(CO)$_4$. IARC considers that there is sufficient evidence for human carcinogenicity associated with exposure to nickel sulfate (NiSO$_4$) and for the combinations of nickel sulfides and oxides encountered in the Ni refining industry. While the carcinogenic effects of Ni (primarily of the lung and nasal cavity) are of predominant concern, other effects on the lung have been reported. Of these, occupational asthma as a result of irritation or allergic response is an important adverse event related to pulmonary Ni exposure (Snow and Costa, 1992; Mastromatteo, 1996; Smialowicz, 1998).

Pulmonary Immunomodulation

Humoral Immunity

The effects of Ni on the humoral immune system have been recently reviewed (Smialowicz, 1998) and the majority of studies examine the effects of Ni following exposures from routes other than inhalation or IT instillation. Although not common, a few cases of occupational asthma have been reported in Ni-sensitive workers and it appears that at least some of these are Type I (immediate) hypersensitivity reactions (Malo et al., 1982; Nieboer et al., 1984). Ni-Specific IgE antibodies have been detected in workers occupationally-exposed to Ni along with positive results in other tests for Ni allergy. In one case report, antibodies to Ni-albumin were found in the serum, and subsequent studies indicated that the Ni was bound to the copper binding site on human albumin (Malo et al., 1985). In general, exposure to Ni by any route (oral, parenteral, inhalation) appears to suppress the primary humoral immune response in animal models. Acute inhalation exposure of mice to nickel chloride (NiCl$_2$; 250 mg/m^3) suppressed the splenic humoral response to the T-dependent SRBC antigen (Graham et al., 1978). Similarly, rats exposed by inhalation to nickel (NiO) particles suppressed the serum anti-SRBC response, although exposure was for 4 mo (Spiegelberg et al., 1984). Bencko and colleagues (1983) examined the concentration of serum IgM, IgG, and IgA on workers exposed occupationally to Ni and found that all three immunoglobulins were statistically significantly decreased when compared to non-exposed controls. Despite the ability to demonstrate Ni-induced humoral immunosuppression in animal models, human data is still somewhat lacking.

Cell-Mediated Immunity

The effects of Ni on cell-mediated immunity have been recently reviewed (Smialowicz, 1998) and the majority of studies examine the effects of Ni following exposures from routes other than inhalation or IT instillation. Occupational and non-occupational exposure to Ni can result in the development of a Type IV (delayed) hypersensitivity response, a response mediated by T-lymphocytes. Nickel has been described as a strong sensitizer in humans (contact sensitivity), although significant animal data has been difficult to generate.

Innate/Non-Specific Immunity

Animals which have been exposed by inhalation or IT instillation to Ni compounds exhibit changes in their ability to defend against pathogenic infection. Generally, Ni exposure appears to suppress macrophage activity, and data from Sunderman and colleagues (1989) suggest that alveolar macrophages are a cellular target for Ni-induced toxicity. In their studies, parenteral exposure to $NiCl_2$ caused activation of alveolar macrophages followed within 2 d by suppressed phagocytic activity and enhanced lipid peroxidation. Similar events occur following pulmonary exposures as well. Rabbits exposed to aerosols of Ni dust for 1 - 6 mo demonstrated gross and histopathological changes in the lung, including an activated appearance of alveolar macrophages which also appeared to have reduced phagocytic capacity (Johansson and Camner, 1980; Johansson et al., 1983; Camner et al., 1984). Dunnick and colleagues (1989) observed an inflammatory response and macrophage hyperplasia in the lungs of mice and rats exposed by inhalation to either nickel subsulfide (Ni_3S_2), NiO, or $Ni(CO)_4$ for 13 wk. Similar results were also noted by other Investigators, who reported the rank order of effect to be $NiSO_4 > Ni_3S_2 = NiO$ (Benson et al., 1989)

Exposure of rats for 4 mo to 25 mg NiO/m^3 aerosol resulted in fewer macrophages obtained from BAL fluid (Spiegelberg et al., 1984). Decreased phagocytosis has also been observed, when mice were exposed for 8 wk to aerosols containing 450 mg/m^3 Ni_3S_2, NiO, or $NiSO_4$ (Haley et al., 1990). Alveolar macrophages from rabbits exposed to $NiCl_2$ aerosol for 1 mo displayed suppressed phagocytic ability and decreased lysozyme levels in the cells and BAL fluid (Wiernik et al., 1983; Lundborg and Camner, 1984). Lysozyme is an enzyme used by macrophages to break down bacterial cell walls, and its activity can be correlated with the ability to clear some of the organisms which infect the lung.

In vitro Ni exposure studies have confirmed many of the observations made *in vivo*. Rabbit and rat alveolar macrophages exposed to Ni have exhibited both reduced phagocytic ability and reduced metabolic capacity (Graham et al., 1975a; Adkins et al., 1979; Castranova et al., 1980). In addition, Lundborg et al. (1987) demonstrated that *in vitro* exposure rabbit alveolar macrophages to $NiCl_2$ produced a similar dose-related decrease in lysozyme activity, confirming the *in vivo* observations of Lundborg and Camner (1984) and indicating a direct effect of Ni ions on the macrophage. *In vitro* suppression of stimulus-induced $\cdot O_2^-$ production by rabbit alveolar macrophages by $NiCl_2$ (IC_{50} ~500 mM) has also been reported (Geertz et al., 1994). Finally, one *in vitro* study compared the effect of six NiO compounds on alveolar macrophages from beagle dogs, mice, and rats (Benson et al., 1988). The study concluded that there was species sensitivity of alveolar macrophages to the effects of Ni with the dog being most sensitive and the rat and mouse being nearly equal. This species sensitivity appeared to correlate with alterations in phagocytic function, with the dog showing the greatest inhibition of activity and the mouse being somewhat less sensitive than the rat.

This apparent inhibition of macrophage activity in the lung appears to be at least partially responsible for the changes in host resistance observed in animals exposed to Ni compounds. For example, mice exposed for 2 hr to aerosols of $NiCl_2$ or $NiSO_4$ and then subsequently challenged with an aerosol of *Streptococcus pyogenes* displayed a decreased resistance to the pathogen. This increase in suscepti-bility to infections in these hosts was correlated with a decreased clearance of the pathogen and a decreased phagocytic ability of their alveolar macrophages. Similarly, hamsters IT-instilled with NiO evidenced increased mortality to a subse-quent challenge with influenza virus (Port et al., 1975).

Natural killer cells are also part of the immunosurveillance team, scouting the body primarily for transformed or virally-infected cells. Parenteral exposure to $NiCl_2$ has been demonstrated to inhibit the ability of mice to clear syngeneic melanoma cells from the lung (Smialowicz et al., 1984a, 1985a, 1987b). Interestingly, while consistent suppression of NK function has been shown following parenteral exposure to Ni, pulmonary exposure has shown variable results. Inhalation exposure of mice followed by challenge with murine cytomegalovirus did not alter either NK activity or host resistance to the virus (Daniels et al., 1987). Similarly, no effects on NK function or host resistance have been shown in mice exposed to either Ni_3S_2 or $NiSO_4$ for 12 d (Benson et al., 1987, 1988) or mice exposed to $NiSO_4$ for 65 d (Haley et al., 1990). In contrast, evaluation of pulmonary NK activity in monkeys instilled with Ni_3S_2 showed enhanced activity (Haley et al., 1987).

Recently, attention has focused on the potential effects of Ni (Ni ions) which may be released from dental alloys. The human macrophage cell line THP-1 was exposed *in vitro* to several metal ions, including Ni, and the release of IL-1 and $TNF\alpha$ (basal and stimulus-induced) was measured. Nickel enhanced both basal and stimulus-induced IL-1 and $TNF\alpha$ at concentrations which had previously been shown to be released from dental alloys both *in vivo* and *in vitro* (Wataha et al., 1996). Recently, Edwards et al. (1998) concluded that Ni can accumulate in the nucleus of human macrophages if sufficient time to eliminate the metal from the cell is not allowed.

Mechanism of Action

A significant amount of information exists regarding the mechanism of Ni-induced contact sensitivity (reviewed in Smialowicz, 1998), and although this is primarily a T-lymphocyte-mediated response, it does require antigen-presenting cells (APC)-like macrophages and other well-known APC. It has been convincingly demonstrated that Ni cations (generated as a result of oxidation of Ni alloys) penetrate the skin and bind as haptens on serum or cellular proteins (Scheper et al., 1989), including the MHC Class II molecule on the surface of APCs. Nickel also causes the increased expression of adhesion molecules on other cell types and induces the release of cytokines which produce local inflammation. While the primary APC for this response is the Langerhan's cells in the skin, these data suggest that similar effects may occur in alveolar macrophages following exposure to Ni.

In some cases, the adverse effects of Ni on a variety of systems can apparently be reversed/prevented by concomitant exposure to other metal ions. Therefore, it has been suggested that Ni toxicity and carcinogenicity may be due to the interference with normal physiological roles of other essential metals such as calcium, magnesium, manganese, and zinc (Nieboer et al., 1984; Fisher et al., 1986; Kasprzak et al., 1986).

REFERENCES

ACGIH: American Conference of Government Industrial Hygienists. *Threshold Limit Values for Chemical Substances and Physical Agents and Biological Exposure Indices.* Cincinnati, OH: ACGIH, 1997.

Adkins B, Jr, Richards JH, Gardner DE. Enhancement of experimental respiratory infection following nickel inhalation. Environ. Res., 1979;20:33-42.

Aranyi C, Bradof JN, O'Shea WJ. Effects of arsenic trioxide inhalation exposure on pulmonary antibacterial defenses in mice. J. Toxicol. Environ. Health, 1985;15:163-172.

Arfsten DP, Aylward LL, Karch NJ. "Chromium." In *Immunotoxicology of Environmental and Occupational Metals*, JT Zelikoff, PT Thomas, eds. London: Taylor and Francis, 1998, pp. 63-92.

Aposhian HV. Biochemical toxicology of arsenic. Rev. Biochem. Toxicol., 1989;10:265-299.

ATSDR: Agency for Toxic Substances and Disease Registry. *Toxicological Profile for Nickel.* Atlanta: U.S. Environmental Protection Agency, 1987.

ATSDR: Agency for Toxic Substances and Disease Registry. *Toxicological Profile for Nickel (Update).* Atlanta: U.S. Department of Health and Human Services, Public Health Service, 1993, ATSDR/TP 92/14.

Bagchi D, Bagchi M, Balmoori J, Ye X, Stohs SJ. Comparative induction of oxidative stress in cultured J774A.1 macrophage cells by chromium picolinate and chromium nicotinate. Res. Comm. Mol. Pathol. Pharmacol., 1997;97:335-345.

Beckman G, Beckman L, Nordenson I. Chromosome aberrations in workers exposed to arsenic. Environ. Health Perspect., 1977;19:145-146.

Benko V, Wagner V, Wagnerova M, Reichrtova E. Immuno-biochemical findings in groups of individuals occupationally and non-occupationally exposed to emissions containing nickel and cobalt. J. Hyg. Epidemiol. Microbiol. Immunol., 1983;27:387-394.

Benko V, Wagner V, Wagnerova M, Batora J. Immunological profiles in workers of a power plant burning coal rich in arsenic content. J. Hyg. Epidemiol. Microbiol. Immunol., 1988;32:137-146.

Benson JM, Zelikoff JT. "Respiratory Toxicology of Metals." In *Toxicology of Metals*, LW Chang, ed. Boca Raton, FL: CRC Press, 1996, pp. 929-938.

Benson JM, Carpenter RL, Hahn FF, Haley PJ, Hanson RL, Hobbs CH, Pickrell JA, Dunnick JK. Comparative inhalation toxicity of nickel subsulfide to F344/N rats and $B_6C_3F_1$ mice exposed for twelve days. Fundam. Appl. Toxicol., 1987;9:251-265.

Benson JM, Burt DG, Carpenter RL. Comparative inhalation toxicity of nickel sulfate to F344/N rats and $B_6C_3F_1$ mice exposed for twelve days. Fundam. Appl. Toxicol., 1988a;10:164-178.

Benson JM, Henderson RF, Pickrell JA. Comparative *in vitro* cytotoxicity of nickel oxides and nickel-copper oxides to rat, mouse, and dog pulmonary alveolar macrophages. J. Toxicol. Environ. Health, 1988b;24:373-383.

Benson JM, Burt DG, Cheng YS, Hahan FF, Haley PH, Henderson RF, Hobbs CH, Pickrell JA, Dunnick, JK. Biochemical responses of rat and mouse lung to inhaled nickel compounds. Toxicology, 1989;57:255-266.

Borella P, Bargellini A. Effects of trace elements on immune system: Results in cultured human lymphocytes. J. Trace Elem. Electrolytes Health Dis., 1993;7:231-233.

Borella P, Manni S, Giardino A. Cadmium, nickel, chromium, and lead accumulate in human lymphocytes and interfere with PHA-induced proliferation. J. Trace Elem. Electrolytes Health Dis, 1990;4:87-95.

Boscolo P, Di Gioacchino M, Cervone M, Di Giacomo F, Bavazzano P, Builiano G. Lymphocyte subpopulations of workers in a plant producing plastic materials (preliminary study). G. Ital. Med. Lav., 1995;17:27-31.

Briggs TM, Owens TW. *Industrial Hygiene Characterization of the Photovoltaic Cell Industry.* Cincinnati, OH: U.S. Department of Health, Education and Welfare, NIOSH Technical Report, DHEW (NIOSH) Publication, 1980, pp. 80-112, .

Broeckaert F, Buchet JP, Huaux F, Lardot C, Lison D. Reduction of the *ex vivo* production of tumor necrosis factor-α by alveolar phagocytes after administration of coal fly ash and copper smelter dust. J. Toxicol. Environ. Health, 1997;51:189-202.

Burns LA. "Arsenic." In *Immunotoxicology of Environmental and Occupational Metals*, JT Zelikoff, PT Thomas, eds. London: Taylor and Francis, 1998, pp. 1-26.

Burns LA, Sikorski EE, Saady J, Munson AE. Evidence for arsenic as the primary immunosuppressive component of gallium arsenide. Toxicol. Appl. Pharmacol., 1991;110:157-169.

Bustamante J, Dock L, Vahter M, Fowler B, Orrenius S. The semiconductor elements arsenic and indium induce apoptosis in rat thymocytes. Toxicology, 1997;118:129-136.

Camner P, Casarett-Bruce M, Curstedt T, Jarstrand C, Wiernikm A, Johansson A, Lundborg M, Robertson B. "Toxicology of Nickel." In *Nickel in the Human Environment*, IARC Scientific Publication No. 53, FW Sunderman, Jr, ed. Lyon, France: International Agency for Research on Cancer, 1984, pp. 267-276.

Casto BC, Meyers J, DiPaolo JA. Enhancement of viral transformation for evaluation of the carcinogenic or mutagenic potential of inorganic metal salts. Cancer Res., 1979;39:193-198.

Castranova V, Bowman L. Reason MJ, Miles PR. Effects of heavy metal ions on selected oxidative metabolic processes in rat alveolar macrophages. Toxicol. Appl. Pharmacol., 1980;53:14-23.

Castranova V, Bowman L, Wright JR, Colby H, Miles PR. Toxicity of metallic ions in the lung: Effect of alveolar macrophages and alveolar Type II cells. J. Toxicol. Environ. Health, 1984;13:845-856.

Chaumard C, Forestier F, Quero AM. Influence of inhaled cadmium on the immune response to influenza virus. Arch. Environ. Health, 1991;41:50-56.

Cifone MG, Procopio A, Napolitano T, Alesse E, Santoni G, Santoni A. Cadmium inhibits spontaneous (NK), antibody-mediated (ADCC) and IL-2-stimulated cytotoxic functions of natural killer cells. Immunopharmacology, 1990;20:73-80.

Cohen MD, Costa, M. "Chromium." In *Environmental and Occupational Medicine, 2nd Edition*, WN Rom, ed. Toronto/Boston/NY: Little, Brown, and Company, 1992, pp. 799-805.

Daniels MJ, Menache MG, Burleson GR, Graham JA, Selgrade MJ. Effects of $NiCl_2$ and $CdCl_2$ on susceptibility to murine cytomegalovirus and virus-augmented natural killer cell and interferon responses. Fundam. Appl. Toxicol., 1987;8:443-453.

Daum JR, Shepherd DM, Noelle RJ. Immunotoxicology of cadmium and mercury on B-lymphocytes. I: Effects on lymphocyte function. Int. J. Immunopharmacol., 1993;15:383-394.

DeFlora S, Camoirano A, Bagnasco M, Bennicelli C, Corbett GE, Kerger BD. Estimates of the chromium (VI) reducing capacity in human body compartments as a mechanism for attenuating its potential toxicity and carcinogenicity. Carcinogenesis, 1997;18:531-537.

Dickerson OB. "Antimony, Arsenic, and Their Compounds." In *Occupational Medicine, 3rd Edition*, C Zenz, OB Dickerson, EP Horvath, Jr., eds. St. Louis: Mosby, 1994, pp. 468-472.

DiPaolo JA, Casto BC. Quantitative studies of *in vitro* morphological transformation of Syrian hamster cells by inorganic metal salts. Cancer Res., 1979;39:1008-1013.

Dunnick JK, Elwell MR, Benson JM, Hobbs CH, Hahn FF, Haley PJ, Cheng YS, Edison AF. Lung toxicity after 13-week inhalation exposure to nickel oxide, nickel subsulfide, or nickel sulfate hexahydrate in F344/N rats and $B_6C_3F_1$ mice. Fundam. Appl. Toxicol., 1989;12:584-594.

Edwards DL, Wataha JC, Hanks CT. Uptake and reversibility of uptake of nickel by human macrophages. J. Oral Rehabil., 1998;25:207.

EPA: Environmental Protection Agency. *Special Report on Ingested Inorganic Arsenic: Skin Cancer and Nutritional Essentiality. Risk Assessment Forum.* Washington, DC: U.S. EPA, 1987.

Farmer JG, Johnson LR. Assessment of occupational exposure to inorganic arsenic based on urinary concentration and speciation of arsenic. Br. J. Ind. Med., 1990;47:342-348.

Fisher GL, McNiell KL, Democko CJ. Trace element interactions affecting pulmonary macrophage cytotoxicity. Am. J. Physiol., 1986;252:C677-C683.

Funkhouser SW, Martinez-Maza O, Vredevoe D. Cadmium inhibits IL-6 production and IL-6 mRNA expression in a human monocytic cell line, THP-1. Environ. Res., 1994;66:77-86.

Gallagher KE, Gray I. Cadmium inhibition of RNA metabolism in murine lymphocytes. J. Immunopharmacol., 1982;3:339-361.

Gallagher K, Mattarazzo WJ, Gray I. Trace metal modification of immunocompetence: II. Effect of Pb^{2+}, Cd^{2+}, and Cr^{3+} on RNA turnover, hexokinase activity, and blastogenesis during B-lymphocyte transformation *in vitro*. Clin. Immunol. Immunopathol., 1979;13:369-377.

Galvin JB, Oberg SG. Toxicity of hexavalent chromium to the alveolar macrophage *in vivo* and *in vitro*. Environ. Res., 1984;33:7-16.

Garcia-Vargas GG, Cebrian ME. "Health Effects of Arsenic." In *Toxicology of Metals*, LW Chang, ed. Boca Raton, FL: CRC Press, 1996, pp. 423-438.

Glaser U, Hochrainer D, Kloppel H, Kuhmen H. Low level chromium(VI) inhalation effects on alveolar macrophages and immune functions in Wistar rats. Arch. Toxicol., 1985;57:250-256.

Glaser U, Hochrainer D, Kloppel H, Oldiges H. Carcinogenicity of sodium dichromate and chromium (VI/III) oxide aerosols inhaled by male Wistar rats. Toxicology, 1986;42:219-232.

Goyer RA. "Toxic Effects of Metals." In *Casarett and Doull's Toxicology: The Basic Science of Poisons,* CD Klaassen, ed. New York: McGraw-Hill, 1996, pp. 691-698,

Graham JA, Gardner DE, Waters MD, Coffin DC. Effect of trace metals on phagocytosis by alveolar macrophages. Infect. Immun., 1975;11:1278-1283.

Graham JA, Miller FJ, Daniels MJ, Payne EA, Gardner DE. Influence of cadmium, nickel, and chromium on primary immunity in mice. Environ. Res., 1978;16:77-87.

Hadley JG, Gardner DE, Coffin DL, Menzel DB. Inhibition of antibody-mediated rosette formation by alveolar macrophages: A sensitive assay for metal toxicity. J. Reticuloendothel. Soc., 1977;22:417-422.

Haley PJ, Bice DE, Muggenburg BA, Hahn FF, Benjamin SA. Immunopathologic effects of nickel subsulfide on the primate pulmonary immune system. Toxicol. Appl. Pharmacol., 1987;88:1-12.

Haley PJ, Shopp GM, Benson JM, Cheng YS, Bice DE, Luster MI, Dunnick JK, Hobbs CH. The immunotoxicity of three nickel compounds following a 13-week inhalation exposure in the mouse. Fundam. Appl. Toxicol., 1990;15:476-487.

Harrison RJ. "Gallium arsenide." In *State of the Art Reviews: Occupational Medicine, Vol. 1*, Philadelphia: Hanley and Belfus, 1986, pp. 49-58.

Hassoun EA, Stohs SJ. Chromium-induced production of reactive oxygen species, DNA single-strand breads, nitric oxide production, and lactate dehydrogenase leakage in J774A.1 cell cultures. J. Biochem. Toxicol., 1995;10:315-321.

Hill AB, Faning EL. Studies in the incidence of cancer in a factory handling inorganic compounds of arsenic: I. Mortality experience in the factory. Br. J. Ind. Med., 1948;5:1-7.

Horiguchi H, Mukaida N, Okamoto S, Teranishi H, Kasuya M, Matsushima K. Cadmium induces interleukin-8 production in human peripheral blood mononuclear cells with the concomitant generation of superoxide radicals. Lymphokine Cytokine Res., 1993;12:421-428.

Huaux F, Lasfargues G, Lauwerys R, Lison D. Lung toxicity of hard metal particles and production of interleukin-1, tumor necrosis factor-α, fibronectin, and cystatin-c by lung phagocytes. Toxicol. Appl. Pharmacol., 1995;132:53-62.

IARC: International Agency for Research on Cancer. *Some Metals and Metallic Compounds, Monograph 23*. Lyon, France: IARC, World Health Organization, 1980.

IARC: International Agency for Research on Cancer. *Overall Evaluation of Carcinogenicity: An Updating of IARC Monographs 1-42, Supplement 7*. Lyon, France: IARC, World Health Organization, 1987.

IARC: International Agency for Research on Cancer. *Chromium, Nickel, and Welding, Monograph 49*. Lyon, France: IARC, World Health Organization, 1990.

IARC: International Agency for Research on Cancer. *Beryllium, Cadmium, Mercury and Exposures in the Glass Manufacturing Industry, Monograph 58*. Lyon, France: IARC, World Health Organization, 1993.

Johansson A, Camner P. Effects of nickel dust on rabbit alveolar epithelium. Environ. Res., 1980;22:510-516.

Johansson A, Camner P. Jarstrand C, Wiernik A. Rabbit lungs after long-term exposure to low nickel dust concentrations. II. Effects on morphology and function. Environ. Res. 1983;30:142-151.

Johansson A, Wiernik A, Jarstrand C, Camner P. Rabbit alveolar macrophages after inhalation of hexa- and trivalent chromium. Environ. Res., 1986a;39:372-385.

Johansson A, Robertson B, Curstedt T, Camner P. Rabbit lung after inhalation of hexa- and trivalent chromium. Environ. Res., 1986b;41:110-119.

Karakaya A, Yulesoy B, Sardas OS. An immunological study on workers occupationally exposed to cadmium. Human Exp. Toxicol., 1994;13:73-75.

Kasprzak KS, Waalkes MP, Porier LA. Antagonism by essential divalent metals and amino acids of nickel (II)-DNA binding *in vitro*. Toxicol. Appl. Pharmacol., 1986;82:336-343.

Kastelan M, Gerencer M, Kastelan A, Gamulin S. Inhibition of mitogen- and specific antigen-induced human proliferation by cadmium. Exp. Cell Biol., 1981;49:15-19.

Koller LD. "Cadmium." In *Immunotoxicology of Environmental and Occupational Metals*, JT Zelikoff, PT Thomas, eds. London: Taylor and Francis, 1998, pp. 41-61.

Koller LD, Brauner JA. Decreased B-cell response after exposure to lead and cadmium. Toxicol. Appl. Pharmacol., 1977;42:621-624.

Kramer CM, Coles RB, Carchman RA. *In vitro* effects of cadmium chloride on calcium metabolism in guinea pig alveolar macrophages: Lack of correlation with superoxide anion release or phagocytosis. In Vitro Toxicol., 1990;3:153-160.

Krzystyniak K, Fournier M, Trotter B, Nadieu B, Chevalier G. Immunosuppression in mice after inhalation of cadmium aerosol. Toxicol. Lett., 1987;38:1-12.

Labedzka M, Gulyas H, Schmidt N, Gercken G. Toxicity of metallic ions and oxides to rabbit alveolar macrophages. Environ. Res., 1989;48:255-274.

La Dou J. Health issues in the microelectronics industry. Ocup. Med., 1986;1:1-11.

Landrigan PJ. "Arsenic." In *Environmental and Occupational Medicine, 2nd Edition*, WN Rom, ed. Toronto: Little, Brown, and Company, 1992, pp. 773-779.

Lantz RC, Parliman G, Chen GJ, Carter DE. Effect of arsenic exposure on alveolar macrophage function I. Effect of soluble As(III) and As(V). Environ. Res., 1994;67:183-195.

Lantz RC, Parliman G, Chen GJ, Barber D, Winski S, Carter DE. Effect of arsenic exposure on alveolar macrophage function II. Effect of slightly soluble forms of As(III) and As(V). Environ. Res., 1995;68:59-67.

Lauwerys RR. "Cadmium and Its Compounds." In *Occupational Medicine, 3rd Edition*, C Zenz, OB Dickerson, EP Horvath, Jr., eds. St. Louis: Mosby, 1994, pp. 481-486.

Lee TC, Oshimura M, Barrett JC. Comparison of arsenic-induced cell transformation, cytotoxicity, mutation, and cytogenetic effects in Syrian hamster embryo cells in culture. Carcinogenesis, 1985;10:1421-1426.

Lee TC, Tanaka N, Lang PW, Gilmol TM, Barrett JC. Induction of gene amplification by arsenic. Science, 1988;24:79-81.

Loose LD, Silkworth JB, Warrington D. Cadmium-induced depression of the respiratory burst in mouse pulmonary alveolar macrophages, peritoneal macrophages, and polymorphonuclear neutrophils. Biochem. Biophys. Res. Commun., 1977;79:326-332.

Loose SD, Silkworth JB, Simpson DW. Influence of cadmium on the phagocytic and microbial activity of murine peritoneal macrophages, pulmonary alveolar macrophages, and polymorphonuclear neutrophils. Infect. Immun., 1978a;22:378-381.

Loose LD, Silkworth JB, Warrington D. Cadmium-induced phagocyte cytotoxicity. Bull. Environ. Contam. Toxicol., 1978b;20:582-588.

Lundborg M, Camner P. Lysozyme levels in rabbit lung after inhalation of nickel, cadmium, cobalt, and copper chlorides. Environ. Res., 1984;34:335-342.

Lundbor M, Johansson A, Camner P. Morphology and release of lysozyme following exposure of rabbit lung macrophages to nickel or cadmium *in vitro*. Toxicology, 1987;46:191-202.

McCabe M, Maguire D, Nowak M. The effects of arsenic compounds on human and bovine lymphocyte mitogenesis *in vitro*. Environ. Res., 1983;31:323-331.

Mabuchi K, Lilienfeld AM, Snell LM. Lung cancer among pesticide workers exposed to inorganic arsenicals. Arch. Environ. Health, 1979;34:312-320.

Malo JL, Cartier A, Doepner M, Nieboer E, Evans S, Dolovich J. Occupational asthma caused by nickel sulfate. J. Allergy Clin. Immunol., 1982;69:55-59.

Malo JL, Cartier A, Gagnon G, Evans S, Dolovich J. Isolated late asthmatic reaction due to nickel antibody. Clin. Allergy, 1985;15:95-99.

Maloney ME. Arsenic in dermatology. Dermatol. Surg., 1996;22:301-304.

Mastromatteo E. "Nickel and Its Compounds." In *Occupational Medicine, 3rd Edition*, C Zenz, OB Dickerson, EP Horvath, Jr., eds. St. Louis: Mosby, 1994, pp. 558-571.

Morgan LG, Usher V. Health problems associated with nickel refining and use. Ann. Occup. Hyg., 1994;38:189-198.

Newman-Taylor AJ. "Cadmium." In *Environmental and Occupational Medicine, 2nd Edition*, WN Rom, ed. Toronto: Little, Brown, and Company, 1992, pp. 767-772.

Nieboer E, Maxwell RI, Stafford AR. "Chemical and Biological Reactivity of Insoluble Nickel Compounds and the Bioinorganic Chemistry of Nickel." In *Nickel in the Human Environment. Proceedings of a Joint Symposium, IARC Scientific Publication No. 53*, FW Sunderman, Jr., ed. Lyon, France: IARC, World Health Organization, 1984a, pp. 439-468.

Nieboer E, Evans SL, Dolovich J. Occupational asthma from nickel sensitivity. II. Factors influencing the interaction of Ni^{2+}, HSA, and serum antibodies with nickel related specificity. Br. J. Ind. Med., 1984b;41:56-63.

NIOSH: National Institute for Occupational Safety and Health. *NIOSH Testimony to U.S. Department of Labor: Comments at the OSHA Arsenic Hearing, July 14, 1982. NIOSH Policy Statement.* Cincinnati, OH: U.S. Department of Health and Human Services, PHS, CDC, NIOSH, 1982.

NIOSH: National Institute for Occupational Safety and Health. *National Occupational Exposure Survey.* Cincinnati, OH: U.S. Department of Health and Human Services, PHS, CDC, NIOSH, 1990, pp. 89-103.

Nordenson I, Beckman G, Beckman L, Nordstrom S. Occupational and environmental risks in and around a smelter in northern Sweden. II. Chromosomal aberrations in workers exposed to arsenic. Hereditas, 1978;88:47-50.

Ohsawa M, Kawai K. Cytological shift in lymphocytes induced by cadmium in mice and rats. Environ. Res., 1981;24:192-200.

Ohyama S, Ishihishi N, Hisanaga A, Yamamoto A. Chronic toxicity, including tumorigenicity, of gallium arsenide to the lung of hamsters. Appl. Organomet. Chem., 1988;2:333-337.

OSHA: Occupational Safety and Health Agency. *Occupational Safety and Health Standards. Code of Federal Regulations, Part 1910.100, Subpart Z - Toxic and Hazardous Substances. 1997.* Washington, DC: U.S. Government Printing Office, 1997.

Petres J, Baron D, Hagedorn M. Effects of arsenic on cell metabolism and cell proliferation: Cytogenetic and biochemical studies. Environ. Health Perspect., 1977;19:223-227.

Port CD, Fenters JD, Ehrlich R, Coffin DL, Gardner D. Interaction of nickel oxide and influenza infection in the hamster. Environ. Health Perspect., 1975;10:268.

Roussin A, Cabec VL, Lonchampt M, De Nadai J, Canet E, Maridonneau-Parini I. Neutrophil-associated inflammatory responses in rats are inhibited by phenylarsine oxide. Eur. J. Pharmacol., 1997;322:91-96.

Sawyer HJ. "Chromium and Its Compounds." In *Occupational Medicine, 3rd Edition*, C Zenz, OB Dickerson, EP Horvath, Jr., eds. St. Louis: Mosby, 1994, pp. 487-495.

Scheper RJ, Blomberg M, Vreeburg KJ, van Hoogstraten IM. "Recent Advances in Immunology of Nickel Sensitization." In *Nickel and the Skin: Immunology and Toxicology*. HI Maibach, T Menne, eds. Boca Raton, FL: CRC Press, 1989, pp. 55-63.

Shenker BJ, Matarazzo WJ, Hirsch RL, Gray I. Trace metal modification of immunocompetence. I. Effect of trace metals in cultures on *in vitro* transformation of B-lymphocytes. Cell. Immunol., 1977;34:19-24.

Sikorski EE, McCay JA, White KL, Jr, Bradley SG, Munson AE. Immunotoxicity of the semiconductor gallium arsenide in female B_6C3F_1 mice. Fundam. Appl. Toxicol., 1989;13:843-858.

Smialowicz RJ. "Nickel." In *Immunotoxicology of Environmental and Occupational Metals*, JT Zelikoff, PT Thomas, eds. London: Taylor and Francis, 1998, pp. 163-194.

Smialowicz RJ, Rogers RR, Riddle MM, Stott GA. Immunologic effects of nickel. I. Suppression of cellular and humoral immunity. Environ. Res., 1984;33:413-427.

Smialowicz RJ, Rogers RR, Riddle MM, Garner RJ, Rowe DG, Leubke RW. Immunologic effects of nickel. II. Suppression of natural killer (NK) activity. Environ. Res., 1985;36:56-66.

Smialowicz RJ, Rogers RR, Riddle MM, Leubke RW, Fogelson LD, Rowe, DG. Effects of manganese, calcium, magnesium, and zinc on nickel-induced suppression of murine natural killer cell activity. J. Toxicol. Environ. Health, 1987;20:67-80.

Snow ET. Metal carcinogenesis: Mechanistic implications. Pharmac. Ther., 1992;53:31-65.

Snow ET, Costa M. "Nickel Toxicity and Carcinogenesis." In *Environmental and Occupational Medicine, 2nd Edition*, WN Rom, ed. Toronto: Little, Brown, and Company, 1992, pp. 807-813.

Snyder CA. Immune function assays as indicators of chromate exposure. Environ. Health Perspect., 1991;92:83-86.

Spiegelberg T, Kordel W, Hichrainer D. Effects of NiO inhalation on alveolar macrophages and the humoral immune system of rats. Ecotoxicol. Environ. Safety, 1984;8:516-525.

Squibb KS, Fowler BA. "The Toxicity of Arsenic and Its Compounds." In *Biological and Environmental Effects of Arsenic,* BA Fowler, ed. Amsterdam: Elsevier Biomedical Press, 1983, pp. 233-269

Steenland K, Loomis D, Shy C, Simonsen N. Review of occupational lung carcinogens. Am. J. Ind. Med., 1996;29:474-490.

Stern RM. "Chromium Compounds: Production and Occupational Exposure." In *Topics in Environmental Health, Vol. 5, Biological and Environmental Aspects of Chromium*, S. Langard, ed. Amsterdam: Elsevier Biomedical Press, 1982, pp. 5-44.

Sunderman FW, Jr. Mechanisms of metal carcinogenesis. Biol. Trace Element Res., 1979;1-63-84.

Sunderman FW, Jr, Hopfer SM, Lin SM, Plowman MC, Stojanivic T, Wong SH, Zaharia O, Ziebka L. Toxicity of alveolar macrophages in rats following parenteral injection of nickel chloride. Toxicol. Appl. Pharmacol., 1989;100:107-118.

Tanaka A, Hisanaga A, Hirata M, Omura M, Makita Y, Inoue N, Ishinishi N. Chronic toxicity of indium arsenide and indium phosphide to the lungs of hamsters. Fukuoka Acta Med., 1996;87:108-115.

Theocharis SE, Souliotis T, Panayiotisid P. Suppression of interleukin-1β and tumor necrosis factor-α biosynthesis by cadmium in *in vitro* activated human peripheral blood mononuclear cells. Arch. Toxicol., 1994;69:132-136.

Tian L, Lawrence DA. Metal-induced modulation of nitric oxide production *in vitro* by murine macrophages: Lead, nickel, and cobalt utilize different mechanisms. Toxicol. Appl. Pharmacol., 1996;141:540-547.

Wang JY, Wicklund BH, Gustile RB, Tsukayama DT. Titanium, chromium, and cobalt ions modulate the release of bone-associated cytokines by human monocytes/macrophages *in vitro*. Biomaterials, 17:2223-2240.

Wataha JC, Ratanasathien S, Hanks CT, Sun Z. *In vitro* IL-1 beta and TNF-alpha release from THP-1 monocytes in response to metal ions. Dental Mater., 1996;12:322-327.

Watrous RM, McCaughey MB. Occupational exposure to arsenic in the manufacture of arsphenamine and related compounds. Ind. Med., 1945;14:639-646.

Webb DR, Sipes IG, Carter DE. *In vitro* solubility and *in vivo* toxicity of gallium arsenide. Toxicol. Appl. Pharmacol., 1984;76:96-104.

Webb DR, Wilson SE, Carter DE. Comparative pulmonary toxicity of gallium arsenide, gallium (III) oxide, or arsenic (III) oxide intratracheally-instilled into rats. Toxicol. Appl. Pharmacol., 1986;82:405-416.

Wiernik A, Johansson A, Jarstrand C, Camner P. Rabbit lung after inhalation of soluble nickel. I. Effects on alveolar macrophages. Environ. Res., 1983;30:129-141.

WHO: World Health Organization. *Environmental Health Criteria: Arsenic, Vol. 19.* Geneva: EHE/EHC, World Health Organization, 1981.

Yamamoto S, Konishi Y, Matsuda T, Murai T, Shibata M, Matsui-Yuasa I, Otani S, Kuroda K, Endo G, Fukushima S. Cancer induction by an organic arsenic compound, dimethylarsenic acid (cacodylic acid), in F344/DuCrj rats after pretreatment with five carcinogens. Cancer Res., 1995, 55:1271-1275.

Yamanaka K, Hasegawa A, Sawamura R, Okada S. DNA strand breaks in mammalian tissues induced by methylarsenics. Biol. Trace Elem. Res., 1989, 21:413-417.

Yamanaka K, Hayashi H, Kato K, Hasegawa A, Okada S. Involvement of preferential formation of apurinic/apyrimidinic sites in dimethylarsenic-induced DNA strand breaks and DNA-protein cross-links in cultured alveolar epithelial cells. Biochem. Biophys. Commun., 1995, 207:244-249.

11. OTHER METALS:
ALUMINUM, COPPER, MANGANESE, SELENIUM, VANADIUM, AND ZINC

Mitchell D. Cohen, Ph.D.
New York University School of Medicine
Department of Environmental Medicine
57 Old Forge Road
Tuxedo, New York 10987

ALUMINUM

While aluminum (Al) is one of the most abundant elements encountered in the environment, and some daily exposure is unavoidable, inhalation of Al by the general population is generally considered to be negligible (i.e., 0.14 mg aluminum dust/d [Jones and Bennett, 1986]). Conversely, smelters, miners, and other workers involved in various metal industries are often acutely exposed, accidentally or intentionally, to much higher levels of ambient Al. For example, one study has indicated that aluminum welders were often exposed to localized atmospheres containing 2.4 mg Al/m^3, resulting in time-weighted-average intakes of >23 mg Al/8 hr shift (Sjogren et al., 1985). In addition, because of its demonstrable role as a prophylactic agent against silicotic lung diseases (Denny et al., 1939; Le Bouffant et al., 1977; Begin et al., 1986, 1987; Brown et al., 1989), many miners were intentionally exposed to airborne mixtures of finely-ground aluminum powder containing ≈15% elemental Al and 85% aluminum oxide (alumina; Al_2O_3) prior to their entering the mineshafts (reviewed in Rifat et al., 1990). Although there is no information describing acute or chronic effects upon human pulmonary health from environmental exposure to airborne Al, there are numerous reports which have described increases in pneumonia, bronchitis, asthma, hard metal pneumoconiosis, lung cancers, and/or pulmonary fibrosis in occupationally-exposed subjects (Gibbs, 1985; Abramson et al., 1989; Chan-Yeung et al., 1989; Larsson et al., 1989; Schwarz et al., 1994; Soyseth et al., 1995; Kilburn, 1998; Sorgdrager et al., 1998).

Al-Containing compounds, the majority of which are the various forms of alumina (i.e., activated, metal-grade, calcined), are widely employed in the chemical, abrasive, ceramic, and refractory industries apart from their use in the primary aluminum production. In the latter, Al is first refined from bauxite ("Bayer process") to isolate alumina and then recovered by electrolytic reduction ("Hall-Heroult process") of an Al_2O_3/cryolite ($3NaF\cdot AlF_3$)-containing melt held in a large reduction pot. During "dry-scrubbing" to minimize fluoride release, virginal (primary) Al_2O_3 is used to adsorb any fluoride emitted and the now-modified (recovery/secondary) alumina is, in turn, added to the reduction pot. However, this remedial process increases the generation of Al-containing dusts within the potroom environment. Not surprisingly, a concurrent increase in the incidence of respiratory symptoms has been noted among those workers in potrooms where dry scrubbing was employed (Gylseth and Jahr, 1975; Edouard and Lie, 1981). While this clearly suggests that Al_2O_3 and AlF_3 are two major airborne compounds to which workers might potentially be at risk for exposure, other Al-containing agents may also be present at significant levels that could pose a health risk. These include aluminum: isopropylate; sulfate; hydroxide; chloride; nitrate; and, phosphide, as well as triethyl aluminum and Al metal itself (NIOSH, 1984).

Measurable levels of Al have been found in the majority of urban atmospheres, with levels in many major metropolitan areas ranging from 0.4 - 10.0 mg Al/m^3 (ATSDR, 1992). In rural areas and those without contributing point sources, average atmospheric Al levels are substantially lower but highly variable. For example, levels in central Canada have been shown to range from 0.27 - 0.39 mg Al/m^3, those in rural Hawaii ranged from 0.005 - 0.032 mg Al/m^3. Although no international guidelines are in place for permissible levels of workplace airborne Al levels, national regulations do exist but vary depending upon the Al compound or generation source involved. For example, the current accepted levels (8 hr time-weighted-average [TWA]) recommended by OSHA for Al metal dust, Al_2O_3, welding fume Al, and soluble Al salts are 15, 10, 5, and 2 mg Al/m^3, respectively (OSHA, 1989).

As evidenced by these varying OSHA standards, it is compound solubility which ultimately is critical to the extent of pulmonary (immuno)toxicities which might evolve following inhalation of Al-containing compounds. However, while many studies have described the post-inhalation distribution/excretion of Al_2O_3 (Rollin et al., 1991), Al flakes (Ljunggren et al., 1991), Al-welding fumes (Sjogren et al., 1985, 1988), or aluminum chlorhydrate (Stone et al., 1979), to date, none have formally compared or characterized the respective half-lives and/or clearance patterns of those Al compounds encountered in occupational settings. In addition, several studies have concluded that pulmonary burdens of Al normally tend to increase with host longevity irrespective of the source of the Al (reviewed in Ganrot, 1986; DeVoto and Yokel, 1994).

Among the studies which have directly examined the effects of inhaled Al compounds upon the development of pulmonary fibrosis and/or asthma ("potroom asthma"), only a few have assessed the histological changes/inflammatory responses that evolved following intrapulmonary deposition of these agents. For example, an examination of the lungs of rats instilled once with either 1 or 5 mg Al_2O_3/kg BW evidenced dose-dependent increased inflammatory responses (as characterized by an increased neutrophil (PMN) influx and total numbers of cells recovered by lavage) and a minimal amount of interstitial inflammation/type II cell hyperplasia over a 2 mo post-exposure period (Lindenschmidt et al., 1990). Unlike what was observed following instillation with a fibrogenic dust containing silica (i.e., the inflammatory response increasing as a function of time post-instillation), the effects from

Al_2O_3 were maximal within the first 7 - 14 d post-exposure and slowly decreased toward normal thereafter. A comparison of the inflammatory effects of virginal Al_2O_3 against those of potroom dust (a mixture of aluminas and AlF_3), indicated that while a single instillation of virginal Al_2O_3 had the capacity to induce pulmonary inflammation and an associated influx of PMN into the lungs, the effects from a similar exposure to potroom dust were often greater (albeit not always dose-dependent) (White et al., 1987). In this study, potroom dust-induced increases in the numbers of PMN in the lungs were often accompanied by concurrent decreases in the amounts of recoverable alveolar macrophages. This is important to note in that in a later study which examined the effects of primary/virginal Al_2O_3 vs those of secondary/recovery Al_2O_3, single instillation of an $\approx$8-fold higher amount (i.e., 40 mg/rat) of virginal Al_2O_3 failed to give rise to any change in total cell numbers and/or PMN recovered at 1 mo post-instillation (Tornling et al., 1993). Conversely, instillation of an equivalent amount of fluoride-adhered secondary Al_2O_3 caused (in this same timeframe) a near doubling in the levels of total cells and of alveolar macrophages, as well as a 10-fold increase in the numbers of PMN, recovered from the lungs of treated hosts. That apparently incongruous results arose among these studies is not surprising given that generation-related variations in the purity/co-contaminant composition of various "alumina" samples are sufficient to induce disparate immunotoxicologic effects in the lungs of treated hosts (Ess et al., 1993). As such, attempting to draw any broad conclusions about pulmonary (immuno)-toxicologic effects from host exposure to "potroom dust" should be done with great caution as relative dust composition (i.e., amounts of primary and secondary Al_2O_3 among any/all Al-containing agents present) can clearly affect the observable outcomes of assays which measure inflammation and associated responses.

A few studies have examined the effects from host inhalation/exposure to Al compounds upon bronchoalveolar fluid (BAL) composition (i.e., to be used as indirect indicator of cell/tissue damage). The studies by Lindenschmidt et al. (1990) demonstrated that a single instillation of Al_2O_3 induced short-lived significant increases in the amounts of lactate dehydrogenase (LDH; an index of lung cell membrane damage), total protein (TP; an index of potential fibrotic activity/vascular damage), and of β-glucuronidase and N-acetylglucosaminidase (BG and NAG; markers of macrophage/PMN membrane damage) in rat BAL. In determining which particular protein(s) in the BAL might comprise this increase in total protein content, Tornling et al. (1993) concluded that Al_2O_3 instillation induced time-dependent increases in the levels of BAL fibronectin, while levels of albumin and hyaluronan remained unaffected.

There have also been a few studies have examined the effects from host inhalation/exposure to Al compounds upon the ability of the lungs of these exposed hosts to resist/clear viable bacterial challenges. In mice which had been exposed for a single 3 hr period to atmospheres containing 0.3 - 0.5 mg Al/m^3 (as aluminum sulfate [$Al_2(SO_4)_3$]) or 0.2 - 0.5 mg Al/m^3 (as aluminum ammonium sulfate [$Al_2(SO_4)_3(NH_4)_2SO_4$]) and then immediately (within 1 hr of Al exposure) infected by a 10 - 15 min exposure to atmospheres containing viable *Streptococcus zooepidemicus*, those hosts exposed to either metal compound displayed increased (albeit not significant) mortality rates compared to air-exposed infected control mice (Drummond et al. 1986). Unfortunately, these results both contradict and support those in earlier studies that examined whether host inhalation of/instillation with Al-containing compounds could modulate host pulmonary responses to infection. In one study, while one 3 hr exposure to increasing amounts of $Al_2(SO_4)_3$ resulted in a significant increase in host mortality due to subsequently-inhaled *S. pyogenes*, inhalation of equivalent (or greater) amounts of $Al_2(SO_4)_3(NH_4)_2SO_4$ never significantly modified host survival against the pathogen (Ehrlich, 1980). Conversely, if the compounds were first instilled and the hosts subsequently infected with *S. pyogenes*, mice that received $Al_2(SO_4)_3(NH_4)_2SO_4$ displayed significantly greater mortality rates than did those that received an approximately equivalent amount of $Al_2(SO_4)_3$ (Hatch et al., 1981). Of even greater interest in this latter study is that when mice were permitted to inhale either compound rather than have either deposited in their lungs by instillation, no significant differences in host mortality (between the two agents or as compared with sham/air-exposed infected controls) were documented.

Very little is known about the effects of Al upon lung immune cell-related functionalities. While one study has demonstrated that aluminum potroom workers display abnormal systemic levels of T-helper (T4)- and T-suppressor (T8)-lymphocyte levels (as well as abnormal T4/T8 ratios) as compared to control populations, a subtyping of lung lymphocyte populations was not performed (Davis et al., 1990). In a study which did examine lung lymphocyte levels in potroom workers who had been exposed to Al-bearing dusts for up to 12 yr, total lymphocyte levels were not significantly different from those in lavages obtained from control workers; again, unfortunately, population subtyping was not performed and so it remains unclear as to whether variations in the relative percentages of T4 and T8 lymphocytes were present in the lungs of the potroom workers (Eklund et al., 1989). Analyses of other immune cells recovered during lavaging also indicated that both the relative percentages in the total populations and the viabilities of both alveolar macrophages and PMN in the lungs of Al-exposed workers were unaffected by the chronic exposures. Though it appeared that the ability of potroom worker macrophages to bind/ingest yeast particles was significantly reduced compared to that by cells obtained from control employees, it was subsequently shown that neither surface opsonin (i.e., OKM1 C3bi) receptor expression nor phagocytic activity itself were modified. A similar lack of effect upon alveolar macrophage viability or phagocytic activity was demonstrated *in vitro* in hamster cells incubated with Al_2O_3 particles (Warshawsky et al., 1994). Using rabbit alveolar macrophages co-cultured with Al_2O_3, it has also been shown that oxidative metabolism (as measured using nitroblue tetrazolium [NBT] reduction in resting and stimulated cells) was only nominally affected (Gusev et al., 1993).

In *in vitro* studies which examined the effects from a soluble Al compound (i.e., aluminum chloride, $AlCl_3$) upon rat alveolar macrophage viability, membrane integrity, oxygen consumption, and reactive oxygen intermediate formation, although the exposure did not induce any alteration in cell membrane integrity or viability it did cause significant reductions in O_2 consumption by either resting or zymosan-stimulated cells, and subsequently, reductions in superoxide anion generation by the latter (Castranova et al., 1980). That soluble $AlCl_3$ did not display an increased cytotoxic effect upon macrophages as compared to insoluble Al_2O_3 agents is somewhat contrary to previously-documented solubility-dependent differences in effects observed with several other trace elements (reviewed in Cohen et al., 1997, 1998). One likely explanation for this similarity in effects engendered by these soluble and insoluble Al agens has been provided by a study which demonstrated that upon $AlCl_3$ deposition in the lungs, there is a rapid sequestration of free Al ions by resident macrophages and type I pneumocytes such that little Al is able to cross the alveolar-capillary barrier (Berry et al., 1988). Once entrapped within cellular lysosomes, complexation with free phosphate ions (generated by the activity of organelle acid phosphatase(s)) then results in the formation of insoluble aluminum (ortho)phosphate ($AlPO_4$). As a result of this process, Al (irrespective of its parent form as an insoluble or soluble compound) remains trapped within the macrophages and pneumocytes and is only slowly cleared from the lungs during alveolar clearance. In light of this somewhat unusual mechanism, it is then not so surprising that the relative toxicities of Al_2O_3 and $AlCl_3$ upon the alveolar macrophages from varying host species did not differ all that much.

COPPER

Unlike some of the metals discussed in this chapter, studies of copper (Cu) immunotoxicity are complicated by the fact that, much in the manner of zinc, Cu is essential to maintenance of immunocompetency. Deficiencies of Cu due to hereditary factors as occur in Menke's (trichopoliodystrophy) and Wilson's disease or poor nutrition give rise to compromised host resistance (Beach et al., 1982; Lukasewycz and Prohaska, 1990) as well as developmental emphysema due to a decrease in the activity of Cu-dependent lung lysyl oxidase (Harris et al., 1974; O'Dell, 1981).

The paucity of epidemiological studies reporting ill effects from exposure to inorganic Cu compounds might suggest that Cu is not particularly hazardous. However, Cu present in several environmental media is unwittingly inhaled by animals and humans and instances of Cu intoxication that give rise to alterations in lung immune cells and their function do occur. The highest air levels of Cu are generally encountered in the vicinity of metallurgical processing plants, iron and steel mills, and around coal-burning power plants. In these workplace atmospheres, Cu is most commonly found as cuprous oxide ($Cu(I)_2O$) and cupric hydroxide ($Cu(II)(OH)_2$) (Peoples et al., 1988); in the presence of atmospheric moisture, the Cu_2O is readily converted to cupric oxide ($Cu(II)O$). As a result of industrial processes, industrial smokes contain particulate Cu at levels often ranging from 50 - 900 μg Cu/m^3; conversely, airborne Cu concentrations in rural/suburban areas average only ≈0.01 - 0.26 μg Cu/m^3 (Purves, 1977; Tsuchiyama et al., 1997). Apart from incidences of occupational exposure, daily Cu intake averages ≈0.02 mg (EPA, 1980).

The major immunologically-based conditions reported to arise from inhalation of inorganic Cu compounds are metal-fume fever (with specific reference to Cu, termed 'copper fever') (McCord, 1960; Cohen, 1974; Piscator, 1976; Nemery, 1990) and pneumofibrosis (vineyard sprayers' lung) (Pimentel and Marques, 1969; Villar, 1974). A few epidemiological studies have indicated that workers in copper smelters have an increased incidence of lung cancers (Kuratsune et al., 1974, Tokudome and Kuratsune, 1976; Ostiguy et al., 1995; Sorahan et al., 1995); although Cu binds with and affects the structure and integrity of DNA and RNA (Sagripanti et al.

1991), a definitive linkage to cancer causation is still not at hand. Interestingly, precisely because Cu ions can enhance DNA damage in several biological systems, it has been combined with select antitumor agents (i.e., bleomycin, streptonigrin, and camptothecin, or irradiation) in efforts to reduce tumor size *in vivo*. (Raisfeld et al., 1982; Sugiura et al., 1984, 1985).

Deficiencies of Cu can give rise to alterations of immune cell numbers and functions. These changes include: a reduction in T-lymphocyte (primarily the $CD4^+,CD8^-$-helper-T [T_H] cell) numbers; inhibition of B- and T-lymphocyte responsiveness to mitogens; decreased antibody formation; decreased phagocytic cell microbicidal activity; increased acute and delayed inflammatory responses; and, enhanced host susceptibility to infections (Newberne et al., 1968; Prohaska and Lukasewycz, 1981; Jones and Suttle, 1983; Lukasewycz and Prohaska, 1983; Vyas and Chandra, 1983; Jones, 1984; Flynn, 1985; Lukasewycz et al., 1985). Although the above *ex vivo* studies, for the most part, utilized systemic immune cells, it is likely that lung immune cells also displayed the same alterations as a result of the Cu deficiency.

While much attention has focused upon the essential role of Cu in proper immune function, immunomodulation arising from Cu toxicosis has also been examined. In studies examining the clearance of soluble cupric sulfate ($CuSO_4$) and insoluble CuO from the lungs of rats, it was shown that the clearance half-time of CuO was more than 5 times that of the soluble sulfate (37 h vs 7.5 h post-instillation) (Hirano et al., 1990, 1993). Although in each case metallothionein formation was induced, it was determined that this Cu-binding protein contributed little to the localized accumulation of Cu originating from either compound. Similarly, both $CuSO_4$ and CuO induced strong pulmonary inflammatory responses (assessed using cytological, biochemical, and elemental indices) that became predominant within 12 hr of instillation and remained significantly elevated for 3 d thereafter. As each agent induced similar effects *in situ*, even though their clearance times strikingly differed, it was concluded that CuO particles undergo rapid dissolution following their deposition in the deep lung.

The proinflammatory effects of Cu may help to explain, at least in part, the observed increase in pulmonary fibrosis following host exposure to copper sulfide (CuS)-bearing dusts or $CuSO_4$-containing Bordeaux mixtures used to treat crops (Lutsenko et al., 1997; Eckert and Jerochin, 1982). Somewhat similar effects have also been observed in the lungs of hosts exposed to more novel compounds like copper gallium diselenide and copper indium diselenide, agents routinely used in the photovoltaic and semiconductor industries (Morgan et al., 1995, 1997). Both of these agents gave rise to strong proinflammatory responses, type II cell hyperplasia, and increases in biochemical indices of fibrosis following instillation. However, unlike with CuS or $CuSO_4$, histological changes indicative of increased pulmonary interstitial fibrosis were not observed.

Apart from inflammation, other effects of Cu upon pulmonary immunocompetence have also been documented. In mice that underwent single or multiple 3 hr exposures to $CuSO_4$ aerosols (over a broad range of concentrations), resistance to challenge with viable *S. zooepidemicus* was drastically reduced (Ehrlich et al., 1978; Ehrlich, 1980; Drummond et al., 1986). In Cu-exposed mice, mortality rates were increased from 54 - 100% (compared to that in air-exposed controls) and changes in mean survival time were decreased from 4 - 11 d (over a 14 d post-infection period). In mice repeatedly exposed to $CuSO_4$ prior to infection, these endpoints were still significantly altered, but not to the same extent as that in mice that simultaneously inhaled Cu and bacteria. Similar effects were also observed in mice instilled with $CuSO_4$ and then challenged with *S. pyogenes* (Group C) (Hatch et al., 1981) and these Investigators determined that the bactericidal activity of alveolar macrophages was impacted upon by the Cu ion with *in situ* killing of bacteria being dose-dependently decreased. This latter effect was in keeping with previous observations that inhaled $CuSO_4$ reduced lung macrophage phagocytic/endocytic activity *in situ* in the absence of a concurrent change in total cell numbers (Skornik and Brain, 1983).

Other studies provide further evidence that alveolar macrophages are a target for the immunotoxic effects of inhaled Cu. Following a single instillation of CuO, rat alveolar macrophages were shown to undergo hypertrophy and to develop increased levels of polymorphic nuclei with margination of the chromatin, crystalloid-like inclusions, concentric and parallel lamellar structures, lattice formations, and degenerative membranous structures (Murthy et al., 1982). A different study indicated that exposure of macrophages to Cu resulted in decreased lysosomal permeability, reduced levels of lysosomal enzymes (without changes in enzyme activities), increased secretion of lysosomal enzymes from vesicles, and increased lysosome labilization (Ludwig and Chvapil, 1981). Alveolar macrophages recovered from rabbits exposed to 0.6 mg/m^3 cupric chloride ($CuCl_2$) every weekday for 1 mo were shown to contain increases in the amounts of lamellated inclusions in the cytoplasm, lysozyme content, and surface blebs (Johansson et al., 1983; Lundborg and Camner, 1984; Camner et al., 1985; Johansson and Camner, 1986). However, unlike the alveolar macrophages of the earlier-described mice exposed to $CuSO_4$, cells from rabbits did not display changes in phagocytic activity, oxidative metabolism, or bactericidal function; it should be noted that these results may differ not because of the host species or the Cu compounds employed but, rather, because these functionality studies were all performed *ex vivo* as opposed to *in situ*. To further confound the issue of the pulmonary immunotoxicity of Cu, the rabbits that inhaled the $CuCl_2$ also failed to evidence any lung inflammation.

It is likely that Cu toxicoses arising from non-inhalation uptake of Cu also affect the function of lung immune cells. For example, dietary Cu oversupplementation increases the severity of infections in animals (Wilhemsen, 1979; Hill, 1980), reduces cell-mediated responses to PHA and lysozyme (Kornegay et al., 1989), and significantly increases autoantibody levels in sera (Pocino et al., 1990, 1991). These studies also indicated specific differences in immunotoxic mechanisms of Cu ions in that splenic T-lymphocyte responses to ConA were inhibited after acute and subchronic host exposure to Cu, while B-lymphocyte responses to LPS were enhanced by acute, but not subchronic, exposures. Similar time-dependent effects were also seen in studies monitoring the onset and strength/duration of delayed-type hypersensitivity reactions. Although it is still not clear whether the same effects occur with lung lymphocytes following inhalation of Cu, these studies provide a basis for predicting what types of non-macrophage-related immunomodulation may occur.

Many of the *in vitro* studies performed to try to delineate the mechanisms by which Cu induces systemic immunomodulation have utilized Cu ions in complexes with serum proteins. This is a logical premise since ≈90% of all extrapulmonary Cu is bound to ceruloplasmin (CP), associated with Cu-dependent enzymes (i.e., ferroxidases, cytochrome oxidase, superoxide dismutase, and amine oxidase), or complexed with free amino acids (i.e., histidine). Unlike systemic leukocytes, however, pulmonary immune cells are more likely to encounter "free" non-complexed Cu (either as solubilized ions or particulates). Nevertheless, these *in vitro* studies do provide information useful for determining how inhaled Cu induces pulmonary immunomodulation.

In several *in vitro* studies, it has been demonstrated that Cu, both as "free" (loosely-associated) ions and in CP-bound forms, inhibited T-lymphocyte function moreso than that of B-lymphocytes or monocytes. These depressions in lymphocyte activities (such as mitogen-induced T-lymphocyte responsiveness, and T_H-directed B-lymphocyte stimulation/antibody synthesis) were apparently related to a Cu-dependent enhanced formation of hydrogen peroxide. An increase in peroxide formation, coupled with a deficiency of peroxide-metabolizing enzymes in T-lymphocytes, caused extensive oxidative damage to these cells that could then lead to reduced functional capacities during cell-mediated or humoral immune responses (Lipsky and Ziff, 1980; Lipsky, 1981, 1984; Anderson and Tomasi, 1984). However, because B-lymphocytes do contain the necessary enzymes which could minimize the impact from any Cu-initiated excess of extracellular peroxide, alternative mechanisms were needed to explain the reduced mitogenic responses of these cells in response to Cu. It was determined that Cu complexes directly oxidized essential surface thiol groups on lymphocytes, and that B-lymphocytes were far more sensitive than T-lymphocytes to this effect (Duncan and Lawrence, 1988). It remains to be determined, however, which immune cell type is the primary target of the Cu-induced alterations in antibody responses.

In vitro studies have also been performed to determine the mechanisms by which Cu-induced modulation of macrophage functions might arise. Some of these studies have shown that Cu treatments can influence directional migration by macrophages (as well as by PMN), and alter macrophage prostaglandin (i.e. PGE_2) and thromboxane (TBX_2) production (Lewis, 1982; Elliott et al., 1987). It has also been demonstrated that Cu could affect the ability of macrophages to act as antigen-presenting cells (Smith and Lawrence, 1988); however, it is still not clear as to which phase of the antigen presentation process (e.g., antigen capture, processing, or presentation in the context of major histocompatibility complex proteins) was primarily altered by Cu treatment.

MANGANESE

Although manganese (Mn) is widely distributed in nature, it normally occurs only in trace amounts in most biologic materials/tissues. It is known that Mn is an essential trace element for all living organisms; in general, it is believed that Mn most often acts as a dissociable co-factor for several enzymes, including one of great importance in the lungs, superoxide dismutase. Similarly, the there are numerous reports in the literature which have described the toxicologic effects from acute or chronic exposures to Mn-containing compounds in the environment and/or the workplace. With respect to toxicities which evolve in the lungs, the two most common clinical presentations have been manganic pneumonitis and croupous pneumonia (Flinn et al., 1940; Davies, 1946; NAS, 1973; Smyth et al., 1973; Saric et al., 1977; Bencko and Crikt, 1984; Saric, 1992).

Mn-Containing compounds are widely employed in alloy steel manufacture for deoxidation and to promote hardenability, in the electric industry production of dry cells, as oxidants in the chemical industry, in the manufacture of fertilizers, as catalysts, wood-impregnating agents, and additives in paints/varnishes, and for the production of glass and glazes (Stokinger, 1981). Of increasing interest to both regulators and toxicologists over the past nearly two decades have been the agents methylcyclopentadienyl manganese tricarbonyl (MMT) and its demethylated counter-part cyclopentadienyl manganese tricarbonyl (CMT), antiknock additives and com-bustion improvers for a variety of fuels. Apart from the introduction of Mn into the environment by the combustion of MMT-/CMT-containing fossil fuels, the major industrial sources which result in substantial release of Mn directly into the environment include several types of mining industries and alloy/steel production facilities (reviewed in Cooper, 1984).

Manganese has been found at measurable levels in the majority of suspended particulate matter (including respirable coal fly ash) in urban environements. On average, while air levels of Mn in many major metropolitan areas containing steel or alloy plants have been found to range from 0.5 - 3.3 μg Mn/m^3 (with levels of up to 10 μg Mn/m^3 sometimes being attained), the majority (>80% of all sites assessed) have levels < 0.1 μg Mn/m^3 (EPA, 1975; Ondov et al., 1982). In those areas without any contributing point sources, average air levels range from 0.03 - 0.07 μg Mn/m^3. Although guidelines for workplace airborne Mn levels vary between that recommended by WHO (i.e., 300 μg Mn/m^3 [TWA]) and by OSHA (i.e., 5000 μg Mn/m^3 [ceiling value]), occupational levels often are on the order of 1 - $\geq$100 μg Mn/m^3 and have even been shown to reach 1 g Mn/m^3 in one ferro-manganese production plant (reviewed in Cooper, 1984).

As with the other metals discussed throughout this Chapter, the effect of compound solubility upon clearance of inhaled Mn-containing particles is critical to the extent of pulmonary immunomodulation which might occur. In studies to compare the clearance of soluble manganous chloride ($MnCl_2$) and insoluble manga-nomanganic (manganese) oxide (Mn_3O_4) from the lungs of rats, it was shown that the rate of clearance of the soluble form of Mn was four times that of the insoluble agent over the first seven days post-exposure (Drown et al., 1986). Regardless of solubility, both agents were cleared in a biphasic manner and the post-translocation disposition of the Mn to distal body organs was similar for both classes of compounds. Similar initially-rapid clearance rates have also been observed in the lungs of mice exposed to Mn_3O_4 (Adkins et al., 1980), and of guinea pigs (Bergstrom, 1977) or mice (Maigetter et al., 1976) that had inhaled fairly high (i.e., 22 - 69 mg Mn/m^3) levels of insoluble manganese dioxide (MnO_2). Inter-experi-mental variations in the lengths of exposures and Mn concentrations employed have made direct comparisons (with regards to the biologic half-times in the lungs) between Mn compounds difficult to achieve. However, among those studies which have examined only the clearance of MnO_2, half-life values for these particles have been reported to range from <1 d in the mouse, to $\approx$2 d in the guinea pig, to 38 d in dogs, to 65 d in humans (Morrow et al., 1964, 1967; Maigetter et al., 1976; Bergstrom, 1977).

While several studies have directly examined the effects of inhaled Mn compounds upon pulmonary immune parameters, the majority have focused upon assessing the histological changes and/or inflammatory responses following intra-pulmonary deposition of Mn-containing agents. For example, histologic examination of the lungs of monkeys that had inhaled either 0.7 or 3.0 mg Mn/m^3 (as MnO$_2$) every day over the course of 10 mo evidenced hyperplasia of interstitial lymphoidal tissues, over-deposistion of dusts in the pulmonary interstitium, bronchiolar retention of exudate, pulmonary emphysema, and atelectasis (Suzuki et al., 1978). Similarly, rats exposed to 0.3 mg MnO$_2$/m^3 for 6 mo displayed perivascular and peribronchial sclerosis of their lung tissues (reviewed in Ulrich et al., 1979). Using mice as the exposure model, it was demonstrated that the inflammatory effects of MnO$_2$ in this species (and likely, as well, in all other mammalian hosts) were a function of both the total surface area and the gravimetric amount of the MnO$_2$ particles in the inhaled atmosphere (Lison et al., 1997).

Studies which have utilized soluble MnCl$_2$ or volatile MMT help to demonstrate the role of solubility upon intrapulmonary effects that arise from exposure to Mn-containing compounds. In a study using rabbits, inhalation of 1 - 4 mg Mn/m^3 (as MnCl$_2$) 5 d/wk for 4 - 6 wk failed to induce any gross histopathologic changes in the lungs and did not induce any inflammatory response as had been observed with the insoluble Mn agents (Camner et al., 1985). In studies which examined the effects from exposure to MMT (an agent selectively toxic to the lungs regardless of the route of administration [Hinderer, 1979]), a single intraperitoneal (IP) treatment of rats, mice, or hamsters with MMT (at their respective $\approx$LC$_{50}$s) resulted in substantial interstitial pneumonitis characterized by interstitial thickening, PMN infiltration, and increased numbers of intraalveolar macrophages in the mice and rats (Hakkinen and Haschek, 1982). Similar acute pneumotoxic effects have also been observed in rats when MMT or CMT was administered subcutaneously (Hakkinen and Haschek, 1982; Blanchard et al., 1996). In all three species, Clara cell necrosis was evident within one day of IP exposure to MMT. In the mouse and hamster, the loss of Clara cells was nearly total in the terminal bronchioles and the murine basement membranes became greatly denuded and partially covered with flattened abnormal-looking ciliated epithelium; Clara cell changes in the rats were the least severe and usually consisted of some flattening and cytoplasmic vacuolation. Except for the parenchymal damage in the treated rats, each of the above-noted pathologies was resolved by the end of a 3-wk post-MMT exposure period; as with MMT, CMT exposure resulted primarily in localized damage to alveolar regions of the lung and only minor bronchiolar damage. While the strong but variable potential for damage to lung cells should engender concern, it must be noted that atmospheric MMT/CMT have very short half-lifes due to photochemical decomposition (Ter Haar et al., 1975) and that it is their combustion byproducts (primarily Mn$_3$O$_4$ as well as MnO$_2$ and manganosite (MnO) [Ter Haar et al., 1975; Loranger and Zayed, 1997]) which potentially present major risks to pulmonary (immuno)compentence.

Other studies have examined the effects from host inhalation/exposure to Mn compounds upon the ability of the lungs to resist and clear viable bacterial/viral challenges. In guinea pigs which had been exposed to atmospheres containing 22 mg MnO_2/m^3 for 24 hr and then immediately infected by a 10 min exposure to atmospheres containing viable *Enterobacter cloacae*, those hosts exposed to the metal displayed significantly lower burdens of the microorganism within the than did their air-exposed counterparts. However, after two hours and continuing over the next 22 hr, burdens of *E. cloacae* remained significantly greater in the lungs of the Mn-treated guinea pigs (Bergstrom, 1977). In mice which had been allowed to inhale one of several concentrations of Mn_3O_4 for 2 hr prior to a 20 min exposure to a *S. pyogenes*-containing atmosphere, the rates of host mortality were found to increase in direct relation to the amounts of Mn deposited in their lungs (Adkins et al., 1980c). Within the lungs of Mn-exposed mice, there was both a delayed clearance and subsequent continuous enhanced growth of the organism over a 4 d period post-infection; budens of *S. pyogenes* in the lungs of air-exposed infected hosts peaked within 2 d of infection and decreased thereafter. Similarly, in mice exposed for 3 hr to MnO_2-containing atmospheres and then challenged 1 - 5 hr post-exposure with airborne *Klebsiella pneumoniae*, host mortality was increased (and mean survival time decreased) in infected hosts compared to that in air-exposed counterparts (Maigetter et al., 1976). In this study, it is interesting to note that while increasing the numbers of daily exposures to MnO_2 prior to infection did not significantly enhance the observed effects upon host resistance to the *K. pneumoniae*, increasing the period between MnO_2 exposure and the initiation of the infection did. When hosts already bearing a lung infection (i.e., mice infected with influenza A/PR/8/34 virus) for 24 or 48 hr prior to exposure to a MnO_2-bearing atmosphere, it was found that a single 3 hr exposure after either lag period was sufficient to significantly reduce host survival against the virus and that the effect of MnO_2 upon this parameter was greater in those animals that already bore greater viral titres. Oddly, if already-infected hosts underwent further daily exposures to MnO_2, host mortality levels actually decreased to levels almost equivalent to those observed with air-exposed control mice.

At the level of lung immune cell-related functionality, the effects of Mn upon alveolar macrophages have been the best studied. In rabbits exposed 6 hr/d for 4 - 6 mo to 1 - 4 mg Mn (as $MnCl_2)/m^3$, although alveolar macrophage numbers and viabilities remained unaffected (Camner et al., 1985) they did display Mn concentration-related increases in their diameters. Oxidative metabolism (measured using NBT reduction in resting and stimulated cells) and phagocytic activity of these cells were similarly unimpaired by the $MnCl_2$ treatments. In mice acutely exposed (2 hr; 532 or 897 µg Mn/m^3) to Mn_3O_4, alveolar macrophage numbers, viabilities, and phagocytic activities were again unaffected; however, the study did indicate that levels of both ATP and the activity of cellular acid phosphatase were significantly increased as a result of the single exposure (Adkins et al., 1980b). These *in vivo* exposure results contrast with those obtained in *in vitro* studies wherein treatment of cultured rabbit alveolar macrophages with $MnCl_2$ resulted in a decrease in cell viability and number, an increased incidence of cell lysis (Waters et al., 1975), and significant reductions in phagocytic activities (Graham et al., 1975).

Far less is known about the effects from insoluble Mn agents upon alveolar macrophages or other immune system cells of the lungs. Lundborg et al. (1984) demonstrated that freshly-harvested rabbit alveolar macrophages were readily able to phagocytize MnO_2 particles (0.1 - 0.5 µm diameter) and that the ingested particles

were subsequently dissolved in a pH-dependent process within macrophage phago-
somes. These studies also demonstrated that these processes were time-dependent
(maximal dissolution occurring within the first 72 hr of exposure) and that the overall percentages of Mn dissolved was both highly dependent upon the amount of
MnO_2 added to each culture and saturable. Interestingly, when rabbit alveolar macrophages were cultured for 48 - 72 hr prior to the exposure to MnO_2 particles, the
capacity to dissolve MnO_2 particles was almost completely ablated. In a separate
study, it has also been demonstrated that incubation of freshly-harvested guinea pig
alveolar macrophages with MnO_2 resulted in an increased formation/release into the
surrounding microenvironment of an active short-lived chemotactic/neutrophil-
recruiting factor within a few hours of initiation of particle uptake (Snella, 1985).
The results of this specific study are important in that they provide a partial
mechanistic basis for the repeatedly-observed time-limited increased influx of
inflammatory cells into the lungs following inhalation of this insoluble agent
(Bergstrom, 1977; Lison et al., 1997).

SELENIUM

As with Cu, the essentiality of selenium (Se) in immunocompetency is well
known. However, there is limited information regarding effects from Se toxicosis
upon the immune system in general, and even less so regarding that in the lungs.
While several reports in the literature describe occurrences of occupational or
personal overexposure to Se-containing compounds, the results presented are mostly
clinical in nature (Buchan, 1947; Clinton, 1947; Glover, 1970; Schecter et al.,
1980; Nantel et al., 1985; Koppel et al., 1986; Holness et al., 1989).

Se-Containing compounds have been widely employed in the production of
glass and glazes, the manufacturing of photoelectric cells, in xerographic equipment,
for production of paints, insecticides, and medicated shampoos, and as an alloying
agent in stainless steel and copper-based alloys. The major industrial sources which
emit Se directly into the environment include the mining, milling, smelting and
refining of several types of metals (i.e., copper, zinc, lead, uranium) and the burning
of fossil fuels. Within the facilities themselves, ambient levels on the order of
several mg Se/m^3 have been commonly measured (WHO, 1987), and, as would be
expected, high (i.e., μg Se/m^3) atmospheric levels of Se are encountered in the
vicinity of these industries. Coal burning serves as the primary source for atmospheric Se in rural and less-industrialized regions, with levels of 0.1 - 10 ng Se/m^3
routinely being encountered (Hashimoto and Winchester, 1967; Zoller and Reamer,
1970; WHO, 1987). Within the coal, the Se is most often found in complexes of
selenide which are converted to selenium dioxide (SeO_2) during combustion. The
SeO_2 itself can then react with water vapor in the air to form selenious acid
(H_2SeO_3) or it can undergo reduction by other ambient pollutants (such as sulfur
dioxide) and revert to elemental selenium metal (Se^0) (Andren et al., 1975).

The impact of compound solubility upon clearance of the particle(s) is critical
to the ultimate extent of pulmonary immunomodulation which can occur; unfortunately, information regarding solubility and clearance of Se compounds is very
limited. In studies examining the clearance of soluble H_2SeO_3 and insoluble Se^0

from the lungs of beagles, it was shown that the clearance of the acid into the blood was slightly faster than that of the metal particle (29 d vs 33 d post-inhalation, respectively) (Weissman et al., 1983); once translocated out of the lungs, the disposition of each compound was similar. In similar studies employing rats, these same trends in solubility-related differences in translocation from the lungs (and post-translocation disposition) were observed (Medinsky et al., 1981a and c). Interestingly, while this study concludes that clearance of each compound from the alveolar regions of the lung (albeit at differing rates) was equivalent, differences in absorption of elemental and acid Se compounds from the upper respiratory tract/nasal regions would eventually give rise to significantly different circulating levels of Se. In a study utilizing rats intratracheally-instilled with sodium selenite (Na_2SeO_3) or selenate (Na_2SeO_4), clearance of each of these relatively soluble agents was found to be equivalent and intermediate between that of the H_2SeO_3 and Se^0, suggesting that valence had no impact upon this parameter (Rhoads and Sanders, 1985); post-translocation distribution of the Se derived from each compound was comparable to that of the highly soluble acid.

Few studies have directly examined the effects of inhaled Se compounds upon immunologic parameters in the lungs. Copper gallium diselenide or copper indium diselenide, agents routinely used in the photovoltaic and semiconductor industries, gave rise to strong proinflammatory responses, type II cell hyperplasia, and increases in biochemical indices of fibrosis (however, without concurrent histologic changes indicative of increased pulmonary interstitial fibrosis) in the lungs of rats following instillation (Morgan et al., 1995, 1997). In guinea pigs intratracheally-instilled with Na_2SeO_3, the formation of the arachidonic acid metabolites thromboxane (TXB_2) and leukotriene C4 (LTC_4) were decreased, while that of PGE_2 was increased (Bell et al., 1997a). The Investigators reported in a separate study that small but significant increases in the levels of lactate dehydrogenase, β-glucuronidase, alkaline phosphatase, and total protein were evident in the lavage fluid obtained from Se-instilled guinea pigs (Bell et al., 1997b).

Although not as yet documented under *in vivo* conditions, one possible mechanism by which Se compounds might impart some of their effects in the immune cells of the lung is through their reaction with cellular thiol residues. This in turn would give rise to bridged R-S-Se-S-R' complexes which could subsequently undergo decomposition to yield R-S-S-R' and elemental Se (Medinsky et al., 1981b). This two-step process of complexation/decomposition likely explains why in *in vitro* studies, H_2SeO_3 is more cytotoxic than Se^0 toward macrophages and why red particles become increasingly evident within macrophage vacuoles after exposure to H_2SeO_3 but not in cells exposed to Se^0.

Other studies have examined the effect of host Se status upon lung immune cell-related functionality. In Se-deficient rabbits provided a diet supplemented with Na_2SeO_3, the survival rate of phagocytizing activated alveolar macrophages and their associated glutathione peroxidase activity levels were significantly increased compared to cells obtained from Se-deficient controls (Oh et al., 1982); however, glutathione reductase and superoxide dismutase activity, as well as macrophage reduced glutathione levels, were unaffected by resupplementation. Host Se-deficiency has also been shown to result in altered formation of arachidonate acid metabolites (primarily leukotriene B_4) by alveolar macrophages (Gairola and Tai, 1985). These alterations which occur *in situ* during changes in Se status have since been shown to modify the pulmonary immunotoxicities of other toxicants, including cigarette smoke (Gairola and Tai, 1986), ozone (Elsayed et al., 1983; Eskew et al., 1986), and the clearance/retention (and as a likely consequence, the immunotoxicity) of mercury (Khayat and Dencker, 1983, 1984; Danielsson et al., 1990).

VANADIUM

Vanadium (V) is one of the more ubiquitous trace metals found in the environment (Nriagu and Pacyna, 1988; reviewed in Cohen, 1998). Because clays and shales can contain >300 ppm V, coals upwards of 1% V (by weight), and petroleum oils 100 - 1400 ppm V depending on the site of recovery, the combustion of fossil fuels is the most readily identifiable source for delivery of V-bearing particles into the atmosphere. Ambient air concentrations of V have been shown to vary from 0.02% by weight in soil-derived aerosols, to 0.02 - 0.20% in automobile-derived fumes, to 0.54 - 0.82% in oil combustion-generated aerosols, depending upon the region under study (Thurston and Spengler, 1985; Watson et al., 1994). Typical V concentrations in rural areas range from 250 pg - 75 ng V/m^3 while those in urban settings are usually higher (i.e., 60 - 300 ng V/m^3) (WHO, 1988); on average, ambient V concentrations in cities are often several µg V/m^3. Seasonal variations, wherein winter urban air V levels are 6-fold greater than summer levels, arise from the increased combustion of V-bearing oils, shales, and coals for heat and electricity generation. At these levels ($\approx$50 ng V/m^3), and based on experimental inhalation studies, it is estimated that $\approx$1 µg V enters the average adult human lung each day (Byrne and Kosta, 1978).

In the workplace, the use of fossil fuels for energy and of V-bearing ores for steel production and chemical processes enhance the risk for occupational exposure to V-bearing gases and particles (ATSDR, 1991). Examples of jobs with a high risk of exposure to V include: mining and milling of V-bearing ores, oil-fired boiler cleaning, and the production of V metal, oxides, and catalysts, with ambient levels of V in these settings sometimes exceeding 30 mg V/m^3 (a value which approximates the established value for immediate danger to life or health (IDLH) of 70 mg V/m^3) (NIOSH, 1985).

The most common form of pentavalent V which is found in air is vanadium pentoxide (V_2O_5), although ferrovanadium, vanadium carbide (VC), and various forms of vanadates (VO_3^- and VO_4^{3-}) can also be readily encountered. Colloidal V_2O_5 can liberate vanadate through the loss of water molecules; the resulting monomeric vanadate ions can be further converted into higher polymeric forms much in the manner that chromate ions are linked during olation. These conversions, and therefore the distribution of V species, in solution are both pH- and V-concentration-dependent events. As a rule, as the number of vanadate units in the polymer increases, overall toxicity declines; however, even large polymers, such as decavanadate can give rise to toxicities.

Once inhaled, V is rapidly transported into the systemic circulation. The initial clearance of V, as either insoluble V_2O_5 or soluble vanadates/vanadyl (VO^{2+}) ions, is fairly rapid, with $\approx$40% cleared within 1 hr (Conklin et al., 1982; Sharma et al.,

1987). However, after 24 hr, the two forms diverge in their ability to be cleared, with V_2O_5 persisting in the lungs (Edel and Sabbioni, 1988). As such, absorption of V compounds (50 - 85% of an inhaled dose) vary as a function of solubility; total clearance of V is never achieved and commonly 1 - 3% of the original dose can persist for extended periods of time (i.e., months --> years; Oberg et al., 1978; Rhoads and Sanders, 1985; Paschoa et al., 1987). Though inhalation is the primary means of delivery of V into the lungs, exposure to V by other routes can also give rise to increased lung V burdens and subsequent toxic manifestations (Hopkins and Tilton, 1966; Kacew et al., 1982).

Pentavalent vanadates and oxides have long been known to alter pulmonary immunity in exposed hosts (reviewed in Zelikoff and Cohen, 1995; Cohen, 1998). Workers exposed to airborne V display an increased occurrence of prolonged coughing spells, tuberculosis, and general irritation of the respiratory tract; post-mortem examinations of these individuals indicated extensive lung damage with the primary cause of death being bacterial infection-induced respiratory failure. Epidemi-ological studies have demonstrated that both acute exposure to high concentrations and/or chronic exposure to moderate levels of V-bearing dusts/fumes by workers resulted in a higher incidence of a wide variety of pulmonary diseases, including: asthma, rhinitis, pharyngitis, ('Boilermakers') bronchitis, and pneumonia (Lees, 1980; Musk and Tees, 1982, Levy et al., 1984, reviewed in Cohen, 1998), as well as increased localized fibrotic foci (Kivuoloto, 1980) and lung cancers (arising from non-V sources; Stocks, 1960; Hickey et al., 1967). More detailed cytological studies with cells from exposed populations showed V-induced disturbances in polymorpho-nuclear cell numbers/cellularity and in plasma cell numbers/immunoglobulin production (Kivuoloto et al., 1979, 1980, 1981). This risk to worker health led to an establishment of acceptable limits (0.14 - 50 µg V/m^3 per 8 - 24 hr period) for workplace V-bearing dusts and fumes (NIOSH, 1985; ACGIH, 1986).

Changes in pulmonary immune function induced by V are reproducible in a variety of animal models. Subchronic and acute exposures of various rodent hosts to pentavalent V agents have been shown to induce: decreased alveolar macrophage phagocytosis and lysosomal enzyme activity/release (Waters et al., 1974; Fisher et al., 1978; Labedzka et al., 1989; Cohen et al., 1997); alterations in lung immune cell population numbers and profiles (Knecht et al., 1985, 1992; Cohen et al., 1996b); modified mast cell histamine release (Al-Laith et al., 1989); reduced cytokine (e.g., interferon [IFN]-γ, interleukin [IL]-6, tumor necrosis factor [TNF]-α) and bactericidal/tumoricidal factor production *in situ* and *ex vivo* (Cohen et al., 1996b, 1997); disturbed macrophage Class II antigen expression/induction by IFNγ and calcium (Ca^{2+}) ion balance (Cohen et al., 1996b); and, increased *in situ* (but not *in vitro*) expression/production of macrophage inflammatory protein-2 (MIP-2) and KC CXC chemokine mRNA by alveolar macrophages and, subsequently, by PMN recruited to the lungs during the inflammatory response (Pierce et al., 1996). This latter study also demonstrated a similar effect when tetravalent vanadyl sulfate ($VOSO_4$) was employed, suggesting that some aspects of the pulmonary immuno-toxicity of V may not necessarily be dependent upon valence.

A few studies, including several recently-reported, have implicated the presence of V as a significant factor in the pulmonary immunomodulation induced following inhalation of urban particulate matter (PM) or residual oil fly ash (ROFA) (Schiff and Graham, 1984; Pritchard et al., 1996; Dreher et al., 1997; Gavett et al., 1997). Many effects, including: intense inflammation, pulmonary eosinophilia, and neutrophilic alveolitis; changes in lung compliance/resistance to acetylcholine; and modified host resistance to pulmonary infection correlated with the levels of V in the particles and were reproduced by exposures of parallel sets of animals to soluble or insoluble V at amounts equivalent to that of the V in the PM/ROFA. An *in vitro* study has subsequently also shown that the V in ROFA likely acts to undermine pulmonary immunocompetency, in part, by inducing dysregulation of cytokine production (Schiff and Graham, 1984; Carter et al., 1997).

Studies of the immunomodulatory effects of V using non-inhalation or instillation exposure regimens have provided information critical for determining the mechanisms which underlie increased host susceptibility to lung infections following inhalation of V. A decreased resistance to/increased mortality from a *Listeria monocytogenes* infection in mice that underwent acute/subchronic IP exposures to ammonium metavanadate (NH_4VO_3) suggested that cell-mediated immunity might primarily be affected (Cohen et al., 1989). The peritoneal macrophages recovered from these mice displayed decreased capacities to phagocytize opsonized *Listeria* and to kill even those few organisms ingested; these defects were attributed to V-induced disturbances in macrophage superoxide anion formation, glutathione redox cycle activity, and hexose-monophosphate shunt activation (Cohen and Wei, 1988). While effects upon pathways critical to maintaining cell energy levels alone might underlie changes in macrophage function/host resistance, other studies indicated that decreased macrophage phagocytic activity and intracellular killing might also be related to decreases in lysosomal enzyme release and activity and/or in surface complement/F_c-receptor expression/binding activity (Cohen et al., 1986; Vaddi and Wei, 1991a and b). *In vitro* studies using V-treated murine macrophage cultures have also indicated that production/release of several monokines critical to the antilisteric response was diminished in conjunction with an increase in the spontaneous formation/release of potentially immunoinhibitory PGE_2 (Cohen et al., 1993).

More recently, research has shown that a critical aspect of the immune response that is very sensitive to V, and may even contribute to the above-described defects in macrophage functionality, is the capacity of V-exposed cells to bind with/respond to IFNγ (Cohen et al., 1996a and b, 1997). Results of *in vitro* exposures of macrophage cell lines with V indicated that surface levels and binding affinities of two classes of surface IFNγR were greatly modified by V treatments. Subsequently, IFNγ-inducible responses (i.e., enhanced: Ca^{2+} influx, Class II antigen expression, and zymosan-inducible reactive oxygen intermediate formation) in all V-treated cells were diminished; similar decrements in macrophage responsiveness to IFNγ were observed with alveolar macrophages recovered from rats subchronically-exposed to NH_4VO_3-bearing atmospheres. Effects upon surface receptor is apparently a common feature of the overall toxicology of V; lymphocytes (and other non-immune cell types) treated *in vitro* with vanadate also have a demonstrated altered affinity for hormones (i.e., epidermal growth factor and insulin) or cytokines (Kadota et al., 1987; Torossian et al., 1988; Evans et al., 1994).

Although the underlying causation for these effects upon IFNγR are not known, it has been suggested that V might directly modify proteins which constitute this and other cytokine/opsonin receptors on macrophages (Cohen et al., 1996a). Modified receptor responses might also be related to V-induced changes in cellular protein kinase and/or phosphatase activities (Swarup et al., 1982; Nechay et al., 1986; Klarlund et al., 1988; Grinstein et al., 1990; Trudel et al., 1991). By this, a V-induced prolonged phosphorylation of receptor proteins as well as cytokine-induced secondary messenger proteins might induce false states of cell activation (Pumiglia et al., 1992; Imbert et al., 1994) which could modulate cytokine receptor expression. In addition, prolonged phosphorylation of cellular proteins could also lead to bypass of normal signal transduction pathways and subsequent activation of cytokine DNA response elements (Igarishi et al., 1993) which, in turn, can lead to downregulation of cytokine receptor expression/functionality. This is best exemplified in a study in which the activation of protein kinases A and C, an event which can result in a downregulated IFNγR expression, was noted both in peritoneal macrophages harvested from V-treated mice and in naive cells exposed *in vitro* (Vaddi and Wei, 1996).

Lastly, it is possible that the effects from V upon receptor expression and/or functionality might be a result of induced effects upon the various processes involved in cytokine-receptor complex handling. Agents (such as V) that are able to disrupt: endocytic delivery of surface receptor-ligand complexes to lysosomes; subsequent complex dissociation; and receptor recycling/*de novo* receptor synthesis, can diminish the magnitude of cytokine-induced responses by macrophages. In macrophages and other cell types, V has been shown to: disrupt the structural integrity of microtubule/microfilaments (Wang and Choppin, 1981; Bennett el at., 1993); induce alterations in local pH due to V polyanion formation (Rehder, 1995); modify lysosomal enzyme release and activity (Vaddi and Wei, 1991b); alter secretory vesicle fusion to lysosomes (Goren et al., 1984), and disrupt cell protein metabolism at the levels of both synthesis and catabolism (Montero et al., 1981; Seglen and Gordon, 1981).

ZINC

Although the pulmonary immunotoxic effects of zinc oxide (ZnO) have been extensively studied due to its long-recognized role in the occupational flu-like illness 'metal fume fever' (Lehman et al., 1910; Drinker et al., 1927), far less has been determined about the immunotoxic properties of the wide variety of other zinc (Zn)-containing compounds commonly encountered in industrial and environmental settings. With regard to the latter agents, the majority of information that is available has been derived from clinical/autopsy reports which described occurrences of occupational or personal overexposure to Zn-containing compounds (Evans, 1945; Matarese and Matthews, 1986; Hjortso et al., 1988; van Netten et al., 1990; Homma et al., 1992).

Zinc metal alone is commonly used as a protective coating for other metals, in alloys (i.e., bronze and brass), and in chemical reduction processes. The various salts of Zn are commonly utilized in photographic paper preparation, wood preservatives, fertilizers, pesticides, textiles, ceramics, and in the vulcanization of rubber. Some Zn salts also have practical applications in medicine, in some cases being used as solubilizing agents for the preparation of pharmaceuticals or used directly for remediation of zinc deficiency. As occurs with several of the other metals discussed in this Chapter, the major industrial processes which result in the greatest emission of Zn directly into the environment include the production of iron or steel as well as operations involved in the smelting/refining of Zn (as well as of copper and lead) (ATSDR, 1994). In regions surrounding smelting operations, mean annual high concentrations approaching 5 μg Zn/m^3 (with 24 hr values ranging from 0.27 - 15.7 μg Zn/m^3) have routinely been encountered (Ragaini et al., 1977; EPA, 1980); other studies have reported even higher levels (Patterson et al., 1977). Galvanized welding represents an additional process in which significant amounts of Zn are released into the atmosphere at a more localized level (Mali and Carter, 1987; Weir et al., 1989; Mali et al., 1993; Contreras and Chan-Yeung, 1997); however, unlike in steel/iron production or major smelting operations, this release is most problematic for the exposed worker rather than for non-workers or the surrounding environment. Lastly, although the burning of fossil fuels (primarily coals) also results in substantial release of Zn into the atmosphere, in these cases, the Zn is just one of many potential toxicants associated with the ash matrix.

Apart from those areas near highly-industrialized regions, ambient concentrations of Zn are overall relatively low and constant. Average values in the United States are < 1 μg/m^3 (EPA, 1980), with the majority of the Zn being derived from automobile exhaust, soil erosion, and local commercial/industrial/construction activities. As expected, levels of Zn are often greater in urban than in rural areas. For example, average atmospheric Zn concentrations for New York City, San Francisco, and northern New Jersey have been measured at 0.29 - 0.38, 0.02 - 0.50, and 0.07 - 0.59 μg Zn/m^3, respectively (Lioy et al., 1978; John et al., 1973; Daisey, 1987). Conversely, rural ambient Zn concentrations have been estimated to range from 0.01 - 0.06 μg Zn/m^3. Unlike what has been observed with several other toxicologically-active inhalable metals (i.e., V) associated with coal fly ash and/or particulate matter, additional burning of coal required for heating during the winter apparently does not correlate with an increase in the concentration of atmospheric Zn relative to those levels in the summer (Barrie and Hoff, 1985; Daisey, 1987).

No international guidelines are in place for permissible levels of workplace airborne Zn levels. National regulations do exist but vary depending upon the Zn compound solubility or generation source involved. For example, the current permissible exposure level (PEL; 8 hr time-weighted-average) recommended by OSHA and the ACGIH for Zn as soluble zinc chloride/hexite ($ZnCl_2$) is 1 mg Zn/m^3 while that for insoluble ZnO is 5 - 10 mg Zn/m^3 (depending upon if it is encountered as total dust or just the respirable fraction). Conversely, the PEL for poorly-soluble carcinogen zinc chromate ($ZnCrO_4$) is 0.1 mg Zn/m^3 (ATSDR, 1994). Several individual states have established tougher atmospheric Zn standards (i.e., 8 hr PEL of < 1 - 20 μg Zn (as $ZnCl_2$)/m^3 and < 1 - 100 μg Zn (as ZnO)/m^3); interestingly, even individual areas within a given state have acceptable ambient Zn standards which differ by more than 1000-fold.

As with other metals, the role of solubility upon clearance of Zn-containing particle(s) is critical to their immunomodulatory potential in the lungs. Unfortunately, information regarding both solubility and clearance of Zn compounds is limited primarily to that for ZnO (i.e., Dinslage-Schlunz and Rosmanith, 1976).

Overall, few studies (apart from those dealing with metal fume fever) have directly examined the effects of inhaled Zn compounds upon immunologic parameters in the lungs. Exposure of humans and animal models to ZnO/hexachloroethane mixtures used for preparation of screening smokes has been shown to generally induce strong proinflammatory responses, type II cell hyperplasia, and increases in fibrosis (primarily interstitial) (Marrs et al., 1983; Karlsson et al., 1986; Hjortso et al., 1988; Brown et al., 1990). Instillation with the $ZnCl_2$ combustion product of the ZnO/hexachloroethane mixture induced similar effects in the lungs; however, the effect from $ZnCl_2$ differed from that to the parent mixture in that an increased amount of lymphocytic infiltration into areas of alveolar damage and in the numbers of foamy macrophage-containing aggregates in the alveolar lumena were also apparent. Except for the presence of foamy macrophages, the immunohistological changes in the lungs observed after host exposure to $ZnCl_2$ are very similar to those observed in the lungs of rats instilled with zinc hydroxide ($Zn(OH)_2$; a colloidal form of zinc under neutral conditions) (Ishiyama et al., 1997). In this latter study, instillation of rats with zinc sulfate ($ZnSO_4$) solution did not induce any changes in the immunologic profile of the lungs.

There have been a few studies have examined the effects from host inhalation/exposure to Zn compounds upon the ability of the lungs of exposed hosts to resist/clear viable bacterial challenges. In one study, one 3 hr exposure of mice to increasing amounts of $ZnSO_4$ resulted in significant increases in host mortality due to subsequently-inhaled *S. pyogenes* as compared to that in air-exposed control mice or mice exposed to equivalent or greater amounts of zinc ammonium sulfate ($Zn_2(NH_4)_2(SO_4)_2$) (Ehrlich, 1980). Unlike what had been observed with similar sulfates/ammonium sulfates of Al, if the Zn compounds were first instilled and the hosts then infected with the *S. pyogenes*, the mice that received $Zn_2(NH_4)_2(SO_4)_2$ did not now display the greater mortality rates (Hatch et al., 1981). Once again, no significant differences in host mortality (between the two agents or as compared with sham/air-exposed infected controls) were evinced when mice in this latter study were permitted to inhale either compound rather than have them instilled.

The effects of Zn agents upon alveolar macrophages among all the lung immune cell types have been the most studied. For example, macrophages recovered from the lungs of rats instilled with a single 5 mg dose of ZnO displayed distinct changes in both size and ultrastructure (Migally et al., 1982). Within 7 d of exposure, both alveolar and interstitial macrophages were found to possess electron-dense structures containing Zn; the Authors indicated that these results were indicative of a likely transfer of the Zn particles from one macrophage type to the other. At the functional level, a single 4 hr exposure of hamsters to increasing amounts of either $ZnSO_4$ or $Zn_2(NH_4)_2(SO_4)_2$ resulted in significant decreases in the phagocytic activity of macrophages recovered in the period between 1 - 48 post-exposure (Skornik and Brain, 1983). When effects upon phagocytic activity are analyzed in the context of exposure concentrations, the results in this study parallel those observed in the mouse antibacterial resistance studies of Ehrlich (1980) in that the effect from $ZnSO_4$ was far greater than that induced by $Zn_2(NH_4)_2(SO_4)_2$. A similar, but apparently species-dependent, effect upon phagocytic activity has been observed in alveolar macrophages recovered 24 hr after guinea pigs were exposed (for 3 hr) to 5 mg ZnO/m^3 (Gordon et al., 1992); no effects were demonstrable in cells recovered from rabbits exposed to the ZnO atmosphere. This latter study showed that while phagocytic activity was impaired by ZnO exposure, the effect was at the level of phagocytic capacity rather than upon phagocytic index. Inhibition of macrophage phagocytic activity has also been demonstrated *ex vivo* using ZnO (Fisher et al., 1986); however, while Zn was observed to be a much weaker inhibitor of phagocytic function than several of the other metals discussed in this chapter (i.e., V and Mn), it was more directly cytotoxic than Mn and had less effect upon cell adherence than did V. Lastly, in a study analyzing macrophages recovered from the lungs of hosts instilled with $Zn(OH)_2$, it was found that the cells displayed significant increases in levels of proliferating cell nuclear antigen (Ishiyama et al., 1997). In addition, when lung slices from naive rats were incubated with $Zn(OH)_2$, oxidative metabolism (as measured using NBT reduction) in each was significantly stimulated. Conversely, treatment of the hosts or naive rat lung slices with $ZnSO_4$ failed to induce any effects upon either the nuclear antigen level or reactive oxygen species formation endpoints.

As noted earlier in this section, the majority of the information known about the pulmonary immunotoxicologic effects of Zn have been obtained from studies of ZnO and its relation with the development of metal fume fever. In one study which examined the effects from 1, 2, or 3 d (3 hr/d) exposures of guinea pigs to atmospheres containing 2.3, 5.9, or 12.1 mg ZnO/m^3, there were consistent dose- and number of exposure-dependent increases in the levels of total protein (indicative of *in situ* fibrotic activity/vascular damage) and the activities of angiotensin-converting enzyme, lactate dehydrogenase (LDH; indicating local cell membrane damage), β-glucuronidase (BG; marker of macrophage/PMN membrane damage), and of alkaline phosphatase (Conner et al., 1988); the activity of acid phosphatase in these BAL samples were only significantly affected by the 2 and 3 d exposures to 12.1 mg ZnO/m^3. Due to the design of this study by Conner et al., it was not clear whether the appearance of effects upon these endpoints in guinea pigs sacrificed after their second exposure to ZnO was an acute-onset response or some latent effect from the first exposure. The studies by Gordon et al. (1992) clarified that it was the latter by showing that both guinea pigs and rats that underwent a single 3 hr exposure to atmospheres containing 2.5 or 5 mg ZnO/m^3, did not evince changes in several of these parameters (i.e., total protein, LDH and BG activities) until anytime from 4 - 24 hr after the exposure. Interestingly, this same study also indicated that interspecies variations in response to ZnO could occur in that rabbits exposed once for 2 hr to the 5 mg ZnO/m^3 atmosphere only evinced a change in BG activity and that the effect was actually a decrement rather than increase in same.

Because metal fume fever is a febrile inhalational syndrome, it is plausible to expect that levels of pyrogenic cytokines in the lungs would be increased following inhalation of ZnO particles. Studies of human volunteers exposed to either galvanized steel welding fumes for 15 - 30 min (Blanc et al., 1993) have indicated that significant amounts of several cytokines which are either pyrogenic (IL-1 and TNFα), chemotactic (IL-8), or anti-inflammatory (i.e., IL-6) are released in time-associated manner over the post-exposure period. In was observed in this study that there was a very rapid (within 3 hr) increase in IL-1, TNFα, and IL-8 levels in the BAL of exposed hosts. With increasing time (i.e., from 8 - 22 hr) post-exposure, the levels of these biomediators decreased while those of IL-6 increased; analysis of other anti-inflammatory cytokines such as IL-4 or IL-10 were either inconclusive or not performed. In studies in which human volunteers were exposed to furnace-generated ZnO particles for 2 hr (2.5 or 5.0 mg ZnO/m^3), levels of IL-6 in the BAL were again shown to undergo continual increases in the period 3 - 6 hr post-exposure (Fine et al., 1997). Similarly, studies in which volunteers were exposed for 10, 15, or 30 min to 33 mg Zn/m3 atmospheres (to yield $\approx$540 mg Zn (as ZnO)·min/m^3 cumulative doses) again displayed significant elevations in IL-8 and TNFα levels in their BAL even 24 hr after termination of the individual exposures (Kuschner et al., 1995). In a follow-up study by these Investigators (Kuschner et al., 1997), analysis of BAL from volunteers 3 hr after $\approx$770 mg Zn (as ZnO)·min/m^3 cumulative dose exposures showed that levels of these cytokines were again elevated. However, while the levels of IL-1, IL-8, and TNFα did undergo a decrease over the 3 - 24 hr post-exposure period similar to that observed in the above-described study of humans exposed to galvanized steel welding fumes, the levels of IL-6 did not significantly increase.

REFERENCES

Abramson MJ, Wlodarczyk JH, Saunders NA, Hensley MJ. Does aluminum smelting cause lung disease? Am. Rev. Respir. Dis., 1989;139:1042-1057.

ACGIH: American Conference of Government Industrial Hygienists, *Documentation of the Threshold Limit Values and Biological Exposure Indices, 5th Edition*. Cincinnati, OH, 1986.

Adkins B, Luginbuhl GH, Gardner DE. Acute exposure of laboratory mice to manganese oxide. Am. Ind. Hyg. Assoc. J., 1980a;41:494-500.

Adkins B, Luginbuhl GH, Gardner DE. Biochemical changes in pulmonary cells following manganese oxide inhalation. J. Toxicol. Environ. Health, 1980b;6:445-454.

Adkins B, Luginbuhl GH, Miller FJ, Gardner DE. Increased pulmonary susceptibility to streptococcal infection following inhalation of manganese oxide. Environ. Res., 1980c;23:110-120.

Al-Bayati MA, Raabe OG, Teague SV. Effect of inhaled dimethylselenide in the Fisher 344 male rat. J. Toxicol. Environ. Health, 1992;37:549-557.

Al-Laith M, Pearce FL. Some further characteristics of histamine secretion from rat mast cells stimulated with sodium orthovanadate. Agents Action, 1989;27:65-67.

Anderson WL, Tomasi TB. Suppression of lymphocyte proliferation by copper-albumin chelates, J. Biol. Chem., 259, 7602-7606, 1984.

Andren HW, Klein DH, Tolmi Y. Selenium in coal-fired steam plant emissions. Environ. Sci. Technol., 1975;9:856-858.

ATSDR: Agency for Toxic Substances and Disease Registry, *Toxicological Profile for Aluminum*. Atlanta: U.S. Department of Public Health and Human Services, 1992.

ATSDR: Agency for Toxic Substances and Disease Registry, *Toxicological Profile for Vanadium*. Atlanta: U.S. Department of Public Heealth and Human Services, 1991.

ATSDR: Agency for Toxic Substances and Disease Registry, *Toxicological Profile for Zinc*. Atlanta: U.S. Department of Public Health and Human Services, 1994.

Barrie LA, Hoff RM. Five years of air chemistry in the Canadian Arctic. Atmos. Environ., 1985;19:1995-2010.

Beach RS, Gershwin ME, Hurley LS. Zinc, copper, and manganese in immune function and experimental oncogenesis. Nutr. Cancer, 1982;3:172-191.

Begin R, Masse S, Rola-Pleszczynski M, Martel M, Desmarais Y, Geoffroy M, LeBouffant L, Daniel H, Martin J. Aluminum lactate treatment alters the lung biological activity of quartz. Exp. Lung. Res., 1986;10:385-399.

Begin R, Masse S, Sebastien P, Martel M, Bosse J, Dubois F, Geoffroy M, Labbe J. Sustained efficacy of aluminum to reduce quartz toxicity in the lung. Exp. Lung. Res., 1987;13:205-222.

Bell RR, Soliman M, Nonavinakere VK, Hammerbeck DM, Early JL. Cadmium and/or selenium effects on guinea pig lung PGE_2, TxB_2, and LTC_4. Res. Commun. Mol. Pathol. Pharmacol., 1997a;97:233-236.

Bell RR, Soliman M, Nonavinakere VK, Hammerbeck DM, Early JL. Selenium- and cadmium-induced pulmonary functional impairment and cytotoxicity. Toxicol. Lett., 1997b;90:107-114.

Bencko V., Cikrt M. Manganese: A review of occupational and environmental toxicology. J. Hyg. Epidemiol. Microbiol. Immunol., 1984;28:139-148.

Bennett PA, Dixon RJ, Kellie S. The phosphotyrosine phosphatase inhibitor vanadyl hydroperoxide induces morphological alterations, cytoskeletal rearrangements, and increased adhesiveness in rat neutrophil leukocytes. J. Cell Sci., 1993;106:891-901.

Bergstrom R. Acute pulmonary toxicity of manganese dioxide. Scand. J. Work Environ. Health, 1977;3(Suppl. 1):7-41.

Berry JP, Meignan M, Escaig F, Galle P. Inhaled soluble aerosols insolubilised by lysosomes of alveolar cells. Application to some toxic compounds: Electron microprobe and ion microprobe studies. Toxicology, 1988;52:127-139.

Blanc PD, Boushey HA, Wong H, Wintermeyer SF, Bernstein MS. Cytokines in metal fume fever. Am Rev. Respir. Dis., 1993;147:134-138.

Blanchard KT, Clay RJ, Morris JB. Pulmonary activation and toxicity of cyclopentadienyl manganese tricarbonyl. Toxicol. Appl. Pharmacol., 1996;136:280-288.

Brown GM, Donaldson K, Brown DM. Bronchoalveolar leukocyte response in experimental silicosis: Modulation by a soluble aluminum compound. Toxicol. Appl. Pharmacol., 1989;101:95-105.

Brown RF, Marrs TC, Rice P, Masek LC. The histopathology of rat lung following exposure to zinc oxide/hexachloroethane smoke of instillation with zinc chloride followed by treatment with 70% oxygen. Environ. Health Perspect., 1990;85:81-87.

Buchan RF. Industrial selenosis. Occup. Med., 1947;3:439-456.

Byrne AR, Kosta L. Vanadium in foods and in human body fluids and tissues. Sci. Total Environ., 1978;10:17-30.

Camner P, Curstedt T, Jarstrand C, Johansson A, Robertson B, Wiernik A. Rabbit lung after inhalation of manganese chloride: A comparison with the effects of chlorides of nickel, cadmium, cobalt, and copper. Environ. Res., 1985;38:301-309.

Camner P, Custedt T, Jarstrand C, Johannsson A, Robertson B, Wiernik A. Rabbit lung after inhalation of manganese chloride: A comparison with the effects of chlorides of nickel, cadmium, cobalt, and copper. Environ. Res., 1985;38:301-309.

Carter JD, Ghio AJ, Samet JM, Devlin RB. Cytokine production by human airway epithelial cells after exposure to an air pollution particle is metal-dependent. Toxicol. Appl. Pharmacol., 1997;146:180-188.

Castranova V, Bowman L, Miles PR, Reasor MJ. Toxicity of metal ions to alveolar macrophages. Am. J. Ind. Med., 1980;1:349-357.

Chan-Yeung M, Enarson DA, MacLean L, Irving D. Longitudinal study of workers in an aluminum smelter. Arch. Environ. Health, 1989;44:134-139.

Chasteen ND. The biochemistry of vanadium. Struct. Bonding, 1983;53:105-138.

Clay RJ, Morris JB. Comparative pneumotoxicity of cyclopentadienyl manganese tricarbonyl and methylcyclopentadienyl manganese tricarbonyl. Toxicol. Appl. Pharmacol., 1989;98:434-443.

Clinton M. Selenium fume exposure. J. Ind. Hyg. Toxicol., 1947;29:225-226.

Cohen MD, Becker S, Devlin R, Schlesinger RB, Zelikoff JT. Effects of vanadium upon polyI:C-induced responses in rat lung and alveolar macrophages. J. Toxicol. Environ. Health, 1997;51:591-608.

Cohen MD, Chen CM, Wei CI. Decreased resistance to *Listeria monocytogenes* in mice following vanadate exposure: Effects upon the function of macrophages. Int. J. Immunopharmacol., 1989;11:285-292.

Cohen MD, Chen LC, Zelikoff JT, Schlesinger RB. Pulmonary retention and distribution of inhaled chromium: Effects of particle solubility and ozone co-exposure. Inhal. Toxicol., 1997;9:843-865.

Cohen MD, Parsons E, Schlesinger RB, Zelikoff JT. Immunotoxicity of *in vitro* vanadium exposure: Effects on interleukin-1, tumor necrosis factor, and prostaglandin E_2 by WEHI-3 macrophages. Int. J. Immunopharmacol., 1993;15:437-446.

Cohen MD, Wei CI, Tan H, Kao KJ. Effect of ammonium metavanadate on the murine immune response. J. Toxicol. Environ. Health, 1986;19:279-298.

Cohen MD, Wei CI. Effects of ammonium metavanadate treatment upon macrophage glutathione redox cycle activity, superoxide production, and intracellular glutathione status. J. Leukocyte Biol., 1988;44:122-129.

Cohen MD, Yang Z, Qu Q, Schelsinger RB, Zelikoff JT. Vanadium affects macrophage interferon-γ binding and -inducible responses. Toxicol. Appl. Pharmacol., 1996a;138:110-120.

Cohen MD, Yang Z, Zelikoff JT, Schlesinger RB. Pulmonary immunotoxicity of inhaled ammonium metavanadate in Fisher 344 rats. Fundam. Appl. Toxicol., 1996b;33:254-263.

Cohen MD, Zelikoff JT, Chen LC, Schlesinger RB. Immunotoxicologic effects of inhaled chromium: Role of particle solubility and co-exposure to ozone. Toxicol. Appl. Pharmacol., 1998;152:30-40.

Cohen MD. "Vanadium." In *Experimental Immunotoxicology*, JT Zelikoff, P Thomas, eds. London: Taylor and Francis, 1998, pp. 207-229.

Cohen SR. A review of the health hazards from copper exposure. J. Occup. Med., 1974;16:621-624.

Conklin AW, Skinner CS, Felten TL, Sanders CL. Clearance and distribution of intratracheally instilled vanadium compounds in the rat. Toxicol. Lett., 1982;11:199-203.

Conner MW, Flood WH, Rogers AE, Amdur MO. Lung injury in guinea pigs caused by multiple exposures to ultrafine zinc oxide: Changes in pulmonary lavage fluid. J. Toxicol. Environ. Health, 1988;25:57-69.

Contreras GR, Chan-Yeung M. Bronchial reactions to exposure to welding fumes. Occup. Environ. Med., 1997;54:836-839.

Cooper WC. The health implications of increased manganese in the environment resulting from the combustion of fuel additives: A review of the literature. J. Toxicol. Environ. Health, 1984;14:23-46.

Daisey JM. "Chemical Composition of Inhalable Particulate Matter - Seasonal and Intersite Comparisons." In *Toxic Air Pollution: A Comprehensive Study of Non-Criteria Air Pollutants*, PJ Lioy, JM Daisey, eds. Chelsea, MI: Lewis Publishing Inc., 1987, pp. 47-63.

Danielsson BR, Khayat A, Dencker L. Fetal and maternal distribution of inhaled mercury vapor in pregnant mice: Influence of selenite and dithiocarbamates. Pharmacol. Toxicol., 1990;67:222-226.

Davies TA. Manganese pneumonitis. Br. J. Ind. Med., 1946;3:111-135.

Davis RL, Milham S. Altered immune status in aluminum reduction plant workers. Am. J. Ind. Med., 1990;18:79-85.

Denny JJ, Robson WD, Irwin DA. The prevention of silicosis by metallic aluminum. Can. Med. Assoc. J., 1939;40:213-228.

DeVoto E, Yokel RA. The biological speciation and toxicokinetics of aluminum. Environ. Health Perspect., 1994;102:940-951.

Dinslage-Schlunz A, Rosmanith J. The course of dust elimination from the rat lung after long-term inhalation of zinc oxide. Beit. zur Silikoseforsch. - Pneumokon., 1976;28:79-89.

Dreher KL, Jaskot RH, Lehmann JR, Richards JH, McGee JK, Ghio AJ, Costa DL. Soluble transition metals mediate residual oil fly ash-induced acute lung injury. J. Toxicol. Environ. Health, 1997;50:285-305.

Drinker P, Thomson RM, Finn JL. Metal fume fever - IV. Threshold doses of zinc oxide, preventive measures, and the chronic effects of repeated exposures. J. Ind. Hyg., 1927;9:331-345.

Drown DB, Obers SG, Sharma RP. Pulmonary clearance of soluble and insoluble forms of manganese. J. Toxicol. Environ. Health, 1986;17:201-212.

Drummond JG, Aranyi C, Schiff LJ, Fenters JD, Graham JA. Comparative study of various methods used for determining health effects of inhaled sulfates. Environ. Res., 1986;41:514-528.

Duncan DD, Lawrence, DA. Four sulfhydryl-modifying compounds cause different structural damage but similar functional damage in murine lymphocytes. Chem.-Biol. Interact., 1988;68:137-152.

Eckert H, Jerochin S. Copper sulfate-mediated changes of the lung. An experimental contribution to pathogenesis of vineyard sprayer's lung. Z. Erkrank. Atm. Org., 1982;158:270-276.

Edel J, Sabbioni E. Retention of intratracheally-instilled and ingested tetravalent and pentavalent vanadium in the rat. J. Trace Elem. Electrolytes Health Dis., 1988;2:23-30.

Eduard W, Lie A. Influence of fluoride recovery alumina on the work environment and the health of aluminum potroom workers. Scand. J. Work Environ. Health, 1981;7:214-222.

Ehrlich R, Findlay JC, Gardner DE. Susceptibility to bacterial pneumonia of animals exposed to sulfates. Toxicol. Lett., 1978;1:325-330.

Ehrlich R. Interactions between environmental pollutants and respiratory infections. Environ. Health Perspect., 1980;35:89-100.

Eklund A, Arns R, Blaschke E, Hed J, Hjertquist SO, Larsson K, Lowgren H, Nystrom J, Skold CM, Tornling G. Characteristics of alveolar cells and soluble components in bronchoalveolar lavage fluid from non-smoking aluminum potroom workers. Br. J. Ind. Med., 1989;46:782-786.

Elliott GR, van Batenburg MJ, Bonta IL. Copper modulation of macrophage cyclooxygenase metabolite synthesis. Prostaglandins, 1987;34:657-667.

Elsayed NM, Hacker AD, Kuehn K, Mustafa MG, Schrauzer GN. Dietary antioxidants and the biochemical response to oxidant inhalation. II. Influence of dietary selenium on the biochemical effects of ozone exposure in mouse lung. Toxicol. Appl. Pharmacol., 1983;71:398-406.

EPA: Environmental Protection Agency. *Ambient Water Quality Criteria for Copper.* Environmental Protection Agency, National Technical Information Service, PB 81-117475, Springfield, VA, 1980.

EPA: Environmental Protection Agency. *Exposure and Risk Assessment for Zinc.* Environmental Protection Agency, EPA 440/4-81-016, Washington, DC, 1980.

EPA: Environmental Protection Agency. *Scientific and Technical Assessment Rreport on Manganese.* Environmental Protection Agency, EPA-600/6-74-002, Washington, DC, 1975.

Eskew ML, Scheuchenzuber WJ, Scholz RW, Reddy CC, Zarkower A. The effects of ozone inhalation on the immunological response of selenium- and vitamin E-deprived rats. 1986;40:274-284.

Ess SM, Steinegger AF, Ess HJ, Schlatter C. Experimental study on the fibrogenic properties of different types of alumina. Am. Ind. Hyg. Assoc. J., 1993;54:360-370.

Evans EF. Casualties following exposure to zinc chloride. Lancet, 1945;2:368-370.

Evans GA, Garcia GG, Erwin R, Howard OM, Farrar WL. Pervanadate simulates the effects of interleukin-2 (IL-2) in human T-cells and provides evidence for the activation of two distinct tyrosine kinase pathways by IL-2. J. Biol. Chem., 1994;269:23407-23412.

Fine JM, Gordon T, Chen LC, Kinney P, Falcone G, Beckett WS. Metal fume fever: Characterization of clinical and plasma IL-6 responses in controlled human exposures to zinc oxide fume at and below the threshold limit value. J. Occup. Environ. Med., 1997;39:722-726.

Fishburn CW, Zenz C. Metal fume fever. A report of a case. J. Occup. Med., 1969;11:142-144.

Fisher GL, McNeill KL, Democko CJ. Trace element interactions affecting pulmonary macrophage cytotoxicity. Environ. Res., 1986;39:164-171.

Fisher GL, McNeill KL, Whaley CB, Fong J. Attachment and phagocytosis studies with murine pulmonary alveolar macrophages. J. Reticuloendothel. Soc., 1978;24:243-252.

Flinn RH, Neal PA, Rinehart WH, Dallavalee JM, Fulton WB, Dooley AE. *Chronic Manganese Poisoning in an Ore-Crushing Mill.* Public Health Bulletin No. 247, 1940.

Flynn A. *In vitro* levels of copper, magnesium, and zinc required for mitogen-stimulated T-lymphocyte proliferation. Nutr. Res., 1985;5:487-495.

Ganrot, PO. Metabolism and possible health effects of aluminum. Environ. Health Perspect., 1986;65:363-441.

Gavett SH, Madison SL, Dreher KL, Winsett DW, McGee JK, Costa DL. Metal and sulfate composition of residual oil fly ash determines airway hyperreactivity and lung injury in rats. Environ. Res., 1997;72:162-172.

Gibbs GW. Mortality of aluminum reduction plant workers, 1950 through 1977. J. Occup. Med., 1985;27:761-770.

Glover J. Selenium and its industrial toxicology. Ind. Med., 1970;39:50-54.

Gordon T, Chen LC, Fine JM, Schlesinger RB, Su W, Kimmel TA, Amdur MO. Pulmonary effects of inhaled zinc oxide in human subjects, guinea pigs, rats, and rabbits. Am. Ind. Hyg. Assoc. J., 1992;53:503-509.

Goren MB, Swendsen SL, Fiscus J, Miranti C. Fluorescent markers for studying phagosome-lysosome fusio. J. Leukocyte Biol., 1984;36:273-282.

Graham JA, Gardner DE, Waters MD, Coffin DL. Effect of trace metals on phagocytosis by alveolar macrophages. Infect. Immun., 1975;11:1278-1283.

Gusev VA, Danilovskaja YV, Vatolkina OY, Lomonosova OS, Velichkovsky BT. Effect of quartz and alumina dust on generation of superoxide radicals and hydrogen peroxide by alveolar macrophages, granulocytes, and monocytes. Br. J. Ind. Med., 1993;50:732-735.

Gylseth B, Jahr J. Some hygienic aspects of working in aluminum reduction potrooms with special reference to the use of alumina from the dry cleaning process of Søderberg potgases. Staub-Reinhalt. Luft, 1975;35:430-432.

Hakkinen PJ, Haschek WM. Pulmonary toxicity of methylcyclopentadienyl manganese tricarbonyl: Non-ciliated bronchiolar epithelial (Clara) cell necrosis and alveolar damage in the mouse, rat, and hamster. Toxicol. Appl. Pharmacol., 1982;65:11-22.

Harris ED, Gonnerman WA, Savage JE, O'Dell BL. Purification and partial characterization of lysyl oxidase from chick aorta. Biochim. Biophys. Acta, 1974;341:332-344.

Hashimoto Y, Winchester JW. Selenium in the atmosphere. Environ. Sci. Technol. 1967;1:338-340.

Hatch GE, Slade R, Boykin E, Hu PC, Miller FJ, Gardner DE. Correlation of effects of inhaled versus intratracheally-injected metals on susceptibility to respiratory infection in mice. Am. Rev. Respir. Dis., 1981;124:167-173.

Hickey RJ, Schoff EP, Clelland RC. Relationship between air pollution and certain chronic disease death rates. Arch. Environ. Health., 1967;15:728-739, 1967.

Hill CH. Influence of time of exposure to high levels of minerals on the susceptibility of chicks to Salmonella gallinarum. J. Nutr., 1980;110:433-436.

Hinderer RK. Toxicity studies of methylcyclopentadienyl manganese tricarbonyl (MMT). Am. Ind. Hyg. Assoc. J., 1979;40:164-167.

Hirano S, Ebihara H, Sakai S, Kodama N, Suzuki KT. Pulmonary clearance and toxicity of intratracheally-instilled cupric oxide in rats. Arch. Toxicol., 1993;67:312-317.

Hirano S, Sakai S, Ebihara H, Kodama N, Suzuki KT. Metabolism and pulmonary toxicity of intratracheally-instilled cupric sulfate in rats. Toxicology, 1990;64:223-233.

Hjortsø E, Qvist J, Bud MI, Thomsen JL, Andersen JB, Wiberg-Jorgensen F, Jensen NK, Jones R, Reid LM, Zapol WM. ARDS after accidental inhalation of zinc chloride smoke. Intensive Care Med., 1988;14:17-24.

Holness DL, Taraschuk IG, Nethercott JR. Health status of copper refinery workers with specific reference to selenium exposure. Arch. Environ. Health, 1989;44:291-297.

Homma S, Jones R, Qvist J, Zapol WM, Reid L. Pulmonary vascular lesions in the adult respiratory distress syndrome caused by inhalation of zinc chloride smoke: A morphometric study. Hum. Pathol., 1992;23:45-50.

Hopkins LL, Tilton BE. Metabolism of trace amounts of vanadium[48] in rat organs and liver subcellular particles. Am. J. Physiol., 1966;211:169-172.

Igarishi K, David M, Larner AC, Finbloom DS. In vitro activation of a transcription factor by gamma interferon requires a membrane-associated tyrosine kinase and is mimicked by vanadate. Mol. Cell. Biol., 1993;13:3984-3989.

Imbert V, Peyron JF, Far DF, Mari B, Auberger P, Rossi B. Induction of tyrosine phosphorylation and T-cell activation by vanadate peroxide, an inhibitor of protein tyrosine phosphatases. Biochem. J., 1994;297:163-173.

Ishiyama H, Ogino K, Sato M, Ogura M, Dan S, Hobara T. Histopathological changes induced by zinc hydroxide in rat lungs. Exp. Toxic. Pathol., 1997;49:261-266.

Johansson A, Camner P, Jarstrand C, Wiernik A. Rabbit alveolar macrophages after inhalation of soluble cadmium, cobalt, and copper: A comparison with the effects of soluble nickel. Environ. Res, 1983;31:340-354.

Johansson A, Camner P. Adverse effects of metals on the alveolar part of the lung. Scan. Electron Microsc., 1986;II:631-637.

John W, Kaifer R, Rahn K. Trace element concentrations in aerosols from the San Francisco Bay area. Atmos. Environ., 1973;7:107-118.

Jones DG, Suttle NF. The effect of copper deficiency on the resistance of mice to infection with Pasturella haemolytica. J. Comp. Pathol., 1983;93:143-149.

Jones DG. Effects of dietary copper depletion on acute and delayed inflammatory response in mice. Res. Vet. Sci., 1984;37:205-210.

Jones KC, Bennett BG. Exposure of man to environmental aluminum - an exposure commitment assessment. Sci. Total Environ., 1986;52:65-82.

Kacew S, Parulekar MR, Merali Z. Effects of parenteral vanadium administration on pulmonary metabolism of rats. Toxicol. Lett., 1982;11:119-124.

Kadota S, Fantus IG, Deragon G, Guyda HJ, Posner BI. Stimulation of insulin-like growth factor II receptor binding and insulin receptor kinase activity in rat adipocytes. Effects of vanadate and H_2O_2. J. Biol. Chem. 1987;262:8252-8256.

Karlsson N, Cassel G, Fangmark I, Bergman F. A comparative study of the acute inhalation toxicity of smoke from TiO_2-hexachloroethane and Zn-hexachloroethane pyrotechnic mixtures. Arch. Toxicol., 1986;59:160-166.

Khayat A, Dencker L. Interactions between selenium and mercury in mice: Marked retention in the lung after inhalation of metallic mercury. Chem.-Biol. Interact., 1983;46:283-298.

Khayat A, Dencker L. Interactions between tellurium and mercury in murine lung and other organs after metallic mercury inhalation: A comparison with selenium. Chem.-Biol. Interact., 1984;50:123-133.

Kilburn KH. "Pulmonary and Neurologic Effects of Aluminum." In *Environmental and Occupational Medicine, 3rd Edition*, W Rom, ed. Philadelphia: Lippincott-Raven Publishers, 1998, pp. 1065-1073.

Kivuoloto M, Pakarinen A, Pyy L. Clinical laboratory results of vanadium-exposed workers. Arch. Environ. Health, 1980;36:109-1130.

Kivuoloto M, Rasanen O, Rinne A, Rissanen A. Intracellular immunoglobulin in plasma cells of nasal biopsies taken from vanadium-exposed workers. Annt. Anz. Jena., 1981;149:446-450.

Kivuoloto M, Rasanen O, Rinne A, Rissanen, M. Effects of vanadium on the upper respiratory tract of workers in a vanadium factory. A macroscopic and microscopic study. Scand. J. Work Environ. Health, 1979;5:50-58.

Kivuoloto M. Observations on the lungs of vanadium workers. Br. J. Ind. Med., 1980;37:363-366.

Klarlund JK, Latini S, Forchhammer J. Numerous proteins phosphorylated on tyrosine and enhanced tyrosine kinase activities in vanadate-treated NIH 3T3 fibroblasts. Biochim. Biophys. Acta, 1988;971:112-120.

Knecht EA, Moorman WJ, Clark JC, Hull RD, Biagini RE, Lynch DW, Boyle TJ, Simon SD. Pulmonary reactivity to vanadium pentoxide following subchronic inhalation exposure in a non-human primate animal model. J. Appl. Toxicol., 1992;12:427-434.

Knecht EA, Moorman WJ, Clark JC, Lynch DW, Lewis TR. Pulmonary effects of acute vanadium pentoxide inhalation in monkeys. Am. Rev. Respir. Dis., 1985;132:1181-1185.

Koppel C, Baudisch H, Beyer KH, Kloppel I, Schneider V. Fatal poisoning with selenium dioxide. J. Toxicol. Clin. Toxicol., 1986;24:21-35.

Kornegay ET, van Heugten PH, Lindemann MD, Blodgett DJ. Effects of biotin and high copper levels on performance and immune response of weanling pigs. J. Anim. Sci., 1989;67:1471-1477.

Kuratsune M, Tokudome S, Shirakusa T, Yoshida M, Tokumitsu Y, Hayano T, Seita M. Occupational lung cancer among copper smelters. Int. J. Cancer, 1974;13:552-558.

Kuschner WG, D'Alessandro A, Wintermeyer SF, Wong H, Boushey HA, Blanc PD. Pulmonary responses to purified zinc oxide fume. J. Invest. Med., 1995;43:371-378.

Kuschner WG, D'Alessandro A, Wong H, Blanc PD. Early pulmonary cytokine responses to zinc oxide fume inhalation. Environ. Res., 1997;75:7-11.

Labedzka M, Gulyas H, Schmidt N, Gercken G. Toxicity of metallic ions and oxides to rabbit alveolar macrophages. Environ. Res., 1989;48:255-274.

Larsson K, Eklund A, Arns R, Lowgren H, Nystrom J, Sundstrom G, Tornling G. Lung function and bronchial reactivity in aluminum potroom workers. Scand. J. Work Environ. Health, 1989;15:296-301.

Lawrence DA. Heavy metal modulation of lymphocyte activities. I. *In vitro* effects of heavy metals on primary humoral immune response. Toxicol. Appl. Pharmacol., 1981;57:439-451.

LeBouffant L, Daniel H, Martin J. "The Therapeutic Action of Aluminum Compounds on the Development of Experimental Lesions Produced by Pure Quartz or Mixed Dust." In *Inhaled Particle IV. Part. I*, WH Walton, ed. Oxford: Pergamon Press, 1977, pp. 361-371.

Lees R. E. Changes in lung function after exposure to vanadium compounds in fuel oil ash. Br. J. Ind. Med., 1980;37:253-256.

Lehman KB. Study of technically and hygienically important gases and steams. XIV. Foundry or zinc fever. Arch. Hyg., 1910;72:358-381.

Levy BS, Hoffman L, Gottsegen S. Boilermakers' bronchitis. Respiratory tract irritation associated with vanadium pentoxide during oil-to-coal conversion of a power plant. J. Occup. Med., 1984;26:567-570.

Lewis AJ. The role of copper in inflammatory disorders. Agents Action, 1982;15:513-519.

Lindenschmidt RC, Driscoll KE, Perkins MA, Higgins JM, Maurer JK, Belfiore KA. The comparison of a fibrogenic and two non-fibrogenic dusts by bronchoalveolar lavage. Toxicol. Appl. Pharmacol., 1990;102:268-281.

Lioy PJ, Wolff GT, Kneip TJ. Toxic airborne elements in the New York metropolitan area. J. Air Pollut. Control Assoc., 1978;28:510-512.

Lipsky PE, Ziff M. Inhibition of human helper T-cell function *in vitro* by D-penicillamine and CuSO4. J. Clin. Invest., 1980;65:1069-1076.

Lipsky PE. Immunosuppression by D-penicillamine *in vitro*: Inhibition of human T-lymphocyte proliferation by copper- or ceruloplasmin-dependent generation of hydrogen peroxide and protection by monocytes. J. Clin. Invest., 1984;73:53-654.

Lipsky PE. Modulation of T-lymphocyte function by copper and thiols. Agents Action, 1981;8 (Suppl.):85-102.

Lison D, Lardot C, Huaux F, Zanetti G, Fubini B. Influence of particle surface area on the toxicity of insoluble manganese dioxide dusts. Arch. Toxicol., 1997;71:725-729.

Ljunggren KG, Lidums V, Sjogren B. Blood and urine concentrations of aluminum among workers exposed to aluminum flake powders. Br. J. Ind. Med., 1991;48:106-109.

Loranger S, Zayed J. Environmental contamination and human exposure to airborne total and respirable manganese in Montreal. J. Air Waste Manage. Assoc., 1997;47:983-989.

Ludwig, JC, Chvapil M. "Effects of Metal Ions on Lysosomes." In *Trace Elements in the Pathogenesis and Treatment of Inflammation, Agents and Actions, Supplement Vol. 8*, KD Rainsford, K Brune, MW Whitehouse, eds. Basel, Switzerland: Birkhauser-Verlag, 1981, pp. 65-84.

Lukasewycz OA, Prohaska JR, Meyer SG, Schmidtke JR, Hatfield SM, Marder P. Alterations in lymphocyte subpopulations in copper-deficient mice. Infect. Immun., 1985;48:644-647.

Lukasewycz OA, Prohaska JR. Lymphocytes from copper-deficient mice exhibit decreased mitogen reactivity. Nutr. Res., 1983;3:335-341.

Lukasewycz OA, Prohaska JR. The immune response in copper deficiency. Ann. N.Y. Acad. Sci., 1990;587:147-159.

Lundborg M, Camner P. Lysozyme levels in rabbit lung after inhalation of nickel, cadmium, cobalt, and copper chlorides. Environ. Res., 1984;34:335-342.

Lundborg M, Lind B. Camner P. Ability of rabbit alveolar macrophages to dissolve metals. Exp. Lung Res., 1984;7:11-22.

Lutsenko LA, Borisenkova RV, Gvozdeva LL, Skriabin S, Ivanova LG. Fibrogenic and general toxic effects of copper and nickel sulfide ore dust. Med. Truda Prom. Ekol., 1997;5:38-43.

Lyle WH, Payton JE, Hui M., Haemodialysis and copper fever. Lancet, 1976;2:1324-1325.

Maigetter RZ, Ehrlich R, Fenters JD, Gardner DE. Potentiating effects of manganese dioxide on experimental respiratory infections. Environ. Res., 1976;11:386-391.

Mali JL, Carter A, Dolovich J. Occupational asthma due to zinc. Eur. Respir. J., 1993;6:447-450.

Mali JL, Carter A. Occupational asthma due to fumes of galvanised metal. Chest, 1987;92:375-376.

Marrs TC, Clifford WE, Colgrave HF. Pathological changes produced by exposure of rabbits and rats to smokes from mixtures of hexachloroethane and zinc oxide. Toxicol. Lett. 1983;19:247-252.

Matarese SL, Matthews JI. Zinc chloride (smoke bomb) inhalational lung injury. Chest, 1986;89:308-309.

McCord CP. Metal fume fever as an immunological disease. Ind. Med. Surg., 1960;29:101-106.

Medinsky MA, Cuddihy RG, Griffith WC, McClellan RO. A simulation model describing the metabolism of inhaled and ingested selenium compounds. Toxicol. Appl. Pharmacol., 1981a;59:54-63.

Medinsky MA, Cuddihy RG, Hill JO, McClellan RO. Toxicity of selenium compounds to alveolar macrophages. Toxicol. Lett., 1981b;8:829-293.

Medinsky MA, Cuddihy RG, McClellan RO. Systemic absorption of selenious acid and elemental selenium aerosols in rats. J. Toxicol. Environ. Health, 1981c;8:917-928.

Migally N, Murthy RC, Doye A, Zambernard J. Changes in pulmonary alveolar macrophages in rats exposed to oxides of zinc and nickel. J. Submicrosc. Cytol., 1982;14:621-626.

Montero MR, Guerri C, Ribelles M, Grisoia S. Inhibition of protein synthesis in cell cultures by vanadate and in brain homogenates of rats fed vanadate. Physiol. Chem. Phys., 1981;13:281-287.

Morgan DL, Shines CJ, Jeter SP, Blazka ME, Elwell MR, Wilson RE, Ward SM Price HC, Moskowitz PD. Comparative pulmonary absorption, distribution, and toxicity of copper gallium diselenide, copper indium diselenide, and cadmium telluride in Sprague-Dawley rats. Toxicol. Appl. Pharmacol., 1997;147:399-410.

Morgan DL, Shines CJ, Jeter SP, Wilson RE, Elwell MR, Price HC, Moskowitz PD. Acute pulmonary toxicity of copper gallium diselenide, copper indium diselenide, and cadmium telluride intratracheally-instilled into rats. Environ. Res., 1995;71:16-24.

Morrow PE, Gibb FR, Gazioglu KM. A study of particulate clearance from the human lungs. Am. Rev. Respir. Dis., 1967;96:1209-1221.

Morrow PE, Gibb FR, Johnson L. Clearance of insoluble dust from the lower respiratory tract. Health Phys., 1964;10:543-555.

Murthy RC, Migally N, Doye A, Zambernard J. Ultrastructural changes in rat alveolar macrophages exposed to oxides of copper and cadmium., J. Submicrosc. Cytol., 1982;14:347-353.

Musk AW, Tees JG. Asthma caused by occupational exposure to vanadium compounds. Med. J. Australia, 1982;1:183-184.

Nantel AJ, Brown M, Dery P, Lefebvre M. Acute poisoning by selenious acid. Vet. Hum. Toxicol., 1985;27:531-533.

NAS: National Academy of Sciences. *Committee on Biologic Effects of Atmospheric Pollutants, Division of Medical Sciences, National Research Council: Manganese.* Washington, DC: National Academy of Sciences, 1973, pp. 101-113.

Nechay BR, Nanninga LB, Nechay PE, Post RL, Grantham JJ, Macara IG, Kubena LF, Phillips TD, Nielsen FH. Role of vanadium in biology. Fed. Proc., 1986;45:123-132.

Nemery B. Metal toxicity and the respiratory tract. Eur. Respir. J., 1990;3:202-219.

Newberne PM, Hurt CE, Young VR. The role of diet and the reticuloendothelial system in the response of rats to Salmonella typhimurium infection. Brit. J. Exp. Pathol., 1968;49:448-457.

Nilsen AM, Mylius EA, Gullvag BM. Alveolar macrophages from expectorates as indicators of pulmonary irritation in primary aluminum reduction plant workers. Am. J. Ind. Med., 1987;12:101-112.

NIOSH: National Institute for Occupational Safety and Health. *National Occupational Exposure Survey (1980-1983).* Washington, DC: United States Department of Health and Human Services, 1984.

NIOSH: National Institutes for Occupational Safety and Health. *Pocket Guide to Chemical Hazards: 5th Edition.* Washington, DC: United States Department of Health and Human Services, 1985, pp. 234-235.

Nriagu JO, Pacyna JM. Quantitative assessment of worldwide contamination of air, waiter, and soils by trace metals. Nature, 1988;333:134-139.

O'Dell BL. Roles for iron and copper in connective tissue biosynthesis. Phil. Trans. Royal Soc. Lond. B, 1981;294:91-104.

Oberg SG, Parker, RD, Sharma, RP. Distribution and elimination of intratracheally-administered vanadium compound in the rat. Toxicology, 1978;11:315-323.

Oh S, Lee M, Chung C. Protection of phagocytic macrophages from peroxidative damage by selenium and vitamin E. Yonsei Med. J., 1982;23:101-109.

Ondov JM, Zoller WH, Gordon Ge. Trace element emissions on aerosols from motor vehicles. Environ. Sci. Technol., 1982;16:3188-328.

OSHA: Occupational Safety and Health Administration. United States Department of Labor, Washington, DC, 1989.

Ostiguy G, Vaillancourt C, Begin R. Respiratory health of workers exposed to metal dusts and foundry fumes in a copper refinery. Occup. Environ. Med., 1995;52:204-210.

Paschoa AS, Wrenn ME, Singh MP, Bruenger FW, Miller SC, Cholewa M, Jones, KW. Localization of vanadium-containing particles in the lungs of uranium/vanadium miners. Biol. Trace Elem. Res., 1987;13:275-282.

Patterson JW, Allen HE, Scala JJ. Carbonate precipitation for heavy metal pollutants. J. Water Pollut. Control Fed., 1977;2397-2410.

Peoples SM, McCarthy JF, Chen, LC, Eppelsheimer D, Amdur MO. Copper oxide aerosol: Generation and characterization. Am. Ind. Hyg. Assoc. J., 1988;49:271-276.

Pierce LM, Alessandrini F, Godleski JJ, Paulauskis JD. Vanadium-induced chemokine mRNA expression and pulmonary inflammation. Toxicol. Appl. Pharmacol., 1996;138:1-11.

Pimentel JC, Marques F. Vineyard sprayer's lung: A new occupational disease. Thorax, 1969;24:678-688.

Piscator M. Health hazards from inhalation of metal fumes. Environ. Res., 1976;11:268-270.

Pocino M, Baute L, Malave I. Influence of the oral administration of excess copper on the immune response. Fundam. Appl. Toxicol., 1991;16:249-256.

Pocino M, Malave I, Baute L. Zinc administration restores the impaired immune response observed in mice receiving excess copper by oral route. Immunopharmacol. Immunotoxicol., 1990;12:697-713.

Pritchard RJ, Ghio AJ, Lehmann JR, Winsett DW, Tepper JS, Park P, Gilmour MI, Dreher KL, Costa DL. Oxidant generation and lung injury after particulate air pollutant exposure increase with the concentrations of associated metals. Inhal. Toxicol., 1996;8:457-477.

Prohaska JR, Lukasewycz OA. Copper deficiency suppresses the immune response of mice, Science, 1981;213:559-561.

Pumiglia KM, Lau L, Huang C, Burroughs S, Feinstein MB. Activation of signal transduction in platelets by the tyrosine phosphatase inhibitor pervanadate (vanadyl hydroperoxide). Biochem. J., 1992;286:441-449.

Purves D. "Trace Element Contamination of the Atmosphere." In *Trace Element Contamination of the Environment*, D Purves, ed. New York: Elsevier, 1977, pp. 62-72.

Ragaini RC, Ralston HR, Roberts, N. Environmental trace metal contamination in Kellogg, Idaho, near a lead smelting complex. Environ. Sci. Technol., 1977;11:773-781.

Raisfeld IH, Chu P, Hart NK, Lane A. A comparison of the pulmonary toxicity produced by metal-free and copper-complexed analogs of bleomycin and phleomycin. Toxicol. Appl. Pharmacol., 1982;63:351-362.

Rehder D. "Inorganic Considerations on the Function of Vanadium in Biological Systems." In *Vanadium and Its Role in Life; Metal Ions in Biological Systems, Vol. 31*, H Sigel, A Sigel, eds. New York: Marcel Dekker, Inc., 1995, pp. 1-44.

Rhoads K, Sanders CL Lung clearance, translocation, and acute toxicity of arsenic, beryllium, cadmium, cobalt, lead, selenium, vanadium, and ytterbium oxides following desposition in rat lung. Environ. Res., 1985;36:359-378.

Rifat SL, Eastwood MR, Crapper McLachlan DR, Corey PN. Effect of exposure of miners to aluminum powder. Lancet, 1990;336:1162-1165.

Rollin HB, Theodorou P, Kilroe-Smith TA. Deposition of aluminum in tissues of rabbits exposed to inhalation of low concentrations of Al_2O_3. Br. J. Ind. Med., 1991;48:389-391.

Sagripanti JL, Goering PL, Lamanna A. Interaction of copper with DNA and antagonism by other metals. Toxicol. Appl. Pharmacol., 1991;110:477-485.

Saric M, Holetic A, Ofner E. Acute respiratory diseases in a manganese contaminated area. Proc. Intl. Conf. Heavy Metals Environ., 1977;3:389-398.

Saric M. Occupational and environmental exposures and non-specific lung disease - A review of selected studies. Isr. J. Med. Sci., 1992;28:509-512.

Schecter A, Shanske W, Stenzler A, Quintilian H, Steinberg H. Acute hydrogen selenide inhalation. Chest, 1980;77:554-555.

Schiff LJ, Graham JA. Cytotoxic effect of vanadium and oil-fired fly ash on hamster tracheal epithelium. Environ. Res., 1984;34:390-402.

Schwarz YA, Kivity S, Fischbein A, Ribak Y, Fireman E, Struhar D, Topilsky M, Greif J. Eosinophilic lung reaction to aluminum and hard metal. Chest, 1994;105:1261-1263.

Seglen PO, Gordon PB. Vanadate inhibits protein degradation in isolated rat hepatocytes, J. Biol. Chem., 1981;256:7699-7703.

Sharma RP, Flora SJ, Brown DB, Oberg, SG. Persistence of vanadium compounds in lungs after intratracheal instillation in rats. Toxicol. Ind. Health, 1987;3:321-329.

Singh j, Kaw JL, Zaidi SH. Early biochemical response of pulmonary tissues to manganese dioxide. Toxicology, 1977;8:177-184.

Sjogren B, Elinder C, Lidums V, Chang G. Uptake and urinary excretion of aluminum among welders. Int. Arch. Occup. Environ. Health, 1988;60:77-79.

Sjogren B, Lidums V, Hakansson M, Hedstrom L. Exposure and urinary excretion of aluminum during welding. Scand. J. Work Environ. Health, 1985;11:39-43.

Skornik WA, Brain JD. Relative toxicity of inhaled metal sulfate salts for pulmonary macrophages. Am. Rev. Respir. Dis., 1983;128:297-303.

Smith KL, Lawrence DA. Immunomodulation of in vitro antigen presentation by cations. Toxicol. Appl. Pharmacol., 1988;96:476-484.

Smyth LT, Ruhf RC, Whitman NE, Dugan T. Clinical manganism and exposure to manganese in the production and processing of ferromanganese alloy. J. Occup. Med., 1973;15:101-109.

Snella M. Manganese dioxide induces alveolar macrophage chemotaxis for neutrophils in vitro. Toxicology, 1985;34:153-159.

Sorahan T, Lister A, Gilthorpe MS, Harrington JM. Mortality of copper cadmium alloy workers with special reference to lung cancer and non-malignant diseases of the respiratory system, 1946-1992. Occup. Environ. Med., 1995;52:804-812.

Sorgdrager B, de Looff AJ, de Monchy JG, Pal TM, Dubois AE, Rijcken B. Occurrence of occupational asthma in aluminum potroom workers in relation to preventive measures. Int. Arch. Occup. Environ. Health, 1998;71:53-59.

Søyseth V, Kongerud J, Aalen OO, Botten G, Boe J. Bronchial responsiveness decreases in relocated aluminum potroom workers compared with workers who continue their potroom exposure. Int. Arch. Occup. Environ. Health, 1995;67:53-57.

Stocks P. On the relations between atmospheric pollution in urban and rural localities and mortality from cancer, bronchitis, pneumonia, with particular reference to 3,4-benzopyrene, beryllium, molybdenum, vanadium, and arsenic. Br. J. Cancer, 1960;14:397-418.

Stokinger HE. "The Metals: Manganese." In Patty's Industrial Hygiene and Toxicology, 3rd Edition, G Clayton, F Clayton, eds. New York: Wiley and Sons, 1981, pp. 1749-1769.

Stone CJ, McLaurin DA, Steinhagen WH, Cavender FL, Haseman JK. Tissue deposition patters after chronic inhalation exposures of rats and guinea pigs to aluminum chlorhydrate. Toxicol. Appl. Pharmacol., 1979;49:71-76.

Sugiura Y, Kuwahara J, Suzuki T. DNA interaction and nucleotide sequence cleavage of copper-streptonigrin. Biochem. Biophys. Acta, 1984;782:254-261.

Sugiura Y, Takita T, Umezawa H. Bleomycin antibiotics: Metal complexes and their biological action. Metal Ions Biol. Syst., 1985;19:81-108.

Suzuki Y, Fujii N, Yano II, Ohkita T, Ichikawa A, Nishiyama K. Effects of the inhalation of manganese dioxide dust on monkey lungs. Tokushima J. Exp. Med., 1978;25:119-125.

Swarup G, Cohen S, Garbers DL. Inhibition of membrane phosphotyrosyl-protein phosphatase activity by vanadate. Biochem. Biophys. Res. Commun., 1982;107:1104-1109.

Ter Haar GL, Griffing ME, Brandt M, Oberding DG, Kapron M. Methylcyclopentadienyl manganese tricarbonyl as an antiknock: Composition and fate of manganese exhaust products. J. Air Pollut. Control Assoc., 1975;25:858-860.

Thurston G, Spengler J. A quantitative assessment of source contributions to inhalable particulate matter pollution in metropolitan Boston. Atmos. Environ., 1985;19:9-25.

Tipton IH, Shafer JJ. Statistical analysis of lung trace element levels. Arch. Environ. Health, 1964;8:56-67.

Tokudome S, Kuratsune M., A cohort study on mortality from cancer and other causes among workers at a metal refinery. Int. J. Cancer, 1976;17:310-317.

Tornling G, Blaschke E, Eklund A. Long-term effects of alumina on components of bronchoalveolar lavage fluid from rats. Br. J. Ind. Med., 1993;50:172-175.

Torossian K, Freedman D, Fantus I.G. Vanadate downregulates cell surface insulin and growth hormone receptor and inhibits insulin receptor degradation in cultured human lymphocytes. J. Biol. Chem., 1988;263:9353-9359.

Trudel S, Paquet MR, Ginstein S. Mechanism of vanadate-induced activation of tyrosine phosphorylation and of the respiratory burst in HL60 cells. Biochem. J., 1986;276:611-619.

Tsuchiyama F, Hisanaga N, Shibata E, Aoki T, Takagi H, Ando T, Takeuchi Y. Pulmonary metal distribution in urban dwellers. Int. Arch. Occup. Environ. Health, 1997;70:77-84.

Ulrich CE, Rinehart W, Busey W, Dorato MA. Evaluation of the chronic inhalation toxicity of a manganese oxide aerosol. II. - Clinical observations, hematology, clinical chemistry, and histopathology. Am. Ind. Hyg. Assoc. J., 1979;40:322-329.

Utter MF. The biochemistry of manganese. Med. Clin. N. Amer., 1976;60:713-727.

Vaddi K, Wei CI. Effect of ammonium metavanadate on the mouse peritoneal macrophage lysosomal enzymes. J. Toxicol. Environ. Health, 1991a;33:65-78.

Vaddi K, Wei CI. Modulation of F_C receptor expression and function in mouse peritoneal macrophages by ammonium metavanadate. Int. J. Immunopharmacol., 1991b;13:1167-1176.

Vaddi K, Wei CI. Modulation of macrophage activation by ammonium metavanadate. J. Toxicol. Environ. Health, 1996;49:631-645.

van Netten C, Teschke KE, Souter F. Occupational exposure to elemental constituents in fingerprint powders. Arch. Environ. Health, 1990;45:123-127.

Villar TG. Vineyard sprayer's lung: Clinical aspects. Am. Rev. Respir. Dis., 1974;110:545-555.

Vyas D, Chandra RK. Thymic factor activity, lymphocyte stimulation response, and antibody producing cell in copper deficiency. Nutr. Res., 1983;3:343-349.

Wang E, Choppin PW. Effect of vanadate on intracellular distribution and function of 10 nm microfilaments. Proc. Natl. Acad. Sci. USA, 1981;78:2363-2367.

Wang R, Wang C, Feng Z, Luo Y. Investigation on the effect of selenium on T-lymphocyte proliferation and its mechanisms. J. Tongji Med. Univ., 1992;12:33-38.

Warshawsky D, Reilman R, Cheu J, Radike M, Rice C. Influence of particle dose on the cytotoxicity of hamster and rat pulmonary alveolar macrophage *in vitro*. J. Toxicol. Environ. Health, 1994;42:407-421.

Waters MD, Gardner DE, Aranyi C, Coffin DL. Metal toxicity for rabbit alveolar macrophages *in vitro*. Environ. Res., 1975;9:32-47.

Waters MD, Gardner DE, Coffin DL. Cytotoxic effects of vanadium on rabbit alveolar macrophages *in vitro*. Toxicol. Appl. Pharmacol., 1974;28, 253-263.

Watson J, Chow J, Lu Z, Fujitas E, Lowenthal D, Lawson D. Chemical mass balance source appointment of PM_{10} during the Southern California air quality study. Aerosol. Sci. Technol., 1994;21:1-36.

Weir D, Robertson A, Jones S, Burge P. Occupational asthma due to soft corrosive soldering fluxes containing zinc chloride and ammonium chloride. Thorax, 1989;44:220-223.

Weissman SH, Cuddihy RG, Medinsky MA. Absorption, distribution, and retention of inhaled selenious acid and selenium metal aerosols in Beagle dogs. Toxicol. Appl. Pharmacol., 1983;67:331-337.

White LR, Steinegger AF, Schlatter C. Pulmonary response following intratracheal instillation of potroom dust from an aluminum reduction plant into rat lung. Environ. Res., 1987;42:534-545.

WHO: World Health Organization. *Selenium. Environmnetal Health Criteria #58.* Geneva: World Health Organization, 1987.

WHO: World Health Organization. *Vanadium. Environmental Health Criteria No. 81.* Geneva: World Health Organization, 1988.

Wilhemsen CL. An immunohematological study of chronic copper toxicity in sheep. Cornell Vet., 1979;69:225-232.

Zelikoff JT, Cohen MD. "Immunotoxicity of Inorganic Metal Compounds." In *Experimental Immunotoxicology*, RJ Smialowicz, MP Holsapple, eds. Boca Raton, FL: CRC Press, Inc., 1995, pp. 189-228.

Zoller WH, Reamer DC. "Selenium in the Atmosphere." In *Proceedings of the Symposium on Selenium-Tellurium in the Environment*, Pittsburgh: Industrial Health Foundation, 1976, pp. 54-66.

12. OZONE

Lisa K. Ryan, Ph.D.
Immunotoxicology Branch
National Health and Environmental Effects
Research Laboratory[1]
United States Environmental Protection Agency
Research Triangle Park, NC 27711

INTRODUCTION

Sources and Encounterable Levels of Ozone Exposure

One of the most important air pollutants that presents a great concern for public health is ozone (O_3) in the trophospheric portion of the Earth's atmosphere. Trophospheric O_3 exists as a key component of photochemical smog. Ozone is a gas that is formed from a series of photochemical reactions involving ultraviolet light, nitrogen oxides (NO_x) and volatile organic compounds (VOC). Ozone also exists in the stratosphere where it protects the Earth from ultroviolet radiation. Some of the O_3 in the trophosphere also comes from the stratosphere, especially in spring when exchange with the trophosphere is highest (Lippman, 1989).

Because of the nature of the chemical reactions generating O_3, ambient air concentrations of O_3 are seasonal and diurnal. Levels of O_3 typically are highest in summer months, begin to rise at mid-morning after rush hour, and peak in the late afternoon (Bascom et al., 1996). Although O_3 levels are usually lowest at night, interregional transport can often result in O_3 remaining elevated throughout the evening and in attaining higher concentrations in suburban regions downwind from large cities (Bascom et al., 1996). In the United States, peak summertime ambient air O_3 concentrations typically range from 0.06 - 0.3 ppm; the highest concentrations occur in Southern California, the Northeast corridor and in other metropolitan areas such as Houston, Texas (Figure 1) (Bascom et al., 1996; EPA, 1998). In Europe, many cities, especially those with a high latitude and little sunshine, have O_3 levels that are less than those outside the city. In these northern cities, NO_x emissions act as an O_3 sink in the formation of NO_2 from NO. However, in

---

[1] Disclaimer: This Chapter has been reviewed by the National Health and Environmental Effects Research Laboratory, U.S. Environmental Protection Agency and approved for publication. Approval does not signify that the contents necessarily reflect the views and policies of the agency, nor does mention of trade names or commercial products constitute endorsement or recommendation for use.

Southern Europe, especially among Mediterranean cities, O_3 exposure can be quite high. A 1989 study estimated that 38 - 56% of the European population lived in areas with levels that exceeded 200 μg O_3/m^3 (0.1 ppm) averaged over 1 hr (Sivertsen and Clench-Aas, 1996). The O_3 levels in Mexico City and Sao Paulo are among the highest in the world. A 1996 report estimated average levels approaching 0.4 - 0.5 mg O_3/m^3 (0.20 - 0.25 ppm) (Schwela, 1996). For comparison internationally, the World Health Organization in 1995 set the Air Quality Guideline for O_3 to be 0.06 ppm (120 μg O_3/m^3), averaged over an 8 hr period (Sivertsen and Clench-Aas, 1996).

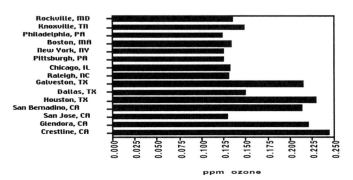

Figure 1. First Maximum Ozone Levels in the United States During 1998. Source: http://www.epa.gov/aqspubl1/annual_summary.html; AIRS Database 1998 Annual Summary Table.

In the United States, the Clean Air Act of 1970 required establishment of National Ambient Air Quality Standards (NAAQS). These included O_3 as one of the six criteria air pollutants used by the Environmental Protection Agency to protect residents against adverse health effects. The NAAQS had set the primary O_3 exposure standard to be 0.12 ppm (235 μg O_3/m^3) over a maximum daily 1 hr average, not to be exceeded more than three times during a 3-yr period. In 1997, the 1 hr standard was replaced with an 8 hr standard and was lowered to 0.08 ppm (157 μg O_3/m^3). The latter standard, which is currently being challenged in the courts, is based on the 3-yr average of the fourth highest daily maximum 8 hr average O_3 concentration measured at central monitors (EPA, 1998). According to an EPA report in 1997, ≈101.6 million people live in areas that have repeatedly exceeded the fourth highest maximum 8 hr NAAQS of 0.08 ppm, emphasizing the importance of this air pollutant as a public health hazard standard (EPA, 1998).

There are many sources that can contribute to formation of O_3 in ambient air besides its exchange from the stratosphere. Combustion sources, such as automobiles and power plants, are major producers of O_3 precursor pollutants. Area sources such as paints and charcoal lighter fluid and even natural sources (e.g. trees producing isoprene and terpene and biogenic decay producing methane) also can contribute by releasing VOC precursors needed for O_3 generation (Lippman, 1989).

Another important source of O_3 occurs in the vicinity of high electric voltage equipment during operation (Stokinger, 1965). It is also widely used as a disinfectant for air and water in bleaches, waxes, textiles, oils, and inorganic synthesis

(Mehlman and Borek, 1987). In addition, welding operations and X-ray machines add other sources of occupational exposure (Stokinger, 1965). The American Congress of Governmental and Industrial Hygienists (ACGIH) adopted threshold limit value-time weighted averages (TLV-TWA; averaged over an 8 hr period) for O_3 of 0.1, 0.08, and 0.05 ppm for light, moderate, and heavy work, respectively. The ACGIH recently changed the recommended TLV-TWA to include 0.2 ppm O_3 for heavy, moderate, or light work, averaged over a period of 2 hr or less (ACGIH, 1999).

Ozone can be an indoor air pollutant as well, although relatively few studies have measured indoor air concentrations and personal exposures. A recent study showed that individual exposure measured by personal monitors may be much lower than area monitors indicate, making personal exposure assessment complex due to the error in the two sampling procedures (Liu et al., 1993). In general, indoor air concentrations of O_3 are much lower than outside. Since most people spend >80% of their time indoors, personal exposure levels may be much lower than estimated from measurements of outdoor O_3 concentrations. However, occupational or excess exposures to indoor sources of O_3 as well as the ventilation of the room can significantly impact upon indoor O_3 concentration and exposure (Lippman, 1989). Sources of O_3 indoors include electrostatic devices (e.g., air purifiers and ion generators) and office equipment with electric motors or ultraviolet lights (e.g., photocopiers) (Bascom et al., 1996). Poor ventilation in the presence of operating equipment in the workplace provides for O_3 exposures that exceed safe levels.

Determinants of Ozone Toxicity

Ozone enters the body primarily via inhalation and reacts with proteins, lipids, and other biomolecules in the airways. Most of the toxic effects occur directly on the lung, although systemic effects also occur. These include effects on red blood cells (Buckley et al., 1975; Hamburger and Goldstein, 1979), the liver (Graham et al., 1983; Laskin et al., 1994), parathyroid (Atwal and Wilson, 1974), and the immune system (reviewed in Jakab et al., 1995). A summary of the pulmonary toxicity of O_3 is listed in Table 1.

The toxicity of O_3 differs with the type (single or repeated, given while exercising or not), duration (acute, subchronic, or chronic) and concentration of exposure. Adaptation responses, genetic susceptibility, diet, sex, age, and health status (i.e. pre-existing respiratory illness) are host factors that play a role in determining the toxic outcome of O_3 exposure and could affect immunotoxicity (reviewed in Jakab et al. [1995] and in Bascom et al. [1996]).

The adaptation phenomenon is defined as a reduced physiological response that occurs on the second day of O_3 exposure, followed by a progressive improvement in response on subsequent days (Bascom et al., 1996). Studies showed that over 4 - 5 d of repeated O_3 exposure, the greatest effect occurred on day 2 of exposure, measured by a reduced decrease in the pulmonary function parameters FEV_1 (forced expiratory volume at 1 sec) and FVC (forced vital capacity) (Hackney et al., 1977; Farrell et al., 1979; Follinsbee et al., 1980; Bedi et al., 1985). Neutrophil influx into the lungs following O_3 exposure also is subject to adaptation, as neutrophil levels in the bronchoalveolar lavage fluid returned to normal in both humans and in rats after

TABLE 1. SUMMARY OF TOXIC EFFECTS OF O_3 IN LUNG

Human Health Effects[**]

Decreased lung function[*]
Increased airway reactivity[*]
Lung inflammation[*]
Increased respiratory symptoms[*]
Decreased exercise capacity[*]
Increased hospitalization of asthmatic patients and others
    with respiratory illness[*]
Increased cough[*]
Reduced athletic performance[*]
Increased respiratory epithelial permeability (0.4 ppm,
    2 hr, intermittent treadmill exercise)
Increased rate of tracheobronchial mucociliary particle
    clearance (0.2 - 0.4 ppm, 2 hr)

[*]These effects were found at ambient O_3 levels at or below 0.12 ppm and were increased with exercise; [**](Lippman, 1989;Bascom et al., 1996)

five consecutive days of exposure (Bascom et al., 1996). Adaptation may also affect host defense functions. In vivo killing of Staphylococcus aureus was impaired in mice 1 - 2 d following a continuous 0.5 ppm O_3 exposure, but returned to normal as the exposure resumed for 7 d (Gilmour et al., 1991b). When rabbits and mice were exposed to O_3 for several days, alveolar macrophages (AM) recovered from the lungs demonstrated no impairment of superoxide anion production or of phagocytic capability; conversely, following a single acute exposure to O_3 (for 1 d or less) impaired these functions (Driscoll et al., 1987; Oosting et al., 1991). There is also a tolerance to the lethal effects of O_3 that occurs upon repeated exposure. One study showed that exposure of rats to nonlethal concentrations of O_3 could tolerize against subsequent lethal doses of O_3 (Stokinger et al., 1956).

The animal model used also may affect the toxicity and outcome of the exposure to O_3. Rats exposed during evening hours when activity is increased have enhanced pulmonary inflammatory responses to a given O_3 concentration compared to that following a daytime exposure to the same concentration of O_3. Conversely, the inflammatory responses during daytime and nighttime O_3 exposures are similar in the guinea pig (van Bree et al., 1992). In another example of a species difference in O_3 response, rats (compared to mice) are relatively resistant to the O_3-induced potentiation of bacterial infection as well as depression of both AM phagocytic function and superoxide anion ($\cdot O_2^-$) production (Goldstein et al., 1978; Kimura and Goldstein, 1981; Ryer-Powder et al., 1988; Gilmour and Selgrade, 1993c).

Genetic factors also play a role in varying responses within an animal species. Both rat strains and inbred mouse strains have varying susceptibility to O_3 (Bascom et al., 1996). In mice, O_3 concentrations which induce lethality vary with strain (Goldstein et al., 1973) as does airway hyperreactivity to acetylcholine [Zhang et al., 1995) and inflammatory responses induced by O_3 (Kleeberger et al., 1990), indicating that genetic susceptibility to O_3 toxicity may be an important factor in determining severity of a response. One study examined genetic susceptibility factors in

C57BL/6J mice, a highly sensitive strain prone to the inflammatory effects of O_3, and C3H/HeJ mice, a highly resistant strain that gives a blunted inflammatory response to O_3 (Kleeberger et al., 1990, 1993a and b). It was found that the inflammatory response, characterized by neutrophil influx into the airways, was controlled by a quantitative trait locus for O_3 susceptibility on chromosome 17 (Kleeberger et al., 1997). The gene itself has yet to be identified, but the tumor necrosis factor-α (TNF) gene, *tnf*, that encodes this proinflammatory cytokine was named as a likely candidate. The two strains also had differential susceptibility to infection with *Streptococcus zooepidemicus* (Group C) and impairment by O_3; C3H/HeJ mice were more susceptible to infection and also more sensitive to impaired AM phagocytosis and corresponding enhanced mortality induced by a 3 hr 0.4 or 0.8 ppm O_3 exposure (Gilmour et al., 1993a). In humans, genetic factors appear to play a role in individual responses to O_3, as large variations exist among responses of individuals to O_3 exposure in chamber and epidemiological studies (Bascom et al., 1996).

Antioxidants in the body can disrupt the oxidant activity of O_3, with dietary antioxidants such as ascorbic acid (Vitamin C), retinol (Vitamin A), and α-tocopherol (Vitamin E) reducing the toxic response to O_3 (Menzel, 1994). Vitamin A deficiencies in genetically-susceptible C57BL/6J and -resistant C3H/HeJ mouse strains enhanced neutrophil influx only in C_3H mice and enhanced epithelial cell loss and lung permeability increases in both strains; replacing this vitamin in these mice partially reversed O_3-induced lung injury (Paquette et al., 1996). Vitamin C-deficient guinea pigs demonstrated enhanced accumulation of protein in bronchoalveolar lavage (BAL) fluid following acute O_3 exposure (Slade et al., 1989). Vitamin E deficiency enhanced lung tissue damage in rats following 1 wk of O_3 exposure [Elsaved et al., 1988). Animals, such as rats and guinea pigs (provided full dietary Vitamin C supplementation), have higher Vitamin C levels compared with humans and are proportionately less sensitive to the effects of O_3 (Slade et al., 1985; Hatch et al., 1986). Vitamin C levels rise in response to repeated O_3 challenges in rats coincident with development of an adaptation response (Tepper et al., 1989). Deficiencies in Vitamin E and selenium modified immune responses of rats, depressing lymphocyte proliferation responses to mitogens and antibody-dependent cell-mediated cytotoxicity events against chicken erythrocytes (Eskew et al., 1986). This study also showed that Vitamin E deficiency led to an enhanced neutrophil and macrophage influx into the lungs following a 2 ppm intermittant O_3 exposure (i.e., 8 hr/d for 4 d, then 2 - 4 d in air, followed by 1 d of O_3 exposure prior to sacrifice). Vitamin C and E also protected rat AM from the suppression of phagocytosis of yeast cells induced by *in vitro* O_3 exposure (Rietjens et al., 1986).

Sensitivity to the inflammatory effects of O_3 appears to decrease with age. Production of prostaglandin E_2 (PGE_2) following an acute 2 - 4 hr 1 ppm O_3 exposure waned from a high amount in BAL fluid in newborn rats to nearly absent in adult rats (Gunnison et al., 1992). Age-related differences in responses also occurred in rats exposed to 0.35 or 0.80 ppm O_3 for 72 hr (Stiles and Tyler, 1988). Younger rats had larger proportions of centriacinar lesions and macrophages while older rats had greater body weight and lung volume changes. Younger animals also appeared to be more sensitive to the inflammatory effects of a 1 or 7 d 12 hr exposure to 0.8 ppm O_3; furthermore, these effects were more pronounced in male than female rats (Dormans et al., 1996). This group also concluded that the age-related differences did not apply when measuring the decrease in host resistance to *Listeria* infection caused by O_3 exposure. However, impaired intrapulmonary killing of *S. zooepidemicus* by 0.8 ppm O_3, which correlated with mortality, was more severe in younger (5-wk-old) than in older (9-wk-old) mice (Gilmour et al., 1993b). An age effect may also apply to humans. Exercising children appear to have a reduction in FEV_1 and FVC while in healthy adults over 50 yrs of age, proportionately smaller changes in these lung function parameters occurred [Bascom et al., 1996b).

Hormonal fluctuations, pregnancy, and lactation also affect the response to O_3 exposure. Pregnancy and lactation in rats induced a metabolic state that increased the sensitivity to the inflammatory effects of O_3 compared with age-matched virgins; BAL levels of total leukocytes, neutrophils, protein, lactate dehydrogenase, and β-glucuronidase activity were all increased to a significantly greater degree (Gunnison et al., 1992). The enhanced sensitivity of lactating rats following exposures to 0.3 - 1.0 ppm O_3 resulted in both an earlier onset and greater intensity and persistence of inflammatory response compared with that in virgin counterparts (Gunnison et al., 1996; Weideman et al., 1996). One factor that contributed to the enhanced sensitivity was the increased ventilation of the lactating rats that resulted in a greater inhaled dose of O_3 (Weideman et al., 1996). Further studies showed that enhanced sensitivity of lactating rats to O_3-induced inflammation also may be due to changes in their immunological status and to the greater propensity of respiratory tract tissues to respond to stimulation by O_3 (Gunnison et al., 1996). In this study, naive lactating rats had a lower concentration of Vitamin C in their airways, a higher number of neutrophils in their circulation and airways, and a higher BAL protein concentration compared with nonlactating virgin rats. These factors may outweigh the apparent effects of progesterone on inhibiting PGE_1 production and decreasing O_3 sensitivity. Nine young adult human females exposed to 0.3 ppm O_3 for 1 hr exhibited a greater decrease in the pulmonary function parameters FEV_1 and FEF_{25-75} (forced expiratory flow between 25 - 75% of FVC) following O_3 exposure during the follicular phase than the mid-luteal phase of their menstrual cycle when progesterone levels were, respectively, lowest and highest (Fox et al., 1993).

Some experiments have incorporated the fact that O_3 does not exist in nature as a pure molecule, but as part of a mixture of other ambient air pollutants. There are two excellent reviews on the toxicity of O_3 that address the details of this issue (Lippman, 1989; Bascom et al., 1996). Many toxicology studies have found that other pollutants, such as acid-coated particles, NO_2, and sulfur dioxide (SO_2) alter the immunotoxicity of O_3. For example, in one study, a 5 d high level exposure of rats to an O_3 + NO_2 mixture induced a depression of AM phagocytosis that was further depressed by diesel exhaust added to the mixture, whereas a 20 d low level exposure of rats to these mixtures induced increased AM phagocytosis (Kleinman et al., 1993). Even inert particles may affect the immunotoxicity of O_3. One study showed that a concomitant 4 hr exposure of Swiss mice to 10 mg inert carbon black particles/m^3 and 1.5 ppm O_3 enhanced suppression of AM F_c-mediated phagocytosis of sensitized sheep erythrocytes compared with suppression induced by O_3 alone (Jakab and Hemenway, 1994). In the same study, O_3-particle interaction also enhanced inflammation as measured by neutrophil influx into the airways of exposed mice; exposure to carbon black particles alone did not affect these parameters.

IMMUNOMODULATORY EFFECTS OF OZONE

Antibody-Mediated Responses

Ozone has variable effects on humoral immune responses, depending on the concentration and when exposure occurs in relation to antigenic stimulation. The nature of the antigen and its route of exposure also will determine the type of effect O_3 has on the response. Whether an antigen initiates a response dependent on the help of T-lymphocytes (termed T-dependent) or not (termed T-independent) and whether a T-helper type 1 (T_H1) or type 2 (T_H2) lymphocyte response is initiated also determines the type of effect O_3 will have on humoral immunity. This fundamental understanding is crucial to comprehending how O_3 can impact upon development of allergy and allergic asthma and the host response to infectious agents that are dependent on humoral immunity.

Effect of Ozone on Protective Antibody Production

Ozone appears to suppress non-allergic antibody production in response to an antigenic stimulation that is strongly dependent on T_H1 lymphocytes. An early study implied a decrease in antibody response to tetanus toxin when mice that had been exposed to O_3 during vaccination against the toxin were shown to be less resistant to subsequent toxin challenge (Campbell and Hilsenroth, 1976). One human study demonstrated suppressed immunoglobulin G (IgG) production in response to lymphocytes exposed to O_3. An *in vitro* study examined the effect of a 2 hr exposure of 0.1, 0.5, and 1.0 ppm O_3 upon human B-lymphocytes (Becker et al., 1991b). These cells were then mixed with control air-exposed T-lymphocytes in a 1:1 ratio and their ability to produce IgG in response to pokeweed mitogen (PWM; T-dependent antigen) and to *Staphylococcus aureus* Cowan I strain (SAC; T-independent antigen) *in vitro* was evaluated. The results showed that there was a significant decrease in IgG production induced by only PWM only at the highest level of O_3 exposure. Conversely, there was a trend toward increased IgG production by B-lymphocytes exposed to SAC; because of the variability of the response from different individuals, this effect was not statistically significant. A similar experiment was conducted with T-lymphocytes being exposed to O_3 and then mixed with air-exposed B-lymphocytes. IgG production was significantly decreased after either PWM or SAC stimulant was added to the cultures containing T-lymphocytes which had been exposed to 0.5 or 1.0 ppm O_3, but not to 0.1 ppm O_3. When T- and B-lymphocytes were mixed in a 1:1 ratio and exposed to similar concentrations of O_3, only PWM-stimulated cells that had been exposed to 1.0 ppm O_3 exhibited a significant decrease in IgG production. It thus appeared that the T-dependent antibody response was more sensitive to O_3 than was the T-independent response.

Ozone also suppresses plaque-forming antibody production (mostly IgM) in mice immunized with a T-dependent antigen such as sheep erythrocytes (SRBC). In one study, BALB/c mice were exposed continuously to 0.8 ppm O_3 for 1, 3, 7 and 14 d prior to immunization with SRBC or the T-independent antigen, DNP (dinitrophenol)-Ficoll (Fujimaki et al., 1984). Results of this study were similar to those obtained by Becker et al. in that there was a depressed plaque-forming cell (PFC) response to the T-dependent, but not T-independent, antigen in mice exposed to O_3 for as little as 1 d prior to immunization. Interestingly, the experiments also showed a trend toward increased IgM antibody production (following immunization with a T-independent antigen) with prolonged (continuously for 14 d.) O_3 exposure. One experiment even showed a significant increase, indicating that O_3 may also have a small enhancing effect on T-independent antigen stimulation of antibody production.

Experiments of similar design showed that mixing 4.0 ppm NO_2 with 0.8 ppm O_3 variably modified BALB/c mouse antibody production (Fujimaki, 1989). Antibody responses to SRBC were equally suppressed by 3, 7, and 14 d exposures to O_3 and the $O_3 + NO_2$ mixture. Utilizing a 56 d exposure to these atmospheres, it was found that while prolonged O_3 exposure significantly suppressed antibody responses to SRBC by 75%, the $O_3 + NO_2$ mixture had no effect. Anti-DNP antibody responses were not suppressed by either atmosphere. Rather, after a 14 d exposure to O_3 or $O_3 + NO_2$ mixture, the antibody response to T-independent DNP-Ficoll was significantly enhanced by 50%. NO_2 alone had no effect on either response. Splenic B-lymphocyte function and macrophage antigen processing, as measured by PFC against SRBC, were unaffected by exposure of CD1 mice to 0.102 ppm O_3 or to a mixture of O_3, 0.1 ppm SO_2, and 1.04 mg/m^3 ammonium sulfate (NH_4SO_4) aerosol for 5 hr/d, 5 d/wk up to 103 d (Aranyi et al., 1983). In this study, mice were immunized against SRBC 3 d prior to the end of exposure.

Ozone suppresses the local humoral immune response in the lung [Gilmour and Jakab, 1991a). Mice were first primed on day 0 with an intraperitoneal (IP) injection of ovalbumin (OA) and boosted via inhalation with aerosolized OA on day 14. Mice were also exposed to either 0.8 or 0.5 ppm O_3 for 23 hr/d for 1, 3, 7 or 14 d (days 20, 17, 14, and 7, respectively). On day 21, 1 wk after the OA aerosol boost and 3 wk after IP injection of OA, the mediastinal lymph node (MLN) and splenic lymphocytes were assayed for proliferative responses to OA antigen, and BAL fluid was assayed for specific IgG and IgA antibody against OA. After 14 d of O_3 exposure, MLN responses to OA were significantly enhanced. With O_3 exposures less than 14 d, no O_3-induced enhancement of OA-stimulated MLN lymphocyte proliferation occurred. OA-Induced splenic (systemic) lymphocyte proliferation was suppressed after 7 and 14 d of exposure; prior to day 7, no suppression of proliferation occurred compared with cells from hosts receiving no O_3 exposure. Pulmonary IgA antibody titers to OA were also lower in BAL from mice exposed to O_3 for 7 or 14 d and IgG in the BAL was suppressed in mice receiving 1, 7, and 14 d of O_3 exposure. The suppression of humoral immune responses occurred at 0.8 ppm and to a lesser degree at 0.5 ppm O_3 (Jakab et al., 1995). Ozone-induced suppression of anti-OA antibodies in serum did not occur at any time, even though OA induced high serum antibody titers in these animals. From the preceding discussion, it appears that O_3 suppresses antibody responses to T-dependent antigens. Both local lung and systemic suppression of these responses were observed, including the suppression of IgM, IgG, and IgA subclasses (see Table 2).

TABLE 2. SUMMARY OF KEY STUDIES SHOWING EFFECTS OF O_3 ON PROTECTIVE ANTIBODY PRODUCTION.

In Vitro Human Cell Exposure Studies

Endpoint	Exposure	Effect
IgG production[a] from B-cells	0.5 - 1.0 ppm (2 hr)	
+ PWM		Suppression of T-dependent Ag-stimulated IgG
+ SAC		Slight enhancement of T-independent Ag-stimulated Ig

In Vivo Exposure Studies in Mice

BALB/c Mice[b]

Plaque assay (mostly IgM)	0.8 ppm O_3, continuous 1 d prior to SRBC immunization	Suppression of PFC to T-dependent Ag
	0.8 ppm O_3, continuous 14 d	Slight enhancement of PFC to T-independent Ag

COBS Mice[c]

Lethal challenge with tetanus toxoid	0.59 ppm O_3, continuous pre-immunization for 5 d, then 36 d post-immunization	Increased mortality

$B_6C_3F_1$ Mice[d]

MLN and Splenic OA-stimulated lymphoproliferation	0.5 - 0.8 ppm, continuous 1 - 14 days	MLN increased splenic suppressed 7-14 d only
Pulmonary OA-specific IgA	0.5 - 0.8 ppm, continuous 1 - 14 days	Depressed levels in BAL on d 7-14 only
Pulmonary OA-specific IgG	0.5 - 0.8 ppm, continuous 1 - 14 days	Depressed levels in BAL on all days

[a]Becker et al., 1991b; [b]Fujimaki et al., 1984; [c]Campbell and Hilsenroth, 1976; [d]Gilmour and Jakab, 1991a

Effect of Ozone on Humoral Immunity in Infectious Disease

Studies on the effects of O_3 on humoral immunity demonstrate that results can be manipulated by changing the exposure regimen. A study using continuous exposure of Swiss mice to 0.5 ppm O_3 during the course of murine influenza A/PR8/34 infection showed that O_3 suppressed humoral immune responses, as evidenced by reduced serum antibody titers and numbers of T- and B-lymphocytes (Jakab and Hmieleski, 1988). However, in a study where several different exposures to O_3 or air were followed by sublethal virus infection and then again by 0.5 ppm O_3 or air, neutralizing antibody titers in BAL and serum were not significantly affected by O_3 (Wolcott et al., 1982). Although there was a decline of serum and BAL antibody titers to influenza A virus in the mice exposed to O_3 after infection with virus, it was not statistically significant. In both studies, it was found that O_3 exposure after infection protected the mice against mortality caused by influenza. In a study where human volunteers were exposed to 0.3 ppm O_3 for 5 d immediately after type 39 rhinovirus inoculation, levels of convalescent serum neutralizing antibody to the virus were not altered by O_3 exposure (Henderson et al., 1988).

Effect of Ozone on Humoral Responses Associated with Atopy

The effect of O_3 on production of allergic antibodies (IgE and IgG_1 in rodents and IgE and IgG_4 in humans) is dependent on the route of antigen exposure, the type of antigen used, the source of the sample, and the method of antibody measurement. The results also may be dependent on whether or not the antigen is given prior to or during exposure to O_3 (Jakab et al., 1995). These effects are summarized in Table 3.

Several studies have suggested that O_3 potentiates IgE production in the lung and subsequent anaphylaxis following inhalation of aerosolized antigens such as OA and hexachloroplatinate. These studies sensitized their animals to the antigen before the end of O_3 exposure and used aerosolized antigen to sensitize the animals. One study showed enhanced sensitization with aerosolized OA in guinea pigs exposed for 30 min to 5 or 10 ppm O_3, but not to 1 ppm O_3 (Matsumura, 1970a). However, when animals were sensitized via IP injection, O_3 had no effect on serum IgE antibody production as measured by passive cutaneous anaphylaxis (PCA). Using a model of OA-induced anaphylaxis in sensitized mice, intermittent exposure to 0.5 - 0.8 ppm O_3 alone or in combination with 1 mg/m^3 sulfuric acid aerosol increased reactivity to subsequent OA aerosol challenge (Osebold et al., 1980). A later study showed that as little as 0.13 ppm O_3 could enhance allergic sensitization in the same animal model (Osebold et al., 1988). When 0.8 or 0.5 ppm O_3 was continuously given to mice for 3 - 4 d prior to aerosolized OA, the increase in the number of cells containing IgE in the lung was 34-fold above that found in normal lung and 3.6-fold compared to the number of IgE-containing cells in lungs of mice sensitized with aerosolized OA but not receiving O_3 exposure (Gershwin et al., 1981). However, when either homologous or heterologous PCA were used to assess serum IgE, none was detected in the serum of most of these animals. Adult male cynomolgus monkeys exposed for 6 hr/d, 5 d/wk for 12 wk to a mixture of 200 $\mu g/m^3$ ammonium hexachloroplatinate and 1 ppm O_3 demonstrated enhanced allergic responses to platinum (Biagini et al., 1986). As allergic responses were better identified using direct skin testing with allergen as compared to radioallergosorbent testing (RAST), this indicated the greater sensitivity of skin testing compared to RAST and PCA.

In addition to the enhancing effect of O_3 upon sensitization to allergens, O_3 enhances susceptibility to antigen *challenge*. When guinea pigs sensitized by IP injection of OA were challenged with an inhaled OA dose designed to induce slight anaphylaxis in control animals, many receiving pre-exposure to >2ppm O_3 (after sensitization but before challenge) had severe dyspneic attacks, with death occurring

TABLE 3. SUMMARY OF KEY STUDIES SHOWING EFFECTS OF O_3 ON HUMORAL RESPONSES ASSOCIATED WITH ATOPY (*IN VIVO* STUDIES).

Endpoint	Exposure	Effect
Serum IgE[a] (PCA in rats)	0.8 ppm, continuous 1, 2, or 4 wk prior to sensitization in BALB/c mice	Suppression of serum of serum IgE, adaptation at wk 2 and 4
Lung anaphylaxis; permeability[b]	5-10 ppm O_3, 30 min guinea pigs during OA sensitization	Enhanced sensitization to OA
Systemic anaphylaxis[c,d]	0.13 - 0.80 ppm O_3 Swiss-Webster mice during OA sensitization	Increased sensitization to OA
IgE-containing cells in lung[e]	0.5 - 0.8 ppm O_3, continuous before and during aerosolized OA sensitization of Swiss-Webster mice	Increased local IgE
Skin Test: methacholine reactivity[f]	1 ppm O_3, 12 wks 5 d/wk, 6 hr/d cynomolgus monkeys	Enhanced allergic responses to platinum
Anaphylaxis[g]	>2 ppm O_3, 30 min guinea pigs	Enhanced response to aerosolized OA challenge
Acetylcholine reactivity[h]	3 ppm O_3, 2 hr *Ascaris*-Ag-sensitized dogs	Increased
Ascaris challenge	3 ppm O_3, 2 hr *Ascaris*-Ag-sensitized dogs	Decreased provocation concentration
Ragweed, grass Ag challenge to decrease FEV_1 by 15% (PC_{15})[i]	0.12 ppm O_3, 1 hr Asthmatic patients allergic to grass, ragweed	Decreased PC_{15}; potentiated allergic response when O_3 is given before Ag

[a]Ozawa et al., 1985; [b]Matsumura, 1970a and b; [c]Osebold et al., 1980; [d]Osebold et al., 1988; [e]Gershwin et al., 1981; [f]Biagini et al., 1986; [g]Matsumura, 1970c; [h]Yanai et al., 1990; [i]Molfino et al., 1991

in several of the animals (Matsumura, 1970c). A 2 hr exposure of *Ascaris* antigen (AAg)-sensitized dogs to 3 ppm O_3 enhanced susceptibility to AAg challenge (Yanai et al., 1990). Compared with AAg-sensitized air-exposed controls, it was found that the O_3-exposed dogs evidenced increases in plasma histamine concentration and airway responsiveness to AAg and decreases in the AAg provocation concentration used to induce respiratory resistance. A similar effect was reported in atopic asthmatic humans exposed to 0.12 ppm O_3 for 1 hr (Molfino et al., 1991). When this realistic ambient O_3 level exposure preceded antigen challenge, the challenge dose required to cause a 15% decrease in FEV_1 was decreased significantly; no significant changes in baseline FEV_1 occurred after exposure to O_3. Thus, even at low O_3 concentrations, bronchial responsiveness to allergens increases in atopic asthmatic subjects and this effect does not seem to be due to a direct effect of O_3 upon baseline airway function.

Epidemiologic studies corroborate these experiments (Bascom et al., 1996b). For example, on days in Atlanta when the 1 hr maximum O_3 exceeded 0.11 ppm, visits by asthmatic children to inner city emergency rooms were significantly higher compared with that on days when O_3 levels were < 0.11 ppm (White et al., 1994). A Japanese study compared two groups of humans exposed to comparable levels of cedar pollen and found that the group that lived closest to a major highway (and, hence, were exposed to higher levels of O_3, NO_2, and other automobile exhaust components) suffered more frequent and severe allergic reactions that did cohorts living five miles from the highway (Ishizaki et al., 1987). This demonstrated that O_3 may contribute to a higher incidence of allergic reactions to airborne pollen.

In contrast to the studies cited above which show an O_3-induced enhancement of local, cytophilic IgE, or an anaphylactic response, one study showed that *serum* IgE antibody production was *suppressed* in O_3-exposed BALB/c mice (Ozawa et al., 1985). This is consistent with studies noted earlier describing O_3-induced suppression of IgG release from human lymphocytes, IgM from mouse spleen cells, and neutralizing antibody against influenza virus (Fujimaki et al., 1984; Jakab and Hmieleski, 1988; Becker et al., 1991b). These experiments are characterized by the fact that O_3 was given *prior to sensitization*. Mice were exposed to 0.8 ppm O_3 for 1, 2, or 4 wk; immediately or 7 d after the end of the O_3 exposure period, the mice were then injected IP with OA and aluminum hydroxide (alum). There were no differences in IgE antibody production (determined by PCA reactions in Wistar male rats) indicating no effect of O_3 upon the primary IgE antibody response of mice immunized IP. Among mice given the same O_3 exposure regimen and then provided aerosolized OA 7 d before IP OA/alum, those exposed to O_3 for 1 wk exhibited a 16-fold suppression on day 7 after this injection; 2- and 4-wk exposures to O_3 did not suppress IgE production on day 7 as much (i.e., only 4-fold); this may have been due to adaptation responses to O_3. It was also shown that induction of T-helper lymphocyte function was suppressed by prior exposure to O_3 for 1 wk if aerosolized OA carrier protein was administered before IP immunization with DNP hapten conjugated to OA. To determine if O_3 affected hapten-specific B-lymphocytes, mice were immunized with carrier protein keyhole limpet hemocyanin (KLH) on day -21, exposed to O_3 for 1 wk on days -14 to -7, treated with aerosolized DNP-OA on day -7 immediately after O_3 exposure, and finally immunized IP on day 0 with DNP-KLH. As neither enhancement nor suppression of DNP-specific IgE production in serum was observed, it was concluded that there was no effect from O_3 exposure on hapten-specific B-lymphocytes. These experiments illustrate the ability of O_3 to suppress IgE responses to T-dependent antigens, provided the exposure to O_3 occurs prior to sensitization with the T-dependent antigen (i.e., OA).

Effects of Ozone in the Absence of Antigenic Stimulation

Without experimental antigenic stimulation, the effect of O_3 on antibody production seems much simpler. In an early study, mice were exposed to 0.5 or 0.8 ppm O_3 for 15 or 24 d, resulting in elevated serum albumin in BAL fluid (peaking on day 4) (Osebold et al., 1979). Exposure to 0.8 ppm O_3 resulted in an increase in the number of IgA-, but not IgG- or IgM-, containing cells in bronchus-associated lymphoid tissue. Increases in IgA, IgG_1, and IgG_2 were found in the BAL of mice exposed to both concentrations of O_3. By plotting the relative changes of Ig content in lavage fluid (independent of serum albumin concentration) in O_3-exposed mice, these Authors hypothesized that the IgA produced was secreted from a local source in the lung and that the IgG source was the serum. It was also hypothesized that the immune cells had received antigenic stimulation of unknown origin. Immuno-globulin levels ceased to be elevated when the mice were removed from the O_3 exposure and placed back in ambient air.

In summary, the effects of O_3 upon humoral antibody responses suggest that the timing and combination of the exposure and stimulation with antigen, whether it be an allergen or infectious agent, is crucial to the outcome of the effect. Under the right circumstances, O_3 can enhance both infectious and allergic diseases.

Cellular Immune Effects

One of the major effects of O_3 is on specific T-lymphocyte-dependent cellular immunity (recently reviewed by Gilmour and Selgrade [1998]). Athymic or nude mice subjected to O_3 exposure evidenced more lung damage than normal mice exposed to O_3 (Dziedzic and White, 1987). T_H (CD4[+])-lymphocytes have been shown to be affected by O_3 (Chen et al., 1995). Immune responses directed against certain pathogens and allergens are regulated by different subpopulations of CD4[+] T-lymphocytes (Mosmann and Sad, 1996). As described above, CD4[+] T_H1 and T_H2 lymphocytes are involved in humoral immune responses and are affected by O_3. CD4[+] T_H1 lymphocytes are also involved in specific cellular immune responses, such as delayed-type hypersensitivity (DTH or Type IV hypersensitivity), contact hypersensitivity, and resistance to intracellular pathogens such as *Listeria monocytogenes* and *Mycobacterium tuberculosis*. CD4[+] T_H1 lymphocytes, along with cytotoxic CD8[+] T-lymphocytes, also are partially involved in resistance to primary influenza virus infections (Graham et al., 1994; Doherty, 1996).

Resistance to Intracellular Pathogens

Ozone enhances infectious diseases that depend on specific cellular immune responses. These diseases also depend on non-specific defense mechanisms, mediated by macrophages (O_3 depression of non-specific mechanisms is described below). However, whether or not O_3 has an effect depends on when the O_3 exposure occurs relative to the initiation of infection. For example, mice infected with *Myco-bacterium tuberculosis* H37Rv or with *Bacillus* Calmette-Guerin (BCG) *after* exposure for 2 mo (4 hr/d, 5 d/wk) to 1.5 ppm O_3 had no alteration in resistance to the bacteria (Thienes et al., 1965). When *M. tuberculosis* R1Rv challenge was given *before* exposure to 1 ppm O_3 for up to 8 wk (3 hr/d, 5 d/wk), resistance was altered (Thomas et al., 1981). In this latter study, in mice sacrificed beginning at 1 or 2 wk after challenge, bacterial titers in the lungs were significantly enhanced after 5 - 8 wk of O_3 exposure; exposure concentrations <1 ppm O_3 had no effect.

Suppressive effects on specific cellular immune responses were also dramatic when exposure to O_3 occurred during an ongoing pulmonary infection to the intra-cellular bacteria *Listeria monocytogenes* (van Loveren et al., 1988). In these experiments, rats were exposed to O_3 concentrations ranging from 0.4 - 1.0 ppm (0.25 - 2.0 mg/m^3) for 1 wk either before or during an intranasal infection with 3.8 x 10^8 *Listeria*. At 1 ppm O_3, mortality and increased numbers of bacteria in the lungs were enhanced. Ozone exposure also increased the severity of the *Listeria*-induced lung lesions and granulomas formed in the infected rats. These studies also showed that the ability of AM to phagocytize and kill the bacteria (a non-specific immune response) was suppressed by O_3. Ozone exposure also suppressed specific cellular immune responses such as the DTH measured by ear swelling 24 hr after *Listeria* antigen challenge and *Listeria* antigen-induced lymphoproliferation by lung draining lymph node and spleen cells. T-/B-Lymphocyte ratios in the O_3-exposed infected rats were also lower compared to that in infected air-exposed controls.

Effect of Ozone on Type IV Hypersensitivity Responses

As in the experiments with *Listeria*, the suppression of specific cellular immunity was also demonstrated in investigations of the DTH response. In one study, O_3 suppressed the DTH response in BALB/c mice exposed to 0.8 ppm O_3 for 1, 3, 7, and 14 d, then immediately immunized with intravenous (IV) SRBC and challenged in the footpad 4 d later with SRBC (Fujimaki et al., 1987). Compared with control mice, the DTH reactions measured by footpad swelling decreased gradually from 1 d to 7 d of O_3 exposure; no decrease was seen at 14 d. If O_3 was provided for 3 d after SRBC immunization and then SRBC challenge was given 4 d later, the DTH response was significantly suppressed compared with that in control mice. This suggested that O_3 affected the growth and differentiation of T-lympho-cytes during their activation by SRBC antigen.

The DTH response was also investigated in the context of hapten-induced pul-monary allergy (Garssen et al., 1997). Unlike pulmonary allergic and anaphylactic responses induced by OA, which are Type I hypersensitivity responses and depend on T_H2 lymphocytes to induce IgE production, low molecular weight compounds (i.e. picryl chloride (PCl), nickel and trimetallic anhydride), induce Type IV allergic responses that are T_H1-dependent (Garssen et al., 1989, 1990, 1991, 1993, 1994, 1997). BALB/c mice were skin-sensitized to PCl and then challenged intranasally 7 d later with picryl sulfonic acid (PSA). Ozone exposure occurred either at night after day 4 (i.e., before PSA challenge) or at night after day 7 (i.e., after PSA challenge) and then tracheas were isolated and exposed to carbachol *in situ* to assess tracheal activity by isometric measurement. No O_3-induced changes on airway reactivity induced by carbachol were noted in nonsensitized mice. However, in sensitized mice, exposure to 0.4 - 1.6 mg O_3/m^3 *prior to challenge* (day 4) inhibited tracheal hyperreactivity induced by carbachol to levels that were comparable to that in O_3-exposed nonsensitized mice. Exposure to all doses of O_3 during the night directly *following intranasal hapten challenge* (day 7) significantly suppressed tracheal hyperresponsiveness in sensitized mice. The induction of tracheal hyperreactivity was also significantly inhibited in mice passively sensitized with donor lymphocytes. The donor mice were exposed to O_3 the night before skin sensitization and were sacrificed for their splenic lymphocytes 7 d later. Even the lowest dose of O_3 was effective in inducing suppression of sensitization in this model, eliminating the possibility of O_3 directly affecting the lung to influence the tracheal hyperreactivity assay. The inflammatory reaction in the lungs, which was maximal at 48 hr after challenge, consisted of mononuclear cells, some neutrophils, and a few eosinophils around the bronchioli and blood vessels. This inflammation was significantly suppressed only by the highest concentration of O_3 given prior to challenge (day 4). Ozone itself did not induce an inflammatory response in control animals. The results showed that O_3 can induce a suppression of pulmonary Type IV hypersensitivity and that the induction of airway hyperresponsiveness is more sensitive to O_3 than the inflammatory component, implying that separate mechanisms exist for inducing these effects.

Effect of Ozone on Autoimmunity

The literature has suggested that the ability of O_3 to react with proteins may induce autoimmunity through modification of antigens recognized by the immune system as "self." Protein antigens could be crosslinked (Buell and Mueller, 1965), denatured (Scheel et al., 1959), or modified by other means (Stokinger and Scheel, 1962). However, little attention has been given to this area of research. Only one study reported a possible autoimmune parathyroiditis in rabbits following 0.75 ppm O_3 inhalation for 48 hr (Atwal et al., 1975). In this study, immunofluorescence and histologic data suggested that O_3 inhalation triggered an autoimmune reaction that caused inflammatory injury to the parathyroid gland. The mechanism of this toxicity was postulated to be due to O_3 reacting with thiol (-SH) groups (especially in basic amino acids) thereby modifying parathyroid hormone (PTH) from a "self" antigen to a "foreign" protein resembling heat-aggregated OA. Hydroxylation of critical tyrosine and tryptophan residues in the parathyroid hormone molecule by O_3 does in fact alter its protein structure, making it crossreactive with heat-aggregated OA (Scheel et al., 1959; Stokinger and Scheel, 1962); this could then trigger an autoimmune reaction and damage to the parathyroid gland.

Non-specific Immune Effects and Effects on Innate Immunity

Effect of Ozone on Immune Organ Weights and Cell Numbers

Studies have shown that in mice, continuous exposure to 0.3 - 0.8 ppm O_3 caused dose-dependent decreases in thymus, spleen, and MLN weights after 1 - 3 d of exposure (Fujimaki et al., 1984; Dziedzic and White, 1986b). In addition, 0.8 ppm O_3 initially depressed the numbers of lymphocytes in the blood, MLN, and spleen (Fujimaki et al., 1987; Gilmour et al., 1991b). Following a prolonged exposure (7 - 14 d), cell numbers returned to normal and increased over control values. These increased numbers were attributed to an increase in T-, but not in B-, lymphocytes (Dziedzic and White, 1986a; Bleavins and Dziedzic, 1990; Li and Richters, 1991a and b). Still, this "adaptation" may not represent a return to entirely normal function since the balance of various lymphocyte subpopulations may be different.

Similarly, lymphocyte responses to phytohemagglutinin (PHA) mitogen and splenocyte tumoricidal activity also were initially decreased (at 3 d post-exposure) but returned to control levels after a prolonged exposure (at 7 d) to 0.8 ppm O_3 (Gilmour et al., 1991b). In addition, thymus weight and number of thymocytes were shown to decrease after 3 d exposure to 0.8 ppm O_3, but then were further depressed after 7 d of exposure (Fujimaki et al., 1987).

Generally, in humans, a suppression of the response to T-lymphocyte mitogens (especially to PHA) occurs upon acute exposure to O_3 (Peterson et al., 1978a). For example, PHA-stimulated lymphocytes from 20 human subjects (exposed to 0.4 ppm O_3 for 4 hr) exhibited depressed mitogenic responses immediately after exposure (Peterson et al., 1978b). The mitogenic response in these subjects did not recover until 2 wk post-exposure. In contrast, prolonged (>1 d) exposure to low levels of O_3 was associated with increased activation of lymphocytes or an enhanced response to concanavalin A in other studies [Ulrich et al., 1980; Foster et al., 1996). Taken together, the data here are consistent with that in previous sections of this Chapter in that T-lymphocytes appear to be a sensitive target for the effects of O_3.

Effect of Ozone on Natural Killer Cell Activity

Natural killer (NK) cell activity is primarily known to target neoplastic and virus-infected cells and is considered an immediate defense mechanism or an innate immune response. NK activity is displayed by cells of different phenotypes, causing lysis of NK-sensitive cell targets via a mechanism not dependent upon the major histocompatibility complex (MHC) (Herberman, 1986). As such, NK activity is defined as a functional, rather than a morphological, characteristic of the cell (van Loveren et al., 1990).

Higher concentrations of O_3 were shown to suppress NK activity in lung homogenates from male rats following a continuous O_3 exposure for up to 7 d. Suppression of NK activitiy against YAC-1 tumor cell targets occurred following exposure for 1, 5, or 7 d to 1 ppm O_3 (Burleson et al., 1989). The effect was most dramatic after 1 d of exposure; 0.5 ppm O_3 (but not 0.1 ppm O_3) also significantly suppressed NK activity on this day. A similar phenomenon occurred after a 7 d continuous exposure to 1.6 mg/m^3 O_3, but NK activity was stimulated at the lower concentrations of 0.4 and 0.8 mg/m^3 (van Loveren et al., 1990).

There again appeared to be adaptation to O_3 exposure over time as exposures longer that 7 d had no effect on pulmonary NK activity. No suppression of NK activity was noted after continuous exposure to O_3 for 10 d (Burleson et al., 1989).

When rats were given an O_3 exposure for 1, 3, 13, 52, or 78 wk that mimicked diurnal exposure patterns frequently seen in worst-case summer environments, no effect was seen even after 1 wk (Selgrade et al., 1990). Daily exposures consisted of a background level of 0.06 ppm for 13 hr, a broad rise and fall over 9 hr ranging from 0.06 - 0.25 ppm, and a 2 hr exposure to ambient air. In contrast to other studies that indicated a suppression of T-lymphocyte mitogen-induced proliferation following O_3 exposure, no suppression of this parameter was demonstrable here.

Exposure to O_3 may impair NK activity in humans. Significant inhibition of human peripheral blood NK activity occurred following an *in vitro* exposure of these cells to O_3 (Harder et al., 1990). Significant suppression of NK activity occurred with as little as 0.18 ppm O_3 and became greater at 0.5 and 1.0 ppm O_3; suppression occurring with 1 ppm O_3 peaked at 6 hr.

Effect of Ozone on Alveolar Macrophages

The immunotoxic effect of O_3 is most dramatic on the phagocytic capacity of AM; as little as 2 hr exposure of rabbits to 0.1 ppm O_3 significantly suppressed AM phagocytosis of bacteria (Driscoll et al., 1987). This suppression of AM phagocytosis was demonstrated following both *in vivo* and *in vitro* exposure in CD-1 mice (Gilmour et al., 1991b), rats (Christman and Schwartz, 1982; Valentine, 1985; van Loveren et al., 1988; Oosting et al., 1991), rabbits (Huber et al., 1971; Driscoll et al., 1987), and humans (Becker et al., 1991a; Devlin et al., 1991a and b). When the effect of O_3 exposure on AM phagocytosis was compared between mice and humans, after correcting for dosimetric differences, the data demonstrated that human and murine AM respond similarly to O_3 exposure, both *in vivo* and *in vitro* (Selgrade et al., 1995).

Reduced phagocytosis can lead to an increased mortality to some infectious agents. A recent study used C57Bl/6J and C_3H/HeJ mice to demonstrate an association between O_3 exposure, reduced AM function, and increased mortality to *S. zooepidemicus* (Group C) (Gilmour et al., 1993a). In mice infected by bacterial aerosol after a 3 hr exposure to either 0.4 or 0.8 ppm O_3 or air, the C_3H/HeJ strain was more susceptible to bacterial infection (with respect to mortality) in all three exposure groups. In both strains of mice exposed to both O_3 concentrations, intrapulmonary killing of the bacteria was impaired. AM function was measured both *in vivo* and *in vitro*, without opsonins on the bacteria, and was found to correlate well with mortality. Further comparisons of the two strains of mice with rats receiving a similar O_3 exposure regimen revealed that inhaled streptococci were cleared normally from the mouse lung by AM, but not in the rat lung, where neutrophils were more important (Gilmour and Selgrade, 1993c).

Human AM complement receptor-mediated phagocytosis of serum-coated *Candida albicans* yeast was suppressed in human subjects exposed during moderate exercise for 6.6 hr to 0.08 or 0.1 ppm O_3 (Devlin et al., 1991a). However, no decreases in F_c receptor-mediated phagocytosis (using IgG-coated yeast) or non-specific receptor-mediated phagocytosis (using unopsonized yeast) were noted in AM isolated from subjects 18 hr after the exposure. When human AM were exposed *in vitro* to 1 ppm O_3, phagocytosis of opsonized *Cryptococcus neoformans* was not affected by O_3 exposure, nor was there a change in the number of AM membrane complement receptors required for phagocytosis of this yeast (Becker et al., 1991a). However, O_3 did induce a decrease in the phagocytosis of antibody-coated SRBC in spite of the lack of effect upon the quantity of F_c and complement receptors.

In addition to depressed phagocytic function, bactericidal mechanisms appear to be depressed following O_3 exposure. Early studies using rabbit AM showed that O_3 exposure (either *in vivo* to 0.5 - 3.0 ppm O_3 or *in vitro*) depressed the activity of intracellular hydrolytic enzymes lysozyme, acid phosphatase, and β-glucuronidase (Hurst et al., 1970; Alpert et al., 1971; Hurst and Coffin, 1971). A depression of cell size and lysozyme content in AM was noted in rats exposed to 0.64 ppm O_3 for 4 wk (Sherwood et al., 1986). Inducing a chronic pulmonary infection with *Pseudomonas aeruginosa* in these rats and then 10 d later exposing them to 0.64 ppm O_3 further reduced AM cell size and lysozyme content. Zymosan, phorbol myristate acetate (PMA), or mezerein-stimulated $\cdot O_2^-$ production was inhibited in mouse AM isolated from animals exposed to as little as 0.11 ppm O_3 for 3 hr suggesting that potentiation of fatal bacterial pneumonia in mice that inhaled O_3 could be due to inhibition of AM $\cdot O_2^-$ production (Ryer-Powder et al., 1988). The results confirmed an earlier observation of decreased $\cdot O_2^-$ formation by PMA-stimulated AM from rats exposed to higher O_3 concentrations and demonstrated that this inhibition could occur at ambient air O_3 levels (Amoruso et al., 1981).

Ozone may also alter AM functions important in tumor surveillance. AM functions were altered in rabbits exposed to 1 ppm for 2 hr/d for 3 d (Zelikoff et al., 1991). Not only did O_3 reduce cell viability of the AM immediately post-exposure, but *ex vivo* AM-mediated cytotoxicity toward mouse 3T12 tumor cell targets was significantly depressed. AM chemotactic migration was also enhanced. Production of $\cdot O_2^-$ in response to zymosan was reduced initially following exposure and then increased 24 hr following exposure. Ozone has also been shown to depress phorbol ester-induced $\cdot O_2^-$ production by human AM and to decrease their expression of the Ia-equivalent cell surface antigen, HLA-DR (Becker et al., 1991a).

In an *in vitro* model system for respiratory syncytial virus (RSV) infection in human AM, O_3 neither altered susceptibility to infection with RSV nor production of interleukin-1 (IL-1), IL-6, or TNF by infected AM (Soukup et al., 1993). However, at low multiplicities of infection, O_3 reduced IL-1 and IL-6 production by infected AM.

Effect of Ozone on Neutrophils

Ozone has also been shown to alter neutrophil or polymorphonuclear leukocyte (PMN) function and morphology. As little as 0.08 ppm O_3 results in an inflammatory response in the lung characterized by the influx of PMN in humans (Koren et al., 1989; Devlin et al., 1991a). This influx has also been demonstrated in many animal models of O_3 exposure and may play a role in the pathophysiology of O_3-induced lung injury (Bascom et al., 1996).

Neutrophil adhesion and motility are increased by exposure to O_3. Rats were exposed to 0.8 ppm O_3 for 2 hr, and circulating PMN were incubated with an epithelial cell line derived from rat lung (ARL-14) or with primary alveolar Type II cell cultures. Compared with the cells obtained from control animals, the PMN from O_3-exposed rats adhered to a greater number of epithelial cells, demonstrated increased motility, and exhibited spontaneous redistribution of actin filaments and cell surface modifications (Bhalla et al., 1993). Signals from O_3-exposed epithelial cells can also increase PMN adherence to epithelial cells. When human airway epithelial cells were exposed *in vitro* to 0.5 ppm O_3 for 2 hr, PMN adherence to them increased significantly after an 8 hr period to a maximum of 75% by 18 - 24 hr, compared with <5% adherence at 2 hr by control PMN (Tosi et al., 1994). PMN adherence was increased further by preactivation with the bacteria cell wall-derived chemoattractant fMLP (*N*-formyl-methionyl-leucyl-phenylalanine). When adhesion molecules, such as intracellular adhesion molecule-1 (ICAM-1), were examined, the adherence of PMN to epithelial cells was shown to be ICAM-1 independent and CD18- and CD11b (Mac-1)-dependent. Since ICAM-1 plays a role in adhesion of PMN to epithelial cells in respiratory virus infections and in endotoxin and cytokine exposure, the O_3 effect upon adherence appears to occur via a different mechanism.

The involvement of the effect of O_3 exposure upon PMN morphology and function in the pathophysiology of O_3 is still unclear. It is unlikely that alterations in PMN function described earlier contribute to the acute injury of pulmonary epithelial cells. When circulating PMN in rats were depleted with rabbit anti-rat PMN serum prior to an 8 hr exposure to 1 ppm O_3, the treatment resulted in total elimination of a PMN influx into the lungs following O_3 exposure but there was no difference in early epithelial cell damage between rats possessing circulating PMN and those rendered PMN-deficient. Using electron microscopy morphometry to assess epithelial cell damage, no significant differences were found in the volume per surface area epithelial cells in central acini between PMN(+) and PMN(-) animals (Pino et al., 1992). Although morphological changes in PMN were noted, PMN from O_3-exposed rats did not alter epithelial permeability (Bhalla et al., 1993). A study that measured spontaneous and phorbol ester-stimulated release of $\cdot O_2^-$, proteolysis of fibronectin, and the ability to damage epithelial cells (by measuring ^{51}Cr release from the A549 epithelial cell line) by cells from rats inhaling 0.2 - 0.8 ppm O_3 for 7 hr/d for 4 d also found that there was no increase in the ability of bronchoalveolar leukocytes to injure epithelial cells, even though leukocyte influx had occurred following exposure (Donaldson et al., 1993). There was some evidence of a depression of the oxidative burst in PMN, but it occurred only upon exposure to the 0.6 ppm concentration of O_3.

Although PMN are brought into the airways by O_3 exposure, their phagocytic function and killing of microorganisms appears to be impaired. A single exposure of human subjects to 0.4 ppm O_3 suppressed the capacity of PMN to phagocytize and kill *Staphylococcus epidermidis in vitro* 72 hr post-exposure (Peterson et al., 1978b); phagocytic function returned to normal only 2 wk later. When rats and mice were exposed to 0.4 or 0.8 ppm O_3 for 3 hr prior to bacterial challenge with *S. zooepidemicus*, PMN were observed in mice 2 or more days post-infection and did not alter the fatal infection. In rats, microbial inactivation was impaired in the O_3-exposed lungs during the first 48 hr after infection and disappearance of the bacteria corresponded with PMN influx into the lung (Gilmour and Selgrade, 1993c). Elimination of PMN with anti-neutrophil antiserum treatment *in vivo* resulted in greater numbers of bacteria in the rat lung following O_3 exposure, impairing the bactericidal activity in these rats even further. These studies illustrate the importance of PMN in clearance of bacterial infection in some species.

Effect of Ozone on Infectious Diseases Dependent on Innate Immune Responses

Epidemiologic studies over the years have suggested that air pollutants increase the frequency of respiratory illness (Bascom et al., 1996b). However, there are no experimental examples of the effect of O_3 on human host defense against bacterial infection other than an association between increases in air pollutant levels and the incidence of pneumonia and bronchitis (Bascom et al., 1996; Gilmour and Selgrade, 1998).

The immunosuppressive effects of O_3 are most effectively demonstrated in animal models of bacterial infection where host recovery depends upon the ability of phagocytic cells to engulf and kill the bacteria. These models have been shown to be among the most sensitive for demonstrating the pulmonary immunotoxicity of O_3. As little as 0.08 ppm O_3 was shown to suppress host defense in two bacterial infection studies (Gardner, 1982; van Loveren et al., 1988). The earliest studies on the immunotoxic effects of O_3 demonstrated a decreased resistance to these types of pulmonary infections in rodents (Miller and Ehrlich, 1959; Coffin and Gardner, 1972; Goldstein et al., 1974; Miller et al., 1978; Gardner, 1982) and were of three types (Jakab et al., 1995). One type utilized highly virulent strains of *Klebsiella pneumoniae* (Miller and Ehrlich, 1959) or *Streptococcus* sp. following various O_3 exposure regimens and monitored mortality of mice over a 14 d period.[2] The *Streptococcus* system was exquisitly sensitive in that mice given as little as one 3 hr 0.1 ppm O_3 exposure simultaneously or 2 hr before infection exhibited significant increases in mortality (Miller et al., 1978). Other studies utilized streptococcal pneumonia model to investigate interactions between O_3 and NO_2 (Ehrlich et al., 1977, 1979; Graham et al., 1987) or O_3, SO_2, and NH_4SO_4 (Aranyi et al., 1983) upon pulmonary antibacterial resistance. Another infection model used aerosols containing *Staphylococcus aureus*, a bacterium that is minimally virulent in mice and does not produce lung injury following infection, to monitor intrapulmonary killing (Goldstein et al., 1974). It was shown that exposure of mice to 0.1 - 1.0 ppm O_3 for 3 hr suppressed intrapulmonary killing of *S. aureus* in a dose-dependent manner if O_3 was provided immediately before the bacterial challenge (Huber and LaForce, 1970; Huber et al., 1977). A similar impairment of host defense against this organism occurred when mice were exposed to O_3 17 hr before or 4 hr after challenge (Goldstein et al., 1971; Warshauer et al., 1974). A third type of bacterial infection model employed the *L. monocytogenes* and *M. tuberculosis* models (described earlier), as resistance to these organisms depends upon non-specific macrophage defense systems in addition to specific cellular immunity (Thomas et al., 1981; van Loveren et al., 1988). In one study, a decrease in AM phagocytosis was noted when rats were infected with *L. monocytogenes* and then exposed to O_3 (van Loveren et al., 1988)

---

[2] Early studies labeled the *Streptococcus* sp. as *S. pyogenes* (Group C), but later it was reclassified as *S. zooepidemicus* based on its ability to ferment trehalose.

There have been a number of studies examining the effect of O_3 exposure on the inflammatory response and viral injury to the lung, and on the mortality of mice infected with influenza virus. Results were varied and appeared to depend upon the strain of mouse used, the virulence of the viral strain, and the exposure regimen of O_3 in relation to infection.

Infection of female CD-1 mice with influenza A/Hong Kong/68 virus after a daily 3 hr 1 ppm O_3 exposure regimen resulted in increased mortality and morbidity (evidenced by increased lung wet weights and both histopathologic and pulmonary function changes) to virus infection without an increase in viral titers in the lung (Selgrade et al., 1988). This result was only seen following a 2 d exposure to 1 ppm O_3 and the doubling effect of O_3 on mortality disappeared when the O_3 concentration was lowered to 0.5 ppm. No differences in mortality were noted when infection began following days 1, 3, 4 and 5 of exposure to 1 ppm O_3.

An increase in lung lesions in O_3-exposed virus-infected animals (compared with infected animals exposed to air) was noted 30 d after female Swiss mice were infected with influenza A/PR8/34 virus and continuously exposed to 0.5 ppm O_3 (Jakab and Bassett, 1990). In this study, O_3 potentiated the post-influenzal alveolitis and structural changes in the lung parenchyma - the residual lung damage that occurs following resolution of the viral infection. This O_3 exposure regimen had a protective effect early in the infection, mitigating by approximately 50% the acute virus-induced lung injury 1 - 15 d post-infection.

In a similar study, continuous exposure of female Swiss mice to 0.5 ppm O_3 (beginning 1 hr after infection) during the course of influenza A/PR8/34 infection reduced the severity of the disease as measured on day 9 post-infection by morphometric measurements of lung injury, lung wet/dry weight ratios, and serum albumin in BAL beginning 6 - 8 d following infection (Jakab and Hmieleski, 1988). Ozone exposure reduced the numbers of lavagable lymphocytes in BAL, T- and B-lymphocytes recovered from lung tissues, and, as mentioned above, serum antibody titers against the virus on day 8 post-infection.

Ozone also protected female Swiss-Webster mice infected with influenza A WSN strain. In mice that underwent a 0.5 ppm O_3 exposure for 2 wk before and 2 wk after infection, 2 wk before infection, or 2 wk after infection, mortality was significantly reduced and survival time prolonged as compared to that observed in virus-infected air-exposed counterparts (Wolcott et al., 1982). However, these effects were only apparent when O_3 was present during the course of infection.

In all four studies, there was no difference in peak viral titer between O_3- and air-exposed mice, indicating that the pathologic effects of O_3 were independent of a classic immunosuppressive effect that would increase the peak viral load and result in more damage to the lung. Instead, the decreased serum antibody titers and reduced numbers of T- and B-lymphocytes that were seen in mice receiving O_3 exposures after infection were hypothesized to be factors in the O_3-related reductions in disease severity (Jakab and Hmieleski, 1988). However, in the study by Wolcott et al. (1982) discussed above, disease severity appeared to be independent of the titers of pulmonary virus, pulmonary interferon, and serum-neutralizing antibody. Another factor that appeared to correlate with the outcome of the severity of infection was the distribution of viral antigen in the lung. Focal collection of viral antigen occurred following O_3 exposure in both studies and was hypothesized to also be a factor in the mechanisms by which O_3 reduced disease severity. An earlier study also showed no difference in lung viral titers when male CD-1 mice were infected with A2/Japan 305/57 influenza virus and subsequently exposed continuously to 0.9 ppm O_3 (Fairchild, 1977). However, here, nasal viral titers were lower in O_3-exposed animals than in air-exposed controls, suggesting that higher concentrations of O_3 may protect nasal epithelial cells against influenza growth. The effect that O_3 has on viral-induced inflammation and inflammatory mediators and the role these have on the pathogenic course of the disease requires further investigation.

Mechanisms of Ozone Immunotoxicity: The Role of Cytokines, Metabolites, and Mediators of Inflammation

The mechanism of O_3-induced immunotoxicity is probably mediated via a combination of direct effects of O_3 upon biomolecules associated with immune and epithelial cells and by inducing these cells to release cytokines and mediators of inflammation. Ozone is a potent oxidant and reacts with biomolecules, particularly those containing thiols, amines, or unsaturated carbon-carbon double bonds (Pryor, 1976). The resultant alteration by O_3 of enzymes involved in intermediary metabolism, glutathione recycling, xenobiotic metabolism, and antiproteinase function and its interactions with unsaturated lipids could lead to cellular injury by creating and supporting the presence of damaging free radicals or toxic intermediate products (e.g., hydrogen peroxide [H_2O_2] and aldehydes) (Bascom et al., 1996).

Ozone can generate arachidonic acid (AA) metabolites (and their byproducts) from membrane lipids of many cell types; these metabolites, in turn, could exhibit immunotoxic properties. In one study, AA degradation products formed in a cell-free environment from AA exposed to O_3, exhibited immunotoxic properties toward human peripheral blood cells *in vitro*, such as depressed T-lymphocyte proliferative responses to PHA or Con A, decreased NK lysis of K562 target cells, and increased polarization of PMN (Madden et al., 1993). In another study, O_3 caused bovine tracheal epithelial cells to release [^{3}H]-labeled products from epithelial cells that had been incubated with [^{3}H]-AA for 24 hr before O_3 exposure (Leikauf et al., 1988). As these products were shown to be cyclooxygenase and lipooxygenase products, such as PGE_2, $PGF_{2\alpha}$, 6-keto-$PGF_{1\alpha}$, and leukotriene B_4 (LTB_4), this indicated that O_3 could augment eicosanoid metabolism in airway epithelial cells. In humans exposed to 0.4 ppm O_3 for 2 hr, PGE_2 in BAL fluid increased 2-fold (Koren et al., 1989); a subsequent study showed that release of PGE_2 occurred early (within 1 hr) following O_3 exposure (Devlin et al., 1996). It was also shown that the human bronchial epithelial cell line BEAS-S6 released thromboxane B_2 (TxB_2), PGE_2, LTC_4, LTD_4, LTE_4, and 12-hydroxyheptadecatrienoic acid following O_3 exposure (McKinnon et al., 1993). Exposure to 0.1 - 1.2 ppm O_3 stimulated both *in vivo* and *in vitro* eicosanoid production (especially PGE_2) by rabbit (Driscoll et al., 1988a) and human AM (Becker et al., 1991a). It is likely these prostanoids play a role in O_3-induced inhibition of phagocytosis (Canning et al., 1991) as it was shown that when the cyclooxygenase inhibitor indomethacin (which inhibits PGE_2 production) was provided, suppression of AM phagocytosis was reduced and enhanced survival of mice infected with *Streptococcus zooepidemicus* occurred (Gilmour et al., 1993b).

Direct effects of O_3 upon immune cells have also been noted in studies where interaction of cells was required. In one study, O_3 exposure interfered with T- and B-lymphocyte rosetting *in vitro*, indicating that a modification of cell membrane receptors had occurred (Savino et al., 1978). In a second study, AM recovered from rats exposed to 0.05 - 0.1 ppm O_3 exhibited increased cell adherence to nylon fiber (Veninga and Evelyn, 1986). Finally, a series of studies indicated that following O_3 exposure, PMN adherence to epithelial cells was increased (Bhalla et al., 1993; Tosi et al., 1994). As some of these effects were noted *in vitro*, it should be appreciated that there may not be much of a direct effect from O_3 upon systemic immune cells. It is more likely that factors such as cytokines and mediators released from cells as a result of O_3 exposure play the greatest role in the systemic immunotoxicity of O_3.

The induction of proinflammatory cytokines and mediators of inflammation also play an important role in regulating many of the immunomodulatory effects of O_3. These, in turn, likely affect disease processes such as the pathogenesis and resultant mortality from infection, the development of atopic asthma, or the triggering of an asthmatic attack. When human subjects were exposed to 0.4 ppm O_3 for 2 hr, IL-6 appeared in the BAL as soon as 1 hr after exposure (Devlin et al., 1996). BAL IL-6 levels were also significantly increased in exercising men exposed for 6.6 hr to 0.08 ppm O_3 (Devlin et al., 1991a). When smokers and non-smoking O_3-responsive and -nonresponsive (with respect to pulmonary function) humans were exposed (during exercise) to 0.22 ppm O_3 for 4 hr, BAL IL-6 and IL-8 levels increased immediately after exposure; these increases subsequently correlated with a later (i.e., 18 hr post-exposure) increase in the numbers of PMN, lymphocytes, eosinophils, and mast cells in BAL and nasal lavage fluids (Torres et al., 1997). This study demonstrated that O_3-induced airway inflammation and cytokine release was independent of smoking status or airway responsiveness to O_3. These results also confirmed those of an earlier study that found no correlation between levels of O_3-induced symptoms, respiratory tract injury and inflammation, and the magnitude of decrements in the pulmonary function parameters FEV_1 and FVC (Balmes et al., 1996). In addition, granulocyte-macrophage colony stimulating factor (GM-CSF), IL-8, and fibronectin were induced by the 4 hr, 0.2 ppm O_3 exposure. Another study by this group demonstrated induction of an additional cytokine, transforming growth factor (TGF)-β, using a similar exposure of asthmatics and normal subjects (Scanell et al., 1996). Inflammatory parameters, including the levels of cytokines released, were increased in asthmatics compared with normal subjects receiving bronchoscopy following the O_3 exposure.

Upon exposure to O_3, several cells in the lung release cytokines that influence the inflammatory response. Epithelial cells play a role in the inflammatory response to O_3 by releasing mediators and cytokines that not only play a dynamic role in governing inflammatory cell interactions in the lung (Leikauf et al., 1995b), but also may influence development of T-lymphocytes to reflect a T_H2 response (Gilmour and Selgrade, 1998). Epithelial cells were shown to release IL-6, IL-8, the CXC chemokine GRO-γ (Devlin et al., 1997), macrophage inflammatory protein-2 (MIP-2), RANTES (Leikauf et al., 1995b), PGE_2 (Leikauf et al., 1988; McKinnon et al., 1992) and nitric oxide (NO') (Punjabi et al., 1994) following O_3 exposure. Conversely, O_3 exposure has been shown to inhibit IL-2 production (Becker et al., 1989) and interferon (IFN) production by tracheal epithelial cells (Ibrahim et al.,

1976). This inhibition may be mediated by PGE_2 as this prostanoid inhibits produc-tion of T_H1 lymphokines IL-2 and IFNγ but does not affect that of T_H2 IL-4 and IL-5 (Betz and Fox, 1991; Snijdewint et al., 1993; Hilkens et al., 1996). IL-6 production stimulates IL-4 formation, and NO· inhibits development of T_H1, but not T_H2, lymphocytes (Barnes and Liew, 1995). In addition, the responsiveness of AM to IFNγ may be compromised by O_3 exposure, thereby affecting the ability of AM to be activated to phagocytize and kill bacterial pathogens (Cohen et al., 1996).

In vitro exposure of the A549 human Type II epithelial-like cell line to 0.1 ppm O_3 induced DNA-binding activity of transcription factors NF-κB, NF-IL-6, and AP-1 (Jaspers et al., 1997). This increase in binding activity was found to be asso-ciated with an increase in IL-8 gene expression and protein release following O_3 exposure. Thus, this study suggests the possibility of a cellular cascade induced by O_3 which culminates in the recruitment of inflammatory cells into the airways and subsequent immunotoxic events. More investigation is needed to discern the precise mechanism by which epithelial cell-derived mediators influence both local mucosal and systemic O_3 immunotoxicities.

Since O_3 is an insoluble oxidizing gas, it penetrates deep into the lung, affecting not only the epithelial cells, but resident AM. Although O_3 causes a decrease in the phagocytizing and bactericidal abilities of the AM, it also activates AM to increase their metabolic activity (Mochitate and Miura, 1989), to form colonies in soft agar culture (Boorman et al., 1979), and to produce TNF, IL-1, IL-6, and IL-8. The induction and secretion of TNF, IL-1, IL-6, and IL-8 in human AM exposed to 0.1 - 1.0 ppm O_3 was demonstrated *in vitro* in a biphasic culture system (Arsalane et al., 1995). While only the lowest concentrations (0.1 - 0.2 ppm) induced TNF in this study, guinea pig AM demonstrated comparatively less sensitivity with regard to an O_3-induced release of TNF and IL-6 (i.e.,only 0.4 and 1 ppm O_3 induced these cytokines). Bovine AM also released TNF and a chemotactic factor for macrophages in response to a 2 hr *in vitro* exposure to 1.0 - 2.5 ppm O_3 (Polzer et al., 1994; Mosbach et al., 1996). These Investigators suggested that pulmonary surfactant may be able to modulate O_3-induced cytokine and mediator release by bovine AM. A study which examined both *in vivo* and *in vitro* AM release of TNF and IL-1 confirmed this inducible effect of O_3. Rats were exposed to 2 ppm for 3 hr and their AM were isolated for use in measuring *ex vivo* cytokine release over time (Pendino et al., 1994). It was observed that TNF was released early at 2 and 4 hr post-exposure and that IL-1 formation occurred 2 - 24 hr post-exposure. Immunohistochemical staining of lung sections revealed that *in situ* release of TNF, IL-1, and fibronectin had also occurred; this indicated that TNF and IL-1 may have a role in oxidant-induced pulmonary inflammation and injury. In rats exposed to 1 - 2 ppm O_3 for 3 hr, release of TNF and IL-1 also occurred in conjunction with increased NO· production and enhanced protein synthesis in the liver, suggesting an induction of an acute phase response (Laskin et al., 1994).

Further research revealed that AM also produced increased amounts of NO· and inducible nitric oxide synthase (iNOS) in response to *in vitro* treatments with LPS or IFNγ following O_3 exposure (Pendino et al., 1993). In this study, H_2O_2 produc-tion by AM also increased, but ·O_2^- production was decreased following inhalation of O_3. Ozone also increased NO· production and iNOS in Type II pneumocytes following this exposure (Punjabi et al., 1994). The increased iNOS in both cell types following O_3 inhalation was associated with activation of the nuclear transcription factor, NF-κB, suggesting that changes in NF-κB signaling activity may be responsible, in part, for the increased response of these cells to cytokines after O_3 inhalation (Laskin et al., 1998).

Chemotactic factors are also released by AM following exposure to O_3. At early timepoints (i.e., 2 - 24 hr post-exposure) following acute O_3 exposure, PMN arrive in the airways due to the presence of chemoattractive molecules produced by both AM and epithelial cells (Leikauf et al., 1995b). In an early study, rabbit AM-conditioned culture medium (following an *in vitro* O_3 exposure of as little as 0.3 ppm for 2 hr) contained chemotactic factors for PMN and monocytes (Driscoll and Schelsinger, 1988b) including MIP-2, which was associated with PMN chemoattraction in a C57Bl/6 mouse model of O_3 exposure (Driscoll et al., 1993). Further experiments showed that H_2O_2 could directly induce MIP-2 gene expression, suggesting that direct exposure of AM to an oxidant was sufficient to induce production of this chemokine and that MIP-2 was more important in attracting PMN into the airways than LTB_4 (which was observed only following *in vitro* exposure to ~4 ppm O_3, but never *in vivo*) (Leikauf et al., 1988, 1995b). In rats, an additional chemokine, cytokine-induced neutrophil chemoattractant (CINC), released following a 3 hr exposure to 1 ppm O_3 was associated with the neutrophilia that resulted from exposure (Koto et al., 1997). Both CINC and MIP-2 transcripts appeared in rats exposed for 6 hr to 1 ppm O_3 and were reduced only by incubation of recovered AM with *both* anti-TNF and anti-IL-1 antibodies (Ishii et al., 1997). These results suggested that O_3-induced IL-1 and TNF were not associated directly with PMN chemoattraction, but indirectly by stimulating MIP-2 and CINC chemokines. Chemokines governing recruitment of other inflammatory cells are also released from AM following *in vivo* O_3 exposure. Brown-Norway rat AM produced eotaxin following a 6 hr exposure to 1.2 ppm O_3, which caused migration of eosinophils *in vitro*; this result suggested that eotaxin may be involved in eosinophil recruitment *in vivo* following O_3 exposure (Ishii et al., 1998). Cells recovered from the lungs of C57BL/6 mice and Wistar rats exposed for 3 hr to 2 ppm O_3 produced monocyte chemoattractant protein-1 (MCP-1) which was shown to mediate the late recruitment of monocytes into the lung following O_3 exposure. NF-κB was involved in the MCP-1 gene expression that occurred 20 - 24 hr post-exposure (Zhao et al., 1998).

Other chemotactic factors besides those derived from AM may be responsible for the O_3-induced leukocyte recruitment. For example, levels of chemotactic serum complement fragment C3 also increased significantly following exposure of humans to 0.4 ppm O_3 for 2 hr (Koren et al., 1989). In addition, as noted earlier, epithelial cells make many chemotactic factors. Taken together, AM and epithelial cells are very important in producing cytokines and mediators that play a role in O_3-induced lung inflammation and injury.

SUMMARY

The immune system is one of the most sensitive targets for O_3 toxicity. Ozone, at levels that exist in ambient air today despite the gradual decline of the average ambient air concentration since 1988, affects the immune system both locally and systemically. These effects are summarized in Table 4.

TABLE 4. SUMMARY OF IMMUNOTOXIC EFFECTS OF O_3 AT AMBIENT AIR CONCENTRATIONS.

Endpoint	Exposure	Hosts
Enhanced allergic sensitization (Osebold et al. 1988)	0.13 ppm	mice
Enhanced bronchial responsiveness to allergens (Molfino et al. 1991)	0.10 ppm	asthmatics
Increased hospital admissions (White et al. 1994)	>0.11 ppm	asthmatics
In vitro suppression of NK cell activity (Harder et al. 1990)	0.18 ppm	human
Suppression of host defense against bacteria (Huber and LaForce 1970; Huber et al. 1977; Miller et al. 1978; Gardner 1982; van Loveren et al. 1988)	0.08 - 0.10 ppm	mice
In vivo suppression of AM phagocytosis of bacteria or yeast (Driscoll et al. 1987; Devlin et al. 1991a)	0.08 - 0.10 ppm	rabbits, humans
Influx of neutrophils into the lung (Koren et al. 1989; Devlin et al. 1991a)	0.08 ppm	humans
Induction of proinflammatory cytokines and mediators (see text for details) (Driscoll et al. 1988a; Becker et al. 1991a; Devlin et al. 1991a; Arsalane et al. 1995; Balmes et al. 1996; Scanell et al. 1996; Jaspers et al. 1997; Torres et al. 1997)	0.08 - 0.22 ppm	rabbits, humans

The most sensitive effects appear to be inhibition of bacterial phagocytosis, recruitment of inflammatory cells, and production of proinflammatory cytokines. These events could influence the development of T_H2 responses, chronic inflammation, and fibrosis in the lung and exacerbate diseases such as asthma and pneumonia. The precise mechanism of O_3-induced inflammation and its subsequent influence on immunotoxicity needs to be further investigated, especially concerning the O_3 dosing regimen and its effect upon these immune system endpoints.

REFERENCES

ACGIH: American Conference of Governmental and Industrial Hygienist. *1999 TLVs and BEIs.* Cincinnati, OH, 1999;53:76.

Alpert SM, Gardner DE, Hurst DJ, Lewis TR, Coffin DL. Effects of exposure to ozone on defensive mechanisms of the lung. J. Appl. Physiol., 1971;31:246-252.

Amoruso MA, Witz G, Goldstein BD. Decreased superoxide anion radical production by rat alveolar macrophages following inhalation of ozone or nitrogen dioxide. Life Sci., 1981;28:2215-2221.

Aranyi C, Vana SC, Thomas PT, Bradof JN, Fenters JD. Effects of subchronic exposure to a mixture of O_3, SO_2, and $(NH_4)_2SO_4$ on host defenses of mice. J. Toxicol. Env. Health, 1983;12:55-71.

Arsalane K, Gosset P, Vanhee D, Voisin C, Hamid Q, Tonnel AB, Wallaert B. Ozone stimulates synthesis of inflammatory cytokines by alveolar macrophages *in vitro*. Am. J. Respir. Cell Mol. Biol., 1995;13:60-68.

Atwal OS, Samagh S, Bhatnagar MK. A possible autoimmune parathyroiditis following ozone inhalation. II. A histopathologic, ultrastructural, and immunofluorescent study. Am. J. Pathol., 1975;79:53-68.

Atwal OS, Wilson T. Parathyroid gland changes following ozone inhalation. Arch. Environ. Health, 1974;28:91-100.

Balmes JR, Chen L, Scannell C, Tager I, Christian D, Hearne PQ, Kelly T, Aris RM. Ozone-induced decrements in FEV_1 and FVC do not correlate with measures of inflammation. Am. J. Respir. Crit. Care Med., 1996;153:904-909.

Barnes PJ, Liew FY . Nitric oxide and asthmatic inflammation. Immunol. Today, 1995;16:128-130.

Bascom R, Bromberg PA, Costa DL, Devlin R, Dockery DW, Frampton MW, Lambert W, Samet JM, Speizer FE, Utell M. Health effects of outdoor air pollution. Am. J. Respir. Crit. Care Med., 1996;153:3-50.

Becker S, Jordan RL, Orlando GS, Koren HS. *In vitro* ozone exposure inhibits mitogen-induced lymphocyte proliferation and IL-2 production. J. Toxicol. Environ. Health, 1989;26:469-483.

Becker S, Madden MC, Newman SL, Devlin RB, Koren HS. Modulation of human alveolar macrophage properties by ozone exposure *in vitro*. Toxicol. Appl. Pharmacol., 1991a;110:403-415.

Becker S, Quay J, Koren H. Effect of ozone on immunoglobulin production by human B-cells *in vitro*. J. Toxicol. Environ. Health, 1991b;34:353-366.

Bedi JF, Drechsler-Parks DM, Horvath SM. Duration of increased pulmonary function sensitivity to an initial ozone exposure. Am. Ind. Hyg. Assoc. J., 1985;46:731-734.

Betz M, Fox BS. Prostaglandin E_2 inhibits production of T_H1 lymphokines but not of T_H2 lymphokines. J. Immunol., 1991;146:108-113.

Bhalla DK, Rasmussen RE, Daniels DS. Adhesion and motility of polymorphonuclear leukocytes isolated from the blood of rats exposed to ozone: Potential biomarkers of toxicity. Toxicol. Appl. Pharmacol., 1993;123:177-186.

Biagini RE, Moorman WJ, Lewis TR, Bernstein IL. Ozone enhancement of platinum asthma in a primate model. Am. Rev. Respir. Dis., 1986;134:719-725.

Bleavins MR, Dziedzic D. An immunofluorescence study of T- and B- lymphocytes in ozone-induced pulmonary lesions in the mouse. Toxicol. Appl. Pharmacol,, 1990;105:93-102.

Boorman GA, Schwartz LW, Wilson FD. *In vitro* macrophage colony formation by free lung cells during pulmonary injury. J. Reticuloendothelial. Soc., 1979;26:867-872.

Buckley RD, Hackney JD, Clark K, Posin C. Ozone and human blood. Arch. Environ. Health, 1975;30:40-43.

Buell GC, Mueller PK. Potential crosslinking agents in lung tissue. Arch. Environ. Health, 1965;10:213-219.

Burleson GR, Keyes LL, Stutzman JD. Immunosuppression of pulmonary natural killer activity by exposure to ozone. Immunopharmacol. Immunotoxicol., 1989;11:715-735.

Campbell KI, Hilsenroth RH. Impaired resistance to toxin in toxoid-immunized mice exposed to ozone or nitrogen dioxide. Clin. Toxicol., 1976;9:943-954.

Canning BJ, Hmieleski RR, Spannhake EW, Jakab GJ. Ozone reduces murine alveolar and peritoneal macrophage phagocytosis: The role of prostanoids. Am. J. Physiol., 1991;261:L277-L282.

Chen X, Gavett SH, Wills-Karp M. CD4$^+$ T-lymphocyte modulation of ozone-induced murine pulmonary inflammation. Am. J. Respir. Cell Mol. Biol., 1995;12:396-403.

Christman CA, Schwartz LW. Enhanced phagocytosis by alveolar macrophages induced by short-term ozone insult. Environ. Res., 1982;28:241-250.

Coffin DL, Gardner DE. Interaction of biological agents and chemical air pollutants. Ann. Occup. Hyg., 1972;15:219-234.

Cohen MD, Zelikoff JT, Qu Q, Schlesinger RB. Effects of ozone upon macrophage-interferon interactions. Toxicology, 1996;114:243-252.

Devlin RB, McDonnell WF, Becker S, Madden MC, McGee MP, Perez R, Hatch G, House DE, Koren HS. Time-dependent changes of inflammatory mediators in the lungs of humans exposed to 0.4 ppm ozone for 2 hr: A comparison of mediators found in bronchoalveolar lavage fluid 1 and 18 hr after exposure. Toxicol. Appl. Pharmacol., 1996;138:176-185.

Devlin RB, McDonnell WF, Mann R, Becker S, House DE, Schreinemachers D, Koren HS . Exposure of humans to ambient levels of ozone for 6.6 hours causes cellular and biochemical changes in the lung. Am. J. Respir. Cell Mol. Biol., 1991a;4:72-81.

Devlin RB, McDonnell WF, Perez R, Becker S, House DE, Koren HS. Prolonged exposure of humans to ambient levels of ozone causes cellular and biochemical changes in the lung. Am. Rev. Respir. Dis., 1991b;140:72-81.

Devlin RB, McKinnon KP, Reed W, Carter JD, Quay JL. "In Vitro Ozone Exposure Induces Increased Expression of mRNAs Coding for IL-6 and α Chemokines, but Not β-Chemokines, in Human Airway Epithelial Cells." In Correlations Between In Vitro and In Vivo Investigations in Inhalation Toxicology, U Mohr, DL Dungworth, KB Adler, CC Harris, CG Plopper, eds. Washington, DC: ILSI Press, 1997, pp. 59-76.

Doherty PC. Cytotoxic T-cell effector and memory function in viral immunity. Curr. Top. Microbiol. Immunol., 1996;206:1-14.

Donaldson K, Brown GM, Brown DM, Slight J, Maclaren W, Davis JM. Characteristics of bronchoalveolar leukocytes from the lungs of rats inhaling 0.2-0.8 ppm of ozone. Inhal. Toxicol., 1993;5:149-164.

Dormans JA, Boere AJF, van Loveren H, Rombout PJ, Marra M, van Bree L . Age-related toxicity in rat lungs following acute and repeated ozone exposure. Inhal. Toxicol., 1996;8:903-925.

Driscoll KE, Vollmuth TA, Schlesinger RB . Acute and subchronic ozone inhalation in the rabbit: Response of alveolar macrophages. J. Toxicol. Environ. Health, 1987;21:27-43.

Driscoll KE, Leikauf GD, Schlesinger RB. Effects of in vitro and in vivo ozone exposure on eicosanoid production by rabbit alveolar macrophages. Inhal. Toxicol., 1988a;1:109-122.

Driscoll KE, Schlesinger RB . Alveolar macrophage-stimulated neutrophil and monocyte migration: Effects of in vitro ozone exposure. Toxicol. Appl. Pharmacol., 1988b;93:312-318.

Driscoll KE, Simpson L, Carter J, Hassenbein D, Leikauf GD . Ozone inhalation stimulates expression of a neutrophil chemotactic protein, macrophage inflammatory protein-2. Toxicol. Appl. Pharmacol., 1993;119:306-309.

Dziedzic D,White HJ. T-Cell activation in pulmonary lymph nodes of mice exposed to ozone. Environ. Res., 1986a;41:610-622.

Dziedzic D, White HJ. Thymus and pulmonary lymph node response to acute and subchronic ozone inhalation in the mouse. Environ. Res., 1986b;41:598-609.

Dziedzic D, White HJ. Response of T-cell deficient mice to ozone exposure. J. Toxicol. Env. Health, 1987;21:57-71.

Ehrlich R, Findlay JC, Fenters JD, Gardner DE. Health effects of short-term inhalation of nitrogen dioxide and ozone mixtures. Environ. Res., 1977;14:223-231.

Ehrlich R, Findlay JC, Gardner DE. Effects of repeated exposures to peak concentrations of nitrogen dioxide and ozone on resistance to streptococcal pneumonia. J. Toxicol. Environ. Health, 1979;5:631-642.

Elsaved NM, Kass R, Mustafa MG, Hacker AD, Ospital JJ, Chow CK, Cross CE. Effect of dietary Vitamin E level on the biochemical response of rat lung to ozone inhalation. Drug Nutr. Interact., 1988;5:373-386.

EPA US. *National Air Quality and Emissions Trends Report, 1997*. Research Triangle Park, NC:U.S. EPA, Office of Air Quality Planning and Standards, Emissions Monitoring and Analysis Division, Air Quality Trends Analysis Group, 1988.

Eskew ML, Scheuchenzuber WJ, Scholz RW, Reddy CC, Zarkower A. The effects of ozone inhalation on the immunological response of selenium and vitamin E-deprived rats. Environ. Res., 1986;40:274-284.

Fairchild GA. Effects of ozone and sulfur dioxide on virus growth in mice. Arch. Environ. Health, 1977;32:28-33.

Farrell BP, Kerr HD, Kulle TJ, Sauder LR, Young JL. Adaptation in human subjects to the effects of inhaled ozone after repeated exposures. Am. Rev. Respir. Dis., 1979;119:725-730.

Follinsbee LJ, Bedi JF, Horvath SM. Respiratory responses in humans repeatedly exposed to low concentrations of ozone. Am. Rev. Respir. Dis., 1980;121:431-439.

Foster WM, Wills-Karp M, Tankersley CG, Chen X, Paquette NC. Bloodborne markers in humans during multiday exposure to ozone. J. Appl. Physiol., 1996;81:794-800.

Fox SD, Adams WC, Brookes KA, Lasley BL. Enhanced response to ozone exposure during the follicular phase of the menstrual cycle. Environ. Health Perspect., 1993;101:242-244.

Fujimaki H, Ozawa M, Imai T, Shimizu F. Effect of short-term exposure to O_3 on antibody response in mice. Environ. Res., 1984;35:490-496.

Fujimaki H, Shiraishi F, Ashikawa T, Murakami M. Changes in delayed hypersensitivity reaction in mice exposed to O_3. Environ. Res., 1987;43:186-190.

Fujimaki H. Impairment of humoral immune responses in mice exposed to nitrogen dioxide and ozone mixtures. Environ. Res., 1989;48:211-217.

Gardner DE. The use of experimental airborne infections for monitoring altered host defenses. Environ. Health Perspect., 1982;43:99-107.

Garssen J, Nijkamp FP, van der Vliet H, van Loveren H. T-Cell-mediated induction of airway hypersensitivity in mice. Am. Rev. Respir. Dis., 1991;144:931-938.

Garssen J, Nijkamp FP, Wagenaar SS, Zwart AA, Askenase P, van Loveren H. Regulation of delayed-type hypersensitivity-like responses in the mouse lung, determined with histological procedures: Serotonin, T-cell suppressor inducer factor, and high antigen dose tolerance regulate the magnitude of T-cell-dependent inflammatory reactions. Immunology, 1989;68:51-58.

Garssen J, van Loveren H, van der Vliet H, Nijkamp FP. An isometric method to study respiratory smooth muscle responses in mice. J. Pharmacol. Meth., 1990;24:209-217.

Garssen J, van Loveren H, van der Vliet H, Bot H, Nijkamp F. T-Cell-mediated induction of airway hyperresponsiveness and altered lung functions in mice are independent of increased vascular permeability and mononuclear cell infiltration. Am. Rev. Respir. Dis., 1993;147:307-313.

Garssen J, Nijkamp FP, van Vugt E, van der Vliet H, van Loveren H. T-Cell-derived antigen binding molecules play a role in the induction of airway hyperresponsiveness. Am. J. Respir. Crit. Care Med., 1994;150:1528-1538.

Garssen J, van Bree L, van der Vliet H, van Loveren H. Ozone-induced impairment of pulmonary Type IV hypersensitivity and airway hyperresponsiveness in mice. Inhal. Toxicol., 1997;9:581-599.

Gershwin LJ, Osebold JW, Zee YC. Immunoglobulin E-containing cells in mouse lung following allergen inhalation and ozone exposure. Int. Arch. Allergy Appl. Immunol., 1981;65:266-277.

Gilmour MI, Jakab GJ. Modulation of immune function in mice exposed to 0.8 ppm ozone. Inhal. Toxicol., 1991a;3:293-308.

Gilmour MI, Hmieleski RR, Stafford EA, Jakab GJ. Suppression and recovery of the alveolar macrophage phagocytic system during continuous exposure to 0.5 ppm ozone. Exp. Lung Res., 1991b;17:547-558.

Gilmour MI, Park P, Selgrade MJK. Ozone-enhanced pulmonary infection with *Streptococcus zooepidemicus* in mice. Am. Rev. Respir. Dis., 1993a;147:753-760.

Gilmour MI, Park P, Doerfler D, Selgrade MJK. Factors that influence the suppression of pulmonary antibacterial defenses in mice exposed to ozone. Exp. Lung Res., 1993b;19:299-314.

Gilmour MI, Selgrade MJK. A comparison of the pulmonary defenses against *Streptococcal* infection in rats and mice following O_3 exposure: Differences in disease susceptibility and neutrophil recruitment. Toxicol Appl. Pharmacol., 1993c;123 211-218.

Gilmour MI, Selgrade MJK. "Modulation of T-Lymphocyte Responses by Air Pollutants." In *T-Lymphocyte Subpopulations in Immunotoxicology*, I Kimber, MJK Selgrade, eds. Chichester, England: John Wiley and Sons, 1998, pp. 253-272.

Goldstein E, Tyler WS, Hoeprich PD, Eagle C. Ozone and the antibacterial defense mechanisms of the murine lung. Arch. Intern. Med., 1971;127:1099-1104.

Goldstein B, Lai LY, Ross SR,Cuzzi-Spada R. Susceptibility of inbred mouse strains to ozone. Arch. Environ. Health, 1973;27:412-413.

Goldstein E, Lippert W, Warshauer D. Pulmonary alveolar macrophage: Defender against bacterial infection of the lung. J. Clin. Invest., 1974;54:519-528.

Goldstein E, Bartlema HC, van der Ploeg M, van Duijn P, van der Stap JG, Lippert W. Effect of ozone on lysosomal enzymes of alveolar macrophages engaged in phagocytosis and killing of inhaled *Staphylococcus aureus*. J. Infect. Dis., 1978;138:299-311.

Graham JA, Menzel DB, Miller FJ, Illing JW, Ward R, Gardner DE. "Influence of Ozone on Xenobiotic Metabolism." In *Advances in Modern Environmental Toxicology Volume 5*, SD Lee, MG Mustafa, MA Mehlman, eds. Princeton, NJ: Princeton Scientific, 1983, pp. 95-117.

Graham JA, Gardner DE, Blommer EJ, House DE, Menache MG, Miller FJ. Influence of exposure patterns of nitrogen dioxide and modifications by ozone on susceptibility to bacterial infectious disease in mice. J. Toxicol. Environ. Health, 1987;21:113-125.

Graham MB, Braciale VL, Braciale TJ. Influenza virus-specific CD4[+] T-helper type 2 T-lymphocytes do not promote recovery from experimental virus infection. J. Exp. Med., 1994;180:1273-1282.

Gunnison AF, Weideman PA, Sobo M, Koenig KL, Chen LC. Age-dependence of responses to acute ozone exposure in rats. Fundam. Appl. Toxicol., 1992;18:360-369.

Gunnison AF, Hatch GE, Crissman K, Bowers A. Comparative sensitivity of lactating and virgin female rats to ozone-induced pulmonary inflammation. Inhal. Toxicol., 1996;8:607-623.

Hackney JD, Linn WS, Mohler JG, Collier CR. Adaptation to short-term respiratory effects of ozone in men exposed repeatedly. J. Appl. Physiol., 1977;43:82-85.

Hamburger SJ, Goldstein BD. Effect of ozone on agglutination of erythrocytes by concanavalin A. I. Studies in rats. Environ. Res., 1979;19:292-298.

Harder SD, Harris DT, House D, Koren HS. Inhibition of human natural killer cell activity following *in vitro* exposure to ozone. Inhal. Toxicol., 1990;2:161-173.

Hatch GE, Slade R, Stead AG, Graham JA. Species comparison of acute inhalation toxicity of ozone and phosgene. J. Toxicol. Environ. Health, 1986;19:43-53.

Henderson FW, Dubovi EJ, Harder S, Seal E, Graham D. Experimental rhinovirus infection in human volunteers exposed to ozone. Am. Rev. Respir. Dis., 1988;137:1124-1128.

Herberman R. Natural killer cells. Ann. Rev. Med., 1986;37:347-352.

Hilkens CM, Snijders A, Vermeulen H, van der Meide PH, Wierenga EA, Kapsenberg MI. Accessory cell-derived IL-12 and prostaglandin E_2 determine the IFN-γ level of activated human $CD4^+$ T-cells. J. Immunol. 1996;156:1722-1727.

Huber GL, LaForce FM. Comparative effects of ozone and oxygen on pulmonary antibacterial defense mechanisms. Antimicrob. Agents Chemother., 1970;10:129-136.

Huber GL, Mason RJ, LaForce M, Spencer NJ, Gardner DE, Coffin DL. Alterations in the lung following the administration of ozone. Arch. Intern. Med. 1971;128:81-87.

Huber GL, LaForce FM, Johanson WG. "Experimental Models and Pulmonary Antimicrobial Defenses." In *Respiratory Defense Mechanisms, Part II*, JD Brain, DF Proctor, LM Reid, eds. New York: Marcel Dekker, 1977, pp. 983-1022.

Hurst DJ, Gardner DE, Coffin DL. Effect of ozone on acid hydrolases of the pulmonary alveolar macrophage. J. Reticuloendothel. Soc., 1970;8:288-300.

Hurst DJ, Coffin DL. Ozone effect on lysosomal hydrolases of alveolar macrophages *in vitro*. Arch. Int. Med., 1971;127:1059-1063.

Ibrahim AL, Zee YC, Osebold JW. The effects of ozone on the respiratory epithelium and alveolar macrophages of mice. I. Interferon production. Proc. Soc. Exp. Biol. Med., 1976;152:483-488.

Ishii Y, Yang H, Sakamoto T, Nomura A, Hasegawa S, Hirata F, Bassett DJ. Rat alveolar macrophage cytokine production and regulation of neutrophil recruitment following acute ozone exposure. Toxicol. Appl. Pharmacol., 1997;147:214-223.

Ishii Y, Shirato M, Nomura A, Sakamoto T, Uchida Y, Ohtsuka M, Sagai M, Hasegawa S. Cloning of rat eotaxin: Ozone inhalation increases mRNA and protein expression in lungs of Brown-Norway rats. Am. J. Physiol., 1998;274:L171-L176.

Ishizaki T, Koizumi K, Ikemori R, Ishiyama Y, Kushibiki E. Studies of prevalence of Japanese cedar pollinosis among the residents in a densely cultivated area. Ann. Allergy, 1987;58:265-270.

Jakab GJ, Hmieleski RR. Reduction of influenza virus pathogenesis by exposure to 0.5 ppm ozone. J. Toxicol. Environ. Health, 1988;23:455-472.

Jakab GJ, Bassett DJ. Influenza virus infection, ozone exposure, and fibrogenesis. Am. Rev. Respir. Dis., 1990;141:1307-1315.

Jakab GJ, Hemenway DR. Concomitant exposure to carbon black particulates enhances ozone-induced lung inflammation and suppression of alveolar macrophage phagocytosis. J. Toxicol. Env. Health, 1994;41:221-231.

Jakab GJ, Spannhake EW, Canning BJ, Kleeberger SR, Gilmour MI. The effects of ozone on immune function. Environ. Health Perspect., 1995;103(Suppl. 2):77-89.

Jaspers I, Flescher E, Chen LC. Ozone-induced IL-8 expression and transcription factor binding in respiratory epithelial cells. Am. J. Physiol., 1997;272 (Lung Cell. Mol. Physiol. 16):L504-L511.

Kimura A, Goldstein E. Effect of ozone on concentrations of lysozyme in phagocytizing alveolar macrophages. J. Infect. Dis., 1981;143:247-251.

Kleeberger SR, Bassett DJP, Jakab G, Levitt RC. A genetic model for evaluation of susceptibility to ozone-induced inflammation. Am. J. Physiol., 1990;258(Lung Cell Mol. Physiol. 2):L313-L320.

Kleeberger SR, Levitt RC, Zhang L-Y. Susceptibility to ozone-induced inflammation. I. Genetic control of the response to subacute exposure. Am. J. Physiol., 1993a;264:L15-L20.

Kleeberger SR, Levitt RC, Zhang L-Y. Susceptibility to ozone-induced inflammation. II. Separate loci control responses to acute and subacute exposures. Am. J. Physiol., 1993b;264:L21-L26.

Kleeberger SR, Levitt RC, Zhang L-Y, Longphre M, Harkema J, Jedlicka A, Eleff SM, DiSilvestre D, Holroyd KJ. Linkage analysis of susceptibility to ozone-induced lung inflammation in inbred mice. Nature Genetics, 1997;17:475-478.

Kleinman MT, Bhalla DK, Ziegler B, Bucher-Evans S, McClure T. Effects of inhaled fine particles and ozone on pulmonary macrophages and epithelia. Inhal. Toxicol., 1993;5:371-388.

Koren HS, Devlin RB, Graham DE, Mann R, McGee MP, Horstman DH, Kozumbo WJ, Becker S, House DE, McDonnell WF, Bromberg PA. Ozone-induced inflammation in the lower airways of human subjects. Am. Rev. Respir. Dis., 1989;139:407-415.

Koto H, Salmon M, Haddad EB, Huang TJ, Zagorski J, Chung KF. Role of cytokine-induced neutrophil chemoattractant (CINC) in ozone-induced airway inflammation and hyperresponsiveness. Am. J. Respir. Crit. Care Med., 1997;156:234-239.

Laskin JD, Pendino KJ, Punjabi CJ, Rodriguez M, Laskin DL. Pulmonary and hepatic effects of inhaled ozone in rats. Environ. Health Perspect., 1994;102:61-64.

Laskin DL, Sunil V, Guo Y, Heck DE, Laskin JD. Increased nitric oxide synthase in the lung after ozone inhalation is associated with activation of NF-κB. Environ. Health Perspect., 1998;106:1175-1178.

Leikauf GD, Driscoll KE, Wey HE. Ozone-induced augmentation of eicosanoid metabolism in epithelial cells from bovine trachea. Am. Rev. Respir. Dis., 1988;137:435-442.

Leikauf GD, Simpson LG, Santrock J, Zhao Q, Abbinante-Nissen J, Zhou S, Driscoll KE. Airway epithelial cell responses to ozone injury. Environ. Health Perspect., 1995b;103 (Suppl. 2):91-95.

Li AF, Richters A. Ambient level ozone effects on subpopulations of thymocytes and spleen T-lymphocytes. Arch. Environ. Health, 1991a;46:57-63.

Li AF, Richters A. Effects of 0.7 ppm ozone exposure on thymocytes: *In vivo* and *in vitro* studies. Inhal. Toxicol., 1991b;3:61-71.

Lippman M. Health effects of ozone. JAPCA, 1989;39:672-695.

Liu L, Koutrakis P, Suh HH, Mulik JD, Burton RM. Use of personal measurements for ozone exposure assessment: A pilot study. Environ. Health Perspect., 1993;101:318-324.

Madden MC, Friedman M, Hanley N, Siegler E, Quay J, Becker S, Devlin R, Koren HS. Chemical nature and immunotoxicological properties of arachidonic acid degradation products formed by exposure to ozone. Environ. Health Perspect., 1993;101:154-164.

Matsumura Y. The effects of ozone, nitrogen dioxide, and sulfur dioxide on the experimentally-induced allergic respiratory disorder in guinea pigs. I. The effect on sensitization with albumin through the airway. Am. Rev. Respir. Dis., 1970a;102:430-437.

Matsumura Y. The effects of ozone, nitrogen dioxide, and sulfur dioxide on the experimentally-induced allergic respiratory disorder in guinea pigs. II. The effects of ozone on the absorption and the retention of antigen in the lung. Am. Rev. Respir. Dis., 1970b;102:438-443.

Matsumura Y. The effects of ozone, nitrogen dioxide, and sulfur dioxide on the experimentally-induced allergic respiratory disorder in guinea pigs. III. The effect on the occurrence of dyspneic attacks. Am. Rev. Respir. Dis., 1970c;102:444-447.

McKinnon K, Noah T, Madden M, Koren H, Devlin R. Cultured human bronchial epithelial cells release cytokines, fibronectin, and lipids in response to ozone exposure. Chest, 1992;101:22S.

McKinnon KP, Madden MC, Noah TL, Devlin RB. *In vitro* ozone exposure increases release of arachidonic acid products from a human bronchial epithelial cell line. Toxicol. Appl. Pharmacol, 1993;118:215-223.

Mehlman MA, Borek C. Toxicity and biochemical mechanisms of ozone. Environ. Res., 1987;42:36-53.

Menzel DB. The toxicity of air pollution in experimental animals and humans: The role of oxidative stress. Toxicol. Lett., 1994;72:269-277.

Miller FJ, Illing JW, Gardner DE. Effect of urban ozone levels on laboratory-induced respiratory infections. Toxicol. Lett., 1978;2:163-169.

Miller S, Ehrlich R. Susceptibility to respiratory infection of animals exposed to ozone. I. Susceptibility to *Klebsiella pneumoniae*. J. Infect. Dis., 1959;103:145-149.

Mochitate K, Miura T. Metabolic enhancement and increase of alveolar macrophages induced by ozone. Environ. Res., 1989;49:79-92.

Molfino NA, Wright SC, Katz I, Tarlo S, Silverman F, McClean PA, Szalai JP, Raizenne M, Slutsky AS, Zamel N. Effect of low concentrations of ozone on inhaled allergen responses in asthmatic subjects. Lancet, 1991;338:199-203.

Mosbach M, Wiener-Schmuck M, Seidel A. Influence of coexposure of ozone with quartz, latex, albumin, and LPS on TNF-α and chemotactic factor release by bovine alveolar macrophages *in vitro*. Inhal. Toxicol., 1996;8:625-638.

Mosmann TR, Sad S. The expanding universe of T-cell subsets: T_H1, T_H2, and more. Immunol. Today, 1996;17:138-146.

Oosting RS, van Golde LM, van Bree L. Species differences in impairment and recovery of alveolar macrophage function following single and repeated ozone exposures. Toxicol. Appl. Pharmacol., 1991;110:170-178.

Osebold JW, Owens SL, Zee YC, Dotson WM, LaBarre DD. Immunological alterations in the lungs of mice following ozone exposure: Changes in immunoglobulin levels and antibody-containing cells. Arch. Environ. Health, 1979;34:258-265.

Osebold JW, Gershwin LJ, Zee YC. Studies on the enhancement of allergic lung sensitization by inhalation of ozone and sulfuric acid aerosol. J. Environ. Pathol. Toxicol., 1980;3:221-234.

Osebold JW, Zee YC, Gershwin LJ. Enhancement of allergic lung sensitization in mice by ozone inhalation. Proc. Soc. Exp. Biol. Med., 1988;188:259-264.

Ozawa M, Fujimaki H, Itami T, Honda Y, Watanabe N. Suppression of IgE antibody production after exposure to ozone in mice. Int. Arch. Allergy Appl. Immunol., 1985;76:16-19.

Paquette NC, Zhang L-Y, Ellis WA, Scott AL, Kleeberger SR. Vitamin A deficiency enhances ozone-induced lung injury. Am. J. Physiol., 1996;270:L475-L482.

Pendino KJ, Laskin JD, Shuler RL, Punjabi CJ, Laskin DL. Enhanced production of nitric oxide by rat alveolar macrophages after inhalation of a pulmonary irritant is associated with increased expression of nitric oxide synthase. J. Immunol., 1993;151:7196-7205.

Pendino KJ, Shuler RL, Laskin JD, Laskin DL. Enhanced production of interleukin-1, tumor necrosis factor-α, and fibronectin by rat lung phagocytes following inhalation of a pulmonary irritant. Am. J. Respir. Cell Mol. Biol., 1994;11:279-286.

Pendino KJ, Meidhof TM, Heck DE, Laskin JD, Laskin DL. Inhibition of macrophages with gadolinium chloride abrogates ozone-induced pulmonary injury and inflammatory mediator production. Am. J. Respir. Cell Mol. Biol., 1995;13:125-132.

Pendino KJ, Gardner CR, Quinones S, Laskin DL. Stimulation of nitric oxide production in rat lung lavage cells by anti-Mac-1b antibody: Effect of ozone inhalation. Am. J. Respir. Cell Mol. Biol., 1996;14:327-333.

Peterson ML, Rommo N, House D, Harder S. *In vitro* responsiveness of lymphocytes to phytohemagglutinin. Arch. Env. Health, 1978a;33:59-63.

Peterson ML, Harder S, Rummo N, House D. Effect of ozone on leukocyte function in exposed human subjects. Environ. Res., 1978b;15:485-493.

Pino MV, Stovall MY, Levin JR, Devlin RB, Koren HS, Hyde DS. Acute ozone-induced lung injury in neutrophil-depleted rats. Toxicol. Appl. Pharmacol., 1992;114:268-276.

Polzer G, Lind I, Mosbach M, Schmidt A, Seidel A. Combined influence of quartz dust, ozone, and NO2 on chemotactic mobility, release of chemotactic factors and other cytokines by macrophages *in vitro*. Toxicol. Lett., 1994;72:307-315.

Pryor WC. Free radical reactions in biology: Initiation of lipid autooxidation by ozone and nitrogen dioxide. Environ. Health Perspect., 1976;16:180-181.

Punjabi CJ, Laskin JD, Pendino KJ, Goller NL, Durham SK, Laskin DL. Production of nitric oxide by rat Type II pneumocytes: Increased expression of inducible nitric oxide synthase following inhalation of a pulmonary irritant. Am. J. Respir. Cell Mol. Biol., 1994;11:165-172.

Rietjens IM, Poelen MC, Hempenius RA, Gijbels MJ, Alink GM. Toxicity of ozone and nitrogen dioxide to alveolar macrophages: Comparative study revealing differences in their mechanism of toxic action. J. Toxicol. Environ. Health, 1986;19:555-568.

Ryer-Powder JE, Amoruso MA, Czerniecki B, Witz G, Goldstein B. Inhalation of ozone produces a decrease in superoxide anion radical production in mouse alveolar macrophages. Am. Rev. Respir. Dis., 1988;138:1129-1133.

Savino A, Peterson ML, House D, Turner AG, Jeffries HE, Baker R. The effect of ozone on human cellular and humoral immunity: Characterization of T- and B-lymphocytes by rosette formation. Environ. Res., 1978;15:65-69.

Scanell C, Chen L, Aris RM, Tager I, Christian D, Ferrando R, Welch B, Kelly T, Balmes JR. Greater ozone-induced inflammatory responses in subjects with asthma. Am. J. Respir. Crit. Care Med., 1996;154:24-29.

Scheel LD, Dobrogorski OJ, Mountain JT, Svirbely JL, Stokinger HE. Physiologic, biochemical, immunologic, and pathologic changes following ozone exposure. J. Appl. Physiol., 1959;14:67-80.

Schwela D. Exposure to environmental chemicals relevant for respiratory hypersensitivity: Global aspects. Toxicol. Lett., 1996;86:131-142.

Selgrade MJK, Illing JW, Starnes DM, Stead AG, Menache MG, Stevens MA. Evaluation of effects of ozone exposure on influenza infection in mice using several indicators of susceptibility. Fundam. Appl. Toxicol., 1988;11:169-180.

Selgrade MJK, Daniels MJ, Grose EC. Acute, subchronic, and chronic exposure to a simulated urban profile of ozone: Effects on extrapulmonary natural killer cell activity and lymphocyte mitogenic responses. Inhal. Toxicol., 1990;2:375-389.

Selgrade MJK, Cooper KD, Devlin RB, van Loveren H, Biagini RE, Luster MI. Immunotoxicity - Bridging the gap between animal research and human health effects. Fundam. Appl. Toxicol., 1995;24:13-21.

Sherwood RL, Lippert WE, Goldstein E. Effect of 0.64 ppm ozone on alveolar macrophage lysozyme levels with rats with chronic pulmonary bacterial infection. Environ. Res., 1986;41:378-387.

Sivertsen B, Clench-Aas J. Exposure to environmental chemicals relevant for respiratory hypersensitivity: European aspects. Toxicol. Lett., 1996;86:143-153.

Slade R, Stead AG, Graham JA, Hatch GE. Comparison of antioxidant levels in humans and laboratory animals. Am. Rev. Respir. Dis., 1985;131:742-746.

Slade R, Highfill JW, Hatch GE. Effects of depletion of ascorbic acid or nonprotein sulfhydryls on the acute inhalation toxicity of nitrogen dioxide, ozone, and phosgene. Inhal. Toxicol., 1989;1:261-271.

Snijdewint FGM, Kalinski P, Wierenga EA, Bos JD, Kapsenberg ML. Prostaglandin E_2 differentially modulates cytokine secretion profiles of human T-helper lymphocytes. J. Immunol., 1993;150:5321-5329.

Soukup J, Koren HS, Becker S. Ozone effect on respiratory syncytial virus infectivity and cytokine production by human alveolar macrophages. Environ. Res., 1993;60:178-186.

Stiles J, Tyler WS. Age-related morphometric differences in responses of rat lungs to ozone. Toxicol. Appl. Pharmacol., 1988;92:274-285.

Stokinger HE. Ozone toxicology: A review of research and industrial experience, 1954-1964. Arch. Environ. Health, 1965;10:819-831.

Stokinger HE, Scheel LD. Ozone toxicity: Immunochemical and tolerance-producing aspects. Arch. Environ. Health, 1962;4:327-334.

Stokinger HE, Wagner WD, Wright PG. Potentiating effects of exercise and tolerance development. AMA Arch. Ind. Health, 1956;14:158-162.

Tepper JS, Costa DL, Lehmann JR, Weber MF, Hatch GE. Unattenuated structural and biochemical alteration in the rat lung during functional adaptation to ozone. Am. Rev. Respir. Dis., 1989;140:493-501.

Thienes CH, Skillen RG, Hoyt A, Bogen E. Effects of ozone on experimental tuberculosis and on natural pulmonary infections in mice. Am. Ind. Hyg. Assoc. J., 1965;26:255-260.

Thomas GB, Fenters JD, Ehrlich R, Gardner DE. Effects of exposure to ozone on susceptibility to experimental tuberculosis. Toxicol. Lett., 1981;9:11-17.

Torres A, Utell MJ, Morow PE, Voter KZ, Whitin JC, Cox C, Looney RJ, Speers DM, Tsai Y, Frampton MW. Airway inflammation in smokers and nonsmokers with varying responsiveness to ozone. Am. J. Respir. Crit. Care Med., 1997;156:728-736.

Tosi MF, Hamedani A, Brosovich J, Alpert SE. ICAM-1-independent, CD18-dependent adhesion between neutrophils and human airway epithelial cells exposed *in vitro* to ozone. J. Immunol., 1994;152:1935-1942.

Ulrich L, Malik E, Hurbankova M, Kemka R. The effect of low-level ozone concentrations on the serum levels of immunoglobulins, α-1-antitrypsin and transferrin and on the activation of peripheral lymphocytes. J. Hyg. Epidemiol. Microbiol., 1980;24:303-308.

Valentine R. An *in vitro* system for exposure of lung cells to gases: Effects of ozone on rat macrophages. J. Toxicol. Environ. Health, 1985;16:115-126.

van Bree L, Marra M, Rombout PJA. Differences in pulmonary biochemical and inflammatory responses of rats and guinea pigs resulting from daytime or night-time, single and repeated exposure to ozone. Toxicol. Appl. Pharmacol., 1992;116:209-216.

van Loveren H, Rombout PJA, Wagenaar SS, Walvoort HC, Vos JG. Effects of ozone on the defense to a respiratory *Listeria monocytogenes* infection in the rat. Suppression of macrophage function and cellular immunity and aggravation of histopathology in lung and liver during infection. Toxicol. Appl. Pharmacol., 1988;94:374-393.

van Loveren H, Krajnc EI, Rombout PJA, Blommaert FA, Vos JG. Effects of ozone, hexachlorobenzene, and bis(tri-*n*-butyltin)oxide on natural killer activity in the rat lung. Toxicol. Appl. Pharmacol., 1990;102:21-33.

Veninga TS, Evelyn P. Activity changes of pulmonary macrophages after *in vivo* exposure to ozone as demonstrated by cell adherence. J. Toxicol. Environ. Health, 1986;18:483-489.

Warshauer D, Goldstein E, Hoeprich PD, Lippert W. Effect of vitamin E and ozone on the pulmonary antibacterial defense mechanisms. J. Lab. Clin. Med., 1974;83:228-240.

Weideman PA, Chen LC, Gunnison AF. Enhanced pulmonary inflammatory response to ozone during lactation in rats: Evaluation of the influence of inhaled dose. Inhal. Toxicol., 1996;8:495-519.

White MC, Etzel RA, Wilcox WD, Lloyd C. Exacerbations of childhood asthma and ozone pollution in Atlanta. Environ. Res., 1994;65:56-68.

Wolcott JA, Zee YC, Osebold JW. Exposure to ozone reduces influenza disease severity and alters distribution of influenza viral antigens in murine lungs. Appl. Environ. Microbiol., 1982;44:723-731.

Yanai M, Aikawa OT, Okayama H, Sekizawa K, Maeyama K, Sasaki H, Takishima T. Ozone increases susceptibility to antigen inhalation in allergic dogs. J. Appl. Physiol., 1990;68:2267-2273.

Zelikoff JT, Kraemer GL, Vogel MC, Schlesinger RB. Immunomodulating effects of ozone on macrophage functions important for tumor surveillance and host defense. J. Toxicol. Environ. Health, 1991;34:449-467.

Zhang LY, Levitt RC, Kleeberger SR. Differential susceptibility to ozone-induced airways hyperreactivity in inbred strains of mice. Exp. Lung Res., 1995;21:503-518.

Zhao Q, Simpson LG, Driscoll KE, Leikauf GD. Chemokine regulation of ozone-induced neutrophil and monocyte inflammation. Am. J. Physiol., 1998;274:L39-L46.

13. SULFUR AND NITROGEN OXIDES

Richard B. Schlesinger, Ph.D, Lung-Chi Chen, Ph.D.,
and Judith T. Zelikoff, Ph.D.
New York University School of Medicine
Department of Environmental Medicine
57 Old Forge Road
Tuxedo, New York 10987

SULFUR OXIDES

Sulfur oxides comprise both gaseous and particulate chemical species. There are four of the former, namely sulfur monoxide, sulfur dioxide, sulfur trioxide and disulfur monoxide. The particulate phase sulfur oxides consist of strongly-to-weakly acidic sulfates, namely sulfuric acid (H_2SO_4) and its products of neutralization with ammonia: letovicite [$(NH_4)_3H(SO_4)_2$], ammonium bisulfate (NH_4HSO_4), and ammonium sulfate [$(NH_4)_2SO_4$]. Most of the toxicologic database for sulfur oxides involves sulfur dioxide (SO_2) and H_2SO_4.

Sources

Sulfur dioxide derives from a number of natural and anthropogenic sources. The single largest natural source is volcanic eruptions, but it is also produced via oxidation of reduced sulfur species released during the biological degradation of organic material containing sulfur compounds, or by sulfate (SO_4^{-2}) reduction under anaerobic conditions. Most anthropogenic SO_2 emissions result from the combustion of sulfur-containing fossil fuels, especially coal and oil. At high temperatures, organic sulfur compounds within the fuel degrade to elemental sulfur or to reduced inorganic sulfur species and these, in turn, are converted to SO_2. A small fraction of the SO_2 so produced may be oxidized to sulfur trioxide (SO_3), which combines with water vapor in the plume, producing ultrafine (<0.1 μm diameter) sulfuric acid particles which are emitted from the emission source as pure droplets, or as a surface coating on mineral ash particles. Another significant source of outdoor SO_2 is industrial processes, such as smelters using non-ferrous sulfide ores, oil refineries and cement manufacturing plants; mobile sources account for a small amount of SO_2 emissions.

Atmospheric transformation of SO_2 to SO_4^{-2} is the major source of particulate phase acidity, with about 20% of the emitted SO_2 undergoing such conversion during its atmospheric lifetime. This "secondarily"-produced acid aerosol generally does not remain in the atmosphere for extended periods of time, since it will become partially or completely neutralized by atmospheric ammonia (NH_3). Because rates of neutralization vary widely, depending upon such factors as NH_3 levels and meteorological conditions, ratios of SO_2 to SO_4^{-2} particles are quite variable, both temporally and spatially. Pure H_2SO_4, often used as the model for acidic sulfates in toxicological studies, rarely exists in ambient outdoor air, and the acidic sulfate component usually consists of NH_4HSO_4, or a mixture of H_2SO_4 with partially-to-completely neutralized sulfates.

In the indoor home environment, a potential major source of sulfur oxides is unvented or improperly vented kerosene space heaters using sulfur-containing fuels. However, more significant indoor exposures to sulfur oxides generally occur in occupational environments, in such diverse settings as petroleum refining, smelting, paper manufacturing, chemical plants, wineries, paper pulp mills, and in the processing and preservation of fruit. Sulfuric acid is the most commonly used industrial chemical and the highest volume chemical produced in the United States, and one to which hundreds of thousands of workers have potential exposure. It is used in such industries as phosphate fertilizer production, iron pickling, petroleum refining, and various manufacturing processes, especially storage batteries.

Exposure Levels

The major environmental exposures for sulfur oxides are regional air pollution episodes, local plume impacts, and acid fogs. Such episodes generally occur in the summertime, and are characterized by high sulfur oxide concentrations occurring over periods ranging from hours to days, often in association with oxidant pollutants, such as ozone. Local plume impacts involve exposures all year and occur in areas downwind of strong point sources, such as power plants or smelters. Acid fogs are associated with cool temperatures, and result from the condensation of water vapor on preexisting fine particles when the atmosphere is saturated.

Current ambient outdoor air levels of SO_2 in the United States are fairly low; the annual average concentration is around 0.021 mg/m^3 (0.008 ppm) (Council on Environmental Quality, 1992). However, short-term peaks (e.g., 10 min averages) in excess of 1.3 mg/m^3 (0.5 ppm) can occur downwind of strong point sources (USEPA, 1982).

There are few data on outdoor concentrations of acidic sulfates. Measurements conducted over limited time periods and generally during only one season report peaks (as H_2SO_4) of 0.02 - 0.04 mg/m^3 occurring over periods of 1 - 12 hr, but annual averages are generally <0.005 mg/m^3 (Lioy and Waldman, 1989).

Indoor exposure levels of SO_2 in homes with kerosene space heaters can range up to 0.15 mg/m^3 (0.06 ppm), and associated concentrations of acidic particles, likely present as pure H_2SO_4, can reach 0.075 mg/m^3 (Leaderer et al., 1990). Peak SO_2 levels in some industrial environments can exceed 15.7 mg/m^3 (6 ppm) (Sandström et al., 1989a). Likewise, occupational exposures to H_2SO_4 can be quite high. While some older studies in battery plants indicated levels ranging from 3 - 35 mg/m^3 (Malcolm and Paul, 1961; El-Sadik et al., 1972), a more recent assessment indicated exposures up to 1.7 mg/m^3 (Jones and Gamble, 1984).

The particle size of sulfate aerosols depends upon their source. Ambient outdoor sulfates occur in two particle size modes. Those resulting from regional pollution events and plumes are generally found in the range between 0.3 - 0.6 μm. On the other hand, acid fog particles are generally between 5 - 30 μm. The median particle size of industrial aerosols is in the range of 3 - 14 μm, depending upon the method of generation.

Dosimetry

Sulfur dioxide is highly water soluble and can be very efficiently scrubbed (>90%) from inspired air within the upper respiratory tract (Dalhamn and Strandberg, 1961; Speizer and Frank, 1966, Frank et al., 1969). However, the extent of removal may depend upon the SO_2 concentration inhaled. Scrubbing efficiency appears to be greater at high SO_2 concentrations (Strandberg, 1964). At levels relevant to ambient outdoor air pollution, however, removal by the upper respiratory tract may be less efficient (Strandberg, 1964; Amdur, 1966), resulting in greater penetration into the lungs. Upon contacting the airway lining fluid, SO_2 rapidly dissolves into the aqueous phase and readily dissociates into bisulfite and sulfite ions, which can then be translocated into the systemic circulation. These are eventually detoxified via sulfite oxidase, and excreted as sulfate in the urine.

Inhaled sulfate aerosols will be subjected to physical deposition mechanisms in the respiratory tract which are dependent upon the specific aerosol particle size distribution and ventilation characteristics. Sulfate particles have properties which may dynamically alter their size following inhalation, namely deliquescence (e.g, for NH_4HSO_4) or hygroscopicity (e.g., for H_2SO_4). This may result in substantial growth in diameter, often by factors of 2 - 4 times, within the airways while the particles are airborne in the high humidity atmosphere of the respiratory tract. Such particles will then deposit on airway surfaces according to their hydrated, rather than their initial dry, size (Morrow, 1986; Martonen and Zhang, 1993).

Inhaled sulfate particles may be further modified while still airborne by under-going neutralization reactions with endogenous respiratory tract NH_3 (Larson et al., 1977, 1982). Since NH_3 concentrations are greater in the mouth than in the nasal passages, the mode of inhalation influences the ratio of specific sulfate species available for deposition in the lungs. Deposited acid sulfate particles may be further modified by the action of buffers present in the fluid lining of the airways (Holma, 1985).

Immunotoxicity

Much of the database involving effects of sulfur oxides on lung defense is concerned with non-specific, i.e., non-immunologic, mechanisms. Limited evidence on pulmonary humoral or cell-mediated immunity provides some equivocal indication of the ability of sulfur oxides to enhance sensitization to antigens, or to modulate the activity of cells involved in allergic responses.

Exposure to SO_2 can facilitate local allergic sensitization in guinea pigs (Riedel et al., 1988). Animals were exposed at 0.1, 4.3, and 16.6 ppm for 8 hr/d for 5 d. On the last 3 days, SO_2 exposure was followed by inhalation of aerosolized ovalbumin (OA) for 45 min. One week later, specific bronchial provocation with inhaled OA followed by measurement of airway obstruction was performed every 2 d during a 2-wk period. The SO_2 exposed groups demonstrated 67 - 100% positive bronchial reactions to inhaled OA compared to 7% in the control group. In addition, the degree of bronchial obstruction was significantly higher in the exposed group, compared to the control group, for all SO_2 concentrations. OA-specific antibodies in serum and bronchoalveolar fluid increased in SO_2-exposed groups as well.

Sulfuric acid exposure has been shown to enhanced histamine release from guinea pig lung mast cells (Fujimaki et al., 1992). When guinea pigs were exposed to H_2SO_4 at 0.3, 1.0, and 3.2 mg/m^3 for 2 and 4 wk, antigen-induced histamine release from isolated mast cells was significantly enhanced by exposure to the two highest levels for 2 wk but not by cells isolated from hosts exposed for 4 wk. These alterations were not accompanied by changes in mast cell numbers in the lungs.

Host resistance studies have been performed to examine the effects of inhaled SO_2 upon both antibacterial and antiviral pathways in the lungs. Studies examining mice and rats exposed to atmospheres of SO_2 alone (100 - 170 μg SO_2/m^3) or in combination with monodispersed ferrous sulfate particles (MMAD = 2.5 μm) either 17 hr prior to, or 4 hr after, infection with aerosols of *Staphylococcus aureus* or *Streptococci* (Group C) revealed no changes in *Staphylococcal* clearance or alveolar macrophage ingestion of the bacteria (Goldstein et al., 1979). Other studies have demonstrated that continuous exposure of mice to ~10 ppm SO_2 for up to 3 wk reduced resistance to infection with *Klebsiella pneumoniae*, thereby leading to increased host mortality and shortened survival times (Azoulay-Dupuis et al., 1982). Inasmuch as Ehrlich (1980) also observed no effect from SO_2 upon host resistance against group C *Streptococci pyogenes*, it seems likely that SO_2 may selectively alter those pulmonary immune defense mechanisms specific for combating Gram-negative, rather than Gram-positive, infections.

Host resistance studies examining the effects of inhaled SO_2 upon viral infection have demonstrated mixed results. While studies by Fairchild (1977) showed that exposure of mice to SO_2 (as high as 6 ppm) produced no effect upon influenza virus replication, other studies (Fairchild et al., 1972; Lebowitz and Fairchild, 1973; Ukai, 1977) have demonstrated either additive or synergistic effects of the two stressors on pulmonary inflammatory responses.

In an experiment using mice as the model, the combination of ozone (O_3) and H_2SO_4 aerosol was found to enhance sensitization to aerosolized OA (Osebold et al., 1980). Mice were exposed continuously to a combination of 0.5 ppm O_3 and H_2SO_4 aerosol (1 mg/m^3) for 4 d, then given aerosolized antigen (2% solution of OA) and adjuvant (*Bordetella pertussis*, intraperitoneally-injected). Following inhalation of OA, the mice were held in ambient air a few days, and then the cycle of air pollutant exposure and aerosolized OA sensitization was repeated four times (a total of 16 pollution exposure days and six contacts with OA). On day 45 of the experiment, anaphylactic shock was induced by intravenous injection of OA. An enhancement of the allergic response in animals exposed to a mixture of H_2SO_4 and O_3 as compared to the animals exposed to either pollutant alone was observed.

The alveolar macrophage is important in various aspects of both antibacterial and antiviral defense pathways in the lungs. Exposure of rats to 2.6 - 52.4 mg SO_2/m^3 (1 - 20 ppm) for 24 hr resulted in increased *in vitro* phagocytosis by alveolar macrophages at ≥13.1 mg SO_2/m^3 (5 ppm) (Katz and Laskin, 1971). On the other hand, macrophages obtained from exercising hamsters exposed to 131 mg SO_2/m^3 (50 ppm) for 4 hr showed decreased motility and phagocytic function (Skornik and Brain, 1990). Examination of lavage fluid of humans exposed to 21 mg SO_2/m^3 (8 ppm) for 20 min during mild exercise showed an increase in lysozyme-positive macrophages (Sandström et al., 1989b), an indicator of cell activation. *In vitro* exposure of human alveolar macrophages to 6.6 - 32.8 mg SO_2/m^3 (2.5 - 12.5 ppm) for 10 min resulted in a dose-dependent increase in spontaneous production of reactive oxygen species; a 30 min exposure of cells to 32.8 mg SO_2/m^3 (12.5 ppm) had a cytotoxic effect, and decreased the production of reactive oxygen species compared to an exposure to 6.6 mg SO_2/m^3 (2.5 ppm) (Kienast et al., 1993).

Some other studies have examined effects of SO_2 on alveolar macrophage function following *in vitro* exposure. These have indicated that SO_2 can alter cell migration via changes in chemotactic mechanisms and decrease production or release of tumor necrosis factor-α (TNFα) and interleukin-1α (IL-1α) (Knorst et al., 1996a and b). The relevance of these results to *in vivo* exposure is, however, unclear.

The effects of H_2SO_4 upon host pulmonary resistance/defense mechanisms have also been examined. Alveolar macrophages recovered from rabbits exposed for 3 hr to H_2SO_4 at 1 mg/m^3 had a reduced ability *ex vivo* to phagocytose and kill ingested *S. aureus* (Zelikoff et al., 1997). While the effects did not appear to be due to alterations in alveolar macrophage F_cR expression, H_2SO_4-induced reductions in superoxide anion ($\cdot O_2^-$) production may have played a role, at least in part, in bringing about these observed effects upon intracellular killing given the importance of reactive oxygen intermediates for killing *Staphylococcal* organisms. In conjunction with the results of earlier studies (Zelikoff and Schlesinger, 1992; Zelikoff et al., 1994), it was concluded that H_2SO_4-induced changes in macrophage $\cdot O_2^-$ and TNFα production may be responsible for the changes observed in host resistance to infection.

At levels ≤1mg/m^3, H_2SO_4 has been shown to affect certain functions of macrophages lavaged from various species following single or repeated inhalation exposure. This includes phagocytic activity, adherence, random mobility, intra-cellular pH and the release or production of certain cytokines (e.g., TNFα and IL-1α) and reactive oxygen species (Schlesinger, 1990; Zelikoff and Schlesinger, 1992; Zelikoff et al., 1994; Chen et al., 1995). Such effects may ultimately be reflected in alterations in the ability of these cells to adequately perform their role in host defenses, including resistance to disease. However, the evidence that H_2SO_4 reduces resistance to bacterial infection is conflicting, and may depend upon the animal model used (USEPA, 1989a; Zelikoff et al., 1994).

Modulation of pulmonary pharmacological receptors may underlie some sulfur oxide-induced responses. β-Adrenergic stimulation downregulates pulmonary macrophage function, and this has been shown to be influenced by short term, repeated exposures of rabbits to H_2SO_4 (McGovern et al., 1993). Any acid-induced enhanced downregulation, by affecting production of reactive oxygen species by macrophages, may create an environment conducive to secondary pulmonary insult, such as bacterial infections, especially in susceptible populations. In addition, alterations in mediator release from the macrophage due to receptor down regulation may, in turn, influence airway responsiveness.

The ability of sulfur oxides to affect pharmacological receptors in the airways was also demonstrated by the blunting of β-adrenergic function in human airway epithelial cells exposed *in vitro* to SO_2 (Fine et al., 1991). This was accompanied by a decrease in cAMP in exposed cells. The likely subsequent upset of the balance between cAMP and cGMP, involved in the pharmacological control of airway muscle tone, may underlie any increased spasmogenic activity of bronchoconstrictor agents, i.e., acetylcholine or histamine, observed following sulfur oxide exposure.

NITROGEN OXIDES

Nitrogen oxides (NO_x) includes nitrogen dioxide (NO_2), nitrous oxide (N_2O), nitric oxide (NO), nitrogen trioxide (NO_3), dinitrogen trioxide (N_2O_3), dinitrogen tetroxide (N_2O_4) and dinitrogen pentoxide (N_2O_5). N_2O occurs naturally due to biological processes occurring in soil; it is commonly used as an anesthetic agent. Except for N_2O, the various NOx are interconvertible, and many of them co-exist in particular atmospheres. From an environmental health perspective, the materials of most concern are NO and NO_2.

Sources And Exposure

Combustion from both mobile and stationary sources is responsible for ambient NO_x emissions. The major mobile source is motor vehicles, while the major stationary source is electric power generation using fossil fuels, with industrial combustion processes a close second. Potential occupational exposures to NO_x are diverse. Operations in which exposure may occur include welding using oxyacetylene flames; glassblowing; working near internal combustion engines, such as in parking lots; underground blasting operations, such as occurs in mining; storage of silage in agricultural operations; manufacture of nitric acid, oxidized cellulose compounds, and rocket propellants and fertilizers; brazing; metal cleaning; textile (rayon) and food bleaching.

Most NO_x initially produced is in the form of NO, but this is generally rapidly oxidized to NO_2. Several reaction pathways are possible. One is simple oxidation involving molecular oxygen (O_2). In polluted smog atmospheres, irradiation by sunlight may result in transformation of NO to NO_2. The conversion rate of NO to NO_2 depends upon various factors including the initial NO concentration and the operating temperature of any combustion process. Because of these factors, distribution of NO and NO_2 generated from industrial sources is variable.

In ambient outdoor air, the pattern of NO_2 concentration in urban areas is characterized by two daily peaks which are related to motor vehicle traffic patterns in the morning and afternoon. In those areas not impacted by mobile sources, NO_2 levels have little variation on an hourly basis throughout the day, unless there is transport of NO_2 into that region. Areas that have significant stationary sources are characterized by baseline NO_2 levels that may be low, but have superimposed upon them, generally on an irregular basis, spikes to higher levels. In most areas of the United States, NO_2 levels are highest in the summer. However, California is characterized by elevated NO_2 throughout the year.

Annual average outdoor concentrations of NO_2 in most regions of the country are in the 0.015 - 0.035 ppm range; levels in major urban areas tend towards the high end, while those in nonmetropolitan or rural areas towards the lower. The median NO_2 annual average is 0.02 - 0.03 ppm, but some areas (e.g., Southern California) may show annual averages >0.04 ppm (USEPA, 1989b, 1990). Short-term levels follow a similar pattern as do yearly averages, with 24-hr average levels generally ≤0.17 ppm, and 1-hr averages ≤0.4 ppm in major metropolitan areas (Sexton et al., 1983). However, hourly averages in most regions often exceed 0.20 ppm at least once during the year.

There are few data for ambient NO levels, but they appear to be higher than those for NO_2. For example, peak concentrations (8-hr average) of 0.12 - 0.16 ppm were noted in one area of Southern California (Appel et al., 1980).

Nitrogen dioxide is also an important indoor pollutant in households. In fact, non-occupational indoor concentrations and exposures are generally higher than those outdoors when significant sources such as gas-fired ranges, kerosene heaters, and improperly or unvented gas space heaters are used. Because the combustion from these tends to be episodic, very high short-term concentrations are possible. Indoor levels vary widely, depending upon cooking pattern, strength of the sources, and degree of ventilation. Nitrogen dioxide levels as high as 0.54 ppm (24-hr average) have been measured in homes using gas-fired stoves, ovens, or heaters, while levels of 0.38 ppm have been found in homes using kerosene heaters (Spengler and Cohen, 1985). Daily (24-hr average) levels in homes with gas stoves may average 0.05 ppm (Keller et al., 1979), while peak instantaneous values NO_2 in such homes may reach 1 ppm (Sexton et al., 1983). Levels of NO associated with gas cooking may reach 0.4 ppm (24 hr average) (Keller et al., 1979), and average about 0.11 ppm.

The amount of NO_x found in occupational settings is very variable, but may reach fairly high values. For example, concentrations of 50 - 250 ppm NO_x (as NO_2) may be produced during high temperature oxyacetylene welding processes due to reactions between atmospheric N_2 and O_2; up to several hundred ppm NO_x may occur in freshly-filled agricultural silos due to fermentation of the stored silage (Norwood et al., 1966; Cummins et al., 1971). More constant, lower level exposures may occur in some occupations. For example, railroad workers may be regularly exposed to up to 0.12 ppm NO_2 derived from diesel exhaust (Woskie et al., 1989).

Dosimetry

A large percentage of NO_2 is removed in the respiratory tract upon inhalation; absorption of up to 90% of the amount inhaled has been reported for both humans and laboratory animals (Wagner, 1970; Postlethwait and Mustafa, 1981; Kleinman and Mautz, 1989). The NO_2 is absorbed along the entire tracheobronchial tree and within the respiratory region, but the major dose to tissue is delivered at the transition zone of the lungs, that is, the junction between the conducting and respiratory (alveolated) airways (Miller et al., 1982; Overton, 1984). Beyond the transition zone, a rapid falloff in dose delivered to tissue occurs due to the rapid increase in lung surface area.

Once deposited, NO_2 may dissolve in lung fluids and/or react with epithelial surface components, and various chemical intermediates may subsequently be formed. However, the initial transformation products have been the subject of some speculation. Dissolution of NO_2 would form nitric and nitrous acids, which could then be responsible for any subsequent toxic effect (Goldstein et al., 1977, 1980). On the other hand, the primary reaction of NO_2 may not be with lung water, but rather with oxidizable tissue components, such as lipids and/or proteins, to produce nitrite ion (Postlethwait and Mustafa, 1981; Saul et al., 1983; Postlethwait and Bidani, 1989, 1990). This would then rapidly pass into the bloodstream, and undergo other chemical reactions in extrapulmonary sites, for example, oxidation to nitrate by interaction with hemoglobin in red blood cells (Case et al., 1979; Kosaka et al., 1979; Oda et al., 1981; Parks et al., 1981). It is likely that both oxidative and non-oxidative mechanisms are involved in the expression of NO_2 toxicity.

Little is known about NO absorption, and even less on its subsequent distribution. Since NO has a lower water solubility and reactivity than NO_2, it should follow that its absorption from inhaled air would be less. However, the available data indicate that NO absorption is quite similar to that for NO_2 (Vaughan et al., 1969; Wagner, 1970; Yoshida et al., 1980; Chiodi and Mohler, 1985). But the lower solubility of NO may, however, result in greater amounts reaching the respiratory region, where it can then diffuse into the blood and react with hemoglobin . In addition, NO has a faster rate of diffusion through tissue than does NO_2 (Chiodi and Mohler, 1985).

Immunotoxicity

Nitrogen oxides have immunotoxic action. Nitrogen dioxide impairs resistance to infectious agents, i.e., bacteria and viruses, in animals exposed to levels as low as 0.5 ppm NO_2 for 3 mo (Ito, 1971; Ehrlich et al., 1979; Rose et al., 1988). A few infectivity studies involved exposure to a baseline level upon which spikes to a higher level were superimposed so as to better mimic some ambient exposure conditions. The relative effect of such spikes is not clear, but seems to depend upon both spike duration and the time between spikes. Miller et al. (1987) noted that mortality due to infection was greater in a spike regimen (to 0.8 ppm) than in the baseline-exposed group (0.2 ppm). Others have also found that the number and amplitude of spikes are of importance in increasing mortality (Gardner et al., 1979; Graham et al., 1987). In fact, effects from such exposures may approach those due to more continuous exposure to a lower concentration.

The effect of NO_2 on bacterial infectivity increases with both exposure duration and concentration, although the latter seems to have more influence than the former (Gardner et al., 1979). Any differences between intermittent and continuous exposure also seem to disappear as the number of days of exposure increase. Other studies suggest that, as concentration increases, a shorter exposure time is needed for intermittent and continuous exposure regimes to produce similar degrees of effect (Ehrlich and Henry, 1968; Ehrlich, 1979). Mortality due to infection is also proportional to exposure duration if the bacterial challenge is given immediately after exposure, but may not be when the challenge is given much later (Ehrlich, 1980; Gardner et al., 1982), suggesting that a critical time between NO_2 and bacterial exposure is needed for increased susceptibility.

The mechanism(s) underlying increased microbial infectivity are not known. However, since exposure levels which alter host resistance do not generally affect physical particle clearance processes (Parker et al., 1990), the response to NO_2 may be due to impaired intracellular killing of microbes, probably due to macrophage dysfunction. For example, macrophages are a source of numerous biochemical mediators that are directly involved in antibacterial action, e.g., $\cdot O_2^-$, and a depression in $\cdot O_2^-$ production has been noted following NO_2 exposure (Amoruso et al., 1981; Suzuki et al., 1986).

Changes in susceptibility to infectious agents are an indirect index of altered immune function. A number of studies have directly examined the effects of NO_2 exposure on specific parameters of both humoral and cellular immunity. While immune suppression has clearly been shown to follow exposure to levels of NO_2 >5 ppm, as evidenced by various endpoints including the response of T- and B-lymphocytes, antibodies, or production of interferon (Valand et al., 1970; Campbell and Hilsenroth, 1976; Holt et al., 1979; Fujimaki and Shimizu, 1981), there are only a few reports of responses to lower concentrations. These suggest that short-term repeated exposures to 0.25 - 0.5 ppm NO_2 may result in reductions in lympho-cyte numbers in the lungs or spleen, or depressed antibody responsivity to particular antigens (Maigetter et al., 1978; Fujimaki et al., 1982; Richters et al., 1988).

Enhanced immune function may follow NO_2 exposure (e.g., Balchum et al., 1965), but this is just as detrimental as suppressed function, through overstimula-tion of response and/or hypersensitivity. The direction of change in immune system responses appears to depend upon exposure concentration. For example, the humoral responses in monkeys chronically-exposed to NO_2 were enhanced at a low concentration (1 ppm) and suppressed at a higher level (5 ppm) (Fenters et al., 1971, 1973). Alveolar macrophages obtained by bronchoalveolar lavage 3.5 hr after exposure to continuous 0.60 ppm NO_2 tended to inactivate influenza virus *in vitro* less effectively than cells collected after air exposure (Frampton, et al., 1989).

The effect of NO_2 upon antimicrobial defenses of the lungs may be modulated by prior infection or by immunosuppression (Jakab, 1988). The intrapulmonary killing of *Staphylococcus aureus* in mice was impaired at 5 ppm NO_2, but this effect was found at 2.5 ppm or less when NO_2 exposure was superimposed on lungs which were immunosuppressed with corticosteroids. Furthermore, the adverse effect of NO_2 occurred at lower concentrations when exposure followed bacterial challenge.

Alveolar macrophages have been found to suffer significant functional deficits in response to NO_2 exposure (Robison et al., 1993). Rats were exposed to 0.5 ppm NO_2 for periods of 0.5 - 10 d and alveolar macrophages obtained by bronchoalveolar lavage (BAL). Exposure to NO_2 produced complex effects upon unstimulated and stimulated alveolar macrophage arachidonate metabolism. Unstimulated synthesis of leukotriene B_4 (LTB_4) was depressed rapidly within 1 d of exposure, and depressed again at 5 d. Production of thromboxane B_2 (TxB_2), LTB_4, and 5-hydroxyeicosa-tetraenoate (5-HETE) in response to stimulation with the calcium ionophore A23187 was acutely depressed within 1 d of exposure; generation of these compounds recovered to air-control levels with longer exposure, while levels of 5-HETE were increased at 10 d. In contrast, alveolar macrophage production of LTB_4 in response to another stimulus, zymosan-activated rat serum (ZAS), was not depressed until following 5 d of exposure and remained slightly lower than air-control levels at 10 d. Levels of TxB_2, prostaglandins E_2 (PGE_2), and $F_{2\alpha}$ ($PGF_{2\alpha}$), and LTB_4 in BAL fluid (BALF) were found to be depressed within 4 hr of exposure, suggesting an acute decrease in pulmonary arachidonate metabolism *in vivo*; however, production of these compounds generally recovered to air-control levels with longer exposure. Alveolar macrophage $\cdot O_2^-$ production stimulated by phorbol myristate acetate (PMA) was decreased rapidly and continuously throughout the study. Thus, exposure to a low concentration of NO_2 acutely depressed activation of alveolar macrophage arachidonate metabolism and $\cdot O_2^-$ production in response to external stimuli, and may impede defense against pulmonary infection.

Nitrogen dioxide also affects respiratory tract susceptibility to viral infection (Rose et al., 1988). CD-1 mice were inoculated intratracheally with murine cyto-megalovirus (MCMV) during exposure to varying concentrations of NO_2. Exposure lasted for 6 hr/d; it began 2 consecutive days prior to instillation of MCMV and continued for 4 d after virus inoculation. Exposure to 5 ppm NO_2 resulted in MCMV proliferation and a mild bronchopneumonia in some animals inoculated with plaque-forming units of virus at a dose that was too low to produce either viral replication or histologic abnormalities in the lungs of air-exposed animals. In addition, the amount of virus required to infect animals exposed to 5 ppm of NO_2 was 100-fold lower than that needed to consistently produce infection in air-exposed animals. Animals exposed to 5 ppm NO_2 also exhibited depressed phagocytosis of colloidal [198]Au *in vivo* as well as diminished macrophage destruction of instilled MCMV compared to air-exposed animals. These results demonstrate that exposure to 5 ppm NO_2, although not associated with evidence of overt lung injury *per se*, is nevertheless capable of predisposing the lower respiratory tract to viral infection.

Nitrogen dioxide has no effect on circulating and BALF human lymphocyte subtypes (Rubinstein et al., 1991). Blood and bronchoalveolar lavage fluid were obtained from five healthy nonsmoking adult volunteers. More than 2 wk later, these volunteers were exposed to 0.60 ppm NO_2 for 2 hr with intermittent light to moderate exercise on 4 separate days within a 6-d period. Neither any single day's exposure nor all four exposures caused a change in symptoms or in the results of tests of pulmonary function. There was no change in the total number of circulating lymphocytes obtained after NO_2 exposure or the proportions of lymphocyte subtypes. In the BALF obtained after NO_2 exposure and in the baseline state, the total number of lymphocytes and the percentages of T-lymphocytes (CD3), B-lymphocytes (CD20), T cytotoxic-suppressor lymphocytes (CD8), T helper-inducer lymphocytes (CD4), and large granular lymphocytes (CD57) did not differ after NO_2 exposure. Only a slightly but significantly greater proportion of natural killer cells (CD16) was found in the BALF obtained after NO_2 exposure.

At a higher concentration of NO_2 (4 ppm), however, repeated exposures (20 min with light exercise, every second day for a total of six exposures) to NO_2 caused decreases in the number of lavaged alveolar macrophages, B-lymphocytes, and natural killer (NK)-cells as well as alteration in the T-helper-inducer/cytotoxic-suppressor cell ratio (Sandström et al., 1992). In addition, lymphocyte numbers in peripheral blood were reduced after exposure.

Alveolar macrophages obtained by lavage from Fischer 344/N rats exposed for 4 hr to 40 ppm NO_2 were significantly more phagocytic to opsonized sheep red blood cells (SRBC), exhibited an increased cytotoxic response toward syngeneic mammary adenocarcinoma cells, and were more sensitive to activation by agents such as lipopolysaccharide, muramyl dipeptide, and macrophage-activating factor, as compared with the response of alveolar macrophages obtained from unexposed control rats (Sone et al., 1983). However, repeated 4 hr/d NO_2 exposures over 7- or 14-d periods usually resulted in alveolar macrophage activity similar to control levels, with some instances of increased phagocytic activity of the alveolar macrophages but not to the extent of that observed for a single 4 hr exposure. There were no significant decreases in the cytotoxic or phagocytic activities of the alveolar macrophages during any of the exposure periods. For the *in vitro* exposures, alveolar macrophages were lavaged from normal rats and then exposed for various periods to 10, 20, or 40 ppm NO_2. A dose-related, time-dependent enhanced cytotoxic response of alveolar macrophages was observed. Maximum alveolar macrophage-mediated cytotoxicity occurred after an *in vitro* exposure to 10 ppm NO_2 for 2 hr. The cyto-toxic response was directed toward syngeneic mammary adenocarcinoma cells but not against syngeneic embryoblast cells, indicating that the alveolar macrophages retained the ability to distinguish between normal and abnormal cells. No inhibitory effects of NO_2 on alveolar macrophage-mediated cytotoxicity were observed. These experiments suggest that the host alveolar macrophage-mediated immune defense of the lung may be modulated by host exposure to inhaled chemicals.

Holt et al. (1979) examined immunological endpoints in mice exposed to 10 ppm NO for 2 hr/d, 5 d/wk up to 30 wk. Leukocytosis was evident by 5 wk of exposure, while a decrease in mean hemoglobin content of red blood cells was found by 30 wk. The ability of spleen cells to mount a graft vs host reaction was stimulated by 20 wk of exposure, but suppressed by 26 wk. When the ability of mice to reject virus-induced tumors was assessed, less of the NO exposed animals survived tumor challenge compared to control; this suggests that NO at high levels may have affected the immunologic competence of the animals. In another study, (Azoulay et al., 1981), mice were exposed continuously to 2 ppm NO for 6 hr up to 4 wk, to assess the effect on resistance to bacterial infection. There was some indication that NO-exposed females showed a significant increase in percentage mortality and a significant decrease in survival time, which was not seen in males.

REFERENCES

Amdur MO. Respiratory absorption data and SO_2 dose-response curves. Arch. Environ. Health, 1966;12:729-732.

Amoruso MA, Witz G, Goldstein BD. Decreased superoxide anion radical production by rat alveolar macrophages following inhalation of ozone or nitrogen dioxide. Life Sci., 1981;28:2215-2221.

Appel BR, Wall SM, Tokiwa Y, Haik M. Simultaneous nitric acid, particulate nitrate and acidity measurements in ambient air. Atmos. Environ., 1980;14:549-554.

Azoulay E, Bouley G, Blayo MC. Effects of nitric oxide on resistance to bacterial infection in mice. J. Toxicol. Environ. Health, 1981;7:873-882.

Azoulay-Dupuis E, Bouley G, Blayo, MC. Effects of sulfur dioxide on resistance to bacterial infection in mice. Environ. Res., 1982;29:312-319.

Balchum OJ, Buckley RD, Sherwin R, Gardner M. Nitrogen dioxide inhalation and lung antibodies. Arch. Environ. Health, 1965;10:274-277

Campbell KI, Hilsenroth RH. Impaired resistance to toxin in toxoid-immunized mice exposed to ozone or nitrogen dioxide. Clin. Toxicol., 1976;9:943-954.

Case GD, Dixon JS, Schooley JC. Interactions of blood metalloproteins with nitrogen oxides and oxidant air pollutants. Environ. Res., 1979;20:43-65.

Chen LC, Wu CY, Qu QS, Schlesinger RB. Number concentration and mass concentration as determinants of biological response to inhaled irritant particles. Inhal. Toxicol., 1995;7:577-588.

Chiodi H, Mohler JG. Effects of exposure of blood hemoglobin to nitric oxide. Environ. Res., 1985;37:355-363.

Council on Environmental Quality. *Environmental Quality*. Washington, DC: Council on Environmental Quality, 1992.

Cummins BT, Ravency FJ, Jesson MW. Toxic gases in tower silos. Ann. Occup. Hyg., 1971;14:275-283.

Dalhamn T, Strandberg L. Acute effect of sulfur dioxide on rate of ciliary beat in trachea of rabbit *in vivo* and *in vitro*, with studies on absorptional capacity of nasal cavity. Int. J. Air Water Pollut., 1961;4:154-160.

Ehrlich R, Findlay J, Gardner DE. Effects of repeated exposures to peak concentrations of nitrogen dioxide and ozone on resistance to streptococcal pneumonia. J. Toxicol. Environ. Health, 1979;5:631-642.

Ehrlich R, Henry MC. Chronic toxicity of nitrogen dioxide: I. Effect on resistance to bacterial pneumonia. Arch. Environ. Health, 1968;17:860-865.

Ehrlich R. "Interaction Between Environmental Pollutants and Respiratory Infections." In *Proceedings of the Symposium on Experimental Models for Pulmonary Research*, DE Gardner, EP Hu, JA Graham, eds., EPA-600/9-79-022, Research Triangle Park, NC: United States Environmental Protection Agency, 1979, pp. 145-163.

Ehrlich R. Interaction between environmental pollutants and respiratory infections. Environ. Health Perspect., 1980;35:89-100.

El-Sadik YM, Osman AH, El-Gazzar RM. Exposure to sulfuric acid in manufacture of storage batteries. J. Occup. Med., 1972;14:224-226.

Fairchild GA, Roan J, McCarroll J. Atmospheric pollutants and the pathogenesis of viral respiratory infection. Arch. Environ. Health, 1972;25:174-182.

Fairchild GA. Effects of ozone and sulfur dioxide on virus growth in mice. Arch. Environ. Health, 1977;32:28-33.

Fenters JD, Ehrlich R, Findlay J, Spangler J, Tolkacz V. Serologic response in squirrel monkeys exposed to nitrogen dioxide and influenza virus. Am. Rev. Respir. Dis., 1971;104:448-451.

Fenters JD, Findlay JD, Port CD, Ehrlich R, Coffin DL. Chronic exposure to nitrogen dioxide: Immunologic, physiologic, and pathologic effects in virus-challenged squirrel monkeys. Arch. Environ. Health, 1973;27:85-89.

Fine JM, Chen LC, Finkelstein M, Finklestein I, Gruenert D, Gordon T. Effects of SO_2 on cAMP levels in human airway epithelial cells. Am. Rev. Respir. Dis., 1991;143:A489."

Frank NR, Yoder RE, Brain JD, Yokoyama E. SO_2 (^{35}S-labeled) absorption by the nose and mouth under conditions of varying concentration and flow. Arch. Environ. Health, 1969;18:315-322.

Fujimaki H, Katayama N, Wakamori K. Enhanced histamine release from lung mast cells of guinea pigs exposed to sulfuric acid aerosols. Environ. Res., 1992;58:117-123.

Fujimaki H, Shimizu F, Kubota K. Effect of subacute exposure to NO_2 on lymphocytes required for antibody response. Environ. Res., 1982;29:280-286.

Fujimaki H, Shimizu F. Effects of acute exposure to nitrogen dioxide on primary antibody response. Arch. Environ. Health, 1981;36:114-119.

Gardner DE, Miller FJ, Blommer EJ, Coffin DL. Influence of exposure mode on the toxicity of NO_2. Environ. Health Perspect., 1979;30:23-29.

Gardner DE, Miller FJ, Illing JW, Graham JA. "Non-respiratory Function of the Lungs: Host Defenses Against Infection." In *Air Pollution by Nitrogen Oxides*, T Schneider, L Grant, eds., New York: Elsevier, 1982, pp. 401-415."

Goldstein E, Goldstein F, Peek NF, Parks NJ. "Absorption and Transport of Nitrogen Oxides." In *Nitrogen Oxides and Their Effects on Health*, SD Lee, ed., Ann Arbor, MI: Ann Arbor Sciences, 1980, pp. 143-160.

Goldstein E, Lippert W, Chang DP, Tarkington B. Effect of near-ambient exposures to sulfur dioxide and ferrous sulfate particles on murine pulmonary defense mechanisms. Arch. Environ. Health, 1979;34:424-432.

Goldstein E, Peek NF, Parks NJ, Hines HH, Steffey EP, Tarkington B. Fate and distribution of inhaled nitrogen dioxide in rhesus monkeys. Am. Rev. Respir. Dis., 1977;115:403-412.

Graham JA, Gardner DE, Blommer EJ, House DE, Menache MG, Miller FJ. Influence of exposure patterns of nitrogen dioxide and modifications by ozone on susceptibility to bacterial infectious disease in mice. J. Toxicol. Environ. Health, 1987;21:113-125.

Holma B. Influence of buffer capacity and pH-dependent rheological properties of respiratory mucus on health effects due to acidic pollution. Sci. Total Environ., 1985;41:101-123.

Holt PG, Finlay-Jones LM, Keast D, Papadimitrou JJ. Immunological function in mice chronically exposed to nitrogen oxides (NOx). Environ. Res., 1979;19:154-162.

Ito K. [Effect of nitrogen dioxide inhalation on influenza virus infection in mice]. Nippon Eiseigaku Zasshi, 1971;26:304-314.

Jakab GJ. Modulation of pulmonary defense mechanisms against viral and bacterial infections by acute exposures to nitrogen dioxide. HEI Research Reports, 1988;20:1-38.

Jones W, Gamble J. Epidemiological-environmental study of lead acid battery workers. I. Environmental study of five lead acid battery plants. Environ. Res., 1984;35:1- 10.

Katz GV, Laskin S. "Effect of Irritant Atmospheres on Macrophage Behavior." In *Pulmonary Macrophages and Epithelial Cells*, C Sanders, R Schneider, G Dagel, HA Ragan, eds., Springfield, VA: NTIS, 1971, pp. 358-375.

Keller MD, Lanese RR, Mitchell RI, Cote RW. Respiratory illness in households using gas and electricity for cooking. I. Survey of incidence. Environ. Res., 1979;19:495-503.

Kienast K, Müller-Quernheim J, Knorst M, Schlegel J, Mang C, Ferlinz R. Realistic *in vitro* study of oxygen radical libaration by alveolar macrophages and mononuclear cells of the peripheral blood after short-term exposure to SO_2. Pneumologie, 1993;47:60-65.

Kleinman MT, Mautz WJ. "Upper Airway Scrubbing at Rest and Exercise." In *Susceptibility to Inhaled Pollutants*, MJ Utell, R Frank, eds., ASTM STP 1024, Philadelphia: American Society for Testing and Material, 1989, pp. 100-110.

Kosaka H, Imaizumi K, Imai K, Tyuma I. Stoichiometry of the reaction of oxyhemoglobin with nitrate. Biochim. Biophys. Acta, 1979;581:184-188.

Larson TV, Covert D, Frank R, Charlson RJ. Ammonia in the human airways: Neutralization of inspired acid sulfate aerosols. Science, 1977;197:161-163.

Larson TV, Frank R, Covert DS, Holub D, Morgan MS. Measurements of respiratory ammonia and the chemical neutralization of inhaled sulfuric acid aerosol in anesthetized dogs. Amer. Rev. Respir. Dis., 1982;125:502-506.

Leaderer BP, Boone PM, Hammond SK. Total particle, sulfate, and acidic aerosol emissions from kerosene space heaters. Environ. Sci. Technol., 1990;24:908-912.

Lebowitz MD, Fairchild GA. The effects of sulfur dioxide and A2 influenza virus on pneumonia and weight reduction in mice: An analysis of stimulus-response relationships. Chem.-Biol. Interact., 1973;7:317-326.

Lioy PJ, Waldman JM. Acidic sulfate aerosols: Characterization and exposure. Environ. Health Perspect., 1989;79:15-34.

Maigetter RZ, Fenters JD, Findlay JC, Ehrlich R, Gardner DE. Effect of exposure to nitrogen dioxide on T- and B-cells in mouse spleens. Toxicol. Lett., 1978;2:157-161.

Malcolm D, Paul E. Erosion of the teeth due to sulphuric acid in the battery industry. Br. J. Ind. Med., 1961;18:63-69.

Martonen TB, Zhang Z. Deposition of sulfate aerosols in the developing human lung. Inhal. Toxicol., 1993;5:165-187.

McGovern TJ, Schlesinger RB, El-Fawal HA. Effect of repeated in vivo ozone and/or sulfuric acid exposures on β-adrenergic modulation of macrophage function. The Toxicologist, 1993;13:49.

Miller FJ, Graham JA, Raub JA, Illing JW, Menache MG, House DE, Gardner DE. Evaluating the toxicity of urban patterns of oxidant gases. II. Effects in mice from chronic exposure to nitrogen dioxide. J. Toxicol. Environ. Health, 1987;21:99-112.

Miller FJ, Overton JH, Myers ET, Graham JA. "Pulmonary Dosimetry of Nitrogen Dioxide in Animals and Man." In Air Pollution by Nitrogen Oxides: Proceedings of the U.S.-Dutch International Symposium, T Schneider, L Grant, eds., Amsterdam: Elsevier Scientific Publishing, 1982, pp. 377-386.

Morrow PE. Factors determining hygroscopic aerosol deposition in airways. Physiol.. Rev., 1986, 66, 330-376.

Norwood WD, Wisehart DE, Earl CA, Adley FE, Anderson DE. Nitrogen dioxide poisoning due to metal-cutting with oxyacetylene torch. J. Occup. Med., 1966;8:301-306.

Oda H, Tsubone H, Suzuki A, Ichinose T, Kubota K. Alterations of nitrite and nitrate concentrations in the blood of mice exposed to nitrogen dioxide. Environ. Res., 1981;25:294-301.

Osebold JW, Gershwin LJ, Zee YC. Studies on the enhancement of allergic lung sensitization by inhalation of ozone and sulfuric acid aerosol. J. Environ. Pathol. Toxicol., 1980;3:221-234.

Overton JH, Jr. "Physicochemical Processes and the Formulation of Dosimetry Models." In Fundamentals of Extrapolation Modeling of Inhaled Toxicants: Ozone and Nitrogen Dioxide, FG Miller, DB Menzel, eds., Washington, DC: Hemisphere Publishing Company, 1984, pp. 93-114.

Parker RF, Davis JK, Cassell GH, White H, Dziedzic D, Blalock DK, Thorp RB, Simecka JW. Short-term exposure to nitrogen dioxide enhances susceptibility to murine respiratory Mycoplasmosis and decreases intrapulmonary killing of Mycoplasma pulmonis. Am. Rev. Respir. Dis., 1990;140:502-512.

Parks NJ, Krohn KA, Mathis CA, Chasko JH, Greiger KR, Gregor ME, Peek NF. Nitrogen-13-labeled nitrate: Distribution and metabolism after intratracheal administration. Science, 1981;212:58-61.

Postlethwait EM, Bidani A. Pulmonary disposition of inhaled NO_2-nitrogen in isolated rat lungs. Toxicol. Appl. Pharmacol., 1989;98:303-312.

Postlethwait EM, Bidani A. Reactive uptake governs the pulmonary air space removal of inhaled nitrogen dioxide. J. Appl. Physiol., 1990;68:594-603.

Postlethwait EM, Mustafa MG. Fate of inhaled nitrogen dioxide in isolated perfused rat lung. J. Toxicol. Environ. Health, 1981;7:861-872.

Richters A, Damji KS. Changes in T-lymphocyte subpopulations and natural killer cells following exposure to ambient levels of nitrogen dioxide. J. Toxicol. Environ. Health, 1988;25:247-256.

Riedel F, Krämer M, Scheibenbogen C, Rieger HL. Effects of SO_2 exposure on allergic sensitization in the guinea pig. J. Allergy Clin. Immunol., 1988;82:527-534.

Robison TW, Murphy JK, Beyer LL, Richters A, Forman HJ. Depression of stimulated arachidonate metabolism and superoxide production in rat alveolar macrophages following *in vivo* exposure to 0.5 ppm NO_2. J. Toxicol. Environ. Health, 1993;38:273-292.

Rose RM, Fuglestad JM, Skornik WA, Hammer SM, Wolfthal SF, Beck BD, Brain JD. The pathophysiology of enhanced susceptibility to murine cytomegalovirus respiratory infection during short-term exposure to 5 ppm nitrogen dioxide. Am. Rev. Respir. Dis., 1988;137:912-917.

Rubinstein I, Reiss TF, Bigby BG, Stites DP, Boushey HA. Effects of 0.60 ppm nitrogen dioxide on circulating and bronchoalveolar lavage lymphocyte phenotypes in healthy subjects. Environ. Res. 1991;55:18-30.

Sandström T, Helleday R, Bjermer L, Stjernberg N. Effects of repeated exposure to 4 ppm nitrogen dioxide on bronchoalveolar lymphocyte subsets and macrophages in healthy men. Eur. Respir. J., 1992;5:1092-1096.

Sandström T, Stjernberg N, Andersson MC, Kolmodin-Hedman B, Lundgren R, Ångström T. Is the short term limit value for sulphur dioxide exposure safe? Effects of controlled chamber exposure investigated with bronchoalveolar lavage. Br. J. Ind. Med., 1989a;46:200-203.

Sandström T, Stjernberg N, Andersson MC, Kolmodin-Hedman B, Lundgren R, Rosenhall L, Ångström T. Cell response in bronchoalveolar lavage fluid after exposure to sulfur dioxide: A time-response study. Am. Rev. Respir. Dis., 1989b;140:1828-1831

Saul RL, Archer MC. Nitrate formation in rats exposed to nitrogen dioxide. Toxicol. Appl. Pharmacol., 1983;67:284-291.

Schlesinger RB. The interaction of inhaled toxicants with respiratory tract clearance mechanisms. CRC Crit. Rev. Toxicol., 1990;20:257-286.

Sexton K, Letz R, Spengler JD. Estimating human exposure to nitrogen dioxide: An indoor/outdoor modeling approach. Environ. Res., 1983;32:151-166.

Skornik WA, Brain JD. Effect of sulfur dioxide on pulmonary macrophage endocytosis at rest and during exercise. Am. Rev. Respir. Dis., 1990;142:655-659.

Sone S, Brennan LM, Creasia DA. *In vivo* and *in vitro* NO_2 exposures enhance phagocytic and tumoricidal activities of rat alveolar macrophages. J. Toxicol. Environ. Health, 1983;11:151-163.

Speizer FE, Frank NR. The uptake and release of SO_2 by the human nose. Arch. Environ. Health, 1966;12:725-728.

Spengler JD, Cohen MA. "Emissions From Indoor Sources." In *Indoor Air and Human Health*, RB Gammage, SV Kaye, eds., Chelsea, MI: Lewis Publishers, 1985, pp. 261-278.

Strandberg LG. SO_2 absorption in the respiratory tract. Arch. Environ. Health, 1964;9:160-166.

Suzuki T, Ikeda S, Kanoh T, Mizoguchi I. Decreased phagocytosis and superoxide anion production in alveolar macrophages of rats exposed to nitrogen dioxide. Arch. Environ. Contam. Toxicol., 1986;15:733-739.

Ukai K. Effects of SO_2 on the pathogenesis of viral upper respiratory infection in mice. Proc. Soc. Exp. Biol. Med., 1977;154:591-596.

USEPA: United States Environmental Protection Agency. *Air Quality Criteria for Particulate Matter and Sulfur Oxides*. Office of Health and Environmental Assessment, Environmental Criteria and Assessment Office, United States Environmental Protection Agency, Research Triangle Park, NC, EPA-600/8-82-029aF-cF. Springfield, VA: NTIS, 1982, PB84-156777.

USEPA: United States Environmental Protection Agency. *An Acid Aerosols Issue Paper: Health Effects and Aerometrics*. Report No. EPA-600/8-88-005f, Office of Research and Development, United States Environmental Protection Agency, Research Triangle Park, NC. Springfield, VA: NTIS, 1989.

USEPA: United States Environmental Protection Agency. National Air Quality and Emissions Trends Report. 1987. EPA-450/4-89-001. United States Environmental Protection Agency, Research Triangle Park, NC. Springfield, VA: NTIS, 1989b."

USEPA: United States Environmental Protection Agency. Progress and Challenges: Looking at EPA Today. EPA J., 1990;16:15-28.

Valand SB, Acton JD, Myrvik QN. Nitrogen dioxide inhibition of viral-induced resistance in alveolar monocytes. Arch. Environ. Health, 1970;20:303-309.

Vaughan TR, Jr, Jennelle LF, Lewis TR. Long-term exposure to low levels of air pollutants. Effects on pulmonary function in the beagle. Arch. Environ. Health, 1969;19:45-50.

Wagner HM. Absorption of NO and NO_2 in MIK and MAK concentrations during inhalation. Staub Reinhalt Luft. (Engl.), 1970;30:25-26.

Woskie SR, Hammond SK, Smith TJ, Shenker MB. Current nitrogen dioxide exposures among railroad workers. Am. Ind. Hyg. Assoc. J., 1989;50:346-353.

Yoshida K, Kasama K, Kitabatake M, Okuda M, Imai M. Metabolic fate of nitric oxide. Int. Arch. Occup. Env. Health, 1980;46:71-77.

Zelikoff JT, Frampton MW, Cohen MD, Morrow PE, Sisco M, Tsai Y, Utell MJ, Schlesinger RB. Effects of inhaled sulfuric acid aerosols on pulmonary immunocompetence: A comparative study in humans and animals. Inhal. Toxicol., 1997;9:731-752.

Zelikoff JT, Schlesinger RB. Modulation of pulmonary immune defense mechanisms by sulfuric acid: Effects on macrophage-derived tumor necrosis factor and superoxide. Toxicology, 1992;76:271-281.

Zelikoff JT, Sisco MP, Yang Z, Cohen MD, Schlesinger RB. Immunotoxicity of sulfuric acid aerosol: Effects on pulmonary macrophage effector and functional activities critical for maintaining host resistance against infectious diseases. Toxicology, 1994;92:269-286.

14. ASBESTOS AND SILICA

David B. Warheit, Ph.D.
DuPont Haskell Laboratory
1090 Elkton Road
Newark, DE 19714

INTRODUCTION

The potential role of altered immunity in the development of asbestos-related tumors has been recognized for several years. This was evident when the Report of the Advisory Committee on Asbestos Cancer to the Director of the International Agency for Research on Cancer recommended that parameters of immunocompetence should be evaluated in asbestos-exposed patients (IARC, 1973). This brief review is organized to discuss some immunological alterations which have been reported to occur in asbestos exposed populations and to relate these findings to possible inflammatory and fibrogenic mechanisms related to pulmonary immunopathological effects in animals exposed to asbestos fibers. The reader is directed to a recent review by Rosenthal et al. (1998) for additional information on new developments on asbestos-related immunotoxicity.

STUDIES IN ASBESTOS-EXPOSED POPULATIONS

Patients with asbestosis are known to have alterations of humoral and cellular immune responses. Some of these changes include an increased prevalence of autoantibodies and rheumatoid factors (Turner-Warwick and Parkes, 1970; Kagan et al. 1977a and b; Lange, 1980; Lange et al., 1980). Other case studies of patients with asbestos-related fibrosis have reported enhanced levels of the immunoglobulins IgA, IgM, IgG, and IgE or of complement components 3 and 4, suggesting common immunologic effects (Kagan et al., 1977a and b; Doll et al., 1983). The mechanisms responsible for the increase in serum immunoglobulins have not been elucidated but could be related to the adjuvant-like action of asbestos or possibly reduced suppressor T-lymphocyte function (Miller et al., 1978).

Numerous reports document the direct and indirect effects of asbestos on the human immune system. The results of several studies suggest that absolute total T-lymphocytes numbers are reduced in patients with tumors, while absolute numbers of total T- and T-helper (T_H) lymphocytes were normal in asbestos workers without

neoplasia. It was interesting to note that the numbers of T-suppressor (T_S) lymphocytes were not different in the mesothelioma population but significantly increased
among asbestos workers without neoplasia. This alteration relative to the normal
balance in T-lymphocyte subsets resulted in a marked reduction in T_H:T_S ratios in
mesothelioma patients and in asbestos workers.

Results from clinical studies on patients exposed to asbestos have shown that
natural killer (NK) cells isolated from peripheral blood demonstrate impaired
activity, thus leading to speculation that such suppression may play a role in the
development of lung cancer in these individuals (Kubata et al., 1985; Tsang et al.,
1988). Regarding the role of NK cells in development of asbestos-related lung cancer
and fibrosis, Rosenthal et al. (1998) have suggested that while circulating NK cells
may play a protective role in controlling development of pulmonary cancer, the
interstitial pulmonary NK cell is a more likely candidate for the local control of
neoplastic development in the lung. Thus, it was postulated that these lung-localized
cells protect the host from altered phenotype expression that may arise during the
pathogenesis of asbestosis. In support of this role, recent studies have shown the
ability of NK cells to lyse tumors of mesothelial origin in mice. In these studies,
the influence of inhaled asbestos on interstitial NK number and function in
C57BL/6 mice was investigated. Mice were exposed to chrysotile asbestos for 3 hr/d
for 3 d and then sacrificed at 7, 28, and 56 d post-exposure. Functional assessment
of the interstitial NK population was measured by their ability to lyse target YAC-1
cells. The ability of pulmonary NK cells to lyse target cells was significantly
suppressed in asbestos-exposed mice on days 7 and 56 post-exposure when compared
to air-exposed animals (Rosenthal et al., 1998). In summary, alterations in cell-
mediated immunity have been reported in many clinical studies of patients with
asbestosis. The results of a variety of clinical and experimental studies on asbestos
exposure suggest a possible connection between diminished T-lymphocyte function
and the development of pulmonary fibrosis and tumors following asbestos exposure.

ROLE OF COMPLEMENT IN CLEARANCE OF INHALED ASBESTOS FIBERS

Complement refers to a system consisting of a group of plasma and cell membrane proteins which play an important role in host defense processes. The
mediation of immunological responses by complement has been widely recognized
for nearly a century (Kunkel et al., 1981; Larsen and Henson, 1983; Snyderman,
1986). Some of the more important functions of the complement system are associated with: (1) lysis of cells, bacteria, and enveloped virus; (2) mediating the process
of opsonization as a prelude to phagocytosis by macrophages and other cell types;
and, (3) generation of peptide fragments that play a role in (a) vasodilation at sites of
inflammation, (b) adherence of phagocytic cells to endothelium and the corresponding egress of these cells from the vasculature, (c) chemotaxis of phagocytic cells
into areas of inflammation, and (d) clearance of infectious agents. Most of the early-
acting plasma proteins found in normal plasma undergo sequential activation along
two distinct but interrelated pathways and can be activated both by immunological
and non-immunologic stimuli. The better known classical pathway may be activated

by IgM- and IgG-containing immune complexes, polynucleotides, and some viruses. In this regard, a single molecule of IgM on an antigenic surface, or two molecules of IgG of appropriate subclass can bind and activate C1 and the corresponding pathway. Activators of the alternative complement pathway do not require antibody for activation and include plant, fungal, and bacterial polysaccharides, organic and inorganic particles, and fibers. Activation of either pathway produces a diversity of biologic effects as described above (Frank, 1997).

Complement proteins are known to be normal constituents of lung fluids and are present in biologically active concentrations on alveolar surfaces (Robertson et al., 1976; Kolb et al., 1981). The functional importance of the complement system in host resistance reactions to infectious agents has been well-recognized. In this regard, complement activation modulates inflammatory and bacterial clearance responses in the lungs of exposed animals and humans. In general, following activation, the smaller fragments, i.e., C5a and C3a, have important immunoregulatory effects on T-lymphocyte function, either stimulating (i.e., C5a) or inhibiting (i.e., C3a) aspects of cell-mediated immunity. C5a is strongly chemotactic for neutrophils and macrophages, inducing their migration along a concentration gradient toward the site of generation and thus mediates the recruitment of macrophages and neutrophils into alveolar regions of the lung (Frank, 1997).

Phagocytic cells such as alveolar macrophages (AM) play an important role in clearing inhaled microorganisms and particulate matter from gas exchange regions of the lung and are critical for maintaining the integrity of the alveolar compartment. Yet, the mechanisms through which AM are attracted to inhaled particulates on alveolar surfaces following deposition of particles are not well understood. Particle identification by macrophages is a necessary prerequisite prior to phagocytosis since particles must be encountered by phagocytes before particle engulfment and clearance can occur.

In studies developed to assess the early events of fiber clearance following short-term inhalation of asbestos it was reported that significant numbers of AM had accumulated selectively at sites of chrysotile asbestos fiber deposition (i.e., alveolar duct bifurcation surfaces) within 2 d following a 1 hr exposure (Figure 1). In contrast, alveolar duct bifurcation surfaces of sham-exposed animals did not contain significant numbers of these cells. Recruitment of macrophages to sites of chrysotile fiber deposition formed a component of an early asbestos-induced lesion which was characterized by an increase in tissue volume on the duct bifurcation surface and was maintained through a 1 mo post-exposure period (Warheit et al., 1984).

In attempting to elucidate mechanisms through which macrophages are selectively recruited to alveolar duct bifurcations, it was hypothesized that AM are recruited to sites of asbestos deposition by a chemotactic factor generated by the inhaled fibers on alveolar surfaces. This postulation was consistent with the *in vitro* finding that incubation of chrysotile asbestos fibers with normal serum activates the alternative complement pathway, resulting in generation of the C5a potent chemotactic factor for neutrophils and macrophages. To test the hypothesis, a variety of correlative *in vitro* chemotaxis experiments and inhalation studies were conducted with rats and mice. Alveolar macrophage chemotaxis was developed as a bioassay for complement activation, using zymosan-activated serum as a positive control.

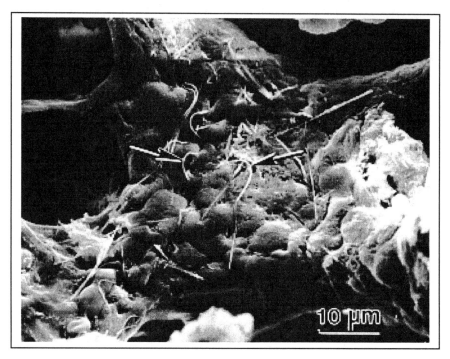

Figure 1. Scanning electron micrograph of lung tissue demonstrating inhaled asbestos fibers (arrows) which have deposited on an alveolar duct bifurcation of a rat.

In support of the hypothesis, the results showed that incubation of chrysotile asbestos fibers with rat serum or unexposed lavaged proteins generated a chemotactic factor leading to an enhanced macrophage chemotactic response compared to controls. Moreover, this finding was confirmed *in vivo* where it was demonstrated that fluids lavaged from the lungs of asbestos-exposed rats contained increased chemotactic activity for macrophages, and this activity was detected in the molecular weight (MW) range of C5a. To further investigate the role of C5 in facilitating the asbestos-induced AM clearance response, congenic strains of complement-normal ($C5^+$) and complement-deficient ($C5^-$) mice, as well as normal and decomplemented rats (i.e. rats treated with cobra venom factor (CVF) to reduce serum complement) were exposed to chrysotile asbestos for a brief period. The numbers of AM that had been attracted to sites of fiber deposition (i.e., alveolar duct bifurcations) were significantly reduced in the $C5^-$ mice and CVF-treated rats (Figures 2 and 3). Time course studies showed that chemoattractant generation preceded the macrophage migration response and that macrophage phagocytosis of inhaled asbestos fibers was reduced in complement-deficient animals. In addition, it was shown that the depletion of measurable levels of hemolytic complement in lung fluids lavaged from asbestos-exposed rats coincided with the presence of a chemotactic factor, suggesting that complement components were consumed during the 3 hr asbestos exposure. The results of these studies demonstrated clearly that asbestos activated complement generating C5a as a byproduct of this reaction, and this chemoattractant induced migration of macrophages to sites of fiber deposition (Warheit et al., 1985, 1986).

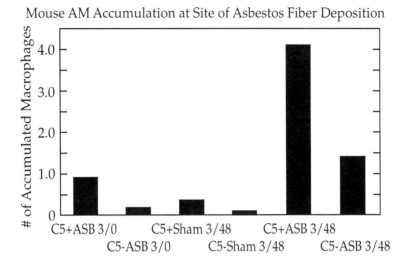

Figure 2. Mouse alveolar macrophage accumulation at sites of asbestos fiber deposition. Inhaled asbestos fibers stimulate a significant macrophage recruitment response to sites of fiber deposition in C5-normal mice by 48 hr post-exposure. In contrast, alveolar macrophages from C5-deficient mice have a reduced macrophage accumulation response, suggesting that C5 plays a role in the recruitment of macrophages following asbestos exposure.

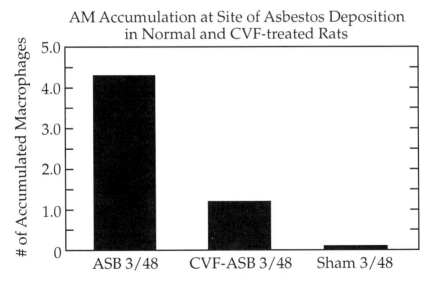

Figure 3. Alveolar macrophage accumulation at sites of asbestos deposition in normal and CVF-treated rats. Inhaled asbestos fibers stimulate a significant macrophage recruitment response to sites of fiber deposition in asbestos-exposed normal rats 48 hr after a 3-hr exposure. In contrast, asbestos-exposed rats depleted of complement via CVF treatment have a reduced macrophage recruitment response.

These findings have been extended to demonstrate that numerous particle or fiber types which activate complement *in vitro* and deposit at alveolar duct bifurcations when inhaled will generate complement-mediated chemoattractants *in vivo* following inhalation exposure (Figure 4). In this regard, it has been reported that a variety of particulates and fiber types such as chrysotile and crocidolite asbestos fibers, glass fibers, iron-coated chrysotile asbestos, wollastonite fibers and carbonyl iron particles activate complement in rat serum to generate chemotactic activity (Figure 5). Conversely, Mount St. Helens volcanic ash particles, incubated in serum through a broad range of concentrations, produced no chemotactic activity. It has been concluded that activation of complement by particulates with a variety of dissimilar chemical characteristics forms an initial step in the process of lung clearance of inhaled materials by providing a mechanism through which macrophages can detect and phagocytose particulates on alveolar surfaces (Warheit et al., 1988).

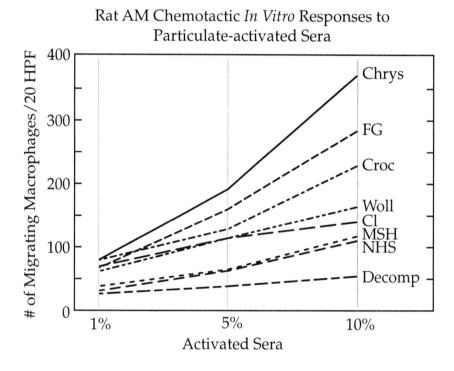

Figure 4. Rat alveolar macrophage *in vitro* response to particulate-activated sera. A variety of particulates were incubated with rat serum and tested for alveolar macrophage chemotactic activity *in vitro*. These included fiberglass (FG), crocidolite asbestos (Croc), chrysotile asbestos (Chrys), wollastonite fibers (Woll) Mount St. Helens particles (MSH), carbonyl iron particles (CI), and appropriate controls - normal heated serum (NHS) and decomplemented sera (Decomp). Chrysotile, fiberglass and crocidolite generated the greatest amount of chemotactic activity.

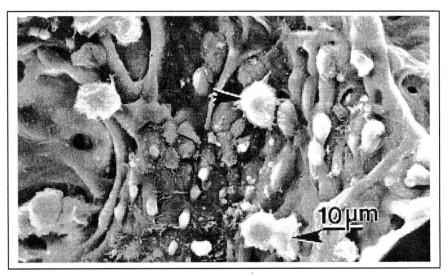

Figure 5. Scanning electron micrograph of lung tissue demonstrating alveolar macrophages (arrowheads) which have been recruited to sites of wollastonite fiber deposition following a brief exposure.

STUDIES ON THE CELLULAR MECHANISMS OF ASBESTOS-INDUCED LUNG INJURY

Animal models of asbestos-induced lung disease have been developed in a variety of rodent species including rats (Wagner et al., 1974; Pinkerton et al., 1984), mice (Bozelka et al., 1983), guinea pigs (Holt et al., 1966), as well as in sheep (Begin et al., 1981, 1983) exposed chronically to the fibers. The long-term inhalation models are particularly useful for identifying the anatomical and cellular patterns of disease. A major shortcoming related to the chronic exposure models, however, is the inability to gauge the time course of initiating pathogenic events. For example, the relationship between initial fiber deposition patterns and the subsequent acute and chronic cellular events that lead to asbestos-induced tissue thickening have not been elucidated. Similarly, the role of macrophages and other inflammatory cell types in the early development of asbestos-related inflammation and fibrosis has not been well established. Therefore, a model of acute asbestos-related lung effects was developed in rats. In this model, rats were exposed for 1 hr to an aerosol of chrysotile asbestos fibers and the development of early pulmonary cellular effects was evaluated at two post-exposure time periods, namely 48 hr and 1 mo post-exposure.

After brief aerosol exposures to chrysotile asbestos by rats, the inhaled fibers were observed by transmission (TEM) and scanning (SEM) electron microscopy to have deposited selectively on alveolar duct bifurcations and were cleared from epithelial surfaces by alveolar macrophage-mediated clearance or fiber translocation (Brody et al., 1981; Warheit et al., 1984). Fibers which had translocated from alveolar regions to the pulmonary interstitium via Type I epithelial cells, were generally phagocytosed, primarily by fibroblasts or by interstitial macrophages

(IM). TEM analysis revealed that the presence of these fibers within fibroblasts induced the formation of intracellular microcalcifications, a form of non-specific cellular injury (Brody et al., 1983). In the alveolar regions distal to the bronchiole-alveolar junction, AM were rapidly recruited to sites of asbestos fiber deposition in order to phagocytose fibers. The mechanisms for macrophage recruitment to alveolar duct bifurcation surfaces are associated with complement activation by the inhaled fibers and consequent generation of chemoattractants as discussed earlier. Light microscopic histological examination of exposed lung tissue indicated that proximal alveolar duct bifurcations appeared to have increased thickness in asbestos-exposed animals sacrificed 2 d after exposure. Using ultrastructural morphometric techniques, Chang et al. (1988) demonstrated that the influx of recruited AM formed a component of an early lesion which was characterized by increases in the volume of the epithelial and interstitial compartments of the bifurcations. In addition, the numbers of AM, IM, and Type I and Type II epithelial cells were all increased. One month after the 1 hr exposure, the numbers of AM on alveolar duct bifurcation surfaces no longer were increased compared to controls, but the tissue volume of the interstitial compartment was still significantly increased by 67% relative to sham controls. This was associated with an increase in the volume of noncellular interstitial matrix along with an accumulation of interstitial cells, including macrophages, myofibroblasts, fibroblasts, and smooth muscle cells. The Investigators concluded that acute structural changes on duct bifurcation surfaces measured at 48 hr after a 1 hr exposure were followed by a progressive interstitial response as evidenced by an increased number of interstitial cells and localized interstitial fibrosis measured at 1 mo post-exposure (Chang et al., 1988).

The documentation of an early lesion of asbestos-related lung injury measured 48 hr and 1 mo after a 1 hr exposure and identification of the target cell types have provided a basis for studying in rodents the cellular mechanisms associated with the development of asbestos-related pathologic response. The finding of alterations in the cellular and noncellular interstitial compartment concomitant with the subsequent development of fibrosis suggests a role for the interstitial-based fibroblasts in proliferating and synthesizing matrix components such as collagen, elastin, and glycosaminoglycans. The rate of collagen accumulation is likely to be balanced by (1) the numbers of fibroblasts and (2) the rate of collagen synthesis by individual cells and degradation by protease-secreting cells. Fibroblast proliferation and connective tissue formation are complex processes and may be independently regulated. Moreover, fibroblasts are known to play an active role in directing the course of fibrosis. Noting the complexity of potential responses in the interstitial microenvironment, it is still attractive to postulate that asbestos-exposed alveolar or interstitial macrophages synthesize and secrete mitogenic factors that enhance the numbers of interstitial cells and generate greater amounts of connective tissue.

Brody and coworkers (1997) have continued to utilize the short-term asbestos exposure model to study the mechanisms associated with the development of fiber-related lung disease. Using this model, Liu et al. (1997) and Lasky et al. (1996) have reported that the gene expression of transforming growth factor (TGF)-α and platelet-derived growth factor (PDGF)-α receptor were upregulated at sites of asbestos fiber deposition following inhalation exposures to asbestos. Brody et al. (1997) postulated that the cellular events leading to the development of fibrosis, namely cellular proliferation and extracellular matrix production, are mediated by three peptide growth factors, i.e., TGFα, TGFβ, and PDGF A- and B-chains. TGFα is known to be a potent mitogen for epithelial cells, while TGFβ reduces fibroblast growth but stimulates extracellular matrix synthesis, and PDGF isoforms are potent growth factors for mesenchymal cells. To test the hypothesis, rats were first exposed to aerosols of asbestos for 3 d and then, at 24 hr post-exposure, gene expression and protein secretion of TGFα, TGFβ, and the PDGF isoforms were assessed. As the results demonstrated that these four genes and their corresponding proteins were upregulated at sites of asbestos fiber deposition within 24 hr post-exposure, this suggested a role of the four in the development of asbestos-related lung disease. More recently, these Investigators have reported that tumor necrosis factor (TNF)-α receptor knockout mice are protected from the fibroproliferative effects of inhaled asbestos and that mice that are refractory to the development of asbestos-induced fibroproliferative lesions have reduced pulmonary gene expression of TNFα and TGF-β1 expression, suggesting that TNFα also plays a central role in the development of asbestosis in mice. The Investigators hypothesized that TNFα is essential for development of asbestos-induced fibrosis and that it mediates its effects through activation of other growth factors such as PDGF and TGFα that regulate cell growth and extracellular matrix production (Liu et al., 1998; Brass et al., 1999).

Recently, the 1 hr asbestos inhalation model has been extended for longer periods in order to address other mechanistic toxicology-related issues. In this respect, an inhalation bioassay, ranging from 3 d exposures to 1 mo aerosol exposures has been developed to assess the toxicity of inhaled particles and fibers. The issue being addressed in this particular study was to compare the pulmonary cellular proliferative effects of inhaled size-separated preparations of chrysotile asbestos fibers with similar aerosol fiber concentrations of p-aramid fibrils in rats. p-Aramid fibrils are an organic fiber-type which have been used as substitutes for asbestos in some commercial applications. In this study, rats were exposed for 2 wk to aerosols of p-aramid fibrils or chrysotile asbestos fibers at design concentrations of 400 and 750 f/cc, and the lungs of control and exposed animals were evaluated over a number of post-exposure periods ranging from 5 d through 1 yr. Two week exposures to p-aramid fibrils produced transient pulmonary inflammatory and transient increases in bromodeoxyuridine (BrdU) cell proliferation responses in terminal bronchiolar and subpleural regions of the respiratory tract. Similar to p-aramid, exposure to chrysotile produced a transient increase in pulmonary inflammatory responses. In contrast to the effects of p-aramid, however, substantial increases compared to controls in pulmonary cell proliferation indices were measured on terminal bronchiolar, parenchymal, subpleural, and mesothelial surfaces immediately after exposures, and some increases persisted for 3 mo post-exposure (Figure 6).

In complementary fiber clearance/biopersistence studies, it was demonstrated that retained p-aramid fibrils were biodegraded (i.e., low biopersistence) in the lungs of exposed rats. In contrast, removal from the lung of short chrysotile asbestos was fast but the clearance of long fibers (i.e. >5 μm) was slow or insignificant, resulting in a pulmonary retention of the long chrysotile asbestos fibers (Warheit et al., 1996). The dimensional changes of asbestos fibers, i.e., the selective retention of long fibers, as well as the pulmonary cell labeling data indicate that chrysotile fibers may have important significance for asbestos-related tumor development. In addition

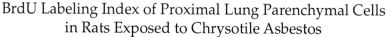

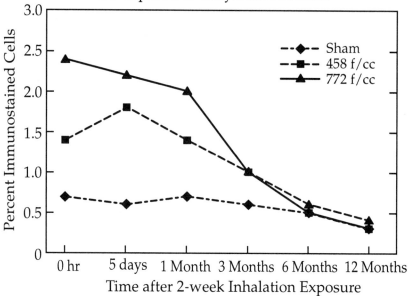

Figure 6. BrdU labeling index of proximal lung parenchymal cells in rats exposed to chrysotile asbestos. Exposure to chrysotile asbestos fibers in rats products a sustained increase in cell proliferative responses vs controls through 3 mo post-exposure.

histopathological evaluation of the exposed lungs revealed that repeated exposures to asbestos fibers for 2 wks resulted in significant alterations occurring in alveolar ducts immediately distal to the terminal bronchiole. The most prominent alterations were increased cellularity and septal wall thickening, which was evident during the first month post-exposure. Thickening of alveolar septal tissue was almost double when compared to controls immediately following exposure, as well as 5 d and 1 mo post-exposure. However, at 3 mo post-exposure, a significant reduction in the volume of affected alveolar tissue was measured in the lungs of exposed animals. These findings suggest that there is significant resolution of the initial tissue changes that were noted up to 1 mo post-exposure. This resolution appears to occur as early as 3 mo post-exposure and does not progress to a more severe state 1 yr after the end of fiber exposure. Thus it is interesting to note the apparent paradox of reversibility of the tissue thickening response concomitant with the lack of clearance of long fibers and sustained increases in cell proliferative responses through 3 mo post-exposure (Pinkerton et al., 1997).

SILICA

Long-term and or acute high-dose exposure to aerosolized crystalline silica (SiO_2) particles results in pulmonary fibrosis and lung tumors in humans and experimental animals. Quartz is a common form of crystalline SiO_2. The mechanism(s) of quartz-related lung disease have not been fully elucidated although the pathogenic effects are known to be associated, in part, with the cytotoxicity of these particles to epithelial and phagocytic cell surfaces and lysosomal membranes. Aerosol exposures to quartz in experimental animals result in the secretion of macrophage- and epithelial-derived cytokines resulting in a persistent lung inflammatory response (Warheit, 1989; Donaldson and Borm, 1998). Generation of reactive oxidants forms a component of the inflammatory response and consequent lung injury (Donaldson and Borm, 1998; Vallyathan et al., 1999). This may occur directly via interactions of SiO_2 particles within pulmonary cell types or may result from activation of recruited leukocytes via macrophage- or epithelial cell-derived cytokines. Continued generation of oxidants is likely to perpetuate the inflammatory cycle and stimulate expression of proinflammatory genes, leading to mutations in epithelial cells. Concomitant with these cellular inflammatory events and growth factors, sustained pulmonary epithelial cell proliferation may progress to epithelial hyperplasia and the consequent development of lung cancer (Donaldson and Borm, 1998; IARC, 1997).

One cytokine that has recently received considerable attention in the development of SiO_2-related pathological effects is TNFα (Huaux et al., 1999). TNFα is regarded as a pivotal mediator in the pathogenesis of SiO_2-induced inflammation. This cytokine is known to have several pro-inflammatory effects including: 1) augmenting inflammatory cell adherence to vascular endothelium as a prerequisite for recruitment to sites of alveolar inflammation; 2) activating the production of mediators by macrophages and neutrophils; 3) facilitating fibroblast proliferation and upregulating gene expression for collagen proteins which contribute to the fibrotic response; and 4) inhibiting collagen breakdown by phagocytic pathways (Donaldson and Borm, 1998; Huaux et al., 1999).

The importance of TNFα in modulating fibrogenic responses to SiO_2 has been demonstrated in studies wherein treatment of rodent hosts with anti-TNFα antibodies resulted in significant reduction of the SiO_2-induced fibrotic response. In these studies, decreased lung total hydroxyproline content relative to those amounts in SiO_2-exposed rats and mice that did not receive antibody were noted (Piquet et al., 1990). These conclusions, highlighting the role of TNFα, have been supported by the recent findings with transgenic animals demonstrating that the expression of a TNFα transgene in mice caused a persistent alveolar inflammatory response which progressed to a fibrotic lesion (Miyazaki et al., 1995). In addition, Ortiz and co-workers (1999) have reported that gene expression of TNFα mRNA in whole lung and bronchoalveolar lavage (BAL) cells was upregulated in mice exposed to quartz particles. Furthermore, the degree of increased TNFα expression in the lung was correlated with the murine strain sensitive to quartz, i.e. C57Bl mice, a sensitive strain, had upregulated TNFα expression and secretion in response to SiO_2 exposure while C3H mice, a strain resistant to effects of SiO_2, did not upregulate TNFα expression and had diminished TNFα secretion in response to SiO_2 (Ortiz et al., 1999). In this study, evidence of SiO_2-induced fibrosis was diminished in SiO_2-exposed double ($p55^{-/-}$, $p75^{-/-}$) TNFα receptor (TNFR) knockout mice as compared to that in normal mouse strains (Ortiz et al., 1999).

The association between TNFα and lung fibrosis has been reported in humans, as evidenced by increased BAL fluid levels of TNFα in patients with pulmonary fibrosis as well as increased release of TNFα by alveolar macrophages in patients with interstitial lung disease (Lasalle et al., 1990). Given the complex nature of the pathogenic response to quartz, it seems clear that many immunopathologic mechanisms are likely to be operative during the development of SiO$_2$-induced pulmonary fibrosis and tumors, and reports on a variety of other immunological mediators is beyond the scope of this brief review. However, the findings reported to date provide strong evidence that TNFα is a central mediator of interstitial pulmonary fibrosis in SiO$_2$-exposed experimental animals and is likely to play a significant role in the pathogenic events in quartz-exposed humans.

SUMMARY

Considerable progress has been made to help explain the mechanisms underlying development of asbestos-induced inflammation and fibrosis in rats following inhalation of chrysotile asbestos fibers. Following deposition, the inhaled fibers deposit at alveolar duct bifurcations and recruit macrophages to these sites. Alveolar macrophages accumulate at the duct bifurcations, phagocytose fibers, and play a significant role in the elaboration of cytokines and growth factors which mediate the growth and proliferation of endothelial, epithelial and mesenchymal cells. Some of the most important cytokines and growth factors appear to be TNFα, TGFα and TGFβ), and PDGF A- and B-chains. The results from acute inhalation studies (i.e., 1 hr exposures) may be different from the findings of longer-term studies or those which evaluate the time course of injury and repair kinetics. In this regard, Pinkerton and coworkers have reported that the asbestos-induced lesions at alveolar duct bifurcations are resolved at 6 and 12 mo post-exposure in the absence of continuing exposure. Thus, it will be necessary to reconcile the studies conducted at early time points relative to later times and to assess the relevance for humans of rodent studies with asbestos fibers, since in rats, cell mediated immunity does not appear to play a major role in the development of asbestos-related lung disease.

REFERENCES

Begin R, Rola-Pleszczynski M, Masse S, Lemaire I, Sirois P, Boctor M, Nadeau D, Drapeau G, Bureau MA. Asbestos-induced injury in the sheep model: The initial alveolitis. Environ. Res., 1983;30:195-210.

Begin R, Rola-Pleszczynski M, Sirois P, Lemaire I, Nadeau D, Bureau MA, Masse S. Early lung events following low-dose asbestos exposure. Environ. Res., 1981;26:392-401.

Bozelka BE, Sestini P, Gaumer HR, Hammad Y, Heather CJ, Salvaggio JE. A murine model of asbestosis. Am. J. Pathol., 1983;112:326-337.

Brass DM, Hoyle GW, Poovey HG, Liu JY, Brody AR. Reduced tumor necrosis factor-α and transforming growth factor β-1 expression in the lungs of inbred mice that fail to develop fibroproliferative lesions consequent to asbestos exposure. Am. J. Pathol., 1999;154:853-862.

Brody AR, Hill LH, Adkins B Jr. O'Connor RW. Chrysotile asbestos inhalation in rats: Deposition pattern and reaction of alveolar epithelium and pulmonary macrophages. Am. Rev. Respir. Dis., 1981;123:670-679.

Brody AR, Hill LH. Interstitial accumulation of inhaled chrysotile asbestos fibers and consequent formation of microcalcifications. Am. J. Pathol., 1982;109:107-114.

Brody AR, Liu JY, Brass D, Corti M. Analyzing the genes and peptide growth factors expressed in lung cells *in vivo* consequent to asbestos exposure and *in vitro*. Environ. Health Perspect., 1997;105 (Suppl. 5):1165-1171.

Chang LY, Overby LH, Brody AR, Crapo JD. Progressive lung cell reactions and extracellular matrix production after a brief exposure to asbestos. Am. J. Pathol., 1988;131:156-170.

Doll NJ, Diem JE, Jones RN, Rodriguez M, Bozelka BE, Stankus RP, Weill H, Salvaggio JE. Humoral immunologic abnormalities in workers exposed to asbestos cement dust. J. Allergy Clin. Immunol., 1983;72:509-512.

Donaldson K, Borm PJA. The quartz hazard: A variable entity. Ann. Occup. Hyg., 1998;42:287-294.

Frank MM. "Complement and Kinin." In *Medical Immunology*, DP Stites, AI Terr, TG Parslow, eds. Stamford, CT: Appleton and Lange, 1997, pp. 169-182.

Holt PF, Mill J, Young DK. Experimental asbestosis in the guinea pig. J. Pathol. Bacteriol., 1966;92:185-195.

Huaux F, Arras M, Vink A, Renauld JC, Lison D. Soluble tumor necrosis factor receptors p55 and p75 and interleukin-10 downregulates TNF activity during the lung response to silica particles in NRMI mice. Am. J. Respir. Cell Mol. Biol., 1999;21:137-145.

IARC: International Agency for Research on Cancer. Biological effects of asbestos. Report of the Advisory Committee on Asbestos Cancers to the Director of the International Agency for Research on Cancer. Ann. Occup. Hyg., 1973;16:9-17.

IARC: International Agency for Research on Cancer. *Monograph on the Evaluation of the Carcinogenic Risk of Chemicals to Humans. Volume 42: Silica and Some Silicates.* Lyon, France: IARC, World Health Organization, 1997

Kagan E, Solomon A, Cochrane JC, Beissmer EK, Gluckman J, Rocks PH, Webster I. Immunological studies of patients with asbestosis. I. Studies of cell mediated immunity. Clin. Exp. Immunol., 1977a;28:261-267.

Kagan E, Solomon A, Cochrane JC, Kuba P, Rocks PH, Webster I. Immunological studies of patients with asbestos II. Studies of circulating lymphoid numbers and humoral immunity. Clin. Exp. Immunol., 1977b;28:268-275.

Kolb WP, Kolb LM, Wetsel RA, Rogers WR, Shaw JO. Quantitation and stability of the fifth component of complement (C5) in bronchoalveolar lavage fluids obtained from non-human primates. Am. Rev. Respir. Dis., 1981;123:226-231.

Kubata M, Kagamimori S, Yokoyama K, Okada A. Reduced killer cell activity of lymphocytes from patients with asbestos. Br. J. Ind. Med., 1985;42:278-280.

Kunkel SL, Fantone JC, Ward PA. Complement-mediated inflammatory reactions. Pathobiol. Ann., 1981;11:127-154.

Lange A, Skibinski G, Garncarek D. The follow-up study of skin reactivity to recall antigens and E- and EAC-RFC profiles in blood in asbestos workers. Immunobiology, 1980;157:1-11.

Lange A. An experimental survey of immunological abnormalities in asbestos. I. Non-organ- and organ-specific autoantibodies. Environ. Res., 1980;22:162-175.

Larsen GL, Henson PM. Mediators of inflammation. Ann. Rev. Immunol., 1983;1:335-359

Lasky JA, Bonner JC, Tonthat B, Brody AR. Chrysotile asbestos induces PDGF-A chain-dependent proliferation in human and rat lung fibroblasts *in vitro*. Chest, 1996;109(Suppl. 3):26S-28S.

Lassalle P, Gosset P, Aerts C, Fournier E, Lafitte JJ, Degreef JM, Wallaert B, Tonnel AB, Voisin C. Abnormal secretion of interleukin-1 and tumor necrosis factor-alpha by alveolar macrophages in coal workers pneumoconiosis: Comparison between simple pneumoconiosis and progressive massive fibrosis. Exp. Lung Res., 1990;16:73-80.

Liu JY, Brass DM, Hoyle GW, Brody AR. TNFα receptor knockout mice are protected from the fibroproliferative effects of inhaled asbestos fibers. Am. J. Pathol., 1998;153:1839-1847.

Liu JY, Morris GF, Lei WH, Hart CE, Lasky JA, Brody AR. Rapid activation of PDGF-A and -B expression at sites of lung injury in asbestos-exposed rats. Am. J. Respir. Cell. Mol. Biol., 1997;17:129-140.

Miller K, Webster I, Handfield RI, Skikne MI. Ultrastructure of the lung in the rat following exposure to crocidolite asbestos and quartz. J. Pathol., 1978;124:39-44.

Miyazaki Y, Araki K, Vesin C, Garcia I, Kapanci Y, Whitsett JA, Piguet PF, Vassalli P. Expression of a tumor necrosis factor-alpha transgene in murine lung causes lymphocytic and fibrosing alveolitis: A mouse model of progressive pulmonary fibrosis. J. Clin. Invest., 1995;96:250-259.

Ortiz LA, Lasky J, Lungarella G, Cavarra E, Martorana P, Banks WA, Peschon JJ, Schmidts HL, Brody AR, Friedman M. Upregulation of the p75 but not the pP55 TNF receptor mRNA after silica and bleomycin exposure and protection from lung injury in double receptor knockout mice. Am. J. Respir. Cell Mol. Biol., 1999;20:825-833.

Pinkerton KE, Elliot AA, Frame SR, Warheit DB. Reversibility of fibrotic lesions in rats inhaling size-separated chrysotile asbestos fibers for 2 weeks. Ann. Occup. Hyg., 1997;41(Suppl. 1):178-183.

Pinkerton KE, Pratt PC, Brody AR, Crapo JD. Fiber localization and its relationship to lung reactions in rats after chronic exposure to chrysotile asbestos. Am. J. Pathol., 1984;117:484-498.

Piquet PF, Collart MA, Gran GE, Sappino AP, Vassalli P. Requirement of tumor necrosis factor for development of silica-induced pulmonary fibrosis. Nature, 1990;344:245-247.

Robertson J, Caldwell JR, Castle JR, Waldman RH. Evidence for the presence of components of the alternative (properdin) pathway of complement activation in respiratory secretions. J. Immunol., 1976;117:900-903.

Rosenthal GJ, Corsini E, Simeonova P. Selected new developments in asbestos immunotoxicity. Environ. Health Perspect., 1998;106(Suppl. 1):159- 169.

Snyderman R. Mechanisms of inflammation and leukocyte chemotaxis in the rheumatic diseases. Med. Clin. N. Amer., 1986;70:217-235.

Tsang PH, Chu FN, Fischbein A, Bekesi G. Impairments in functional subsets of T-suppressor (CD8) lymphocytes, monocytes, and natural killer cells among asbestos-exposed workers. Clin. Immunol. Immunopathol., 1988;47:323-332.

Turner-Warwick M, Parkes WR. Circulating rheumatoid and anti-nuclear factors in asbestos workers. Br. Med. J., 1970;3:492-495.

Vallyathan V, Shi X, Dalal NS, Irr W, Castranova V. Generation of free radicals from freshly fractured silica dust: Potential role in acute silica-induced lung injury. Am. Rev. Respir. Dis., 1988;138:1213-1219.

Wagner JC, Berry G, Skidmore JW, Timbrell V. The effects of inhalation of asbestos in rats. Br. J. Cancer, 1974;29:252-269.

Warheit DB, Chang LY, Hill LH, Hook GER, Crapo JD, Brody AR. Pulmonary macrophage accumulation and asbestos-induced lesions at sites of fiber deposition. Am. Rev. Respir. Dis., 1984;129:301-310.

Warheit DB, George G, Hill LH, Snyderman R, Brody AR. Inhaled asbestos activates a complement-dependent chemoattractant for macrophages. Lab. Invest., 1985;52:505-514.

Warheit DB, Hartsky MA, Frame SR. Pulmonary effects in rats inhaling size-separated chrysotile asbestos fibers or p-aramid fibrils: Differences in cellular proliferative responses. Toxicol. Lett., 1996;88:287-292.

Warheit DB, Hill LH, George G, Brody AR. Time course of chemotactic factor generation and the corresponding macrophage response to asbestos inhalation. Am. Rev. Respir. Dis., 1986;134:128-133.

Warheit DB, Overby LH, George G, Brody AR. Pulmonary macrophages are attracted to inhaled particles through complement activation. Exp. Lung Res., 1988;14:51-66.

Warheit DB. Interspecies comparisons of lung response to inhaled particles and gases. CRC Crit. Rev. Toxicol., 1989;20;1-29.

15. WOODSMOKE, KEROSENE HEATER EMISSIONS, AND DIESEL EXHAUST

Judith T. Zelikoff, Ph.D
New York University School of Medicine
Department of Environmental Medicine
57 Old Forge Road
Tuxedo, New York 10987

WOODSMOKE

Introduction

Over the past two decades, due to rising energy costs and the uncertain availability of petroleum and natural gas, homeowners in the United States have increasingly turned to the use of wood as an alternate heating fuel. This trend has been especially striking in the northeastern and northwestern United States (National Research Council, 1981). For example, the Washington State Department of Ecology estimated that wood was burned in 60% of all Washington households, with about 2.2 million cords of wood consumed annually (Pierson et al., 1989). In Massachusetts, studies showed that 65% of all households surveyed used wood-burning devices, and of those, 68% were in use 24 hr a day (Tuthill, 1984). While sales of woodstoves have been recently declining, the current numbers of woodstoves in the United States is thought to exceed 11 million. It is estimated that approximately 10% of the total space-heating input for the United States is from firewood (Lipfert and Dungan, 1983).

Increased usage of woodburning devices has resulted in greater public exposure to indoor and outdoor pollutants generated during combustion and an increased concern by local residents for the health effects associated with woodsmoke (WS) effluents. These concerns focus on the fact that woodburning stoves, furnaces, and fireplaces emit significant quantities of toxic, as well as some carcinogenic compounds, including respirable particulate matter with diameters <10 μm (PM_{10}), carbon monoxide (CO), nitrogen and sulfur oxides (NO_x/SO_x), aldehydes, polycyclic aromatic hydrocarbons (PAH), and free radicals (Cooper, 1980; Dasch, 1982; Lao, 1983; Tuthill, 1984; Sexton et al., 1986; Traynor et al., 1987; Koenig and Pierson, 1991; Hildemann et al., 1991; Koenig et al., 1993). Citizen concern has prompted the legislation of laws mandating the reduction of residential WS levels. Although these legislative actions appear promising, only a few states are currently affected (Koenig et al., 1988). Meanwhile, those most vulnerable to health impairment secondary to air pollutants, including very young children and the elderly (Koenig and Pierson, 1991), will continue to be exposed both indoors and outdoors for prolonged periods of time to high concentrations of WS-generated particulates and the other smoke-associated pollutants.

Under normal usage conditions, woodburning devices create both indoor and outdoor pollution (Lao, 1983; Sexton et al., 1986; Samet et al., 1987, 1988; Traynor et al., 1987). Even though woodburning stoves and fireplaces are vented directly to the outside, many circumstances including improper installation, negative indoor air pressure, and down drafts, facilitate re-entry of incomplete combustion products directly indoors (Pierson et al., 1989). Traynor et al. (1987) reported that both airtight and non-airtight stoves released PM_{10}, CO, and PAH within the indoor environment; concentrations of PM_{10} as great as 820 $\mu g/m^3$ (current EPA standard for PM_{10} is 150 $\mu g/m^3$ for a 24 hr period) were measured indoors from non-airtight stoves over a 24 hr period. In addition to that amount released directly indoors, a large percentage of outdoor WS from chimneys actually re-enters the home (Pierson et al., 1989). Since individuals typically spend 60 - 70% of their out-of-work time at home (Szalar, 1972; Chapin, 1974; Sexton et al., 1986), indoor WS represents a major source of human exposure.

Woodburning devices also contribute to outdoor air pollution (Butcher and Sorenson, 1979; Cooper, 1980; Koenig et al., 1988; Koenig and Pierson, 1991; Koenig et al., 1993). For example, aldehydes released into the ambient environment from woodburning have been measured at levels comparable to those emitted from power plants and automobiles (Lipari et al., 1984). Studies by Cooper (1980) have shown that on a moderately cold winter day, 51% of the respirable air particulates in the Portland, Oregon area were from residential wood combustion sources. Koenig et al. (1988) reported that residential WS in the Olympia, Washington area accounted for 50% (on clear days) to 85% (on polluted days) of the airborne PM_{10}. Other studies in the same geographic area have demonstrated that 80 - 90% of the PM measured in the ambient air (150 $\mu g/m^3$) was due to use of woodburning devices during nighttime hours (Larson et al., 1992). Such studies have led to the conclusion that WS represents a more significant source of ambient PM, volatile organic compounds, and CO, than the sum total of all industrial point sources in the state of Washington (Koenig et al., 1988).

As previously noted, WS is a complex mixture of several gases, as well as respirable fine particles of varying inorganic/organic composition and diameter (Pierson et al., 1989). Individually, most of these compounds have been shown to produce acute and chronic biologic effects and/or cause deleterious physiologic responses in exposed humans (Anderson et al., 1973; Spiezer et al., 1980; Ramage et al., 1987; Koenig et al., 1988; Pierson et al., 1989; Koenig and Pierson, 1991; Pope et al., 1991; Schwartz, 1993). Carbon monoxide, at levels measured in homes using woodstoves (in the range of 1.2 - 43 ppm, compared to the indoor level of 5 ppm recommended by the American Society for Heating, Refrigeration, and Air Conditioning Engineers [ASHRAE] and the current outdoor standard of 9 ppm for an 8 hr period), has been shown to produce carboxyhemoglobin and increase the incidence of angina among persons with cardiac disease (Anderson, 1973). Nitrogen oxides (primarily nitrogen dioxide and nitric oxide) bind to hemoglobin to produce methemoglobin and hematologic aberrations, affect the activity of several enzyme systems, cause vascular membrane injury and leakage leading to edema, produce bronchoconstriction in asthmatics at low levels, alter pulmonary macrophage function, and increase host susceptibility to pulmonary infections and respiratory illness in children (Speizer et al., 1980; Samet et al., 1987; Pierson et al., 1989). Polycyclic aromatic hydrocarbons, incomplete combustion products from woodburning that are released into the environment adsorbed onto emitted PM_{10}, are immunosuppressive in laboratory animals (White et al., 1994) as well as suspected and/or proven carcinogens in animals and, possibly, in humans (Koenig et al., 1988; Pierson et al., 1989). Aldehydes (mostly in the form of formaldehyde and acrolein) measured indoors at concentrations ranging from 0.3 to 1.0 ppm (compared to ASHRAE-recommended indoor levels of 0.1 ppm) are associated with upper airway irritation, headaches and other neurophysiologic dysfunctions, exacerbation of bronchial asthma and, possibly, cancer (Kerns et al., 1983). In addition to the aforementioned effects, other chronic health effects in humans have also been associated with WS exposure. These include: increased airway resistance, decreased vital capacity, increased respiratory symptoms in children (i.e., cough, wheeze, dyspnea), exacerbation of asthma symptoms, and increased respiratory infections in children (Koenig et al., 1988; Butterfield et al., 1989).

Although the health effects associated with exposure to whole WS effluents are not as well studied as its individual components, a number of adverse health consequences have been demonstrated. For example, exposure of laboratory animals to woodsmoke decreases ventilatory frequency and ventilatory response to CO_2 (Wong et al., 1984), increases microvascular permeability and produces pulmonary edema (Nieman et al., 1988), causes necrotizing tracheobronchial epithelial cell injury (Thorning et al., 1982), increases angiotensin-1-converting enzyme in the lungs (Brizio-Molteni et al., 1984), compromises pulmonary immune defense mechanisms important in host resistance against infectious agents (Fick et al., 1984; Loke et al., 1984; Thomas and Zelikoff, 1999), and, possibly, increases the lung cancer incidence in mice (Liang et al., 1988). In humans, health effects are related to host age at the time of smoke exposure. In adults, prolonged inhalation of WS contributes to chronic bronchitis (Rajpandey, 1984), causes chronic interstitial pneumonitis and fibrosis (Ramage et al., 1987), produces cor pulmonale, interstitial lung disease, pulmonary arterial hypertension (Sandoval et al., 1993), and alters pulmonary immune defense mechanisms by increasing the levels of pulmonary immunoglobulins (Ig) and immune cell population densities and functions (Demarest et al., 1979; Ramage et al., 1988).

While effects of repeated WS exposures on adults are notable, young children appear to be the most susceptible subpopulation. Inhalation exposure by preschool children living in homes heated with woodburning stoves: decreases pulmonary lung function in young asthmatics (Koenig et al., 1993), increases the incidence of acute bronchitis and the severity/frequency of wheezing and coughing (Butterfield, et al., 1989) and increases the incidences, duration and, possibly, severity of acute respiratory infections (Honicky et al., 1983; Rajpandey, 1984; Honicky et al., 1985; Collings et al., 1990; Morris, et al., 1990; Honicky and Osborne, 1991). Even in those few epidemiological studies which fail to correlate WS exposure with respiratory disease/symptoms (Anderson, 1978; Browning et al., 1990), the Authors conclude that WS pollution may aggravate symptoms of respiratory disease and should not be disregarded as a possible contributing factor to increased respiratory infections in young children.

IMMUNOTOXICOLOGY OF WOODSMOKE: EFFECTS AND MECHANISMS OF RESPONSE

Many respirable pollutants found in WS have been shown to alter pulmonary immune defense mechanisms. For example, Hatch et al. (1981) demonstrated that 100 µg of urban air particles (0.4 µm) intratracheally (IT)-instilled into the lungs of mice reduced host resistance to bacterial infection. Jakab (1993) demonstrated that mice co-exposed to carbon black (10 mg/m^3, 4 hr/d, for 4 d) and acrolein (2.5 ppm) had suppressed intrapulmonary killing of *Staphylococcus aureus*, impaired elimination of *Listeria monocytogenes* and Influenza A virus, and altered intrapulmonary killing of *Proteus mirabilis*. The investigators suggested that the biologic effect was due to carbon particles acting as vehicles to carry acrolein into the deep lung. Studies by this laboratory investigating the immunotoxicity of inhaled ambient PM demonstrated the ability of air particulates concentrated from New York City air to exacerbate an ongoing *Streptococcus pneumoniae* infection in exposed rats (Thomas and Zelikoff, 1999; Zelikoff et al., 1999). Besides affecting host resistance, WS-associated components have been shown to alter pulmonary macrophage activity (Burrell et al., 1992; Jakab and Hemenway, 1993). For example, rats exposed nose-only to an iron oxide aerosol (mass median diameter = 1.6 µm) for 2 hr had delayed increases in $F_{c\gamma}RI$-mediated phagocytosis.

While only a handful of animal and human studies have been performed in the past examining the effects of whole WS effluents on pulmonary immunity, suppressive effects on host defense and/or immune cell function similar to those produced by individual WS components have been observed (Demarest et al., 1979; Fick et al., 1984; Loke et al., 1984; Thomas and Zelikoff, 1999). Laboratory investigations by Fick et al. (1984) indicated that a single exposure of rabbits to smoke from the pyrolysis of Douglas fir wood produced an increase in the total number of recovered pulmonary macrophages, and a transitory decrease in macrophage adherence to glass. In addition, acute inhalation of WS also decreased macrophage uptake of the Gram-negative bacterial pathogen, *Pseudomonas aeruginosa*. Woodsmoke-induced effects occurred without changes in macrophage viability and in the absence of an inflammatory response. In another laboratory study, a single inhalation exposure of Douglas fir-generated WS altered rabbit macrophage morphology and increased the total cell number recovered by bronchopulmonary lavage (Loke et al., 1984). These results are similar to those reported by Demarest et al. (1979) who demonstrated an increase in the number of cells obtained by bronchopulmonary lavage from smoke-exposed human subjects; in this study, macrophages also displayed a decreased migration in response to chemoattractants. While these studies have provided some evidence that inhaled WS alters pulmonary immune defense mechanisms and that the lung macrophages, a primary defense of the deep lung that provides a link between the non-specific and specific defense systems of the respiratory tract (see Chapter 2 for further details), is the likely primary target for the immunotoxic effects of inhaled WS, the most compelling evidence demonstrating the immunotoxic potential of inhaled WS comes from experimental studies performed in this laboratory.

Very recent studies have demonstrated that short-term exposure of rats to a relevant indoor WS concentration progressively inhibited bacterial clearance from their lungs even in the absence of any histopathological changes, lung cell damage, or inflammation; the suppressed clearance began as early as 3 hr post-exposure and persisted for almost 2 wk (Thomas and Zelikoff, 1999). Interestingly, similar dramatic effects were not observed in rats exposed to particle-free WS effluents (Zelikoff et al., 1997). Ambient particles of the size emitted during woodburning (respirable WS particulates are composed primarily of a relatively equal mixture of ultrafine/fine [0.02 - 2.5 μm d_a] and course [2.5 - 3.5 μm d_a] particles) (Sexton et al., 1986; Traynor et al., 1987; Hildemann et al., 1991) are likely the most striking aspect of indoor and outdoor pollution due to woodburning (Sexton et al., 1986; Traynor et al., 1987; Pierson et al., 1989; Hildeman et al., 1991). These respirable particles can penetrate into the deep lung and remain indefinitely, producing a variety of morphological and biochemical changes, as well as exacerbation of preexisting medical conditions (Schwartz, 1991). Although still controversial, an increased incidence and rate of infectious respiratory diseases (e.g., bronchitis) has also been associated with inhaled particulates (Pope, 1991). Thus, the importance of particulates in bringing about the observed time-related effect of WS upon host antibacterial responses is not surprising.

Results from *ex vivo* macrophage functional assays also performed in this Author's laboratory showed that exposure to WS suppressed production of superoxide anion (Figure 1), an oxyradical critical for the intracellular killing of *S. aureus*. Alterations in this parameter likely played a role in the observed decrease in host resistance. Moreover, inasmuch as macrophages recovered from WS-exposed rats had an impaired ability to kill *S. aureus in vitro* (compared to control), it appears that short-term WS exposure acts via direct effects upon pulmonary macrophage function rather than indirectly through neuroimmune mechanisms to induce changes in pulmonary antibacterial defense. Overall, these findings as well as those from earlier studies demonstrate that inhaled WS can alter pulmonary immune defense mechanisms and that the lung macrophage is a sensitive target for immunotoxic effects of WS. These findings may help to explain the increased incidence of respiratory infections observed worldwide in children exposed to woodburning emissions (Honicky et al., 1983, 1985; Morris et al., 1990).

KEROSENE HEATER EMISSIONS

Introduction

Kerosene heaters are one type of combustion space heater used to supply supplemental heat (MMWR, 1997). Between 1988 and 1994 unvented combustion

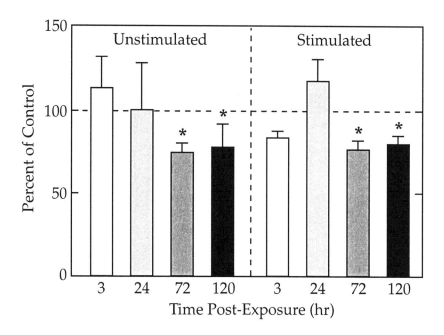

Figure 1. Effect of short-term repeated inhalation of woodsmoke upon superoxide anion production by rat pulmonary macrophages. Each bar is the mean (±SEM) from 12 animals per exposure group. Values significantly (p<0.05) different from those of the air controls are indicated (*).

space heaters (i.e., kerosene- and propane-fueled space heaters, some gas-fueled log sets, and cooking devices used improperly for heating) were used for home heating by an estimated 13.7 million adults. Five million kerosene heaters were sold in 1982 in the United States alone (de la Vega et al., 1990). Usage proves more common among rural-dwelling adults than among those residing in urban environments (i.e., 10 compared with 3.7 million). Moreover, utilization appears greatest among individuals with an annual household income of ≤$20,000 and by African-Americans (11%) more than Caucasians (7%).

While home heating is the primary purpose of kerosene heaters in developed countries, in developing nations it is also heavily relied upon for lighting and cooking. In a developing economy such as that of India, a substantial proportion of its population continues to use kerosene as a domestic fuel in view of its easy availability and subsidized rates; moreover, many industries in these nations also use kerosene oil (a colorless petroleum derivative consisting of saturated and unsaturated hydrocarbons and trace amounts of sulfur-derived and nitrogen-derived impurities) as an organic solvent. In this situation, incomplete combustion of the kerosene oil used for cooking and illuminating purposes generates large volumes of soot, byproducts of incomplete combustion such as PAHs and aliphatic compounds (IARC, 1985; Arif et al., 1997). Exposure of humans/animals to kerosene oil or its soot has been shown to cause biochemical changes in certain tissues which, in some cases, may lead to serious pulmonary or cardiopulmonary disorders (Noa and Illnait, 1987; Goodwin et al., 1988; Arif et al., 1991).

Depending upon operating conditions, heater type, and fuel composition, kerosene heaters can release into small closed indoor settings toxic combustion products including CO, CO_2, NO_x, SO_x, aldehydes, particulates, and carcinogenic and/or non-carcinogenic hydrocarbons often at levels above those considered safe for long-term exposure (Ragland et al., 1985; Arashidani et al., 1996). For example, a study performed in Japan by Arashidani et al. (1996) using both convective (i.e., units with exposed flames) and radiant/reflective (i.e., heater with a metal mesh assembly covering the wick and a reflector to direct radiant heat out the front) kerosene heaters, demonstrated that under non-ventilating conditions NO_x concentrations in the room increased with usage reaching a maximum level of 2.4 ppm after 1 hr. Under non-ventilating conditions, concentrations of CO and CO_2 increased simultaneously, reached a steady state level of 11 ppm within 1 hr after heating, and declined to baseline after 3 hr. Suspended particulate matter, generated by both the convection and reflective heaters, were highly respirable ranging between 0.3 and 0.5 μm in diameter; levels remained below the PM_{10} NAAQS (i.e., ~40 μg/m^3 vs 150 μg/m^3). Polycyclic aromatic hydrocarbon levels (measured as benzo(a)pyrene [BaP]) from both types of kerosene heaters increased with burning time and reached a maximum level between 0.43 and 1.94 ng/m^3; for both heater types, levels of NO_2 exceeded the NAAQS (i.e., 0.05 ppm). In general, the afore-mentioned study demonstrated that ventilation was an effective method for decreasing indoor kerosene-generated pollutant concentrations. Woodring et al. (1985) using unvented kerosene heaters under a controlled-laboratory setting demonstrated similar findings and concluded that either NO_2 or SO_2 were likely the first pollutants to exceed national standards if ventilation was insufficient. The Authors also concluded that increased indoor pollutant levels could arise as a result of some actions taken by homeowners in an attempt to lower operating costs of kerosene heaters which include the use of poor quality fuel, operation in closed rooms, and failure to provide dilution ventilation.

Despite the potential risk of exposure to unacceptably high concentrations of kerosene heater-generated toxic/carcinogenic combustion products, use of unvented combustion heaters has increased in the United States since 1992. This may be due to the fact that in many states regulations prohibiting the use of these devices have been rescinded. As of 1997, only five states (i.e., Alaska, Massachusetts, Minnesota, Colorado, and Utah) prohibit use of unvented gas-fueled or liquid-fueled heaters.

Overall, relatively few studies have been performed investigating the toxic effects of kerosene. Of these, the majority of studies have focused upon the effects of accidental kerosene ingestion on the respiratory tract (Baldachin and Malmed, 1964; Rai and Singh, 1980). In such cases, health impacts including pneumonitis, atelectasis, and pulmonary edema have been observed; in some cases, chronic cough, recurrent bronchitis and asthma have also been noted (Brunner et al., 1964). More studies are clearly needed to better understand human health risks associated with kerosene exposure.

IMMUNOTOXICOLOGY OF KEROSENE HEATER EMISSIONS AND/OR FUMES: EFFECTS AND MECHANISMS OF RESPONSE

Evidence from a few epidemiologic studies infer an association between kerosene exposure and asthma. In one study, de la Vega et al. (1990) reported a correlation between daily exposure to kerosene and asthma severity in a study of 286 asthmatic patients followed for five years. Of those individuals exposed to kerosene fumes, 15.5% improved clinically, while 43.9% of those who failed to improve had daily contact with the fuel (compared to unexposed controls). In a similar study carried out in Havana, Cuba, the prevalence of asthma was found to be twice as high in women exposed to kerosene fumes compared to those in the general population (Diaz et al., 1982). However, since both studies were poorly-explained and some important controls appeared to be lacking, any conclusions reached from these studies should be approached with caution.

Unlike the rather weak epidemiologic data, a number of *in vivo/ex vivo* animals studies investigating the effects of inhaled kerosene on airway hyper-responsiveness have provided compelling evidence relating exposure to kerosene with bronchoconstriction (Casaco et al., 1985a and b and c; Mesa et al., 1988a and b; de la Vega et al., 1990). For example, Casaco et al. (1982) described increased airway resistance in rabbits in response to kerosene inhalation. In another airway reactivity study by the same investigator (Casaco et al., 1985), inhalation of aerosolized kerosene by guinea pigs one hour before exposure to acetylcholine potentiated the response of isolated tracheal strips to the agonist and induced a decrease in the lethal dose of acetylcholine. Inhibition of tracheal acetylcholin-esterase activity and an ensuing increase of acetylcholine activity on airway smooth muscle was thought by the Authors to explain, at least in part, respiratory symptoms that frequently follow kerosene intoxication. A later study by Casaco et al. (1990) supported the aforementioned findings and demonstrated that a sharp rise in airway resistance in rabbits followed a 30 min inhalation of smoke generated from a kerosene kitchen range; a cholinergic response was suggested. since atropine pretreatment prevented the effect. Studies by Mesa et al. (1988a) investigating possible underlying mechanisms by which acute inhalation of kerosene aerosol might induce bronchoconstriction demonstrated that inhalation of kerosene (20.4 ppm) by guinea pigs for 5 min resulted in increased airway resistance in the absence of increased blood histamine levels; on the other hand, bronchoconstrictive effects of kerosene in guinea pigs and rabbits was not modified by administration of the histamine H_1 antagonist mepyramine or the steroidal anti-inflammatory drug triamcinolone, respectively. From this study, the Authors concluded that acute kerosene-induced bronchoconstriction was not mediated by stimulation of histamine H_1 receptors or by the release of chemical mediators. Other studies by these same Investigators (Mesa et al., 1988b) suggested that changes in lysosomal enzyme levels in the lungs of kerosene-exposed animals might explain the inflammatory response observed in animals exposed by inhalation to kerosene vapors or its combustion fumes.

The immunotoxic effects of instilled kerosene soot (a combustion product of kerosene oil) on recovered rat alveolar macrophages was examined in two studies by Arif et al. (1993, 1997). In the earlier study, the Investigators demonstrated that a single IT instillation of kerosene soot at a concentration of 5 mg was cytotoxic to lavaged macrophages and that the observed cytotoxicity appeared to be mediated by production of reactive oxygen intermediates (ROI). In this study, oxidative stress was indicated by increased production of hydrogen peroxide (H_2O_2) and thiobarbituric acid (TBA)-reactive substances from macrophages recovered from soot-exposed rats at different timepoints post-instillation; decreased antioxidant activity was also noted in these animals. The Authors concluded that macrophage injury by soot depended upon ROI formation and, thus, soot exposure could result in serious health problems in exposed human populations. In the later study, the relationship between kerosene, its soot, and asbestos-mediated toxicity was investigated using rats IT-instilled with either a single bolus of: chrysotile asbestos (i.e., 5 mg); kerosene oil (50 µl); kerosene soot (i.e., 5 mg); or asbestos in combination with a single kerosene product. Rats were sacrificed at different timepoints post-exposure and effects upon lavaged cell numbers, biochemical markers of lung cell injury, alveolar macrophage-mediated production of H_2O_2, TBA substances, and antioxidant activity were examined. Co-exposure of rats to both chrysotile asbestos and kerosene oil/soot resulted in increased numbers of alveolar macrophages, elevated levels of H_2O_2 and TBA, and depletion of macrophage glutathione levels (compared to rats exposed to either agent alone). The Authors suggested that the resulting oxidative stress may have altered cell membrane permeability and damaged the phagolysosomal membrane leading, ultimately, to the extracellular release of acid phosphatase and lactate dehydrogenase. They concluded that such injury to exposed macrophages could trigger events leading to lung fibrosis and/or malignancies in exposed animals.

Studies examining the incidence of pulmonary infections in kerosene-exposed-children also infer that kerosene might possess immunotoxic properties. While the majority of these studies found no clear association between kerosene exposure and infectious respiratory disease (Tominaga and Itoh, 1985; Azizi and Henry, 1991; Behera and Jindal, 1991; Awasthi et al., 1996), it has been suggested by Sharma et al. (1998) that exposure to kerosene from cooking increased the incidence of acute respiratory infections in very young children in developing nations. In this study, infants living in an urban slum in northern India, in which the families cooked with kerosene fuel, demonstrated a higher incidence of acute lower respiratory infections than those children living in a less-polluted area. Clearly, additional studies which consider nutritional status and child-rearing habits are needed before a definitive conclusion regarding the role of inhaled kerosene on lower respiratory tract infection can be reached.

DIESEL EXHAUST

Introduction

Diesel exhaust (DE) is a pollutant mixture to which millions of people are exposed (Henderson and Mauderly, 1994). Estimates of DE concentrations to which human populations are exposed span three orders of magnitude and range from 1 - 10 µg/m^3 in the general urban environment to >1 mg/m^3 in some underground mining operations (Cass and Gray, 1995). NIOSH estimates that more than one million workers in the United States are exposed occupationally to DE. Industrial hygiene studies have demonstrated that mean exposure concentrations in the work environment (generally measured over a work shift and estimated using different measurement and analysis methods) range from 4 $\mu g/m^3$ for truck drivers to 1740 $\mu g/m^3$ for underground miners. While considerably fewer data exist concerning exposure of the general population, it is clear that DE is a major pollutant in the non-occupational urban environment.

Diesel exhaust is composed of a particulate and gaseous phase (Henderson and Mauderly, 1994). The particulate phase can be broken down to a fine particulate carbon soot phase and an organic phase adsorbed onto the soot. These particles have a mass median diameter of 0.05 - 1.0 μm, a size that renders them highly respirable and capable of depositing into the deep lung (Snipes, 1989). Diesel exhaust particles (DEP) consist of a carbonaceous core with a large surface area to which various hydrocarbons (HC) are adsorbed, including PAHs and nitro-PAHs which have elicited the most concern in regards to human health. The gaseous phase contains a variety of combustion products and HC, including some of the same PAHs found in the particle phase (Cohen and Nikula, 1999). As in many pollutant mixtures, the composition of DE varies considerably with the conditions under which it is produced (i.e., condition and type of diesel engine, fuel burned, time during the driving period, and load on the engine). Once generated, components of DE undergo atmospheric transformation in ways that may be relevant to human health. For example, nitro-PAHs created by the reaction of directly emitted PAHs with hydroxyl radicals in the atmosphere can be more potent mutagens and carcinogens, and more bioavailable, than their precursors (Winer and Busby, 1995). Compared with gasoline engine exhaust, DE is lower in CO, hydrocarbons, and CO_2, but higher in NO_x and respirable particulate matter (Cohen and Nikula, 1999).

The toxicity of inhaled DE has been studied by numerous investigators worldwide and a number of excellent reviews currently exist on the topic (Mauderly, 1992; Henderson and Mauderly, 1994; Nel et al., 1998; Cohen and Nikula, 1999). The most active area of DE research has perhaps been in the area of carcinogenesis. While certain limitations in the data exist, epidemiological evidence suggests that long-term employment in jobs with substantial DE exposure (i.e., railroad workers, truck drivers, and bus garage workers) is associated with a 20 - 50% increase in risk for lung cancer (relative risk of 1.2 - 1.5%). Laboratory studies using animal models support this finding. Perhaps the most compelling data comes from a study by Mauderly et al. (1987) in which F344/N rats exposed chronically to high levels of DE developed lung tumors after 24 mo of exposure; the prevalence of lung tumors was increased significantly (above control prevalence) at the high and medium dose levels. Though convincing evidence exists demonstrating an association between DE and cancer, some questions still remain including: which specific DE constituent(s) is/are responsible for producing cancer; the role particle burden may play in carcinogenesis; underlying cellular/molecular mechanisms of effect; and whether an exposure threshold exists. Data from several inhalation studies have led to the suggestion that the carcinogenic response of rats to inhaled DE may be non-specific and related to the heavy lung burden of particulate material rather than to a unique carcinogenic property of the chemical. McClellan (1996) suggests that DE induces tumors in rats through non-genotoxic or indirectly genotoxic mechanisms which involve inflammation and/or cell proliferation.

Non-cancer respiratory pathologies have also been associated with inhalation of DE. A number of epidemiologic and animal studies have demonstrated changes in pulmonary function following chronic DE exposure (Gross, 1981; Gamble et al., 1983; Mauderly et al., 1988; Brunekreef et al., 1997; reviewed in Cohen and Nikula, 1999); both restrictive and obstructive patterns of pulmonary dysfunction have been reported. Decreases in lung volumes (Mauderly et al., 1988) and lung compliance (Heinrich et al., 1986) suggest restrictive disease and are compatible with parenchymal fibrosis; decreases in flow rates (Lewis et al. 1986) and increases in airway resistance (Heinrich et al., 1986) suggest obstructive disease. The earliest response to inhaled DE appears to be pulmonary inflammation which in many cases is detected earlier than alterations in pulmonary function. Chronic exposure of rats to DE at a level that induces pulmonary inflammation eventually leads to pulmonary fibrosis and pulmonary function changes, which accompany the structural change to a "stiffer lung".

IMMUNOTOXICOLOGY OF DIESEL EXHAUST EMISSIONS (DEE): EFFECTS AND MECHANISMS OF RESPONSE

Perhaps the most widely studied area of DEE immunotoxicity concerns its role in inducing/exacerbating allergic inflammation and airway hyperresponsiveness (Fujimaki et al., 1997; Lovik et al., 1997; Takano et al., 1997, 1998; Miyabara et al., 1998; Nel et al., 1998). Through this mechanism, particulates from such emissions as DE may be an important contributor to the increased prevalence and morbidity of asthma and allergic rhinitis observed worldwide.

The potential role of DE in enhancing allergic rhinitis and asthma has been well-investigated using laboratory animal models. Animal studies have provided definitive data that DEP (or extracts made from DEP) affect IgE production (Nel et al., 1998). A review by Nel et al. (1998) reports that DEP may enhance IgE production by a variety of mechanisms, including effects on cytokine and chemokine production as well as upon activation of macrophages and other mucosal cell types. In other studies, Kobayashi and Ito (1995) have shown that intranasal exposure of guinea pigs to DE increases nasal airway resistance and augments nasal airway resistance and secretions in response to histamine. Experimental studies have also focused upon the possible adjuvant activity of DE. A number of studies have demonstrated that DEP increases the production of ovalbumin (OA)-specific IgE after repeated intranasal or IT instillation in OA-sensitized and challenged mice (Takafuji et al., 1987; Takano et al., 1997). Diesel exhaust particles also enhance antigen-specific IgE responses after repeated intraperitoneal (IP) injection of mice with DEP plus OA or pollen (Muranaka et al., 1986). Intranasal instillation of DEP and OA causes increased *in vitro* cell proliferation in response to OA and increased interleukin-4 production in cells from the cervical lymph nodes compared with mice instilled with OA alone (Fujimaki et al., 1994, 1995); inhalation of DE has been shown to enhance production of antigen-specific IgE antibody in mice through an alteration of the cytokine network (Fujimaki et al., 1997).

A number of studies have been carried out to determine the particular constituent(s) responsible for the adjuvant activity of DEP. Lovik and colleagues (1997) demonstrated that both DEP and carbon black injected into the footpad of mice in conjunction with OA can have an adjuvant effect on poplitel lymph node inflammation and systemic OA-specific IgE response. While this study suggests that the particle core contributes to the adjuvant activity of DEP, studies by Suzuki et al. (1993) showed that both pyrene and DEP have adjuvant activity on allergen-specific IgE antibody production when mice are immunized by IP injection of OA or pollen allergen with pyrene. These findings suggest that chemical compounds contained in diesel soot can also have adjuvant activity in mice. At the present time, the relative contributions of the particle core versus various adsorbed chemicals to the adjuvant activity of DEP are unresolved. Also unsettled is the relative T-helper-2 lymphocyte (T_H2) adjuvant potency or potential contribution of DEP compared with other inhaled particles in enhancing allergic respiratory disease. Finally, it remains unclear whether inhaled DE at environmental or occupational concentrations would have significant T_H2 adjuvant effects.

A number of *in vitro* studies have also been performed demonstrating the immunomodulating effects of DEP. In studies by Thomas et al. (1995), human peripheral monocytes and alveolar macrophages exposed to suspended DEP had reduced phagocytosis and increased tumor necrosis factor-α release. In another study, human airway epithelial cells exposed to DEP *in vitro* showed increased release of granulo-cyte-monocyte colony stimulating factor, a cytokine that influences eosinophil activity (Ohtoshi et al., 1994). Organic extracts from DEP and phenanthrene, one of the major PAHs adsorbed to DEP, have been shown to enhance *in vitro* production of IgE from IgE-secreting purified human B-lymphocytes and from an Epstein-Barr virus-transformed human B-lymphocyte line (Takenaka et al., 1995; Tsien et al., 1997). Pfutzner et al. (1995) investigated the ability of DEP to influence the number and activity of IgE-secreting cells *in vitro*. In this study, peripheral blood mononuclear cells, isolated from individuals with atopic eczema and from healthy, non-allergic individuals, were exposed to DEP *in vitro*. The number of IgE-secreting cells was increased in some, but not all, of the DEP-exposed cell isolates from allergic eczema patients, and in none of the DEP cell isolates from controls; IgE production paralleled the number of IgE-secreting cells.

The effects of inhaled DE on aspects of pulmonary immunity other than inflammation and humoral-mediated immune functions have also been studied. A number of studies have been performed to assess effects of DE on host susceptibility to respiratory tract infections. For the most part, results of these studies are inconclusive. For example, Campbell et al. (1981) reported that acute and subacute exposure of mice to DE caused an increase in host mortality in response to *S. pyogenes* infection, but not following challenge with influenza virus. Hahon and colleagues (1985) reported no increase in mortality or other measures of influenza viral infection after one month of exposure to DE; after a longer exposure (i.e., 3 and 6 mo), pulmonary consolidation and virus growth were greater in DE-exposed animals while interferon-γ and hemagglutinin-antibody levels were depressed.

SUMMARY

This Chapter has provided an overview of the health effects and immunological impacts associated with three different types of combustion emissions: woodsmoke, kerosene heater effluents, and diesel exhaust. While a number of differences exist between emissions, all three share some similarities concerning chemical constituents and overall health effects. In general, combustion emissions from these sources are increasing worldwide . Increased usage of devices which produce these indoor and outdoor airborne pollutants has resulted in greater public exposure and increased concern by exposed individuals. While more studies need to be performed regarding the underlying mechanisms responsible for the observed effects and the particular constituent(s) responsible, it is clear that inhalation of combustion products from wood, kerosene, and diesel exhaust can have a significant impact on pulmonary homeostasis and/or exacerbation of ongoing disease processes. Standards regulating output of these potentially dangerous effluents need to be considered, particularly for those members of our population deemed most susceptible (i.e., elderly, children, asthmatics, and individuals with ongoing cardiopulmonary disease.

REFERENCES

Anderson EW, Andelman RJ, Strauch JM. Effect of low-level carbon monoxide exposure on onset and duration of angina pectoris - A study of ten patients with ischemic heart disease. Ann. Intern. Med., 1973;79:46-50.

Anderson HR. Respiratory abnormalities in Papua New Guinea children: The effects of locality and domestic wood smoke pollution. Int. J. Epidemiol., 1978;7:63-72.

Arashidani K, Yoshikawa M, Kawamoto T, Matsuno K, Kayama F, Kodama Y. Indoor pollution from heating. Ind. Health, 1996;34:205-215.

Arif JM, Khan SG, Ahmad I, Joshi LD, Rahman Q. Effect of kerosene and its soot on the chrysotile-mediated toxicity to the rat alveolar macrophages. Environ. Res., 1997;72:151-161.

Arif JM, Khan SG, Ashquin M, Rahman Q. Modulation of macrophage-mediated cytotoxicity by kerosene soot: Possible role of reactive oxygen species. Environ. Res., 1193;61:232-238.

Arif JM, Khan SG, Aslam M, Mahmood M, Joshi LD Rahman Q. Early biochemical changes in kerosene exposed rat lungs. Chemosphere, 1991;22:705-712.

Awasthi S, Glick HA, Fletcher RH. Effect of cooking fuels on respiratory diseases in preschool children in Lucknow, India. Am. J. Trop. Med. Hyg., 1996;55:48-51.

Azizi BHO, Henry RL. The effects of indoor environmental factors on respiratory illness in primary school children in Kuala Lumpur. Int. J. Epidemiol. 1991;20:144-150.

Baldachin BJ, Malmed PM. Clinical and therapeutic aspects of kerosene poisoning. A series of 200 cases. Br. Med. J., 1964;2:28-30.

Behera D, Jindal SK. Respiratory symptoms in Indian women using domestic cooking fuels. Chest, 1991;100:385-388.

Behera D, Sood P, Singh S. Passive smoking, domestic fuels and lung function in north Indian children. Indian J. Chest Dis. Allied Sci., 1998;40:89-98.

Brizio-Molteni LG, Piano G, Rice PL, Warpeha R, Fresco R, Solliday N, Molteni A. Effect of wood combustion smoke inhalation on angiotensin-1 converting enzyme in the dog. Ann. Clin. Lab. Sci., 1984;14:381-389.

Browning KG, Koenig JQ, Checkoway H, Larson TV, Pierson WE. A questionnaire study of respiratory health in areas of high and low ambient woodsmoke pollution. Pediat. Asthma Allergy Immunol., 1990;4:183-191.

Brunekreef B, Janssen NA, de Hartog J, Harssema J, Knape M, van Pliet P. Air pollution from truck traffic and lung function in children living near motorways. Epidemiology, 1997;8:293-303.

Brunner S, Rovsing H, Wulf M. Roentgenographic changes in the lungs of children with kerosene poisoning. Am. Rev. Respir. Dis., 1964;29:250-254.

Burrell R, Flaherty DK, Sauers LJ. (eds.) Toxicology of the Immune System: A Human Approach. New York: Van Nostrand Reinhold Publishers, 1992, pp. 1-240.

Butcher SS, Sorenson EM. A study of wood stove particulate emissions. Air Poll. Control Assoc., 1979;29:724-728.

Butterfield P, Edmunson E, LaCava G, Penner J. Woodstoves and indoor air. J. Environ. Health, 1989;59:172-173.

Campbell KL, George EL, Washing IS. Enhanced susceptibility to infection in mice after exposure to dilute exhaust from light duty diesel engines. Environ. Int., 1981;5:377-382.

Casaco A, Carvajal D, Noa M, Gonzalez R, Garcia M, de la Vega AR. Effects of kerosene on airway sensitization to egg albumin in guinea pig. Allergol. Et Immunopathol., 1985a;13:235-239.

Casaco A, Garcia M, Gonzalez R, de la Vega AR. Induction of acetylcholinesterase inhibition in the guinea pig trachea by kerosene. Respiration, 1985b;48:46-49.

Casaco A, Gonzalez R, Arruzazabala L, Garcia M, de la Vega AR. Studies on the effects of kerosene aerosol on airways of rabbits. Allerg. Immunol., 1982;10:361-366.

Casaco A, Gonzalez R, Arruzazabal L, Garcia M, de la Vega AR. Kerosene aerosol induces guinea-pig airway hyperreactivity to acetylcholine. Respiration, 1985c;47:190-195.

Cass GR, Gray HA. "Regional Emissions and Atmospheric Concentrations of Diesel Engine Particulate Matter: Los Angeles as a Case Study." In Diesel Exhaust: A Critical Analysis of Emissions, Exposure, and Health Effects. Cambridge, MA: Health Effects Institute, 1995, pp. 125-137.

Chapin FS. (ed.) Human Activity Patterns in the City. New York: Wiley-Interscience, 1974, pp. 1-30.

Cohen AJ, Nikula K. (eds.) "The Health Effects of Diesel Exhaust: Laboratory and Epidemiologic Studies." In Air Pollution and Health, San Diego: Academic Press, 1999, pp. 707-745.

Collings DA, Martin KS, Sithole SD. Indoor woodsmoke pollution causing lower respiratory disease in children. Trop. Doctor, 1990;20:151-155.

Cooper JA. Environmental impact of residential wood combustion emissions and its implications. Air Poll. Control Assoc., 1980;30:855-861.

Dasch JM. Particulate and gaseous emissions from woodburning fireplaces. Environ. Sci. Technol., 1982;16:639-645.

de la Vega AR, Casaco A, Garcia M, Noa M, Carvajal D, Arruzazabala L, Gonzalez R. Kerosene-induced asthma. Ann. Allergy, 1990;64:362-363.

Demarest GM, Hudson LD, Altman LC. Impaired alveolar macrophage chemotaxis in patients with acute smoke inhalation Am. Rev. Respir. Dis., 1979;119:279-286.

Diaz A, Ponmereck C, Barcelo C. Kerosene, gas, and asthma in housewives in homes in the city of Havana [Keroseno, gas y asma en amas de casas de Ciudad de la Habana]. Librio de Resumenes VIII Seminario Cientifico del CENIC, La Habana, 1982;266:1-5.

Fick RB, Pau ES, Merrill WW, Reynolds HY, Lake JS. Alterations in the antibacterial properties of rabbit pulmonary macrophage exposed to woodsmoke. Am. Rev. Respir. Dis., 1984;129:76-81.

Fujimaki H, Nohara O, Ischinose T, Watanabe N, Saito S. IL-4 production in mediastinal lymph node cells in mice intratracheally-instilled with diesel exhaust particulates and antigen. Toxicology, 1994;92:261-268.

Fujimaki H, Saneyoshi K, Nohara O, Shiraishi F, Imai T. Intranasal instillation of diesel exhaust particulates and antigen in mice modulated cytokine productions in cervical lymph node cells. Int. Arch. Allergy Immunol., 1995;108:268-273.

Fujimaki H, Saneyoshi K, Shiraishi F, Imai T, Endo T. Inhalation of diesel exhaust enhances antigen-specific IgE antibody production in mice. Toxicology, 1997;116:227-233.

Gamble J, Jones W, Hudak J. An epidemiological study of salt miners in diesel and non-diesel mines. Am. J. Ind. Med., 1983;4:435-438.

Goodwin SR, Berman LS. Kerosene aspiration: Immediate and early pulmonary and cardiovascular effects. Vet. Hum. Toxicol., 1988;30:521-524.

Gross KB. Pulmonary function testing of animals chronically-exposed to diluted diesel exhaust. J. Appl. Toxicol., 1981;1:116-123.

Hahon N, Booth JA, Green F, Lewis TR. Influenza virus infection in mice after exposure to coal dust and diesel engine emissions. Environ. Res., 1985;37:44-60.

Hatch GE, Slade R, Boykin E, Hu PC, Miller FJ, Gardner DE. Correlation of effects of inhaled versus intratracheally-injected metals on susceptibility to respiratory infection in mice. Am. Rev. Respir. Dis., 1981;124:167-173.

Heinrich U, Muhle H, Takenaka S, Ernst H, Fuhst R, Mohr U, Pott F, Stober W. Chronic effects on the respiratory tract of hamsters, mice, and rats after long-term inhalation of high concentrations of filtered and unfiltered diesel engine emission. J. Appl. Toxicol., 1986;6:383-395.

Henderson RF, Mauderly JL. "Diesel Exhaust: An Approach for the Study of the Toxicity of Chemical Mixtures." In *Toxicology of Chemical Mixtures*, RSH Yang, ed., San Diego: Academic Press, 1994, pp. 119-133.

Hildemann LM, Markowski GR, Jones MC, Cass GR. Submicrometer aerosol mass distributions of emissions from boilers, fireplaces, automobiles, diesel trucks and meat cooking operations. Aerosol Sci. Tech., 1991;14:138-152.

Honicky RE, Akpom CA, Osborne JS. Infant respiratory illness and indoor air pollution from a woodburning stove. Pediatrics, 1983;71:126-128.

Honicky RE, Osborne JS Akpom CA. Symptoms of respiratory illness in young children and the use of woodburning stoves for indoor heating. Pediatrics, 1985;75:587-593.

Honicky RE, Osborne JS. Respiratory effects of wood heat: Clinical observations and epidemiologic assessment. Environ. Health Perspect., 1991;95:105-109.

IARC: International Agency for Research on Cancer. *Polynuclear Aromatic Compounds, Part 4: Bitumens, Coal-tars and Derived Products, Shale Oils, and Soots, Vol. 35.* Lyon, France: IARC, World Health Organization, 1985, pp. 219-240.

Jakab GJ Hemenway DR. Inhalation co-exposure to carbon black and acrolein suppresses alveolar macrophage phagocytosis and TNFα release and modulates peritoneal macrophage phagocytosis. Inhal. Toxicol., 1993;5:265-279.

Kerns WD, Parkov KL, Donofrio DJ, Gralla EJ, Swenberg JA. Carcinogenicity of formaldehyde in rats and mice after long-term inhalation exposure. Cancer Res., 1983;43:4382-4392.

Kobayashi T, Ito T. Diesel exhaust particulates induce nasal mucosal hyperresponsiveness to inhaled butamine aerosol. Fundam. Appl. Toxicol., 1995;27:295-301.

Koenig JQ, Covert DS, Larson TV, Maykut N, Jenkins P, Pierson WE. Woodsmoke: Health effects and legislation. Northwest Environ. J., 1988;4:41-54.

Koenig JQ, Larson TV, Hanley QS, Rebolledo V, Dumler K, Checkoway H, Wang SZ, Lin D, Pierson WE. Pulmonary function changes in children associated with fine particulate matter. Environ. Res., 1993;63:26-38.

Koenig JQ, Pierson WE. air pollutants and the respiratory system: Toxicity and pharmacologic interventions. Clin. Toxicol., 1991;29:401-411.

Lao YJ. Particulate emissions from wood stoves in a residential area. J. Environ. Health, 1983;46:115-117.

Larson TV, Yucn P, Maykut N. "Weekly Composite Sampling of $PM_{2.5}$ for Total Mass and Trace Elements Analysis." In *Transactions of the Air and Waste Management Association Specialty Conference on PM$_{10}$ Standards and Nontraditional Particulate Source Controls*, J Chow, ed., ISBN 0-92-32-04-09, Pittsburgh, 1992, pp. 1-200.

Lewis TR, Green FHY, Moorman WJ, Burg JAR, Lynch DW. "A Chronic Inhalation Toxicity Study of Diesel Engine Emissions and Coal Dust, Alone and Combined." In *Carcinogenic and Mutagenic Effects of Diesel Engine Exhaust*, N Ishinishi, A Koizumi, RO McClellan, W Stober, eds., New York: Elsevier Publishers, 1986, pp. 361-380.

Liang CK, Quan NY, Cao SR, He XZ, Ma F. Natural inhalation exposure to coal smoke and wood smoke induces lung cancer in mice and rats. Biomed. Environ. Sci., 1988;1:42-50.

Lipari F, Dasch JM, Scruggs WF. Aldehyde emissions from woodburning fireplaces. Environ. Sci. Technol., 1984;18:326-330.

Lipfert FW, Dungan JL. Residential firewood use in the United States. Science, 1983;219:1425-1427.

Loke J, Paul E, Virgulto JA, Smith W. Rabbit lung after smoke inhalation. Arch. Surg., 1984;119:956-959.

Lovik M, Hogseth AK, Gaardner PI, Hagemann R, Eide I. Diesel exhaust particles and carbon black have adjuvant activity on the local lymph node response and systemic IgE production to ovalbumin. Toxicology, 1997;121:165-178.

Mauderly JL, Jones RF, Henderson RF. "Relationship of Lung Structural and Functional Changes to Accumulation of Diesel Exhaust Particles." In *Inhaled Particles VI*, J Dodgson, RI McCallum, MR Bailey, DR Fisher, eds., Oxford: Pergamon Press, 1988, pp. 659-669.

Mauderly JL, Jones RK, Griffith WC, Henderson RF, McClellan RO. Diesel exhaust is a pulmonary carcinogen in rats exposed chronically by inhalation. Fundam. Appl. Toxicol., 1987;9:1-13.

Mauderly JL. "Diesel Exhaust." In *Environmental Toxicants: Human Exposures and Their Health Effects*, M Lipmann, ed., New York: Van Nostrand Reinhiold, 1992, pp. 119-162.

McClellan RO. Lung cancer n rats from prolonged exposure to high concentrations of carbonaceous hazard and human lung cancer risk. Inhal. Toxicol., 1996;8:193-226.

Mesa MG, Parada AC, Valmana LA, Alvarez RG, de la Vega AR. Role of chemical mediators in bronchoconstriction induced by kerosene. Allergol. Et Immunopathol., 1998a;16:421-423.

Mesa MG, Alvarez RG, Parada AC. Biochemical mechanisms in the effects of kerosene on airways of experimental animals. Allergol. Et Immunopathol., 1988b;16:363-367.

Miyabara Y, Ichinose T, Takano H, Sagai M. Diesel exhaust inhalation enhances airway hyperresponsiveness in mice. Int. Arch. Allergy Immunol., 1998;116:124-131.

MMWR- Morbidity Mortality Weekly Reports. Use of unvented residential heating appliances--United States, 1988-1994. Morbidity Mortality Weekly Reports, 1997;46:51:1221-1224.

Morris K, Morganlander M, Coulehan JL, Gahajgen S, Arena VC. Wood-burning stoves and lower respiratory tract infection in American Indian children. Am. J. Dis. Child., 1990;144:105-108.

Muranaka M, Suzuki S, Koizumi K. Adjuvant activity of diesel-exhaust particulates for the production of IgE antibody in mice. J. Allergy Clin. Immunol., 1986;77:613-623.

National Research Council. *Indoor Pollutants. Committee on Indoor Pollutants*. Washington, DC: National Academy Press, 1981.

Nel AE, Diaz-Sanchez D, Ng D, Hiura T, Saxon A. Enhancement of allergic inflammation by the interaction between diesel exhaust particles and the immune system. J. Allergy Clin. Immunol., 1998;102:539-554.

Nieman GF, Clark WR, Goyette D, Hart KE, Bredenberg CE. Woodsmoke inhalation increases pulmonary microvascular permeability. Surgery, 1989;105:481-487.

Noa M, Illnait J. Induction of aortic plaques in guinea pigs by exposure to kerosene. Arch. Environ. Health, 1987;42:31-36.

Ohtoshi T, Takizawa H. Sakamaki C. Cytokine production by human airway epithelial cell and its modulation. Nippon Kyobu Shikkan Gakkai Zasshi - Japan. J. Thor. Dis., 1994;32:23-29.

Pfutzner W, Thomas P, Przybilla B. Influence of suspended diesel exhaust particles (DEP) on number and activity of IgE-secreting cells *in vivo*. J. Allergy Cln. Immunol., 1995;95:226-227.

Pierson WE, Koenig JQ, Bardona EJ. Potential adverse health effects of woodsmoke. West. J. Med., 1989;151:339-342.

Pope CA. respiratory hospital admissions associated with PM_{10} pollution in Utah, Salt Lake, and Cache Valleys. Arch. Environ. Health, 1991;46:90-97.

Ragland KW, Andren AW, Manchester JB. Emissions from unvented kerosene heaters. Sci. Total Environ., 1985;46:171-179.

Rai NC, Singh TSK. Cardiopulmonary changes in mongrel dogs after exposure to kerosene smoke. Indian J. Exp. Biol., 1980;18:1263-1266.

Rajpandey M. Domestic smoke pollution and chronic bronchitis in a rural community of the Hill Region of Nepal. Thorax, 1984;39:337-339.

Ramage JE, Roggli VL, Bell DY, Piantadosi CA. Interstitial lung disease and domestic woodburning. Am. Rev. Respir. Dis., 1988;137:1229-1232.

Samet JM, Marbury MC, and Spengler JD., Health effects and sources of indoor air pollution. Part I. Am. Rev. Respir. Dis., 1987;136:1486-1508.

Samet JM, Marbury MC, and Spengler JD. Health effects and sources of indoor air pollution. Part II. Am. Rev. Respir. Dis., 1988;137:221-242.

Sandoval J, Salas J, Martinez-Guerra ML, Gomez A, Martinez C, Portales A, Palomar A, Villegas M, Barrios R. Pulmonary arterial hypertension and cor pulmonale associated with chronic domestic woodsmoke inhalation. Chest, 1993;103:12-20.

Schwartz J. Particulate air pollution and chronic respiratory disease. Environ. Res., 1993;62:7-13.

Sexton K, Liu KS, Treitman RD, Spengler JD, Turner WA. Characterization of indoor air quality in woodburning residencies. Environ. Int., 1986;12:265-278.

Sharma, S, Sethi GR, Rohtagi A, Chaudhary A, Shankar R, Bapna JS, Joshi V, Sapir DG. Indoor air quality and acute lower respiratory infection in Indian urban slums. Environ. Health Perspect., 1998;106:291-297.

Snipes MB. Long-term retention and clearance of particles inhaled by mammalian species. CRC Crit. Rev. Toxicol., 1989;20:175-211.

Speizer FE, Ferris B, Bishop YM, Spengler J. Respiratory disease rates and pulmonary function in children associated with NO_2 exposure. Am. Rev. Respir. Dis., 1980;121:3-10.

Suzuki T, Kanoh T, Kanbayashi M. The adjuvant activity of pyrene in diesel exhaust on IgE antibody production in mice. Arerugi - Japan. J. Allergology, 1993;42:963-968.

Szalar A. (ed.) *The Use of Time: Daily Activities of Urban and Suburban Populations in Twelve Countries*. The Hague, Netherlands: Mouton, 1972, pp. 1-680.

Takafuji S, Suzuki S, Koizumi K. Diesel-exhaust particulates inoculated by the intranasal route have an adjuvant activity for IgE production in mice. J. Allergy Clin. Immunol., 1987;79:639-645.

Takano, H, Yoshikawa T, Ichinose T, Miyabara Y, Imaoka K, Sagai M. Diesel exhaust particles enhance antigen-induced airway inflammation and local cytokine expression in mice. Am. J. Respir. Crit. Care Med., 1997;156:36-42.

Takenaka H. Zhang K, Diaz-Sanchez D et al. Enhanced human IgE production results from exposure to the aromatic hydrocarbons from diesel exhaust: Direct effects on B-cell IgE production. J. Allergy Clin. Immunol., 1995;95:103-115.

Thomas P, Maerker J, Riedel W, Przybilla B. Altered human monocyte/macrophage function after exposure to diesel exhaust particles. Environ. Sci. Pollut. Res. Int., 1995;2:69-72.

Thomas PT, Zelikoff JT. "Air Pollutants: Modulators of Pulmonary Host Resistance Against Infection." In *Air Pollution and Health*, ST Holgate, JM Samet, HS Koren, RL Maynard, eds., San Diego: Academic Press, 1999, pp. 357-379.

Thorning DR, Marianne LH, Hudson LD, Schumaher RL. Pulmonary responses to smoke inhalation: Morphologic changes in rabbits exposed to pine woodsmoke. Human Pathol., 1982;13:355-364.

Tominaga S., Itoh K. Relationship between parental smoking and respiratory diseases of three year old children. Tokai J. Exp. Clin. Med., 1985;395-399.

Traynor GW, Apte MG, Carruthers AR, Dillworth JF, Grimsrud DT, Gunder LA. Indoor air pollution due to emissions from woodburning stoves. Environ. Sci. Technol., 1987;21:691-697.

Tsien A, Diaz-Sanchez D, Ma J, Saxon A. the organic component of diesel exhaust particles and phenanthrene, a major polyaromatic hydrocarbon constituent, enhances IgE production by IgE-secreting EBV-transformed human B-cells *in vitro*. Toxicol. Appl. Pharmacol., 1997;142:256-263.

Tuthill RW. Woodstoves, formaldehyde, and respiratory disease. Am. J. Epidemiol., 1984;120:952-955.

White KL, Kawabata TT, Ladics GS. "Mechanisms of Polycyclic Aromatic Hydrocarbon Immunotoxicity." In *Immunotoxicology and Immunopharmacology*, JH Dean, MI Luster, AE Munson, I Kimber Jr., eds., New York: Raven Press, 1994, pp. 123-146.

Winer AM, Busby WF. "Atmospheric Transport and Transformation of Diesel Emissions." In *Diesel Exhaust: A Critical Analysis of Emissions, Exposure, and Health Effect.* Cambridge, MA: Health Effects Institute, 1995.

Wong KL, Stock Mf, Malek DE, Alarie Y. Evaluation of the pulmonary effects of woodsmoke in guinea pigs by repeated CO_2 challenges. Toxicol. Appl. Pharmacol., 1984;175:69-80.

Woodring JL, Duffy TL, Davis JT, Bechtold RR. Measurements of combustion product emission factors of unvented kerosene heaters. Am. Ind. Hyg. Assoc. J., 1985;46:350-356.

Zelikoff JT, Nadziejko C, Fang K, Gordon T, Premdass C, Cohen, MD. "Short-term, Low-dose Inhalation of Ambient Particulate Matter Exacerbates Ongoing Pneumococcal Infections in Streptococcus pneumoniae-infected Rats." In *Proceedings of the Third Colloquium on Particulate Air Pollution and Human Health*, R Phalen, Y Bell, eds., 1999, 8-94-8-101.

Zelikoff JT, Li Y, Nadziejko C, Chen LC, Cohen MD. PM_{10} generated from wood burning may be responsible for increased pulmonary infections: A toxicological model. Fundam. Appl. Toxicol., 1997;36:7.

16. TOBACCO SMOKE

Donald E. Gardner, Ph.D.
Inhalation Toxicology Associates, Inc.
P.O. Box 97605
Raleigh, NC 27624-7605

INTRODUCTION

Much is known about the biological responses associated with the inhalation of tobacco smoke. The scientific literature on the health effects of tobacco smoke has grown rapidly and today comprises thousands of articles.

While this enormous database on smoking has provided substantial evidence linking potential health hazards to the inhalation of tobacco smoke, there is still a significant lack in understanding of the mechanistic relationship between smoke and certain disease processes. Chemicals inhaled in tobacco smoke, from either voluntary or involuntary exposure, have been reported to be responsible for increased incidence of numerous health effects. Acute exposure can result in irritation of the eyes, nose, and throat and result in cough, increased sputum production, headaches, and possibly increases in the incidence of lower respiratory illnesses. Many of these effects are transitory discomforts that are influenced by both the smoke concentration and individual differences in sensitivity to the smoke. Long-term exposure has been associated with increased risk of cancer of the lungs, larynx, and possibly other organs (i.e. bladder, pancreas, kidney, and digestive tract); increased risk of chronic obstructive lung disease; and decreases in cardiovascular and respiratory function. Tobacco smoke is now being considered to be associated with unsuccessful pregnancies and certain reproductive and/or prenatal effects. These research studies have been critically evaluated in numerous review documents (DHHS, 1986, 1989; IARC, 1986, 1987; National Research Council, 1986; EPA, 1990, 1992; Federal Register, 1994)

Numerous explanations have been proposed as the mechanism for many of the effects, including genetic susceptibility, tissue damage from repeated irritation, necrosis, cell turnover, increased ability for penetration of certain toxicants in the smoke from the surface of the lung into the lung tissue, metabolic activation, detoxification, and interference with immune competence.

The awareness of the possible effects of tobacco smoke on the immune system has become evident from human clinical and epidemiological studies, animal toxicology studies, and *in vitro* models. Because the lung is the primary route of exposure to tobacco smoke, as well as the portal of entry for many other airborne chemicals and infectious agents, it is important to understand the response of this pulmonary defense system to inhaled tobacco smoke. Recent concerns about the immunotoxicity of inhaled tobacco smoke have prompted this review. The following chapter is divided into three sections. The first reviews the exposure sources, levels of exposure, composition of tobacco smoke, and various approaches being used to reduce exposure. The second section describes the various immunomodulatory events from exposure to tobacco smoke. The final section is devoted to the interpretation and summary of the results and where possible a discussion of the underlying mechanisms by which tobacco smoke may perturb these host defense mechanisms, resulting in an increase risk of disease.

It is hoped that this review stimulates the reader to speculate on critical questions concerning the immunotoxicity of inhaled tobacco smoke, including: (1) what components present in tobacco smoke can interfere with the immune competence of the respiratory system; (2) under what conditions (e.g., exposure levels, health status, age, gender) can tobacco smoke adversely affect the immune system; (3) what are the consequences and risks associated with such exposures to human health; and, (4) what are the gaps in the available database necessary to identify the possible mechanism(s) of action of the immunomodulation effects associated with such responses?

EXPOSURE SOURCES AND CHARACTERIZATION

The chemical composition of tobacco smoke is quite complex. Tobacco, when burned, produces a variety of chemical components due to the incomplete combustion of the tobacco leaf, the wrapper, the many curing agents, additives, and fillers present in the tobacco product.

The generation of tobacco smoke is governed by the particular characteristics of the smoker, (i.e. puff volume, duration, frequency, and inhalation pattern of the smoker), the particular tobacco product being used and whether or not a filter is present. The combustion of tobacco leads to the formation of mainstream smoke (MS) and sidestream smoke (SS). Exposure to MS results during puff-drawing in the burning cone of a tobacco product. The smoke travels through the tobacco column and is inhaled into the oral cavity and the respiratory tract of the smoker. Undeposited smoke, which is exhaled by the smoker, is also considered to be MS. Sidestream smoke is a combination of the smoke emitted into the air during the burning of the tobacco product between puffs, smoke escaping from the smoldering product, and any vapor phase components that might diffuse thought the cigarette paper. It has been estimated that one half of the smoke emitted from a burning tobacco product is SS smoke.

Tobacco smoke affects not only people who smoke but also nonsmokers who are exposed to the smoke. This smoke has been referred to as either passive or secondhand smoke but the more technical term is Environmental Tobacco Smoke (ETS) which is a common component of indoor air pollution (DHHS, 1986; National Research Council, 1986; Samet et al., 1987; EPA, 1990; Rodgman 1992). ETS is defined as that smoke that contaminates the air surrounding a smoker. It is a combination of SS and MS smoke, exhaled by the smoker. Studies have indicated that no inhaled component of the MS is 100% absorbed and retained in the smoker's respiratory system therefore every inhaled MS component must also be considered to be present in exhaled MS, as well (Rodgman, 1992). Passive smoking is generally considered to be the inhalation of ETS produced by someone else's smoking. The term passive smoking also refers to the exposure of a fetus to tobacco smoke products and/or their metabolites from an active or passively smoking mother.

Tobacco smoke has been estimated to be a major source of indoor air pollution. This can be illustrated by the prediction by the EPA (1989) that in the United States ≈50 million smokers annually smoke about 600 billion cigarettes, 4 billion cigars and the equivalent of 11 billion pipefuls of tobacco. Since people spend ≈90% of their time indoors, this means that there is a potential for about 465,000 tons of tobacco being burned indoor each year.

PHYSICAL/CHEMICAL COMPOSITION OF TOBACCO SMOKE

Understanding the physicochemical composition and the identification of the biologically active compounds in a tobacco smoke mixture is useful for understanding the human health risk associated with smoking. Tobacco smoke, whether MS, SS, or ETS, consists of two major phases: a gaseous phase and a particulate phase. However, MS, SS, and ETS differ both chemically and physically from each other, presumably due to chemical transformation that occurs as the mixture is diluted and aged in the atmosphere (Sterling and Kobayashi, 1982; National Research Council, 1986; EPA, 1989)

Fresh MS smoke contains a variety of organic and inorganic particles and gases, too numerous to discuss individually. The gaseous components in fresh smoke include carbon monoxide, carbon dioxide, nitrogen oxide, volatile aldehydes, hydrogen sulfide, isoprene, acetone, toluene, ammonia, hydrazine, toluene, ketones, acetic acid, formic acid, benzene, acetaldehyde, and hydrogen cyanide, as well as an additional 400 - 500 other minor constituents. Components of the particulate phase include nicotine, phenol, catechol, hydroquinone, aniline, 2-toluidine, 2-naphthyl-amine, benz(a)anthracene, benzo(a)pyrene, N-nitrosonornicotine, 4-methylnitroso-amino-1-(3-pyridinyl)-1-butanone, N-nitrosodiethanolane, polonium-210, copper, iron, cadmium, nickel, zinc, chromium, arsenic, and numerous other trace components (Schmeltz and Hoffmann, 1977; Kuller et al., 1986; EPA, 1989, 1992, 1993; Stohs et al., 1997). Bioassays of tobacco smoke have identified more than 43 known carcinogens, co-carcinogens, or suspected carcinogens in tobacco smoke. These include, for example, benzene, nickel, polonium-210, 2-napthylamine, 4-aminobiphenyl, dimethylnitrosamine, nitrosopyrrolidine, formaldehyde, various N-nitrosamines, benzo(α)anthracene, and benzo(α)pyrene. While the majority of the carcinogenic and genotoxic agents in tobacco smoke reside in the particulate phase of the smoke, some components of the vapor phase also contains known human and animal carcinogens (Hoffmann and Wynder, 1986; EPA, 1989, 1992, 1993).

While these compounds are initially found in MS smoke, they also can be expected to occur in SS and ETS, but will differ significantly in quantities of the individual constituents. Table 1 list some examples of the measured constituents both in the gaseous and particulate phase of MS and SS smoke and indicates a ratio that can be expected between SS/MS (Federal Register, 1994).

TABLE 1. CONSTITUENTS OF TOBACCO SMOKE

VAPOR PHASE COMPONENTS	AMOUNT IN MS	RATIO IN SS/MS
carbon monoxide	10 - 23 mg	2.5 - 4.7
carbon dioxide	20 - 40 mg	8.0 - 11.0
carbonyl disulfide	12 - 42 µg	0.03 - 0.13
benzene	12 - 48 µg	5.0 - 10.0
toluene	100 - 200 µg	5.6 - 8.3
formaldehyde	70 - 100 µg	0.1 - 50.0
acrolein	60 - 100 µg	8.0 - 15.0
acetone	1002 - 250 µg	2.0 - 5.0
pyridine	16 - 40 µg	6.5 - 20.0
3-methylpyridine	12 - 36 µg	3.0 - 13.0
3-vinylpyridine	11 - 30 µg	20.0 - 40.0
hydrogen cyanide	400 - 500 µg	0.10 - 0.25
hydrazine	32 ng	3.0
ammonia	50 - 130 µg	3.7 - 5.1
methylamine	11.5 - 28.7 µg	4.2 - 6.4
dimethylamine	7.8 - 10.0 µg	3.7 - 5.1
nitrogen oxides	100 - 600 µg	4.0 - 10.0
N-nitrosopyrrolidine	6 - 30 ng	6.0 - 30.0
acetic acid	330 - 810 µg	1.9 - 3.6
1,3-butadiene	69.2 µg	3.0 - 6.0

PARTICULATE PHASE COMPONENTS	AMOUNT IN MS	RATIO IN SS/MS
nicotine	1.0 - 2.5 mg	2.6 - 3.3
anatabine	2 - 20 mg	< 0.01 - 0.5
phenol	60 - 140 µg	1.6 - 3.0
catechol	100 - 360 µg	0.6 - 0.9
hydroquinone	110 - 300 µg	0.7 - 0.9
aniline	360 ng	30.0
2-toluidine	160 ng	19.0
2 naphthylamine	1.7 ng	30.0
4-aminobiphenyl	4.6 ng	31.0
benz(α)anthracene	20 - 70 ng	2.0 - 4.0
benzo(α)pyrene	20 - 40 ng	2.5 - 3.5
cholesterol	22 µg	0.9
butyrolactone	10 - 22 µg	3.6 - 5.0
quinoline	0.5 - 2 µg	3.0 - 11.0
N-nitrosonornicotine	200 - 3000 ng	0.5 - 3.0
NNK	100 - 1000 ng	1.0 - 4.0
N-nitrosodiethanolamine	20 - 70 ng	1.2

TABLE 1. CONSTITUENTS OF TOBACCO SMOKE (cont'd)

PARTICULATE PHASE COMPONENTS	AMOUNT IN MS	RATIO IN SS/MS
cadmium	110 ng	7.2
nickel	20 - 80 ng	13.0 - 30.0
zinc	60 ng	6.7
benzoic acid	14 - 28 µg	0.67 - 0.95
lactic acid	63 - 174 µg	0.5 - 0.7
succinic acid	110 - 140 µg	0.43 - 0.62

Adapted from Federal Register (1994)

Many of these difference maybe due to the lower temperature during burning, the limited availability of oxygen inside the burning cone during SS formation as well as possible influence of different pHs during the generation of the smoke. For example, SS are generated at distinctly different temperatures (600° for SS and 900° for MS), SS smoke originates from a hydrogen-enriched, strongly-reducing atmosphere and since it is being formed as a result of oxygen deficiency and thermal cracking it will contain more combustion products than MS. At pH levels above 6, there is increasing amounts of unprotonated nicotine present in the generated smoke. SS smoke, which is more basic than MS (pH 6.0 vs. 7.5) could thus be expected to contains more free nicotine in the gas phase than MS. In addition, SS formation involves generation of larger quantities of reaction products of nitrates. In the vapor phase of the SS smoke the levels of volatile nitrosamines are considerably higher than MS (Adams et al., 1987; Spitzer et al., 1990). The particulate phase of SS contains higher levels of aromatic amines than does MS and also higher levels of polycyclic aromatic hydrocarbons. However, it is important to remember that SS/ETS will be inhaled only after considerable dilution, whereas MS is not. Table 2 indicates the expected or estimated MS/ETS dilution factor of MS components inhaled by a smoker vs. ETS components inhaled by a nonsmoker.

TABLE 2. ESTIMATED DILUTION FACTORS[a]

TOBACCO SMOKE COMPONENT	MS/ETS DILUTION FACTOR
Acetone	240 - 2000
Acrolein	1500 - 20,833
Benzene	112 - 7167
Benzo(a)pyrene	68 - 40,740
Nicotine	57,333 - 7,200,000

Adapted from National Research Council (1986).
[a]The dilution factor is MS components inhaled by a smoker vs ETS components inhaled by a nonsmoker breathing ETS-containing indoor air.

The aerosol particles in SS, which is the major component of ETS are somewhat smaller (0.2 vs 0.5 µm) than MS (Oberdorster and Pott, 1987). Exhaled particles from MS have been reported to be, on the average, 20 - 25% larger than inhaled particles from MS. This is because the inhaled particles can be expected to absorb water, during passage in the respiratory tract, thus increasing the size (Ingebrethsen, 1989). However, during dilution of the ETS in the environment the exhaled MS particles will quickly tend to lose this absorbed water as well as other volatile particulate components, (e.g., nicotine, amines, etc.) resulting in particle sizes of ETS that may actually decrease to about 0.15 - 0.20 µm (Rodgman, 1992.) For example, within the ETS, the nicotine will tend to evaporate from smoke particles such that nicotine is then found almost exclusively in the vapor phase of the smoke (Hammond et al., 1987). However, if the cigarette has a filter, the levels of many of the components in MS smoke are significantly reduced compared with nonfiltered cigarettes, but the levels of these compounds in SS smoke would not be expected to be affected by the filter (Spitzer et al., 1990).

ETS is not chemically nor physically equivalent to either MS or SS smoke (Lofroth et al., 1989; Spitzer et al., 1990; Rodgman, 1992). As these smoke components undergo aging and dilution with ambient air, several things happen over time due to a drop in temperature, condensation of vapors onto existing particles, and various chemical reactions in the atmosphere (e.g. exposure of the smoke to ultraviolet light will affect the chemistry of the various nitrogen-based compounds). Also, changes in the ambient temperatures and relative humidity during the burning process will affect the ultimate composition of the smoke produced, evaporation of volatile vapors, coagulation of fresh tobacco particles, sedimentation and adsorption onto surfaces within the room, and the formation of new particles (Phalen, 1984, Spitzer et al., 1990). These very rapid changes in the chemical composition can also result in changes in its biological activity of the inhaled smoke (Dube and Green, 1982). Even though ETS is initially made up of MS (15%) and SS (85%) smoke, its toxicological properties can be expected to continue to change over time due to the temporal and spatial patterns of smoking, the dimension of the room, the presence or absence of people in the room, and the room's ventilation rate (Benner et al., 1989; Eatough et al., 1989). Thus, attempts to try to extrapolate exposure levels from a single air measurement of ETS to a predicted total dose reaching the respiratory tract of an individual involves many variables and is likely to be inaccurate (Spitzer et al., 1990).

Siegel (1993) provided some indication of the relative concentration of ETS in various indoor environments. When ETS was measured in bars, restaurants and homes, it was reported that the mean restaurant ETS concentrations were between 1.6 - 2.0 times higher than those in the office workplace, and 1.5 times higher than levels in homes having at least one smoker present. However, one must be cautious in comparing ETS exposure in the home and workplace based only on ETS concentration, because the duration of exposure can be significantly different in each of these indoor environments. In reevaluating their data, based on the average amount of time spent by individuals in the various environments, the results were that the total exposure to ETS is likely to be at least 1.5 times higher for restaurant workers than for a person who lives with a smoker and at least 4.4 times higher for bar workers than for individuals with only domestic ETS exposure. Table 3 provides examples of the mean ETS constituents concentrations found in certain indoor air environments.

TABLE 3. ETS CONCENTRATIONS IN CERTAIN INDOOR ENVIRONMENTS

Constituent	Concentration Range	Weighted Means[a]
Carbon monoxide (ppm)		
Bars	3.1 - 17.0	11.6
Offices	1.0 - 3.3	3.0
Restaurants	0.5 - 9.9	5.1
Nicotine ($\mu g/m^3$)		
Bars	7.4 - 65.5	19.7
Offices	0.6 - 22.1	4.1
Restaurants	3.4 - 34.0	6.5
Residences	1.6 - 21.0	4.3
Particulates ($\mu g/m^3$)		
Bars	75.0 - 1320.0	348.0
Offices	6.0 - 256.0	57.0
Restaurants	27.0 - 690.0	117.0
Residences	32.0 - 700.0	78.0

Adapted from Siegel (1993).

[a]Weighed average of individual study mean concentrations for all measurements taken. Weights used were the number of restaurants, bars, office, or homes samples.

The components of ETS most often measured in field studies are nicotine, Respirable particles and carbon monoxide. Measurement of these three components represents a minimal effort for characterization of an environment containing ETS. For locations, such as residences, public building etc. large variations in ranges among the different locations have been reported (CIAR, 1990; Federal Register, 1994).

MEASUREMENTS OF EXPOSURE

Inhalation toxicologists seldom measure dose directly. The two most common used surrogates for dose are exposure and biological fluid from the exposed individual which can provide evidence that absorption has taken place or that biologically effective doses have been delivered. Assessing exposure and its resulting dose is a difficult and complex problem for inhalation studies. The estimated number of different chemical components in tobacco smoke exceeds 4000 (Stedman, 1968; Schmeltz and Hoffmann, 1977; DHHS, 1989; EPA, 1989). For health effects studies, the measurement of all of these components in tobacco smoke is not feasible; therefore, it has become more practical to utilize certain biological markers, which when measured, will give an estimate of the frequency, duration, dose-response relationships, and magnitude of the exposure to the inhaled smoke.

A number of such markers have been used including thiocyanate, carboxy-hemoglobin (COHb), nicotine, and cotinine (CIAR, 1990; National Research Council, 1992; Rose and Margolick, 1992; Coggins, 1996). Thiocyanate is non-specific and lacks sensitivity to ETS. Carboxyhemoglobin is non-specific for ETS, and long-term exposures result in only small changes in COHb. Nicotine, which can also be present in an individual's diet, and COHb both have short circulating half-lifes thereby, limiting their utility. Cotinine, a major nicotine metabolite, is currently the best individual index of exposure to tobacco smoke in adults, children, and infants. When measured, it can be related to the number of cigarettes smoked actively or the amount of tobacco smoke inhaled passively. Cotinine not only has a relatively long half-life (14 - 18 hr) and is specific to tobacco exposure, but it can be measured at relatively low concentrations in a variety of biological fluids (saliva, plasma, and urine). In order to improve the accuracy of predicting total inhaled smoke exposure, it has been recommended that other metabolites of nicotine, such as trans-3'-hydroxycotinine, nornicotine, isomethylnicotinum, also be measured. The measurement of all nicotine metabolites should account for 80 - 90% of total inhaled dose of nicotine (CIAR, 1990). Even so, there are limitations to the utilization of biomarkers (such as cotinine) as a surrogate for measuring the total delivered dose of MS, SS, or ETS delivered to the respiratory system. These include possible individual differences in the systemic absorption and metabolism of nicotine. In animal studies, biomarkers may not necessarily reflect the route of uptake of the smoke. Care has to be taken in determining what dose is being deposited via the inhalation route versus the dose that might be taken up by other routes. For example, in animal studies that use whole body exposure various tobacco components can be expected to be deposited on the fur and then be taken up either transdermally or orally by grooming. Chen et al. (1995) concluded that whole-body exposure increased the dose to the gastrointestinal tract by a factor of two, compared to nose-only exposure. Similar differences in dose levels have also been reported by Mauderly et al. (1989) who found that the nicotine concentration in plasma and urine of rats was five- to six-fold higher with whole-body MS exposure, compared to nose-only exposure.

REGULATIONS AND RISK PREVENTION

Smoking control policies encompass a diverse group of actions in both the public and private sectors. They share the common potential for reducing the burden of tobacco induced illnesses by decreasing the prevalence and intensity of tobacco smoking. From a public health perspective, the issue of smoking is quite complex and a host of ethical and social issues are raised by such efforts. Efforts to curtail smoking have focused on the smokers, in an attempt to get smokers to change their behavior based on additional health risks. Governments have placed numerous restriction on certain tobacco advertising, mandated warning labels on tobacco products, restricted smoking in airplanes and in public areas for many types of buildings, banned the distribution free tobacco samples, limited cigarette vending machines, and increased anti-smoking education specifically focused on teenagers and children. Because nonsmokers can be at risk from exposure to ETS, most current and proposed policy initiatives have focused on reducing exposure by restricting smoking in public places and making individuals aware of the existing evidence about the harmful effects of breathing ETS. Since there are no established, health-based acceptable limits for exposure to ETS, the EPA has only recommended that exposure to ETS be minimized, wherever possible, and has supported this effort by encouraging effective way to reduce exposure. Table 4 summarizes the array of tobacco control policies that have been considered or adopted to impact on tobacco usage.

TABLE 4. TOBACCO CONTROL POLICIES BEING CONSIDERED OR ADOPTED

Direct Restraints on Tobacco Use	Economic Incentives	Education and Information
Restrict sales by age and outlets (vending machines)	Increased taxation	Require disclosure of constituents (e.g., nicotine, tars, additives)
Limit smoking in certain places (e.g., schools, hospitals, public places)	Establish legal liability	Issue Government reports; mandate education programs; restrict advertising
Regulate product composition	Mandate insurance incentives (e.g., premium price differentials)	Fund smoking research and education programs. Require health warnings (e.g., on packages and in advertising)

Adapted from DHHS (1989).

The public has become increasingly sensitized to these health issues associated with ETS. For example, the public has been made aware that the EPA (1992) and the Surgeon General (DHHS, 1986, 1989) have determined that ETS is a major source of indoor air pollution and may be responsible for lung cancer deaths of ≈3,000 nonsmokers in the United States each year. With this database, the EPA has identified ETS as a known human carcinogen (Group A). The Agency based this decision on the analogy to MS smoke, as well as the results of numerous epidemiological studies. Much of the public is also aware that exposure to ETS can cause significant other health problems in children and those other individuals who may be particularly sensitive to atmospheric tobacco smoke due to certain respiratory disorders (e.g., asthma, emphysema, and respiratory infections).

IMMUNOMODULATORY EFFECTS OF TOBACCO SMOKE

The lung is an active immunologic organ which, when exposed to toxicants, can elicit a variety of immunotoxic effects. Pulmonary immunity has been defined as all of the physiological mechanisms that enable an individual's body to recognize materials as foreign and to neutralize, eliminate, or metabolize them without injury to its own tissue. Considerable evidence both from humans and animal studies have indicated that the pulmonary immune system is a vulnerable target for the toxic elements present in smoke.

The mammalian pulmonary immune/defense system is a complex system that is dependent on the integration of and cooperative interaction between a network of different cells types with certain recognition and response functions and various soluble factors. The primary function of these defenses includes the identification and removal of inhaled substances, destruction or containment of viable infectious agents, control and removal of neoplastic cells, and the repair of damaged tissue.

Immune functions are performed through certain non-specific mechanisms (innate or natural) as well as through a variety of specific mechanisms (acquired immunity). The immune system in the lungs can be viewed as having three distinct compartments, each responsible for certain pulmonary defenses and each containing immunocompetent cells. Morphologically, these three compartments include cells associated with the (1) bronchoalveolar airspaces, (2) submucosal or secretary antibody system within the tissue of the tracheobronchial tree, and (3) complex network of lymphatic ducts and lymph nodes associated with the tracheobronchial tree. Tobacco smoke has been shown to alter the delicate balance of these protective mechanisms in a number of different ways that adversely affects the functioning of the pulmonary defense system by: (1) reducing the normal immune response (immunosuppression), possibly resulting in an increased incidence of pulmonary infections or tumors; (2) over-activating or exaggerating the immune system, possibly resulting in certain hypersensitivity reactions; or, possibly, (3) triggering an autoimmune reaction, which could result in the failure of the body to distinguish between "self" and "nonself". While this section will not critically review all of the numerous studies available, it will provide the reader with examples of each of the types of immunological modulations that has been reported, a discussion of the proposed mechanisms associated with these changes, and provide an extensive listing of references that can be used to obtain more in-depth explanations. The focus will be on the effects of tobacco smoke on the three principal components of the pulmonary defense system: (1) humoral immunity that involves the synthesis and secretion of antibodies; (2) cell-mediated immunity that involves the production of "sensitized" lymphocytes which are the effectors of cellular defense; and, (3) other non-specific defenses of the pulmonary system (e.g., mucociliary clearance, certain macrophage functions).

Evidence of Effects of Tobacco Smoke on Humoral Immunity

When the functions of the humoral immune system are adversely affected, the most severe consequences may be an increase incidence and severity of respiratory infections and/or an increase in the incidence and progression of malignancies. Such effects have been reported to be associated with exposure to tobacco smoke (DHHS, 1989; EPA, 1990; Spitzer et al., 1990). This literature also indicates that there is convincing evidence that exposure of young children, exposed in homes to ETS, is causally associated with an increase risk of lower respiratory illnesses (pneumonia, bronchitis, and bronchiolitis), resulting in approximately twice as many hospitalizations and acute upper respiratory tract infections (i.e., colds, sore throats) as unexposed children.

Exposure to tobacco smoke plays a role in systemic alterations, such as changes in circulating immunoglobulins (Ig), the primary effector for the humoral immunity. In a recent review (Sopori et al., 1994), there seem to be substantial evidence, although somewhat conflicting, that tobacco smoke does have an effect on immunoglobulin levels. Miller et al. (1996) reported that cigarette smoke caused a significant reduction in IgA levels in rats which persisted after post exposure recovery, suggesting a selective defect in immune response initiated at the mucosal level of the lung. This immunoglobulin forms an important element of the mucosal immune system and its reduction may indicate a possible mechanism whereby the immune system can be deregulated by tobacco smoke. In humans, similar effects have been reported (Masada, 1993). Reduced levels of secretary IgA have been reported both in the saliva and bronchoalveolar lavage fluid of smokers (Sato et al., 1989; Barton et al., 1990).

Other Investigators have found that inhaled cigarette smoke caused a significant reduction in serum levels of other immunoglobulins, including IgG, IgA, and IgM (Gulsvik and Fagerhol, 1979; Hersey et al., 1983; Robertson et al., 1984). However, other reports have indicated somewhat different responses, such as a significant effect on IgG, but not on IgA and IgM, levels (Ferson et al., 1979). Mill et al. (1991) reported a negative dose-response correlation between cigarette smoking and Ig levels. Whether or not these alterations in Ig levels are biologically significant is questionable, but if such Ig deficiency exists and this effect is also associated with some other immune deficiencies, then the exposed individual's resistance to respiratory infections may be compromised. Such a hypothesis is reasonable, given the association between smoking and increased respiratory infections.

Within the respiratory tract, local immunity is also mediated through the action of IgE. IgE serves as an antigen-specific receptor on the surface of inflammatory cells that can release a variety of proinflammatory factors, including vasoactive amines (histamines) and products of the arachidonic acid cascade (prostaglandins and leukotrienes). Alterations in epithelium permeability caused by tobacco smoke has been shown to result in a migration of macrophages, neutrophils, other cell types, and proteins into the lung during exposure. Serum concentrations of IgE are significantly higher in smokers than nonsmokers (Bahna et al., 1980, 1983a; Burrows et al., 1981; Zetterstrom et al., 1981; Warren et al., 1982)

Parental smoking is also associated with an increase in the prevalence of childhood asthma. Asthma has the characteristic features of airway smooth muscle contraction, edema and fluid accumulation, changes initiated presumably by the degranulation of local mast cells and the release of inflammatory mediators such as histamine and leukotrienes. It has been suggested that cell-mediated immune processes may act in concert with, or independently of, IgE antibody in the elicitation of such respiratory reactions. Chronic inflammation, which may be associated with smoking, may play an important role in asthma and is also associated with an accumulation in bronchial mucosa of leukocytes, mucus production, the destruction and sloughing of airway epithelial cells.

There is evidence of a relationship between IgE levels and rates of diagnosed asthma, wheeze, and chronic cough and/or sputum in smokers (Burrows et al., 1981). Asthma is invariably associated with the activity of T-lymphocytes and eosinophils and that in those cases where asthma results from allergic sensitization an important initiating event is IgE-mediated mast cell degranulation, which triggers the accumulation of inflammatory cells. Lehrer et al. (1980) demonstrated tobacco antigen-specific IgE in some people, but the frequency was very low and was not significantly increased in smokers. Thus, the specific role of tobacco smoke-induced IgE in the pathogenesis of pulmonary disease in smokers is not clear. There appears to be no correlation between tobacco smoking, IgE levels, and skin-test reactivity. In fact, smokers exhibit significantly lower skin-test reactivity for a given value of IgE as compared to nonsmokers (Sopori et al., 1994)

Cigarette smokers showed a biphasic response to changes in serum IgD concentrations. There were increases in IgD levels in light and moderate smokers but these levels decreased in heavy smokers (Thomas et al., 1973, 1974a and b, 1975; Bahna et al., 1983b; Sopori et al., 1994). The function of IgD in the lung seems to be limited to its role as a receptor on B-lymphocytes and it cannot mediate any of the effector functions known to be associated with the other classes of Ig.

Immunoglobulin levels in cord blood have been examined in infants from smoking mothers. IgA, IgM, IgD, IgE, and IgG were shown to be elevated in these infants when compared to infants with nonsmoking mothers (Cedarqvist et al., 1984; Magnusson 1986; Magnusson and Johansson, 1986; Spitzer et al., 1990). These Investigators postulated that such changes might indicate sensitization and mucosal damage or could possibly predispose the individual to premature rupture of maternal membranes or endometritis (Cedarquist et al., 1984; Spitzer et al., 1990).

In animal studies, exposure to cigarette smoke results in a reduction in the numbers of antibody plaque-forming cells in the spleen. Splenic lymphocytic mitogen-induced blastogenesis was also depressed in tobacco-exposed mice (Dean and Muray, 1991). However, in most cases the Authors were not able to find a significant relationship between the number of cigarettes smoked and the observed effect.

Investigations of possible autoimmune effects of cigarette smoke are very limited. In one study, autoantibodies, such as antinuclear antibodies and rheumatoid factors, have been reported to be more frequent in smokers than nonsmokers, but these responses have not been confirmed by other studies (Mathews et al., 1993).

Evidence of Effects of Tobacco Smoke on Cell-Mediated Immunity

The cell-mediated immune system, of the respiratory tract, is responsible for providing the defense against intracellular organisms and certain malignancies by controlling spontaneous arising tumors and limiting the growth of established neoplasms.

Because cigarette smoking has been implicated with increases in a variety of malignant diseases a number of studies have evaluated the effect of smoking on natural killer (NK) cell activity. Several Investigators (Ferson et al., 1979; Hersey et al., 1983; Hughes et al., 1985; Tollerud et al., 1989a) reported that NK cell activity of human peripheral blood leukocytes, against melanoma cells, was significantly lower in smokers. Tollerud et al. (1991) observed that while white smokers had lower NK cell activity, black smokers and nonsmokers had comparable NK activity.

Cellular immunity, as measured by the phytohemagglutinin (PHA)-induced lymphoproliferative response or development of tumor-specific cytotoxic T-lymphocytes, was initially increased, but with continued exposure to tobacco smoke, there was a significant reduction in this response (Thomas et al., 1973; Chalmer et al., 1975; Holt et al., 1976). It appeared that cigarette smoke containing the higher levels of tar and nicotine induced changes to a greater degree than smoke containing lower levels of these components. The age of the smoker may also play a role because in young smokers, such changes in lymphocyte functions appear to be elevated, whereas in older smokers, the same effects were significantly decreased (Vos-Brat and Rumke, 1969; Silverman et al., 1975). Other studies have reported that the relationship between cigarette smoking and lymphocyte response to PHA and Concanavalin A (ConA) or to alloantigens, in a mixed-lymphocyte reaction, were not significantly different between smokers and nonsmokers (Suciu-Foca et al., 1974; Whitehead et al., 1974; Daniele et al., 1977; Hugles et al., 1985). In one study with monkeys, there was a decreased response to the T-lymphocyte mitogen ConA following long-term exposure to smoke (Sopori et al., 1985). These Authors also reported a significant decrease in plaque forming cell responses to both T-lymphocyte-dependent (i.e., sheep red blood cells) and T1 antigens in rats following chronic smoke exposure. Chang et al. (1990) observed a T-lymphocyte functional defect in lung-associated lymphoid tissue, but not in anatomically-distant lymph nodes, suggesting that cigarette smoke may have variable effects in different lymphoid organs. In heavy smokers, total T $(CD3^+)$-lymphocytes were increased, but the percentage of $CD4^+$ T-lymphocytes was decreased and the percentage of $CD8^+$ T-lymphocytes were increased, resulting in a significant decrease in the $CD4^+/CD8^+$ ratio (Ginns et al., 1982; Miller et al., 1982; Costabel et al., 1992). One study found that ethnic background may be an important factor. In a white study group, a significant dose-related decrease in $CD4^+$ T-lymphocytes was reported; there was no change in $CD4^+/CD8^+$ ratio among Black smokers (Smart et al., 1986; Tollerud et al., 1989b, 1991).

In general, studies that measured the effects of cigarette smoke on B-lymphocytes seem more consistent. Several Investigators have reported increases in the total number of B-lymphocytes following exposure to tobacco smoke (Burton et al., 1983; Hughes et al., 1985; Mill et al., 1991).

Evidence of Effects of Tobacco Smoke on Non-Specific Immunity

A number of non-specific defense mechanism are operative in the respiratory tract that protect the individual from assault by inhaled airborne chemicals and infectious agents. The inhalation of tobacco smoke have been shown to alter a number of these defenses possibly increasing the individual's risk of a variety of respiratory diseases (Holt and Keast 1972; Holt, 1987; Johnson et al., 1990; Sopori et al., 1994).

In humans and other mammals, a healthy airway epithelium and its secretions serve as a protective barrier to the passage of both autologous and foreign materials across the airway mucosa and provide a means of transporting substances out of the lung by the mucociliary escalator. Tobacco smoke has been shown to adversely affect the functioning of this clearance mechanism through such effects in humans, rats, rabbits, and donkeys (Holma, 1969; Dalhamn and Rylander, 1970; Albert et al., 1974; Park et al., 1982). By using electron microscopy, it has been shown that exposure to smoke damages ciliated cells in the trachea and lung epithelial cells resulting in a denuded and/or ulcerated epithelium. This results in a reversible loss of cilia and the hyperplasia of secretory cells (IARC, 1986; Adelkofer 1988). Defects in this mucociliary transport system can be brought about by changing the chemical nature of the mucus secretions, by paralyzing the ciliary beating movements, or by producing focal lesions in the ciliary epithelium.

Tobacco smoke stimulates the goblet cells to produce excessive mucus discharge, inhibits epithelial ion movement, and initiates biochemical changes in the composition of the mucus. Initiation of these changes can begin after the first exposure (Abraham, 1984). Inhalation to tobacco smoke results in hypertrophy of submucosal glands and hyperplasia of mucin-secreting goblet cells in the airway (Miller et al., 1996). This, in turn, can alter the rheological properties of the airway secretions and result in an inhibition of mucociliary clearance.

Within the gaseous exchange region of the lung, the first line of cellular defense against viable and nonviable particles is the alveolar macrophage. The efficiency of the phagocytic and lytic system of the macrophage determines the sterility of the lung and is essential for health.

Human and animal studies have shown that cells obtained by lavage differ significantly, between smokers and nonsmokers, with respect to the number of recovered cells, their composition, and other various characteristics. Smoking increases epithelial permeability and stimulates the migration of inflammatory cells (neutrophils, lymphocytes, and mast cells) into the airway lumen. Such effects may predispose smokers to infections and increased risk of chronic lung disease (Jeffery and Reid, 1981; Plowman, 1982; Hunninghake and Crystal, 1983; Gairola, 1986; IARC, 1986; Pride, 1987; Gardner, 1994). There is evidence that tobacco smoke can activate the neutrophils, producing a shift in favor or proteolysis by releasing elastase from lysosomal granules and generating oxidants through the NADPH oxidase system and myeloperoxidase. It has been postulated that such a proteolytic imbalance may cause lung tissue destruction, leading to emphysema. Alveolar macrophages, which account for most of the increase in lavagable cells from smokers, exhibit altered morphological, functional, and biochemical characteristics (Brody and Craighead, 1975; Martin and Warr, 1977; Finch et al., 1982; Fisher et al., 1982; Miller et al., 1996). Included among such characteristics of the macrophages of smokers are their ability to release increasing amounts of superoxides and elevated levels of lysosomal enzymes (Cantrell et al., 1973; Martin, 1973; Harris et al., 1975; Hoidal et al., 1981; Hoidal and Niewoehner, 1982); impaired protein and RNA synthesis (Yeager 1969; Low, 1974; Ando et al., 1984); and depression of lymphoproliferative response to antigen and an increase in migration and chemotactic responsiveness (Warr and Martin 1974; Laughter et al., 1977). Animal studies have shown that the ultimate consequence of such alterations is an increased survival of inhaled viable bacteria in the lungs, resulting in excess deaths due to bacterial infections (Henry et al., 1970; Henry and Kouri, 1984; Gardner, 1994).

Smoking also suppresses the local immunity of the lung by decreasing the ability of the isolated human macrophage to release certain cytokines such as inter-leukin (IL)-1 and IL-6 (Yamaguchi et al., 1989, Soliman and Twigg, 1992). Macro-phages from rats exposed to cigarette smoke were unresponsive to lipopolysac-charide stimulation *in vitro*, resulting in a reduced expression of cytokine IL-1β mRNA as compared with controls (Miller et al., 1996). The cytokine 1L-1β is also expressed in macrophages, but these cells, from smoke-exposed animals, were shown to have reduced responsiveness to induction of 1L-1β gene expression by bacterial LPS. Keast and Ayres (1981) also demonstrated that lymphocytes from smokers gave a reduction in response to the mitogen PHA or LPS, also indicating a reduced ability by the immune system to synthesize and release cytokines after tobacco smoke exposure.

There are conflicting reports regarding the phagocytic and bactericidal activity of alveolar macrophages following smoke exposure. Some studies have detected little difference in the ability of macrophages from smokers and nonsmokers to ingest and kill microorganisms, while others have reported a significant depression in phagocytosis and/or bactericidal activity of macrophages isolated from smokers (Harris et al., 1970; Martin and Warr 1977; Park et al., 1982; Green, 1985; McLeod et al., 1985; Johnson et al., 1987). King et al. (1988) reported that macrophages from smokers were able to phagocytize microbes normally but had a deficiency in their ability to kill the engulfed microorganisms.

INTERPRETATION AND SUMMARY

Tobacco smoke is probably one of the best characterized risk factors that adversely affect human health. Over the past decade, data have been accumulated to clearly substantiate cases in which lung immunoregulatory functions of the humoral and/or cell-mediated immunity has been compromised by the inhalation of tobacco smoke.

While the mechanisms for some of the specific classes of immunomodulation can be proposed in this section, insofar as is known, along with the presentation of the response, there is still much uncertainty regarding the specific mechanism actually responsible for most of the response observed.

In 1990, a panel reviewed nearly 3000 articles on the biological, clinical, and epidemiological evidence for the health effects of tobacco smoke (Spitzer et al., 1990). One of their conclusion was that the available literature would support the hypothesis that smoking can affect the immune system, resulting in human health risk. Although the scientific literature concerning tobacco smoke and its associated human health risk continues to grow at a fast pace, the difficulty in clearly inter-preting the results of these studies leaves many important questions unanswered. The current literature is fraught with inconsistencies concerning the immunotoxicity of smoke. For example, while some Investigators observed that smoking inhibits clearance, other reported that it may actually accelerate clearance, or that it did not adversely affect clearance rates (Albert et al., 1971; Binns, 1975; Bohning et al., 1982; Cross et al., 1982).

At least two factors that may lead to clearance impairment by tobacco smoke need to be considered; (1) possible toxic effects on the cells and (2) pulmonary particle overload. Pulmonary overloading impairs lung clearance and, possibly, redistribution of particles into the interstitium, in addition to inflammation, macrophage recruitment, and cell proliferation (Conference Proceedings, 1996). Impaired clearance increases pulmonary retention which, in turn, increases the dose of the various components of the tobacco smoke to the respiratory system. This increased dose may be the basis for some of the immunological effects of inhaled tobacco smoke.

In a review by Sterling (1983), there was some evidence presented suggesting that low doses of tobacco smoke may actually stimulate certain functions of the immune system (e.g. accumulation of macrophages, stimulation of mucus production and possible acceleration of respiratory clearance) possibly providing some degree of "protection" for the smoker. However, such responses could also be pathologic (e.g., stimulation of mucus production in a person with chronic bronchitis or emphysema).

The increase in the number of macrophages may be the result of increased production of IL-1 by the resident macrophages, which causes the influx of the other inflammatory cells (polymorphonuclear leukocytes cells and peripheral blood mononuclear cells) into the lung. The concentration-dependent leukocytosis (increased number of T- and B-lymphocytes) appears to be different in smokers compared to nonsmokers. However, the question of whether this is a relationship between smoke and lymphocyte function is debatable. While animals studies have demonstrated that tobacco smoke suppresses antibody responses, causes biphasic lymphoproliferative capacity, enhanced susceptibility to murine sarcoma virus, and influences virus proliferation, such animal studies can not precisely replicate the human exposure conditions because of the route of exposure and the rapid chemical changes that are known to occur to the tobacco smoke once generated.

It appears that nicotine present in tobacco smoke has an effect on the developing fetus. Evidence supports the hypotheses that *in utero* exposure to mother's smoke and postnatal exposure to tobacco smoke enhances the process of allergic sensitization and predisposes children to early respiratory illness and influences the severity of asthma in children. Passive tobacco smoke increases the number of asthma attacks in children who have the disease as well as increases the number of new cases of asthma in children who did not previously exhibit symptoms. For individuals having allergies and asthma, the inhalation of tobacco smoke can either worsen or produce a reappearance of the illness (Muramatsu et al., 1983). Second-hand smoke is suspected of playing a role in conversion of the respiratory immune responses of children from normal T-helper-1 (T_H1) to allergic T_H2 responses.

While exposure to tobacco smoke alone may not be immediately life-threatening, the association with other environmental stresses, such as infections, could prove critical in the promotion or exacerbation of a pulmonary disease (Bonham and Wilson 1981; Aronson et al., 1982). Smoking is a risk factor for various respiratory infections such as influenza, bronchitis, and pneumonia. Although the exact mechanism and the specific tobacco smoke components associated with these health effects are not known with certainty, such modulation of the pulmonary immunological system must be considered.

It seems, from the available data, that it is not possible, at this time, to clearly identify all of the specific mechanisms of action that tobacco smoke has on the pulmonary immune system. The science of immunotoxicology is complicated by several elements: the complex nature of the immune system; the age of the subject; genetic background; and, a variety of external risk factors. This is especially true when trying to understand the effects of such complex mixtures like tobacco smoke.

The pulmonary immune system is also a very complex toxicological target. It's response to chemical insults such as tobacco smoke has features that complicate interpretation of the experimental findings. For instance, in animal studies, tobacco smoke can show variations in the impact on the test subject that could be attributed to species, strain, or sex differences. Exposure to tobacco smoke also reveals complex dose-response relationships; tobacco smoke can stimulate the immune system at one dose yet suppress it at another. Since the immune system has a reserve capacity, problems with interpreting such dose-response relationships also manifest as problem with evaluating the reserve or redundant capacity of the system. Tests that measure impairment of one component of the immune system may not indicate overall immunotoxicity, since other immune components may compensate for the impairment. It is difficult to determine accurately whether exposure to an airborne substance that reduces one cell type function will actually impair the total immune capacity or whether test agents will actually stimulate another component of the immune system that will then compensate for the impaired component.

Acute and chronic exposure to tobacco smoke can render different results. The age or maturational status of the exposed individual has been shown to affect the results. Often with prenatal or neonatal exposure provoking a greater response than the adult exposure. Finally, a most difficult problem is trying to interpret and evaluate the clinical significance of an altered immune response seen in animals to humans. What is the biological relevance of a moderate and/or transient perturbation of the humoral, cell mediated or non-specific immunity system?

This Chapter provided a review of issues and the status of our present knowledge on the effects of mainstream tobacco smoke, sidestream smoke and environmental tobacco smoke on the immune system in animals and humans. The data would indicate that there are probably several distinct mechanisms, interacting, that induces the various immunomodulations. Because tobacco smoke is such a complex mixture of substance, many of which alone can have the potential to adversely effect the immune system, future research will be necessary to better understand the cellular and molecular mechanisms through which tobacco smoke mixture affects these pulmonary defense systems.

REFERENCES

Abraham WM. "Effects of Inhaled Materials on Airway Secretions." In *Seminars in Respiratory Medicine. Vol. 5.* New York: Publisher Theme-Stratton Inc., 1984, pp. 324-328.

Adams JD, O'Mara-Adams KJ, Hoffmann D. Toxic and carcinogenic agents in undiluted mainstream smoke and sidestream smoke of different types of cigarettes. Carcinogenesis, 1987;8:729-731.

Adelkofer F, Sherer G, Wenzel-Hartung R, Brune H, Thomas C. "Exposure of Hamsters and Rats to Sidestream Smoke of Cigarettes." In *Indoor and Ambient Air Quality*, R Perry, PW Kirk, eds. London: Selper Ltd., 1988, pp. 252-258.

Albert RE, Berger J, Sanborn K, Lippmann M. Effects of cigarette smoke components on bronchial clearance in the donkey. Arch. Environ. Health, 1974;29:96- 101.

Albert RE, Lippmann M, Peterson HT. "The Effects of Cigarette Smoking on the Kinetics of Bronchial Clearance in Humans and Donkeys." In *Inhaled Particles*, WH Walton, ed. Old Woking, England: Unwin, 1971, pp 165-182.

Ando M, Sugimoto M, Nishi R. Surface morphology and function of human pulmonary alveolar macrophages from smokers and nonsmokers. Thorax, 1984;39:850-856.

Aronson MD, Weiss ST, Ben RL, Komaroff AL. Association between cigarette smoking and acute respiratory illness in young adults. J. Am. Med. Assoc., 1982;248:181-183.

Bahna SL, Myhre BA, Heiner DC. IgE elevation and suppression by tobacco smoking. J. Allergy Clin. Immunol., 1980;65:231-236.

Bahna SL, Heiner DC, Myhre BA. Immunoglobulin E pattern in cigarette smokers. Allergy, 1983;38:57-64.

Bahna SL, Heiner DC, Myhre BA. Changes in serum IgD in cigarette smokers. Clin. Exp. Immunol., 1983;51:624-630.

Barton JR, Riad MA, Gaze MA, Moran AG, Ferguson A. Mucosal immunodeficiency in smokers and in patients with epithelial head and neck tumors. Gut, 1990;31:378-382.

Benner CL, Bayona JM, Caka FM, Tang H, Lewis L, Crawford J, Lamb JD, Lee ML, Lewis EA, Hanson LD, Eatough DJ. Chemical composition of environmental tobacco smoke: Particulate-phase compounds. Environ. Sci. Technol., 1989;23:688-699.

Bice DE, Muggenberg DA. Lung response to antigen. Semin. Respir. Med., 1984;5:217-227.

Binns R. Animal inhalation studies with tobacco smoke. Rev. Environ. Health, 1975;2:81-93.

Bohning DE, Atkins HL, Cohn SH. Long-term clearance in man: Normal and impaired. Ann. Occup. Hyg., 1982;26:259-271.

Bonham GS, Wilson RW. Children's health in families with cigarette smoke. Am. J. Publ. Health, 1981;71:290-293.

Brody AR, Craighead JE. Cytoplasmic inclusions in pulmonary alveolar macrophages by cigarette smoke. Lab. Invest., 1975;32:125-132.

Burrows B, Halonen M, Barbee RE, Lebowitz MD. The relationship of serum immunoglobulin E to cigarette smoking. Am. Rev. Respir. Dis., 1981;124:526-525.

Burton RC, Ferguson P, Gary M, Hall J, Hayes M, Smart YC. Effects of age, gender, and cigarette smoking on human immunoregulatory T-cells subsets: Establishment of normal ranges and comparison with patients with colorectal cancer and multiple sclerosis. Diag. Immunol., 1983;1:216-233.

Cantrell ET, Warr GA, Busbee DL, Martin RR. Induction of aryl hydrocarbon hydroxylase in human pulmonary alveolar macrophages by cigarette smoking. J. Clin. Invest., 1973;52:1881-1884.

Cedarqvist LL, Eddey G, Abdel-Latif N, Litwin SD. The effect of smoking during pregnancy on cord blood and maternal serum immunoglobulin levels. Am. J. Obstet. Gynecol., 1984;148:1123-1126.

Chen BT, Benz JV, Finch GL, Mauderly JL, Sabourin PJ, Yeh HC, Snipes MB. Effects of exposure mode on amount of radiolabeled cigarette particles in lungs and gastrointestinal tract of F344 rats. Inhal. Toxicol., 1995;7:1095-1108.

Chalmer J, Holt PG, Keast D. Cell-mediated immune response to transplanted tumors in mice chronically exposed to fresh cigarette smoke. J. Natl. Cancer Inst., 1975;52:1129-1134.

Chang JC, Distler SG, Kaplan AM. Tobacco smoke suppresses T-cell but not antigen-presenting cells in the lung-associated lymph nodes. Toxicol. Appl. Pharmacol., 1990;102:514-523.

CIAR: Center for Indoor Air Research. *Workshop on Environmental Tobacco Smoke Generation and Exposure.* April, 1990.

Conference Proceedings. *Particle Overload in the Rat Lung and Lung Cancer: Implications for Human Risk Assessment.* Proceedings of Conference held at Massachusetts Institute of Technology. March 29-30, 1995. Inhal. Toxicol., 1996;8:1-298.

Coggins CR. The OSHA review of animal inhalation studies with environmental tobacco smoke. Inhal. Toxicol., 1996;8:819-830.

Costabel U, Maier K, Teschler H, Wang YM. Local immune components in chronic obstructive pulmonary disease. Respiration, 1992;59:17-19.

Cross FT, Palmer RF, Filipy RE, Dagle RE, Stuart BO. Carcinogenic effects of radon daughters, uranium ore dust and cigarette smoke in beagle dogs. Health Phys., 1982;42:323-52.

Dalhamn T, Rylander R. Ciliotoxicity of cigar and cigarette smoke. Arch. Environ. Health, 1970;20:252- 253.

Daniele RP, Dauber J, Altose MD, Tawlands, DT, Gorenberg DJ. Lymphocyte studies in asymptomatic cigarette smokers. Am. Rev. Respir. Dis., 1977;116:997-1005.

Dean J, Muray R. "Toxic Response of the Immune System." In *The Health Consequences of Smoking: Toxicology.* Washington, DC: United States Department of Health and Human Services, 1991.

Department of Health and Human Services. *The Health Consequences of Involuntary Smoking. A Report of the Surgeon General.* Washington, DC: U.S. Government Printing Office, DHHS Publ. No. 87-83398, pp. 134-137, 1986.

Department of Health and Human Services. *A Report of the Surgeon General. Reducing the Health Consequences of Smoking: 25 Year Progress.* Washington, DC: U.S. Government Printing Office, 1989.

Dube MF, Green CR. Methods of collection of smoke for analytical purposes. Rec. Adv. Tobacco Sci, 1982;8:42-102.

Eatough DL, Benner CL, Bayona JM, Richards G, Lamb JD, Lee ML, Lewis EA, Hansen LD. Chemical composition of environmental tobacco smoke: Gas-phase acids and bases. Environ. Sci. Technol., 1989;23:679-687/

EPA: Environmental Protection Agency. *Indoor Air Facts: Environmental Tobacco Smoke. June 1989.* Washington, DC: U.S. Government Printing Office, 1989.

EPA: Environmental Protection Agency. *Environmental News Report. January 7, 1993.* Washington, DC: U.S. Government Printing Office, 1993.

EPA: Environmental Protection Agency. *Health Effects of Passive Smoking: Assessment of Lung Cancer in Adults and Respiratory Disorders in Children.* Washington, DC: U.S. Government Printing Office, EPA 600/6-90/006A, May 1990.

EPA: Environmental Protection Agency. *Respiratory Health Effects of Passive Smoking.* Washington, DC: U.S. Government Printing Office, EPA 600/6-90-006F, 1992.

Federal Register, Department of Labor, Vol. 59, No. 65, Part II, April 5, 1994. pp. 15968- 16013.

Ferson M, Edwards A, Lind A, Milton GW, Hersey P. Low natural killer-cell activity and immunoglobulin levels associated with smoking in human subjects. Int. J. Cancer, 1979;23:603-609

Finch GL, Fisher, GL, Hayes TL, Golde DW. Surface morphology and functional studies of human alveolar macrophages from cigarette smokers and nonsmokers. J. Reticuloendothel. Soc., 1982;32:1-23.

Fisher GL, McNeill KL, Finch GL, Wilson FD, Golde DW. Functional evaluation of lung macrophages from cigarette smokers and nonsmokers. J. Reticuloendothel. Soc., 1982;32:311-321.

Gairola CG. Free lung cells response of mice and rats to mainstream cigarette smoke exposure. Toxicol. Appl. Pharmacol., 1986;84:567-575.

Gardner DE. "Direct and Indirect Injury to the Respiratory Tract." In *Air Pollution and Lung Disease in Adults*, P Witorsch, SV Spagnolo, eds. Boca Raton, FL: CRC Press, 1994, pp. 19-47.

Ginns LC, Goldenheim PD, Miller LG. T-Lymphocyte subsets in smoking and lung cancer: Analysis by monoclonal antibodies and flow cytometry. Am. Rev. Respir. Dis., 1982;126:265-269.

Green GM. Mechanism of tobacco smoke toxicity on pulmonary macrophage cells. Eur. J. Respir. Dis., 1985;66:82-85.

Gulsuik A, Fagerhol MK. Smoking and immunoglobulin levels. Lancet, 1979;1:449-451.

Hammond SK, Leaderer BP, Roche AC, Schenker M. Collection and analysis of nicotine as a marker for environmental tobacco smoke. Atmos.. Environ., 1987;21:457-462.

Harris JO, Swenson, EW, Johnson JE, III. Human alveolar macrophages: Comparison of phagocytic ability, glucose utilization and ultrastructure in smokers and nonsmokers. J. Clin. Invest., 1970;49:2086-2096.

Harris JO, Olson GN, Castle JR, Maloney AS. Comparison of proteolytic enzyme activity in pulmonary alveolar macrophages and blood leukocytes in smokers and nonsmokers. Am. Rev. Respir. Dis., 1975;111:579-586.

Henry MC, Kouri RE. *Smoke Inhalation Studies in Mice. Final Report for the Council of Tobacco Research, Contract CTR-0030*. Bethesda, MD: Microbiological Associates, 1984.

Henry MC, Spangler J, Findlay J, Ehrlich R. "Effects of Nitrogen Dioxide and Tobacco Smoke on Retention of Inhaled Bacteria." In *Inhaled Particles III. Vol. 1*, WH Walton, ed. Surrey, England: Unwin, 1970, pp. 527-533.

Hersey P, Predergost D, Edwards A. Effects of cigarette smoking on the immune system: Follow-up studies in normal subjects after cessation of smoking. Med. J. Australia, 1983;2:425-429.

Hoffmann D, Wynder EL. *Chemical Composition and Bioactivity of Tobacco Smoke. IARC Scientific Publication No. 74*. Lyon, France: World Health Organization, 1986, pp. 145-165.

Hoidal JR, Niewoehner DE. Lung phagocyte recruitment and metabolic alterations induced by cigarette smoke in humans and in hamsters. Am. Rev. Respir. Dis., 1982;126:548-552.

Hoidal JR, Fox RB, LeMarbe PA, Perri R, Repine JE. Altered oxidative metabolic responses *in vitro* of alveolar macrophages from asymptomatic cigarette smokers. Am. Rev. Respir. Dis., 1981;23:85-87.

Holma B. The acute effects of cigarette smoke on the initial course of lung clearance in rabbits. Arch. Environ. Health, 1969;18:171-176.

Holt PG, Chalmer J, Roberts LM, Papadimitrious JM, Thosma WR, Keast D. Low-tar high-tar cigarettes: Comparison of effects in mice. Arch. Environ. Health, 1976;31:258-265.

Holt PG. Immune and inflammatory function in cigarette smokers. Thorax, 1987;42:241-249.

Holt PG, Keast D. Environmental-induced changes in immunological function: Acute and chronic effects of inhaled tobacco smoke and other atmospheric contaminants in man and experimental animals. New Engl. J. Med., 1972;307:1042-1046.

Hughes DA, Haslam PL, Townsend PI, Turner-Warwick M. Numerical and functional alteration in circulatory lymphocytes in cigarette smokers. Clin. Exp. Immunol., 1985;61:459-466.

Hunninghake CW, Crystal RG. Cigarette smoking and lung destruction: Accumulation of neutrophils in the lungs of cigarette smokers. Am. Rev. Respir. Dis., 1983;128:833-838

IARC: International Agency for Research on Cancer. *Working Group on the Evaluation of the Carcinogenic Risk of Chemicals to Humans. IARC Monograph Vol. 38*. Lyon, France: IARC, World Health Organization, 1986, pp. 127-198.

IARC: International Agency for Research on Cancer. *Committee on Smoking and Health. London, Her Majesty's Stationery Office, Environmental Carcinogens: Methods of Analysis and Exposure Measurements. Vol. 9. IARC Scientific Publications No. 81*. Lyon, France: IARC, World Health Organization, 1987.

IARC: International Agency for Research on Cancer. *Monograph on the Evaluation of Carcinogenic Risk of Chemicals to Humans. Vol. 38. Tobacco Smoking.* Lyon, France: IARC, World Health Organization, 1986.

Ingebrethsen BJ, Sears SB. Particle evaporation of sidestream smoke in a stirred tank. J. Colloid Interface Sci., 1989;131:526-536.

Jeffery PK, Reid L. The effect of tobacco smoke with and without phenylmethyloxadiazole in rat bronchial epithelium. J. Pathol., 1981;133:341-358.

Johnson S, Musher DM, Lawrence EC. Phagocytosis and killing of *Haemophilus influenzae* by alveolar macrophages: No differences between smokers and nonsmokers. Eur. J. Respir. Dis., 1987;70:309-315.

Johnson JD, Hauchens DP, Kluwe WM, Craig DK, Fisher GL. Effects of mainstream and environmental tobacco smoke on the immune system in animals and humans. CRC Crit. Rev. Toxicol., 1990;20:369-395

Keast D, Ayres DJ. Effect of chronic tobacco smoke exposure on immune responses in mice. Arch. Environ. Health, 1981;36:201-207.

King TE, Savici D, Campbell PA. Phagocytosis and killing of *Listeria monocytogenes* by alveolar macrophages: Smokers versus nonsmokers. Am. Rev. Respir. Dis., 1988;158:1309-1316.

Kuller LH, Garfinkel L, Correa P, Haley N, Hoffmann D, Preston-Martin S, Sandler D. Contribution of passive smoking to respiratory cancer. Environ. Health Perspect., 1986;70:57-69.

Laughter AH, Martin RR, Twomey JJ. Lymphoproliferative responses to antigens mediated by human pulmonary alveolar macrophages. J. Lab. Clin. Med., 1977;89:1326-1332.

Lehrer SB, Wilson MR, Karr RM, Salvaggio JE. IgE antibody response of smokers, nonsmokers, and "smoke-sensitive" persons to tobacco leaf and smoke antigen. Am. Rev. Respir. Dis., 1980;121:168-170.

Lofroth G, Burton RM, Forehand L, Hammond KS, Seila RL, Zweidinger RB, Lewtas J. Characterization of environmental tobacco smoke. Environ. Sci. Technol., 1989;23:L 610-614.

Low RB Protein biosynthesis by the pulmonary alveolar macrophages. Am. Rev. Respir. Dis., 1974;110:466-477.

Magnusson CG. Maternal smoking influences cord serum IgE and IgD levels and increases the risk for subsequent infant allergy. Allergy Clin. Immunol., 1986;78:898-904.

Magnusson CG, Johansson SG. Maternal smoking leads to increased cord serum IgG_3. Allergy, 1986;41:302-307.

Martin RR, Warr GA. Cigarette smoking and human pulmonary macrophages. Hosp. Pract., 1977;86:97-104.

Martin RR. Altered morphology and increased acid hydrolase content of pulmonary macrophages from cigarette smokers. Am. Rev. Respir. Dis., 1973;107:596-601.

Masada M. Determination of human immunoglobulin A and secretary immunoglobulin A in bronchoalveolar lavage fluids by solid phase enzyme immunoassay. Clin. Chem. Acta, 1993;220:145-156.

Mathews JD, Hooper BM, Whittingham S, Mackay IR, Stenhouse NS. Association of autoantibodies with smoking cardiovascular morbidity and death in a Busselton population. Lancet, 1993;2:754-758

Mauderly JL, Bechtold WE, Bond JA, Brooks AL., Chen BT, Cuddihy RG, Harkema JR, Henderson RF, Johnson NF, Rithidech K, Thomassen DG. Comparison of three methods of exposing rats to cigarette smoke. Exp. Pathol., 1989;37:194-197.

McLeod R, Mack DG, McLeod EG, Campbell EJ, Estes RG. Alveolar macrophage function and inflammatory stimuli in smokers with and without obstructive lung disease. Am. Rev. Respir. Dis., 1985;131:377-384.

Mill F, Flanders WD, Boring JR, Annest JL, Destefano AF. The association of race, cigarette smoking, and smoking cessation to measures of immune system in middle-aged man. Clin. Immun. Immunopathol., 1991;59:187-200.

Miller K, Hudspith B, Cunningham M, Prescott C, Meredith B. Effects of cigarette smoke exposure on biomarkers of lung injury in the rat. Inhal. Toxicol., 1996;89:803-817.

Miller LG, Goldstein G, Murphy M, Ginns LC. Reversible alterations in immunoregulatory T-cells in smoking: Analysis by monoclonal antibodies and flow cytometry. Am. Rev. Respir. Dis., 1982;126:265-269.

Muramatsu T, Weber A, Muramatsu S, Akermann F. An experimental study on irritation and annoyance due to passive smoking. Int. Arch. Occup. Environ. Health, 1983;51:305-317

National Research Council. *Environmental Tobacco Smoke: Measuring Exposure and Assessing Health Effects.* Washington, DC: National Academy Press, 1986.

National Research Council. *Biological Markers in Immunotoxicology.* Washington, DC: National Academy Press, 1992.

Oberdorster G, Pott F. Extrapolation from rat studies with environmental tobacco smoke to humans: Comparison of particle mass deposition and of clearance behavior of ETS components. Toxicol. Lett., 1987;35:107-112.

Park SS, Kikkawa Y, Goldring IP, Daly MM, Zeletsky M, Shim CH, Spierer M, Warner CP, Holford-Strevens V, Wong C, Manfreda J. The relationship between smoking and total immunoglobulin E levels. J. Allergy Clin. Immunol., 1982;69:370-375

Phalen RF (ed.). *Inhalation Studies: Foundation and Techniques.* Boca Raton, FL: CRC Press, 1984.

Plowman PN. The pulmonary macrophage population of human smokers. Ann. Occup. Hyg., 1982;25:3933-3405

Pride NB. Epidemiology of bronchial hypersecretion: Recent studies. Eur. J. Respir. Dis., 1987;71:13-18.

Robertson MD, Boyd JE, Collins HP, Davis JM. Serum immunoglobulin levels and humoral immune competence in coal workers. Am. J. Ind. Med., 1984;6:387-393

Rodgman A. Environmental tobacco smoke. Reg. Toxicol. Pharmacol., 1992;16:223-244.

Rose NR, Margolick JB. (eds.) "The Immunological Assessment of Immunotoxic Effects in Man." In *Clinical Immunotoxicology,* New York: Raven Press, 1992, pp. 449-530.

Samet JM, Marbury MC, Spengler JD. Health effects and sources of Indoor air pollution. Am. Rev. Respir. Dis., 1987;136:1486-1508

Sato T, Ohmto AK, Chikuma M, Kado M, Nagai S, Izumi T, Takeyama M, Sopori ML, Cherian S, Chilukun R, Shopp GM. Cigarette smoking causes inhibition of the immune response to intratracheally-administered antigen. Toxicol. Appl. Pharmacol., 1989;97:489-499.

Schmeltz I, Hoffmann D. Nitrogen-containing compounds in tobacco and tobacco smoke. Chem. Rev., 1977;77:295-311.

Siegel M. Involuntary smoking in the restaurant workplace. J. Am. Med. Assoc., 1993;270:490-493.

Silverman NA, Potvin C, Alexander JC, Jr., Chretien PB. *In vitro* lymphocyte reactivity and T-cell levels in chronic cigarette smokers. Clin. Exp. Immunol., 1975;22:285-292.

Smart YC, Cox J, Roberts TK, Brinsmead MW, Burton RC. Differential effects of cigarette smoking on recirculating T-lymphocyte subsets in pregnant women. J. Immunol., 1986;137:1-3.

Soliman DM, Twigg H. Cigarette smoking decreases bioactive interleukins-6 secretion by macrophages. Am. J. Physiol., 1992;263:471-478.

Sopori ML, Goud NS, Kaplan AM. "Effects of Tobacco Smoke on the Immune System." In *Immunotoxicology and Immunopharmacology. 2nd Edition,* JH Dean, MJ Luster, AE Munson, I Kimber, eds. New York: Raven Press, 1994, pp. 412-434.

Sopori ML, Gairola C, DeLucia AJ, Bryant LR, Cherian S. Immune responsiveness of monkeys exposed chronically to cigarette smoke. Clin. Immunol. Immunopathol., 1985;36:338-344.

Spitzer WO, Lawrence V, Dales R, Hill G, Archer MC, Clark P, Abehnaim L, Hardy J, Sampalis J, Pinfold SP, Morgan P. Links between passive smoking and disease: A best-evidence synthesis. A report of the Working Group on Passive Smoking. Clin. Invest. Med., 1990;13:17-42.

Stedman RL. The chemical composition of tobacco and tobacco smoke. Chem. Rev., 1968;68:153-207.

Sterling TB, Kobayashi D. Indoor by product levels of tobacco smoke: A critical review of the literature. J. Air Pollut. Control Assoc., 1982;32:250-259.

Sterling TD. Possible effects on occupational lung cancer from smoking-related changes in the mucus content of the lung. J. Chron. Dis., 1983;36:669-684.

Stohs SJ, Bagchi D, Bagchi M. Toxicity of trace elements in tobacco smoke. Inhal. Toxicol., 1997;9:867-890.

Suciu-Foca N, Molinaro A, Buda J, Rectsma K. Cellular immune responsiveness in cigarette smokers. Lancet, 1974;1:1062.

Thomas WR, Holt PG, Keast D. Humoral immune response of mice with long-term exposure to cigarette smoke. Arch. Environ. Health, 1975;30:78-80.

Thomas WR, Holt PG, Papadimitriou JM, Keast D. The growth of transplanted tumors in mice after chronic inhalation of fresh cigarette smoke. Br. J. Cancer, 1974a;30:459-462.

Thomas WR, Holt PG, Keast D. Antibody production in mice chronically-exposed to fresh cigarette smoke. Experientia, 1974b;30:1469-1470.

Thomas WR, Holt PG, Keast D. Effect of cigarette smoking on primary and secondary humoral responses in mice. Nature, 1973a;243:240-241.

Thomas WR, Holt PG, Keast D. Cellular immunity in mice chronically-exposed to fresh cigarette smoke. Arch. Environ. Health, 1973b;27:372-375.

Tollerud DJ, Clark JW, Brown LM. Association of cigarette smoking with decreased number of circulating natural killer cells. Am. Rev. Respir. Dis. 1989a;139:194-198.

Tollerud DJ, Clark JW, Brown LM. The effects of smoking on T-cell subsets: Population-based survey of healthy caucasians. Am. Rev. Respir. Dis., 1989b;139:1446-1451.

Tollerud DJ, Brown LM, Blattner WA, Man DL, Pankiw-Trost L, Hoover RN. T-Cell subsets in healthy black smokers and nonsmokers. Evidence for ethnic group as an important response modifier. Am. Rev. Respir. Dis., 1991;144:612-616.

Vos-Brat LC, Rumke PH. Immunoglobuline Concentraties, PHA Reacties van Lymfocyten in vitro en Enkele Antistof Titers van Gezonde Rokers. Jaarb Kanker Nederland, 1969;19:49-53.

Warr GA, Martin RR. Chemotactic responsiveness of human alveolar macrophages: Effects of smoking. Infect. Immun., 1974;8:222-227.

Warren CP, Holford-Stevens V, Wong C, Manfreda J. The relationship between smoking and total immunoglobulin E levels. J. Allergy Clin. Immunol., 1982;69:370-375.

Whitehead RH, Hooper Be, Ginshaw DA, Hughes LE. 1974. Cellular immune responsiveness in cigarette smokers. Lancet, 1974;1:1232-1233.

Yamaguchi E, Okazaki N, Itoh A, Obe S, Kawakami Y, Okuiyama H. Interleukin-1 production of macrophages is decreased in smokers. Am. Rev. Respir. Dis., 1989;14:397-402.

Yeager H. Alveolar cells: Depressant effect of cigarette smoke on protein synthesis. Proc. Soc. Exp. Biol., 1969;131:247-250.

Zetterstrom O, Osterman K, Machado L, Johnsson SG. Another smoking hazard: Raised serum IgE concentration and increased risk of occupational allergy. Br. Med. J., 1981;283:1215-1217.

17. APPLYING PULMONARY IMMUNOTOXICITY DATA TO RISK ASSESSMENT

MaryJane K. Selgrade, Ph.D.
National Health and Environmental Effects
Research Laboratory,
United States Environmental Protection Agency,
Research Triangle Park, NC 27711

INTRODUCTION

From the preceding Chapters, it is clear that a number of contaminants in our environment can modulate pulmonary immune responses. These same immune responses represent the first line of defense against many microorganisms for which the respiratory tract is the primary portal of entry. Suppression of these responses may increase the risks associated with these infections. On the other hand, stimulation of the immune response in the lung can lead to hypersensitivity reactions that can range from mild to life-threatening. In order to protect the public from potentially adverse health effects, it is important to understand the risks associated with modulation of these pulmonary immune responses. This Chapter will review the various components of risk assessment, describe some of the approaches that can be used in applying pulmonary immunotoxicity data to the risk assessment process, and discuss the inherent uncertainties and research which may help to reduce these uncertainties.

The National Research Council (1983) of the National Academy of Sciences has defined risk assessment as comprising some or all of the following components: hazard identification, dose-response assessment, exposure assessment, and risk characterization. Hazard identification is a qualitative assessment and addresses the question, "What effects does a chemical cause?" In this case, "Are immune responses in the lung altered as a result of exposure and does this lead to disease?" Dose-response assessment is more quantitative and, broadly interpreted, addresses the question, "Under what conditions does a chemical cause these effects?" Certainly the biggest concern is the effective dose, but of course this can vary based on other factors such as age, genetics, and presence of underlying disease. Exposure assessment addresses the issue of what portion of the population is exposed to a particular toxicant and at what level. Risk characterization uses all of this information to assess the risk that might result from exposure to an agent. This risk characterization is then used along with other information to make risk management (regulatory) decisions. Risk assessments have to be based on available data, and because ideal data sets are rare (if not impossible to achieve) decisions with respect to risk necessarily include some uncertainties.

Three types of data will be discussed in this Chapter: Data demonstrating immune suppression and enhancement of infection as a result of exposure to air pollutants; data demonstrating allergic sensitization to proteins or low molecular weight (MW; hapten) chemicals associated with asthma; and data demonstrating enhanced responses to common allergens as a result of exposure to air pollutants. In each case issues and uncertainties associated with hazard identification, extrapolation from animal data to human health effects, intraspecies variability, and extrapolation from acute to subchronic to chronic exposures will be discussed.

AIR POLLUTANTS AND ENHANCED RISK OF INFECTIOUS DISEASE

Hazard Identification

In the study of systemic immune suppression, effects of chemicals on immune function, typically at the cellular level, have generally been used to test chemicals (hazard identification). Occasionally, these have been followed by tests to determine the effects of a chemical on host resistance to challenge with an infectious agent or tumor cells (reviewed by Selgrade, In Press). The study of chemical exposures on local pulmonary effects has generally taken the opposite approach starting with an assessment of effects of inhaled compounds on resistance to challenge with an infectious agent (or in some cases tumor cells) and then exploring the underlying immunologic mechanisms. For example, studies by a number of Investigators have shown that ozone (O_3; see Chapter 12) and nitrogen dioxide (NO_2; see Chapter 13) impair resistance to challenge with Gram-positive bacteria in both mice and rats (reviewed by Selgrade and Gilmour, 1994). This approach eliminates the uncertainty that arises from having to extrapolate from effects at one level of biologic organization to another, that is, from effects at the cellular (immune function) level to effects at the whole organism or population (disease susceptibility) level. However, although host resistance models may be applied in a research setting, they are not amenable to large-scale testing. Hence, when the United States Environmental Protection Agency proposed a test rule for 25 hazardous air pollutants (USEPA, 1996), an assessment of alveolar macrophage (AM) phagocytic function rather than resistance to challenge with bacteria was requested. In addition to the above-mentioned uncertainty associated with crossing levels of biologic organization, interpretation of the data from this test rule will also require dealing with the uncertainty associated with extrapolating from animal data to human health effects. As described below, work done with ozone suggests that it is possible to extrapolate both from suppression of macrophage function to disease susceptibility and from mouse to human data.

Dose Response and Extrapolating from Animal Data to Human Health Effects

Parallelograms have often been used to make comparisons between effects in laboratory rodents and humans (Figure 1) (Selgrade et al., 1995a; Selgrade, In Press). In these models, different species are represented by the vertical sides. Interspecies extrapolations are made by moving horizontally across the diagram; moving vertically down the models addresses problems associated with extrapolation from *in vitro* data to *in vivo* effects and extrapolation across levels of biologic organization from effects at the cellular level (e.g., immune function) to effects at the organism or population level (e.g., disease susceptibility). Each circle represents a measurable parameter. Models can be expanded by stacking parallelograms to make ladders (Figure 1). For the most part this approach has only been used to make qualitative comparisons; however, more quantitative comparisons could be made by developing dose-response curves for two or more circles in the model, obtaining a function (or equation to describe these dose responses, and then making quantitative comparisons from one circle to another.

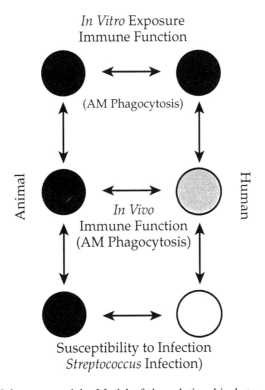

Figure 1. Parallelogram models. Model of the relationship between suppression of human and rodent immune cell (alveolar macrophage) function following *in vitro* and *in vivo* exposure to a toxicant (upper) and the relationship between suppression of immune function and susceptibility to infection (lower). Solid circles represent data that can be readily obtained, cross-hatched circle represents data that can sometimes be obtained, and open circle represents data that can rarely be obtained.

Data from studies demonstrating enhanced susceptibility to *Streptococcus zooepidemicus* in mice following O_3 exposure and suppression of AM phagocytosis as a result of O_3 exposure can be applied to this parallelogram model. In experiments designed to assess effects of O_3 exposure on susceptibility to this bacteria, mice were exposed to O_3 or filtered air for 3 hr. Both treatment groups were then put in the same chamber and challenged with an aerosol of *S. zooepidemicus* which results in deposition of 200 - 4000 bacteria (Miller et al., 1978) in the lung. Few if any air controls died; however, significant increases in mortality have been observed at O_3 levels similar to the National Ambient Air Quality Standard which is 0.08 ppm (based on the maximum 8 hr average concentration) (Coffin and Gardner, 1972). In air controls, bacteria were cleared from the lung by 24 hr post-exposure, delayed clearance and then replication of bacteria were observed in mice exposed to O_3. Different susceptibilities to O_3-enhanced infection occur as a result of age and strain differences (Gilmour et al., 1993a and 1993b.). When percent suppression of AM phagocytosis for different ages and strains of mice was plotted against percent mortality, strains most resistant to O_3-enhanced infection could tolerate the greatest decrease in phagocytosis, whereas the most susceptible strain could tolerate the least suppression of phagocytosis. In fact, in the most susceptible strain, some deaths occurred in air controls without any suppression of phagocytosis; there was no reserve capacity. Similar strain differences have been observed with NO_2 and phosgene enhancement of streptococcal infection (unpublished data) suggesting that the macrophage is also an important target for other pollutants. Hence, AM phagocytic function can be used as the parameter of interest for the top left-hand corner of the lower parallelogram in Figure 1 when susceptibility to *S. zooepidemicus* is used as the parameter of interest for the lower left-hand corner. For at least some mice species, there appears to be a one-to-one relationship between these two endpoints. Studies with other infectious agents and air pollutants suggest this applies to Gram-positive bacterial infections in general (Selgrade and Gilmour, 1994).

Suppression of AM phagocytic activity has also been demonstrated in human subjects following O_3 exposure (Devlin et al., 1991) and can be applied to the upper right-hand corner of the parallelogram (Figure 1, lower half). To date, effects of O_3 on human AM phagocytosis have only been tested using one exposure regimen, 0.08 ppm O_3 for 6.6 hr while undergoing intermittent moderate exercise. Table 1 shows a comparison of mouse and human AM sensitivity to *in vivo* O_3 exposure. Dosimetric differences between mice and humans have been reported by Slade et al. (1997) based on a comparison of ^{18}O deposition in humans and mice exposed to $^{18}O_3$. When one makes adjustments based on this dosimetric data and assumes that the response of AM to O_3 is linear over some portion of the dose-response curve, the sensitivity of human and mouse AM to *in vivo* O_3 exposure appears to be similar (Selgrade et al., 1995a). With this limited amount of data, qualitative comparisons can be made between the top corners of the lower parallelogram in Figure 1. Similarities have also been demonstrated between rabbits and humans with respect to effects of sulfuric acid on macrophage function, including phagocytosis, viability, and production of reactive oxygen intermediates (Zelikoff et al., 1997). This data might also be applied in a parallelogram model.

Data from controlled human studies following *in vivo* exposure to toxicants are rare; O_3 and sulfuric acid are exceptions. It is, however, possible to obtain AM from humans and expose them to toxicants *in vitro*. When human and mouse AM obtained from bronchoalveolar lavage were exposed to 0.8 ppm O_3 for 3 hr, the percent suppression in phagocytic activity was 29 and 21%, respectively (Table 1). These values were not statistically different (Selgrade et al., 1995a). Dose response for these endpoints (represented by the two upper corners of the ladder model) (Figure 1) would allow us to develop quantitative relationships for O_3. These numbers may be useful in interpreting data for more toxic air pollutants, such as phosgene, for which the mechanisms of immune suppression appear to be similar, but human *in vivo* data is not available.

TABLE 1. EFFECT OF OZONE EXPOSURE ON ALVEOLAR MACROPHAGE PHAGOCYTOSIS

Treatment	Mice	Humans
In vitro Air	369.2 (26.4)[a] n=6	386.7 (50.5) n=6
In vitro Ozone (0.8 ppm; 3 hr)	291.7 (17.4)[b] n=6	275.0 (45.1)[b] n=6
% Suppression	21%	29%
In vivo Air	330.6 (10.4)a n=4	714.9 (46.1) n=10
In vivo Ozone[c]	194.0 (19.7)[b] n=4	539.2 (22.3)[b] n=10
% Suppression	42%	25%
% Suppression corrected[d] for dosimetric difference	28%	25%

[a]Mean (±SE) of phagocytic index (number fluorescent particles ingested/100 cells).
[b]Significantly different from air control (p<0.05; Student's t-test)
cMice were exposed to 0.8 ppm O_3 for 3 hr; humans were exposed to 0.08 ppm O_3 for 6.6 hr while undergoing intermittent exercise.
dBased on studies using ^{18}O, AM of mice exposed to 0.8 ppm O_3 for 3 hr receive ≈1.5 times more O_3 than those of humans exposed to 0.08 ppm O_3 for 6.6 hr while exercising moderately (See text).
Reproduced from Selgrade et al., 1995a.

What is Adverse?

The question of what constitutes an adverse effect is usually raised during any risk assessment. With respect to immunosuppression, two components are required: exposure to an immunosuppressive agent and exposure to an infectious agent or neoplastic cell. In the above example, suppression of AM function is not a problem unless infectious bacteria are also present. Two major determinants of adversity are the baseline immune response of any particular individual (i.e., how much reserve capacity does an individual have) and the dose and pathogenicity of the infectious agent or tumor cells. Baseline immune responses in the population are probably represented by a bell-shaped curve with some individuals at much more risk than others (Figure 2). Dose and pathogenicity of the organisms determine what portion of the population is at risk. It is clear that even individuals who appear to have competent immune systems can get sick given a sufficient dose of a sufficiently pathogenic organism. In the above example, AM function varied as a result of age and strain of mice. At least one strain of mouse had no reserve capacity; hence, any degree of immune suppression resulted in enhanced disease. Therefore, suppression of immune function represents an adverse effect. Compensatory mechanisms and redundancy in immune responses are also frequently issues of debate with respect to what constitutes an adverse effect. Specifically, one could ask whether other components of the immune system are able to compensate for the loss in AM function. Certainly this is not the case in mice where increased mortality due to infection is observed following exposure to O_3. However, it has been demonstrated that the *S. zooepidemicus* infectivity model can be applied to rats (Gilmour and Selgrade, 1993). Exposure to O_3 suppressed AM phagocytic function and delayed clearance of bacteria from the lung. However, rats did not die as a result of infection, but eventually cleared the bacteria from their lungs. Recovery from infection appeared to be due to a more rapid influx of neutrophils (PMN) in the rat (peak at 24 hr) as compared to the mouse (peak 2 - 4 d). Rat PMN may also be more efficient at killing bacteria than mouse PMN.

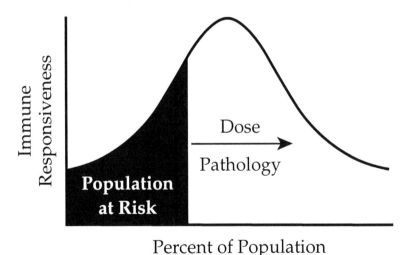

Figure 2. Influence of baseline immune response and dose pathogenicity of the infectious agent on the population at risk.

PMN influx in rats might be considered to be a compensatory mechanism. However, although rats did not die, they did experience delayed clearance of bacteria and an inflammatory response in the lung which did not occur with bacteria or O_3 (at the exposure levels used) alone. This enhanced pathogenesis, although not as drastic as mortality, is certainly adverse.

Intraspecies Variability

A number of factors influence the response of the immune system to toxicant exposure. In addition to genetics, gender and age, factors such as stress, infections, medications, sun exposure, nutrition, and in women the timing of the menstrual cycle affect the immune response. Some of these factors undoubtedly contribute to the wide variability seen in human control groups. Many of these variables are controlled in animal studies, which are generally conducted with inbred, well-nourished, pathogen-free, young adult animals of the same sex, exposed only to the chemical of interest. These idealized conditions reduce variability in experimental data, but are not representative of the human condition. The problem of intraspecies variability is not unique to immunotoxicity, and for purposes of risk assessment is generally dealt with by applying an uncertainty factor of 10. In the O_3 example above, the mice that were most resistant to challenge with *S. zooepidemicus* following O_3 exposure (CD-1, 9-wk-old) differed from the least resistant mice (C_3H/HeJ) by a factor of nine (Selgrade and Gilmour, 1994), suggesting that the 10-fold uncertainty factor may not be unrealistic.

Intraspecies variability is also a confounder in epidemiologic studies and may, in part, account for discrepancies among studies designed to demonstrate enhanced risk of infection in populations exposed to air pollutants. For example, Ware et al. (1984) found no correlation between respiratory illness and indoor exposure to NO_2 generated by combustion of gas for cooking, and Samet et al. (1993) reported no association between indoor NO_2 exposure and respiratory illness in infants <2 years of age. However, in other studies, the incidence of acute respiratory illness was greater in children from households with gas stoves (Speizer et al., 1980) and a 15 ppb increase in household annual NO_2 exposure was associated with an increase of ≈20% in cumulative incidence of lower respiratory tract symptoms in children (Neas et al., 1991). Also, an association was reported between NO_2 exposure and increased incidence of croup in children <2 years of age (Schartz et al., 1991). Of course, a number of other variables may also account for these differences, including differences in exposure to pathogenic organisms during the course of different studies.

Acute vs Subchronic and Chronic Effects

Frequently, a 10-fold uncertainty factor is incorporated into a risk assessment when there is no information on chronic exposure. Very few chronic studies have been done to assess pulmonary immunotoxicity. Such studies are complicated by the fact that immune responsiveness decreases (as does resistance to infection) in aging animals. Alveolar macrophage phagocytic function recovers rapidly after exposure to air pollution ceases. At least in the case of phosgene, as described below, results of acute and subchronic studies were comparable. Exposure to phosgene enhances infection in both mice and rats to *Streptococcus* infectivity in a manner similar to that described in a previous section for O_3. In both mice and rats enhanced susceptibility was observed when infection occurred shortly after a single phosgene exposure (Selgrade et al., 1989; Yang et al., 1995). If infection was delayed until 24 hr after exposure, susceptibility to infection was not enhanced, suggesting that AM function recovers fairly rapidly following phosgene exposure.

Because recovery of AM function following a single phosgene exposure is rapid, the opportunity for enhanced infection is short-lived. However, recent studies demonstrated enhanced susceptibility to infection in rats exposed to 0.1 or 0.2 ppm phosgene, for 6 hr/d, 5 d/wk, for 4 or 12 wk and infected shortly after the last exposure (Selgrade et al., 1995b). The enhanced susceptibility was equal to that observed after a single, acute exposure suggesting that adaptation did not occur nor were effects of repeated exposure additive. Presumably increased risk of infection existed for a short period of time (4 - 6 hr) following each phosgene exposure in this chronic study, greatly expanding the opportunity for enhanced infection as compared to a single exposure. It is clear that recovery time, exposure regimen, and potential adaptation are all factors that can influence the outcome of infection.

Clearly the paucity of subchronic and chronic studies contributes a major uncertainty in using pulmonary immunotoxicity data in risk assessment. It should be noted, however, that if an acute exposure to an immunotoxicant and an encounter with an infectious agent or neoplastic event happen to coincide, enhanced severity of infection might result in permanent damage to the lung, or immune suppression might result in the spread of an agent (infectious or neoplastic) and damage to extrapulmonary organs/tissues. Risk assessors have generally focused on establishing safe "lifetime exposures"; only recently have methodologies for establishing acute reference doses been developed (Guth, 1994).

Implications for Risk Assessment

The preceding provides information on one type of pulmonary immunotoxicity, suppression of AM phagocytic function. From the studies cited it would appear that suppression of AM phagocytosis represents an adverse effect associated with enhanced susceptibility to gram positive bacterial infections. The severity of this effect will depend on the magnitude of suppression, the dose and pathogenicity of bacteria encountered during the time when AM phagocytosis is suppressed, and the genetic makeup of the individual. Effects of O_3 on human and mice AM were quantitatively similar, suggesting that perhaps the 10-fold uncertainty factor for intraspecies variability could be reduced or eliminated. However, considerable intraspecies variability exists, and a 10-fold uncertainty factor is warranted in this case. AM phagocytic function appears to be fairly resilient, and in the case of phosgene there was no quantitative difference between acute and subchronic exposure, suggesting again that the 10-fold uncertainty factor usually applied for extrapolating from subchronic to chronic may be unwarranted. Also, methodologies for establishing acute reference doses may also need to be applied for this endpoint.

ALLERGIC SENSITIZATION TO PROTEINS OR LOW MW (HAPTEN) CHEMICALS ASSOCIATED WITH ASTHMA

Hazard Identification

Proteins and certain low MW compounds (MW<3000, equivalent weight <1000) have the potential to cause allergic sensitization and responses characterized by pulmonary inflammation, mucus secretion, hyperreactivity following challenge with the offending allergen, and hyperresponsiveness following challenge with non-specific agonists such as methacholine (Chan-Yeung, 1990a; Bernstein, 1992). These symptoms are generally considered hallmarks of allergic asthma and as previous Chapters indicated, represent a significant health hazard. All proteins are not equally allergenic. For instance, enzymes that vary by just a few amino acids may behave very differently with respect to allergenicity (Kimber et al., In Press). Low MW compounds must be reactive enough to form conjugates with host cell proteins. The hapten protein conjugate may then induce allergic sensitization, and subsequent challenge with the hapten may elicit allergic responses. However, all isocyanates, for example, do not induce respiratory allergy (Karol, 1995).

The goal for hazard identification is to identify compounds that have the potential to cause allergic sensitization. As in the case of assessment of pulmonary immune suppression, the approach to assessing chemicals for the potential to cause allergic sensitization initially focused on actual disease endpoints rather than immune responses. Guinea pigs were exposed by inhalation or intratracheal (IT) instillation to the agent in question for some period of time. Following a rest period during which immune responses developed, animals were challenged with the agent, again by the respiratory route, and immediate-onset responses assessed based on changes in respiratory rate. Late-onset responses could be assessed also by monitoring respiratory rate (for 24 hr after challenge) and body temperature (Karol et al., 1985). Because this approach is prohibitively expensive (i.e.,>\$100,000/chemical; Selgrade et al., 1994), efforts have been made to develop methods using mice and to identify immunologic biomarkers of effect that could be used to screen chemicals. Antibody responses, e.g., IgG_1 in guinea pigs (assessed by passive cutaneous anaphylaxis) and IgE and IgG_1 (assessed by ELISA) in mice, have received the most attention to date (Sarlo and Clark, 1992; Ritz et al., 1993; Kawabata et al., 1996; Robinson et al., 1996) and have been used to set safe exposure levels in occupational settings (see "Dose Response and Extrapolation from Animal Data to Human Health Effect" below).

Because experimental exposure of laboratory rodents via inhalation is an expensive procedure and successful methods for exposing animals to low MW compounds via IT or intranasal routes have not as yet been developed, efforts have been made to devise methods for screening these compounds for allergenic effects without using respiratory sensitization. Several methods, i.e., the use of structure-activity relationships (SAR), in vitro tests for chemical reactivity with protein, and the mouse IgE test have received significant attention (Kimber et al., 1996). Structure-activity is frequently used in the initial evaluation of low MW compounds (Sarlo and Clark, 1992; Selgrade et al., 1994); two chemical classes, diisocyanates and acid anhydrides, trigger concern. Many compounds that have been identified as respiratory allergens possess electrophilic functionalities that enable the chemical to react covalently with carrier proteins. Electrophilicity has been used in the analysis of SARs for respiratory allergy (Agius et al., 1991). However, electrophilicity is neither definitive nor sufficient for conferring on chemicals the capacity to induce respiratory sensitization. Programs have been developed to identify low MW compounds with potential to induce respiratory sensitization based on the chemical structures of known allergens (Klopman, 1992; Karol et al., 1996); however, the drawback to this approach is the limited number (39) of known low MW allergens on which to base these SARs (Kimber et al., 1996). Also, although SARs may be useful for hazard identification, other approaches have to be used to obtain dose-response information needed for more quantitative risk assessment. *In vitro* tests to assess the capacity of a chemical to react with protein have also been used as a preliminary screen (Wass and Belin, 1990, Sarlo and Clark, 1992). A positive test strongly suggests that a chemical has the potential to act as a hapten and requires further study. Chemicals that do not react *in vitro* are not likely to pose a risk, but again this test is not sufficient to identify a respiratory allergen. Hence, *in vitro* testing represents a preliminary screening procedure and, of course, does not provide information on dose response. In the mouse, IgE test compounds are applied to the skin in a manner similar to that used to assess contact sensitivity. Respiratory sensitizers are distinguished by an increase in total serum IgE. Although this test has been used successfully by some laboratories, attempts at an intra-laboratory validation suggested that more work is needed before it can be adopted outside the research arena (Dearman et al., 1998). Again, the mouse IgE test is primarily a means for hazard identification. Information on the effective dose via the respiratory route needed for respiratory sensitization cannot be derived from this test. However, sensitization via the dermal route can result in respiratory sensitization (Nemery and Lenaerts, 1993; Botham et al., 1989). To the extent that the dermal route of exposure is of concern, the mouse IgE test might be used in a more quantitative fashion. To date, the mouse IgE test has not been used to assess allergenicity of proteins because of problems associated with dermal absorption.

Although there is little doubt that induction of specific cytophilic antibodies (IgG_1 and IgE) are indicative of the potential to induce allergic sensitization, other mechanisms may be responsible for some allergic reactions (reviewed by Bernstein and Bernstein, 1994). In studies of dust mite (DM) allergy, Lambert et al. (1998) transferred serum from DM-sensitized brown Norway rats to naive rats that subsequently exhibited immediate bronchoconstriction responses to allergen challenge, but did not exhibit immune-mediated pulmonary inflammation or hyperreactivity to a non-specific agonist, suggesting that the antibody response plays an important role in the first but not in the latter two responses. Conversely, when they transferred cells from sensitized to naive rats, they observed immune-mediated inflammation and hyperresponsiveness to the non-specific agonist, but did not observe immediate reactions to DM challenge. Clinically, it is possible to have late-onset responses in the absence of immediate-onset types, particularly in the case of low MW compounds (Bernstein, 1992). Mechanisms underlying late-onset responses are uncertain, but it has been suggested that they may be mediated by IgG (Zeiss et al., 1977) or CD-4 cells (Karol, 1995). In many instances, IgE antibodies specific for low MW compounds have not been detected in patients exhibiting a clinical picture consistent with allergic asthma induced by these compounds (Chan-Yeung, 1990a). Hence, the use of cytophilic antibody as a screen for low MW compounds may not identify all compounds with allergenic potential. As research continues to improve our understanding of the mechanisms associated with allergic sensitization and subsequent pulmonary responses, other biomarkers that may be useful in hazard identification may be identified. Certainly the presence of inflammatory cells in the lung, particularly eosinophils, and cytokine responses associated with the eosinophilic inflammation are possible candidates. Until we have a more complete understanding of the mechanisms underlying respiratory hypersensitivity, our ability to assess this hazard will be limited. At the present time there are no standard guidelines for assessing chemicals for respiratory allergenicity. Therefore, although the previously mentioned test rule for 25 hazardous air pollutants proposed by the EPA included tests for immune suppression, tests for allergenicity are not included.

Dose Response and Extrapolating from Animal Data to Human Health Effects

The first dilemma that risk assessors face with respect to allergenic agents is that of which dose response to consider. The dose required to initially sensitize an individual is generally higher than that required to trigger responses in a previously-sensitized individual. In the example below, the detergent industry worked with the dose required to sensitize an individual. Although this choice may be appropriate for the occupational exposures where sensitized individuals can be excluded from work-related exposures, the choice is more difficult when environmental exposures that affect populations at large are under consideration. Another concern that confounds dose response is the possibility of concurrent exposure to agents which are not anti-genic themselves, but act as adjuvants. For example, detergent matrix and common air pollutants (such as diesel and tobacco smoke) have both been shown to act as adjuvants (Burrows et al., 1981, Zetterstrom et al., 1981; Takafuji et al., 1987; Kawabata et al., 1996) and influence the allergenicity of detergent enzymes and aller-genic air pollutants, respectively. Also, proteolytic detergent enzymes enhance the allergic antibody responses of guinea pigs to nonproteolytic enzymes in mixtures (Sarlo et al., 1997a). Exposure to mixtures of agents also presents possibilities for crossreactivity. Finally, as alluded to above, respiratory sensitization may occur by routes other than inhalation, including dermal (Botham et al., 1989; Nemery and Lenaerts, 1993) and gastrointestinal (Björkstén, 1996). All of these issues present complications in any move from a qualitative to a quantitative assessment of risk.

Nevertheless, the detergent industry has been able to use data from the guinea pig IT test to set occupational exposure guidelines (OEG) that protect workers from risks associated with allergic sensitization to enzymes (Sarlo et al., 1997b). In their studies, guinea pigs were IT-dosed with different levels of enzyme protein, and sera from the animals were titered for allergic antibody to the enzyme. The amount of antibody produced to an enzyme was compared to the amount of antibody produced to the same protein dose of subtilisin A (Alcalase). Subtilisin was chosen as the reference allergen because the American Conference of Governmental and Industrial Hygienists (1990) developed a threshold limit value in the workplace for subtilisin A of 60 ng protein/m^3 based on historic human data (Flindt, 1969; Pepys et al., 1969, 1973). Also, Proctor and Gamble has set an internal OEG of 15 ng protein/m^3. In their experience, this guideline has minimized new sensitizations to the enzyme and eliminated asthma and allergies caused by such exposures (Sarlo, 1994). The dose responses of new enzymes were compared to that of Alcalase and used to determine relative potency factor differences. For less potent and equivalent enzymes, the OEG were set at the same level used for Alcalase; for more potent enzymes, the OEG were lowered according to the potency factor derived by comparing the two dose-response curves. This approach was used to set OEG for several new enzymes. Workers were monitored for new sensitization via annual or semi-annual skin-prick tests and results were similar to that observed for Alcalase (no more than 0 - 3% new sensitizations/yr and, as with Alcalase, allergic symptoms were not observed). The data show that a slightly different type of parallelogram (Figure 3) was used to successfully extrapolate in a quantitative fashion from animal data to human health effects. Preliminary studies suggest that a mouse intranasal exposure model in which IgG_1 is assessed demonstrates relative potencies to Alcalase that are similar to those observed with the guinea pig (Robinson et al., 1996). Conceivably, Alcalase could be used as a reference when evaluating proteins other than detergent enzymes, for example, those associated with microbial pesticides (Ward et al., 1998). It has also been suggested that a relative potency approach might be useful in evaluating low MW compounds (Selgrade et al., 1994).

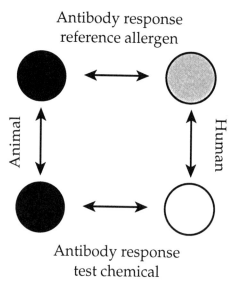

Figure 3. Parallelogram model applied to respiratory allergy. Model of relationship between the production of allergic antibodies to a reference and test allergen in laboratory animal tests and human occupational exposures. Solid circles represent data obtained experimentally. Cross-hatched circle represents data obtained using occupational data for a known sensitizer. Open circle is extrapolated from hatched circle using a factor obtained by comparing dose responses for the two solid circles.

Intraspecies Variability

Clearly genetics (and other host factors such as age, infection, and nutritional status) play an important role in susceptibility to allergic disease because only a portion of individuals receiving the same dose of an allergen develop disease. For example, 10 - 45% of workers exposed to proteolytic enzymes develop asthma (Chan-Yeung, 1990b). In contrast, only 7% of workers exposed to trimellitic anhydride develop allergic disease (Zeiss et al., 1992). Atopy, defined by skin reactivity in prick tests with common allergens such as pollens and house dust mite is usually associated with a family history of allergy (see Chapter 4). Atopy is a risk factor for occupational asthma caused by protein agents, but there is not a strong association between atopy and asthma caused by low MW compounds (Venables, 1989). It is likely that alternative combinations of various sets of genes act independently or synergistically to produce the phenotype of allergic asthma (Rosenwasser, 1996). The incidence of atopic disease in general and asthma in particular has been increasing at a rate that suggests factors other than the gene pool are involved (Weitzman et al., 1992; Vogel, 1997). Animal models of asthma have generally used strains or species that are particularly prone to this type of disease (e.g., guinea pigs, Brown-Norway rats, and BALB/c mice) and conditions can be made even more conducive to development of allergic responses when adjuvants are used (Mizushima et al., 1979; Sedgwick and Holt, 1984; Herz et al., 1996). Allergy and asthma are associated with T-helper (T_H)-2 lymphocyte responses and a number of situations have been described that may either favor or inhibit T_H2 responses (Romagnani, 1996). Susceptibility may very well be described by a bell-shaped curve similar to that in Figure 2. A portion of the population is clearly at greater risk. However, the increasing incidence of allergic disease indicates that this portion of the population may expand under the right conditions. Presently, however, it is not clear what those conditions might be. Both human and animal studies tend to focus on geneti-cally-susceptible individuals; hence, one could argue that a 10-fold uncertainty factor is unnecessary. Were genetics the only factor contributing to susceptibility, this would be true. Certainly other factors are involved, however, and so little is known about these that it is difficult to say what an appropriate uncertainty factor would be.

Acute vs Subchronic and Chronic Effects

Allergic sensitization is an example of an acute exposure that has long-lasting effects in that the immune system is, thereafter, primed to respond to the allergen and trigger respiratory responses. It is more difficult to judge the impact of chronic exposure to doses below that which results in acute sensitization. Little information is available with respect to influence of frequency and duration of exposure on the induction of sensitization. Most animal studies have relied on relatively short-term sensitization regimens ranging from a few days to several weeks. These regimens are driven in part by the desire to develop test methods that fit within a feasible time frame and, in part, by concern that there be a clear separation between sensitization and challenge (i.e., one could begin to see bronchial responses typical of elicitation in a longer-term sensitization protocol). Also, it has been difficult to develop animal models that mimic chronic asthma because repeated exposure of laboratory animals to low-dose antigen eventually results in tolerance (Sedgwick and Holt, 1984). The little evidence available suggests that, at least for low MW compounds, exposure concentration rather than frequency and duration of exposure is of greater importance for sensitization. Karol (1983) found that exposure of guinea pigs for 3 hr/d for 5 consecutive days to 0.61 ppm toluene diisocyanate (TDI)-induced respiratory hypersensitivity in $\approx25\%$ of test animals, but exposure for 6 hr/d for 70 d over a 4 mo period to 0.02 ppm, an equivalent cumulative dose, failed to cause sensitization.

Relatively high concentrations of allergen, in one or a few exposures, are believed to be important in induction of sensitization (Karol, 1989). However, repeated challenge (which could be considered boosts to the original sensitization) was built into the Karol protocol and recommended when initial challenge failed to elicit a response (Karol et al., 1985). For the guinea pig IT test used by Proctor and Gamble in setting safe protein exposure levels as described above, guinea pigs were dosed once per week for 10 - 12 wk (Sarlo et al., 1997b). Depending on allergen potency, seropositive animals were observed as early as 4 wk and the percentage of positive animals increased with successive weeks up to 10 wk. In the mouse intranasal test, currently under development as an alternative to the guinea pig IT test (Robinson et al., 1996), mice exposed three times over a 10 d period exhibited IgG_1 antibody responses, but IgE responses were weak. A stronger IgE response was observed with a more prolonged dosing regimen, suggesting that exposure duration may have some role to play in induction of sensitization. At present there is not enough information with respect to sensitization to either refute of support use of a 10-fold uncertainty factor to account for lack of data following chronic (or even subchronic) exposure. Once sensitization is established, challenge exposures are always an acute event.

Implications for Risk Assessment

From the preceding, it is clear that a lot remains to be learned about allergic asthma. Research is needed to increase understanding of the mechanisms underlying this disease in order to develop better methods for hazard identification, particularly if the goal is to identify potential allergens and set safe exposure levels before allergic sensitization occurs in exposed populations. From studies cited here, it appears that production of cytophilic antibodies represents one useful biomarker which can be assessed in both experimental animals and human subjects. The relative potency approach used to test detergent enzymes provides one means of relating laboratory animal data to potential human effects and setting exposure standards to minimize risk. Because there have been other cases where occupational exposures have provided information on safe and unsafe exposure levels, it may be possible to expand this approach to other situations.

AIR POLLUTANTS AND ENHANCED RESPONSES TO COMMON ALLERGENS

Hazard Identification

For a majority of children with asthma and for $\approx 50\%$ of adult asthmatics, allergic sensitization to common allergens (particularly indoor allergens like dust mite [DM] and cockroach) is a driving stimulus for chronic airway inflammation and bronchial hyperresponsiveness (Henderson et al., 1995; Rosenstreich et al., 1997). Substantial epidemiological evidence has suggested an association between increased levels of ambient air pollutants (including O_3, sulfur dioxide, NO_2, and respirable particulate matter) and increased morbidity due to asthma (ATS, 1996). Modulation of immune responses to common allergens is one of several possible means by which air pollution might influence asthma. Modulation of immune responses could occur at either time of sensitization, challenge, or both. It has also been suggested that air pollution may not only contribute to exacerbation of existing asthma, but also may in part account for the previously-mentioned increased incidence of asthma that has occurred over the last two decades, particularly because the increased incidence corresponds to a time when indoor air pollution may have increased as a result of efforts to make buildings more airtight and, hence, more economical to heat. At least one study argues against a role for ambient air pollutants in the increased incidence of asthma. When the respiratory health status of West and East German children were compared shortly after reunification, atopy, airway reactivity, and asthma episodes were found to be more common in the West Germans even though pre-unification air quality in East Germany was thought to be substantially worse (von Mutius et al., 1994). Nevertheless, several animal studies have shown that exposure to air pollutants (diesel and residual fly ash particles, NO_2) can enhance sensitization to allergens (Takafuji et al., 1987; Gilmour et al., 1996; Lambert et al., In Press).

Our lab has used enhanced IgE (both DM specific and total) levels in serum and bronchoalveolar lavage fluid (BALF), DM-specific proliferation of cells from the mediastinal lymph node, immune-mediated inflammation demonstrated by increased inflammatory cells in BALF, as well as immediate bronchoconstriction responses to DM, as indicators of enhanced allergenic responses (Gilmour et al., 1996; Lambert et al., In Press). Analyses of cytokine responses in BALF that might provide clues as to mechanisms underlying the enhanced sensitization observed in these studies have also been undertaken (Lambert et al., In Press). One hypothesis is that air pollutants stimulate respiratory tract cells to produce cytokines that create a local environment that favors the T_H2 response associated with allergy. For example, in vitro studies have shown that following O_3 exposure, respiratory epithelium produces nitric oxide (Punjabi et al., 1994), prostaglandin E_2 (Leikauf et al., 1988; McKinnon et al., 1993), and IL-6 (Koren et al., 1991) all of which favor T_H2 responses (Betz and Fox, 1991; Hilkens et al., 1996; Barnes and Liew, 1995). Alveolar macrophages also produce these mediators in response to O_3. Similarly, NO_2 (Devalia et al., 1993), diesel particles (Nel et al., 1998), and residual oil fly ash (Carter et al., 1997) have been shown to induce cytokines which favor T_H2 responses. Introduction of an allergen into such an environment might well result in more active sensitization than would occur in the absence of these air pollutants.

Extrapolation from Animal Data to Human Health Effects

The ability to demonstrate cytokine responses associated with allergic disease following exposure of respiratory cells in vitro may provide the means to bridge the gap between animal research and human health effects by again using a double parallelogram model (Figure 4). In the case of O_3 and diesel particles, cytokines were measured in BALF or nasal lavage following in vivo exposure and in respiratory epithelium cultures exposed in vitro. Hence, it would be possible to approach this problem in a manner similar to that described in Figure 1 for immune suppression.

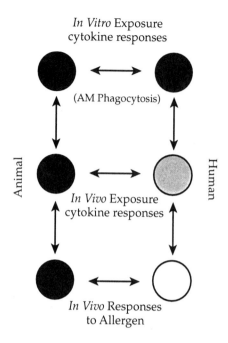

Figure 4. Potential model of relationship between stimulation of cytokines favoring T_H2 response in respiratory epithelium following *in vitro* and *in vivo* exposure to a toxicant and allergic sensitization associated with asthmatic responses. Solid circles represent data that can readily be obtained; cross-hatched circle represents data that can sometimes be obtained; and, open circles represent data that can rarely be obtained.

Intraspecies Variability and Acute vs Chronic Effects

Most of the animal studies to date have been done in genetically-susceptible animals and, hence, do not provide information with respect to the possibility that pollutants might favor a response in an animal that would ordinarily not be susceptible. Human studies have been mixed in that, while some have focused on atopic individuals, others have not. Atopic individuals appear to be more at risk. In addition to genetics, age is clearly a risk factor since allergic asthma usually develops during childhood. Infectious disease and immunizations may also affect the sensitization process. Again, as for the preceding example, the information which is available is insufficient to determine an appropriate uncertainty factor for intraspecies variability. Also, as in the last example, an acute exposure that alters the sensitization process to a common allergen can have long-lasting effects in terms of response to that allergen forever after. Although there is one study in which exposure to diesel particles took place over 15 wk (Ichinose et al., 1998), most of the experimental data currently available have come from acute exposure studies. It is likely that short-term exposure to an agent that primes the lung for a T_H2 response is sufficient to alter sensitization, provided there is sufficient exposure to allergen. However, as in the phosgene example cited earlier, the longer the pollutant exposure period the greater the opportunity for allergen exposure to occur also. Again, more information is needed in order to understand the consequences of chronic exposure.

Implications for Risk Assessment

Although we have a lot to learn with respect to the impact of air pollutants on allergic sensitization, some of the approaches applied in the previous two examples may be useful in approaching this problem. Currently, animal models suggest that air pollutants have the capacity to influence allergic sensitization. There are both animal and epidemiologic data that indicate that air pollutants do enhance the expression of allergic responses in sensitized individuals. Once the mechanisms responsible for these effects become clear, models to provide more quantitative information may be developed.

SUMMARY

In conclusion, it is clear that results from experimental studies can be used to reduce some of the uncertainties typically encountered in the risk assessment process. With respect to pulmonary immunotoxicity, most progress has been made in understanding the relationships between suppression of alveolar macrophage function and susceptibility to infectious disease in mice and men. However, the parallelogram approach can also be used to improve our ability to predict the risk of allergic lung disease. More research is needed in this area. In addition, although experimental data has provided some insight with respect to intraspecies variability and acute vs chronic exposure, there are still many unanswered questions which remain problematic when it comes to risk assessment. As we continue to make progress in understanding how toxicants affect pulmonary immune responses, it is important to keep the risk assessment process in mind so that decisions can be made based on the best possible scientific information.

REFERENCES

Agius RM, Nee J, McGovern B, Robertson A. Structure activity hypotheses in occupational asthma caused by low molecular weight substances. Ann. Occup. Hyg., 1991;35:129-137.

ACGIH: American Conference of Governmental Industrial Hygienists. *Threshold Limit Values and Biological Exposure Indices*. Cincinnati, OH, 1990, p. 38.

ATS: American Thoracic Society. Committee of the Environmental and Occupational Health Assembly. Health effects of outdoor air pollution. Am. J. Respir. Crit. Care Med., 1996;153:3-50.

Barnes PJ, Liew FY. Nitric oxide and asthmatic inflammation. Immunol. Today, 1995;16:128-130.

Bernstein DI. Occupational asthma. Clin. Allergy, 1992;76:917-934.

Bernstein JA, Bernstein IL. "Clinical Aspects of Respiratory Hypersensitivity to Chemicals." In *Immunotoxicology and Immunopharmacology*, JH Dean, MI Luster, AE Munson, I Kimber, eds. New York: Raven Press, 1994, pp. 617-642.

Betz M, Fox BS. Prostaglandin E_2 inhibits production of T_H1 lymphokines but not of T_H2 lymphokines. J. Immunol., 1991;146:108-113.

Björkstén, B. The role of the gastrointestinal tract in the development of respiratory hypersensitivities. Toxicol. Lett., 1996;86:85-88.

Botham PA, Rattray NJ, Woodcock DR, Walsh ST, Hext PM. The induction of respiratory allergy in guinea-pigs following intradermal injection of trimellitic anhydride: A comparison with the response to 2,4-dimitrochlorobenzene. Toxicol. Lett., 1989:47:25-29.

Burrows B, Halonen M, Barbee, RA, Lebowitz MD. The relationship of serum immunoglobulin E to cigarette smoking. Am. Rev. Respir. Dis., 1981;124:523-525.

Carter JD, Ghio AJ, Samet J, Devlin RB. Cytokine production by human airway epithelial cells after exposure to an air pollution particle is metal-dependent. Toxicol. Appl. Pharmacol., 1997;146:180-188.

Chan-Yeung M. A clinician's approach to determine the diagnosis, prognosis, and therapy of occupational asthma. Med. Clin. N. Amer., 1990a;74:811-822.

Chan-Yeung M. Occupational asthma. Chest, 1990b;98:148S-161S.

Coffin D, Gardner DE. Interaction of biological agents and chemical air pollutants. Ann. Occup. Hyg., 1972;15:219-234.

Dearman, RJ, Basketter DA, Blaikie L, Clark ED, Hilton J, House RV, Ladics GS, Loveless SE, Mattis C, Sailstad DM, Sarlo K, Selgrade MJK, Kimber I. The mouse IgE test: Inter-laboratory evaluation and strain comparisons. Toxicol. Meth., 1998;8:1-17.

Devlin RB, McDonnell WF, Mann R, Becker S, House DE., Koren HS. Exposure of humans to ambient levels of ozone for 6.6 hours causes cellular and biochemical changes in the lung. Am. J. Respir. Cell Mol. Biol., 1991;4:72-81.

Devalia JL, Campbell AM, Sapsford RJ, Rusznak C, Quint D, Godard P, Bousquet J, Davies R.J. Effect of nitrogen dioxide on synthesis of inflammatory cytokines expressed by human bronchial epithelial cells *in vitro*. Am. J. Respir. Cell Mol. Biol., 1993;9:271-278.

Flindt ML. Pulmonary disease due to inhalation of derivatives of *Bacillus subtilis* containing proteolytic enzyme. Lancet, 1969;1:1177-1181.

Gilmour MI, Park P, Doerfler D, Selgrade MJK. Factors that influence the suppression of pulmonary anti-bacterial defenses in mice exposed to ozone. Exp. Lung Res., 1993a;19:299-314.

Gilmour MI, Park P, Selgrade MJK. O_3-Enhanced pulmonary infection with *Streptococcus zooepidemicus* in mice: The role of alveolar macrophage function and capsular virulence factors. Am. Rev. Respir. Dis., 1993b;147:753-760.

Gilmour MI, Selgrade MJK. A comparison of the pulmonary defenses against *Streptococcal* infection in rats and mice following O_3 exposure: Differences in disease susceptibility and neutrophil recruitment. Toxicol. Appl. Pharmacol., 1993;123:211-218.

Gilmour MI, Park P, Selgrade MJK. Increased immune and inflammatory responses to dust mite antigen in rats exposed to 5 ppm nitrogen dioxide. Fundam. Appl. Toxicol., 1996;31:65-70.

Guth DJ. Comparison of quantitative exposure-response models: Application to risk assessment of acute exposure to toluene or tetrachloroethylene. The Toxicologist 1994;14:400.

Henderson FW, Henry MM, Ivins SS, Morris R, Neebe EC, Leu SY, Stearts PW. Correlates of recurrent wheezing in school-age children. The Physicians of Raleigh Pediatric Associates. Am. J. Respir. Crit. Care Med., 1995;151:1786-1793.

Herz U, Lumpp U, Da Palma JC, Enssle K, Takatsu K, Schnoy N, Daser A, Kottgen E, Wahn U, Renz H. The relevance of murine animal models to study the development of allergic bronchial asthma. Immunol. Cell Biol., 1996;74:209-217.

Hilkens CM, Snijders A, Vermeulen H, van der Meide PH, Wierenga EA, Kapsenberg ML. (1996) Accessory cell-derived IL-12 and prostaglandin E_2 determine the IFN-γ level of activated human CD4$^+$ T-cells. J. Immunol., 1996;156:1722-1727.

Ichinose T, Takano H, Miyabara Y, Sagai M. Long-term exposure to diesel exhaust enhances antigen-induced eosinophilic inflammation and epithelial damage in the murine airway. Toxicol. Sci., 1998;44:70-79.

Karol MH. (Concentration-dependent immunologic response to toluene diisocyanate (TDI) following inhalation exposure. Toxicol. Appl. Pharmacol., 1983;68:229-241.

Karol MH. "Immunologic Responses of the Lung to Inhaled Toxicants." In *Concepts in Inhalation Toxicology*, RO McClellan, RF Henderson, eds. New York: Hemisphere Publishing, 1989, pp. 403-413.

Karol MH., Stadler J, Magreni C. Immunotoxicologic evaluation of the respiratory system: Animal models for immediate- and delayed-onset pulmonary hypersensitivity. Fundam. Appl. Toxicol., 1985;5:459-472.

Karol MH. "Predictive Testing for Respiratory Allergy." In *Allergic Hypersensitivities Induced by Chemicals*, JG Vos, M Younes, E Smith, eds. New York: CRC Press, 1995, pp. 125-137.

Karol MH., Graham C, Gealy R, Macina OT, Sussman N, Rosenkranz HS. Structure-activity relationships and computer-assisted analysis of respiratory sensitization potential. Toxicol. Lett., 1996;86:187-191.

Kawabata TT, Babcock LS, Horn PA. Specific IgE and IgG$_1$ responses to *subtilisin Carlsberg* (Alcalase) in mice. Development of an intratracheal exposure model. Fundam. Appl. Toxicol., 1996;29:238-242.

Kimber I, Bernstein IL, Karol MH, Robinson MK, Sarlo K, Selgrade MJK. Workshop overview: Identification of respiratory allergens. Fundam. Appl. Toxicol., 1996;33:1-10.

Kimber I. Kerkvliet NI, Taylor SL, Astwood JD, Sarlo K, Dearman RJ. Toxicology of protein allergenicity: Prediction and characterization. Toxicol. Sci., In Press.

Klopman G. MULTICASE. I. A hierarchical computer automated structure evaluation program. Quant. Struct-Act. Relat., 1992;11:176-184.

Koren HS, Devlin RB, Becker S. "Ozone-Induced Inflammatory Response in Pulmonary Cells." In *Xenobiotics and Inflammation: Roles of Cytokines and Growth Factors*, LB Schook, DL Laskin, eds. New York: Academic Press, 1991, pp. 249-281.

Lambert AL, Winsett DW, Costa DL, Selgrade MJK, Gilmour MI. Differential transfer of allergic airway responses with serum and lymphocytes from house dust mite sensitized rats. Am. Rev. Respir. Crit. Care Med., 1998;157:1991-1999.

Lambert AL, Dong W, Winsett DW, Selgrade MJK, Gilmour MI. Residual oil fly ash (ROFA) exposure enhances allergic sensitization to house dust mite (HDM). Toxicol. Appl. Pharmacol., In Press.

Leikauf GD, Driscoll KE, Wey HE. Ozone-induced augmentation of eicosanoid metabolism in epithelial cells from bovine trachea. Am. Rev. Respir. Dis., 1988;137:435-442.

Mizushima Y, Mori T, Ogita T, Nakumara T. Adjuvant for IgE antibody and islet activating protein in *Bordetella pertussis*. Int. Arch. Allergy Appl. Immunol., 1979;58:426-429.

Miller FJ, Illing JW, Gardner DE. Effects of urban ozone levels on laboratory induced respiratory infections. Toxicol. Lett., 1978;2:163-169.

McKinnon K, Madden MC, Noah TL, Devlin RR. *In vitro* ozone exposure increases release of arachidonic acid products from a human bronchial epithelial cell line. Toxicol. Appl. Pharmacol., 1993;18:215-223.

National Research Council. *Risk Assessment in the Federal Government: Managing the Process*. Washington, DC: National Academy Press, 1983, pp. 17-83.

Nel LM, Dockery DD, Ware JH, Spengler JD, Speizer FE, Ferris BG. Association of indoor nitrogen dioxide with respiratory symptoms and pulmonary function in children. Am. J. Epidemiol., 1991;134:204-218.

Nels AE, Diaz-Sanchez E, Ng D, Hiura T, Saxon A. Enhancement of allergic inflammation by the interaction between diesel exhaust particles and the immune system. J. Allergy Clin. Immunol., 1998;102:539-554.

Nemery B, Lenaerts L. Exposure to methylene diphenyl diisocyanate in coal mines. Lancet, 1993;341:318.

Pepys J, Longbottom JE, Margreave FE, Faux J. Allergic reactions of the lungs to enzymes of Bacillus subtilis. Lancet, 1969;1:1181-1184.

Pepys, J., Wells, I.D., D'Souza MF, Greenberg M. Clinical and immunological responses to enzymes of *Bacillus subtilis* in factory workers and consumers. Clin. Allergy, 1973;3:143-160.

Punjabi CJ, Laskin JD, Pendino KJ, Goller NL, Durham SK, Laskin DL. Production of nitric oxide by rat type II pneumocytes: Increased expression of inducible nitric oxide synthase following inhalation of pulmonary irritant. Am. J. Respir. Cell Mol. Biol., 1994;11:165-172.

Ritz HL, Evans BLB, Bruce RD, Fletcher ER, Fisher GL, Sarlo K. Respiratory and immunological responses of guinea pigs to enzyme containing detergents: A comparison of intratracheal and inhalation modes of exposure. Fundam. Appl. Toxicol., 1993;21:31-37.

Robinson MK, Babcock LS, Horn PA, Kawabata TT. (1996) Specific antibody responses to *subtilisin Carlsberg* (Alcalase) in mice: Development of an intranasal exposure model. Fundam. Appl. Toxicol., 1996;34:15-24.

Romagnani S. T$_H$1 and T$_H$2 in human diseases. Clin. Immunol. Immunopathol., 1996;80:225-235.

Rosenstreich DL, Eggleston P, Kattan M, Baker D, Slavin RG, Gergen P, Mitchell H, McNiff-Mortimer K, Lynn H, Ownby D, Malveaux F. The role of cockroach allergy and exposure to cockroach allergen in causing morbidity among inner-city children with asthma. N. Engl. J. Med., 1997;336:1356-1363.

Rosenwasser LJ. Genetics of asthma and atopy. Toxicol. Lett., 1996;86:73-77.

Samet JW, Lambert WE, Skipper BJ, Cushing AH, Hunt WC, Young SA, McLaren LC, Schwab M, Spengler JD. Nitrogen Dioxide and respiratory illness in infants. Am. Rev. Respir. Dis., 1993;148:1258-1265.

Sarlo K. "Human Health Risk Assessment: Focus on Enzymes." In *Proceedings of the 3rd World Conference on Detergents*, A Cahn, ed. Chicago: American Oil Chemists Society Press, 1994, pp. 54-57.

Sarlo K, Clark ED. A tier approach for evaluating the respiratory allergenicity of low molecular weight chemicals. Fundam. Appl. Toxicol., 1992;18:107-114.

Sarlo K, Ritz HL, Fletcher ER, Schrotel KR, Clark ED. Proteolytic detergent enzymes enhance the allergic antibody responses of guinea pigs to nonproteolytic detergent enzymes in a mixture: Implications for occupational exposure. J. Allergy Clin. Immunol., 1997a;100:480-487.

Sarlo K, Fletcher R, Gaines WG, Ritz HL. Respiratory allergenicity of detergent enzymes in the guinea pig intratracheal test: Association with sensitization of occupationally exposed individuals. Fundam. Appl. Toxicol., 1997b;39:44-52.

Schwartz J, Spix C, Wichmann HE, Malin E. Air pollution and acute respiratory illness in five German communities. Environ. Res., 1991;56:1-14.

Sedgwick JD, Holt PG. Suppression of IgE responses in bred rats by repeated respiratory tract exposure to antigen: Responder phenotype influences isotope specificity of induced tolerance. Eur. J. Immunol., 1984;14:893-897.

Selgrade MJK, Starnes DM, Illing JW, Daniels MJ, Graham JA. Effects of phosgene exposure on bacterial viral and neoplastic lung disease susceptibility in mice. Inhal. Toxicol., 1989;1:243-259.

Selgrade MJK, Gilmour MI. "Effects of Gaseous Air Pollutants on Immune Responses and Susceptibility to Infectious and Allergic Disease." In *Immunotoxicology and Immunopharmacology*, JH Dean, MI Luster, AE Munson, I Kimber, eds. New York: Raven Press, 1994, pp. 395-411.

Selgrade MJK, Zeiss CR, Karol MH, Sarlo M, Kimber K, Tepper JS, Henry MC. Workshop on status of test methods for assessing potential of chemicals to induce respiratory allergic reactions. Inhal. Toxicol., 1994;6:303-319.

Selgrade, M.J.K., Cooper, K.D., Devlin, R.B., van Loveren, H., Biagini, R. E., and Luster, MI. Immunotoxicity - Bridging the gap between animal research and human health effects. Fundam. Appl. Toxicol., 1995a;24:13-21.

Selgrade MJK, Gilmour MI, Yang YG, Hatch GE, Burleson GR. Pulmonary host defenses and resistance to infection following subchronic exposure to phosgene. Inhal. Toxicol., 1995b:7:1257-1268.

Selgrade MJK. Use of immunotoxity data in health risk assessments: Uncertainties and research to improve the process. Toxicology, In Press.

Slade R, Watkinson WP, Hatch GE. Mouse strain differences in ozone dosimetry and body temperature changes. Am. J. Physiol., 1997;272:L73-L77.

Speizer FE, Ferris B, Bishop YM, Spengler J. Respiratory disease rates and pulmonary function in children associated with NO_2 exposure. Am. Rev. Respir. Dis., 1980;121:3-10.

Takafuji S, Suzuki S, Koizumi K, Tafokoro K, Miyamoto T, Ikemori R, Muranaka M. Diesel-exhaust particulates inoculated by the intranasal route have an adjuvant activity for IgE production in mice. J. Allergy Clin. Immunol., 1987;79:639-645.

USEPA: Environmental Protection Agency. *Proposed Test Rule for Hazardous Air Pollutants*. Federal Register 61, 33177-33200, 1996.

Venables KM. Low molecular weight chemicals, hypersensitivity, and direct toxicity: The acid anhydrides. Br. J. Ind. Med., 1989;46:222-232.

Vogel G. Why the rise in asthma cases. Science, 1997;276:1645.

von Mutius E, Martinez FD, Fritzsch C, Micolai T, Roell G, Thiemann HH. Prevalence of asthma and atopy in two areas of West and East Germany. Am. J. Respir. Crit. Care Med., 1994;149:358-364.

Ward MDW, Sailstad DM, Selgrade MJK. The biopesticide, *Metarhizium anisopliae*, elicits allergic response in BALB/c mice. Toxicol. Sci., 1998;45:195-203.

Ware JH, Dockery DW, Spiro A, Speizer FE, Ferris BG. Passive smoking, gas cooking, and respiratory health of children living in six cities. Am. Rev. Respir. Dis., 1984;129:366-374.

Wass U, Belin L. An *in vitro* method for predicting sensitizing properties of inhaled chemicals. Scand J. Work Environ. Health, 1990;16:208-214.

Weitzman M, Gortmaker SL, Sobol AM, Perrin JM. Recent trends in the prevalence and severity of childhood asthma. J. Am. Med. Assoc., 1992;268:2673-2677.

Yang YG, Gilmour MI, Lange R, Burleson GR, Selgrade MJK. Effects of acute exposure to phosgene on pulmonary host defenses and resistance to infection. Inhal. Toxicol., 1995;7:393-404.

Zeiss CR, Patterson R, Pruzansky JJ, Miller MM, Rosenberg M, Levitz D. Trimellitic anhydride-induced airway syndromes: Clinical and immunologic studies. J. Allergy Clin. Immunol., 1977;60:96-103.

Zeiss CR, Mitchell JH, van Peenan PRD, Kavich D, Collins MJ, Grammar L, Shaughnessy M, Levitz D, Henderson J, Patterson R. A clinical and immunologic study of employees in a facility manufacturing trimellitic anhydride. Allergy Proc., 1992;13:193-198.

Zelikoff JT, Frampton MW, Cohen MD, Morrow PE, Sisco M, Tsai Y, Utell MJ, Schlesinger RB. Effects of inhaled sulfuric acid aerosols on pulmonary immunocompetence: A comparative study in humans and animals. Inhal. Toxicol., 1997;9:731-752.

Zetterstrom O, Osterman K, Machado L, Johansson SG. Another smoking hazard: Raised serum IgE concentration and increased risk of occupational allergy. Br. Med. J., 1981;283:1215-1217.

Disclaimer: This Chapter has been reviewed by the National Health and Environmental Effects Research Laboratory, U.S. Environmental Protection Agency and approved for publication. Approval does not signify that the contents necessarily reflect the views and policies of the agency, nor does mention of trade names or commercial products constitute endorsement or recommendation for use.

18. BIOMARKERS OF PULMONARY IMMUNOLOGIC RESPONSES

Andrea K. Hubbard, Ph.D
Department of Pharmaceutical Sciences
School of Pharmacy,
University of Connecticut
Storrs, CT 06269-2092

INTRODUCTION

Biomarkers of pulmonary immune/inflammatory responses have been measured in the lungs of humans and animals injured by exposure to occupational toxicants (e.g. silica, asbestos, beryllium, isocyanates) as well as to environmental toxicants (e.g. ozone, nitrogen dioxide). Detection of different populations of inflammatory cells, cell proliferation, protein mediators, reactive oxygen/nitrogen species, and activated transcription factors may indicate acute or chronic inflammation as well as antigen-dependent or -independent immune responses. This Chapter will briefly summarize pulmonary defense in the upper and lower airways, changes in pulmonary immune/inflammatory markers measured in different xenobiotic-induced pulmonary diseases, and the potential significance of these changes as biomarkers of immunologic and inflammatory responses in the lung.

PULMONARY IMMUNE RESPONSE

The lungs are repeatedly exposed even in ambient air to levels of xenobiotics which may elicit an immune response or, alternatively, modulate pulmonary immunity. In response, the lungs are protected by defense mechanisms in the upper airways and bronchi and by mechanisms in the alveolar or lower regions of the lung.

Defense in the Upper Airways

The innate immune response in the upper airways includes anatomic barriers, secreted proteins/mediators and activated phagocytic cells. Anatomic barriers include impaction on the mucosal surface and cough. Cough is triggered by a wide variety of stimuli (Widdicombe, 1995) producing enough turbulence and shearing forces in the major bronchi and trachea to extrude foreign material. The mucociliary apparatus, a layer of mucus overlying ciliated epithelium, is the site of deposition (and thus removal) of about 90% of inhaled particles with a diameter >2 - 3 μm. Mucus, composed of mucoglycoproteins and proteoglycans, also contains other important elements which contribute to lung defense. Bactericidal properties of mucus are also mediated by lysozyme, lactoferrin, peroxidase, and secretory IgA (dimeric). Secretory IgA is especially important in the neutralization of toxins and viruses and in blocking entry of bacteria across the epithelium (Underdown and Schiff, 1986). Epithelial cells (e.g., ciliated, Clara cells) in the upper airways also actively participate in host defense against inflammation. They produce chemokines, substances which cause the directed migration of inflammatory cells. These chemotactic factors include products of arachidonic acid metabolism (i.e., 15-HETE, LTB_4) (Holtzman, 1992), and proteins such as interleukin (IL)-8 (humans) (Bedard et al., 1993), MIP-1, MIP-2, MCP-1, RANTES, and eotaxin. Epithelial cells and macrophages also release growth factors collectively known as colony stimulating factors (CSF) (Sallusto and Lanzavecchia, 1993), and evidence increased expression of adhesion molecules such as intercellular adhesion molecule-1 (ICAM-1) following stimulation with cytokines. Epithelial cells are also capable of increased expression of major histocompatability complex (MHC) Class II molecules in response to cytokines like interferon-γ (IFNγ). Another cell type with high MHC Class II levels on the cell surface are pulmonary dendritic cells which are capable of presenting antigen to lymphocytes during an acquired immune response. Pulmonary dendritic cells are intimately associated with epithelial basal cells forming a network to capture inhaled antigens (Sertl et al., 1986). Having internalized inhaled antigen, and in the presence of such cytokines as tumor necrosis factor-α (TNFα) and granulocyte-monocyte CSF (GM-CSF), they migrate via the lymphatics to regional lymph nodes to initiate an immune response (Macatonia et al., 1987).

Acquired immunity in the upper airways is centered in bronchus-associated lymphoid tissue (BALT). These tissues, at airway branching points, are important afferent pathways of local immunity as both antibody and cell-mediated immune responses against inhaled antigens are initiated at these sites. T-Helper (T_H) CD4+ and T-cytotoxic/suppressor CD8+ lymphocytes are also present in the lungs. These populations primarily possess αβ T-lymphocyte receptors (TCR), but can also be γδ T-lymphocytes as well. Both αβ and γδ T-lymphocytes can differentiate into T_H1 and T_H2 subpopulations depending upon the pathogen encountered and the type of cytokine which predominates (IFNγ and IL-12 vs IL-4 and IL-10) (Ferrick et al., 1995). In addition, both natural killer (NK) cells and γδ T-lymphocytes participate in a type of innate immunity in that they can react to pathogens in the absence of preliminary priming.

Defense in the Lower Airways

Alveolar and interstitial macrophages in the lower airways of the lungs can participate in innate immunity as well as in acquired. Alveolar macrophages (AM) reside on the alveolar epithelial surface and provide phagocytic defense through the release of cytokines, proteolytic enzymes, biologically active lipids, oxygen and/or nitrogen metabolites, antioxidants, glycoproteins and coagulation factors (Sibille and Reynolds, 1990). These cells release large amounts of reactive oxygen species (ROS) and nitric oxide (NO), TNFα, and IFNγ upon stimulation (Franke-Ullmann et al., 1996), but probably do not participate in antigen presentation as they express only small amounts of MHC Class II protein. Alveolar macrophages also release chemokines, such as IL-8, MIP-1, RANTES, and MCP-1, and products of arachidonic acid metabolism. In addition, they also release growth factors (i.e., GM-CSF, G-CSF, M-CSF) and express adhesion molecules including ICAM-1 (Nario and Hubbard, 1997). Interstitial macrophages are found in the interstitial spaces of the lung parenchyma (Crapo et al., 1983), in the peribronchial and perivascular spaces (Sorokin and Brain, 1975), in the lymphatic channels (Lehnert, 1992), and in the visceral pleural region (Mariassy and Wheeldon, 1983). These type of macrophages maintain a lower state of activation than AM (Rouzer et al., 1982) but participate more actively as antigen-presenting cells since they appear to release significant amounts of IL-1 and IL-6 and express more MHC Class II protein (Franke-Ullmann et al., 1996). These pulmonary macrophages are also involved in tissue remodeling with the release of metalloelastases, collagenase, and metalloproteases, as well as fibroblast growth factors such as transforming growth factor-β (TGFβ) or platelet-derived growth factor (PDGF). TGFβ is also known to attract macrophages but to suppress lymphocyte and epithelial cell proliferation.

Neutrophils (PMN), a type of polymorphonuclear leukocyte, normally represent <2% of the cells in the lung. However, upon infection or introduction of toxic substances, PMN are rapidly recruited from the circulation into interstitial and alveolar spaces. Neutrophil emigration begins with a deceleration and rolling of the cells along the endothelial cell (EC) surface followed by upregulation of β2 integrin adhesion molecules such as LFA-1 and MAC-1; the latter recognizes ICAM-1 as one of its ligands. ICAM-1 is constitutively-expressed on lung epithelial and endothelial cells and is upregulated by cytokines such as IL-1, IFNγ, and TNFα (Dustin et al., 1986; Pober et al., 1986). This adhesion to EC ICAM-1 precedes the transendothelial migration of these inflammatory cells from the capillary bed into lung tissue (Springer, 1990). Retention of PMN at the epithelial surface appears due to the increased presence of epithelial cell ICAM-1 induced by inflammatory cytokines (Wegner et al., 1992; Mulligan et al., 1995). It should be noted that there is still some controversy about the requirement for all described adhesion molecules in all types of lung injury. For example, depending upon the type of inflammatory stimuli, PMN emigration out of the pulmonary circulation may or may not depend upon expression of MAC-1 (Doerschuk et al., 1990) or of ICAM-1 (Bullard et al., 1995). Other adhesion molecules have been implicated in ICAM-1-*independent* PMN adherence (Doerschuk et al., 1990; Mulligan et al., 1992; Muller, 1995) and space constraints within the capillary vessels of the pulmonary circulation may not allow the leukocytes to roll, thereby decreasing the dependence upon adhesion molecules (Doerschuk et al., 1993). Neutrophils migrate following chemotactic gradients (Albelda et al., 1994) of chemokines (such as IL-8, MIP-2, GROα, GROβ, and LTB4) and eliminate foreign substances by phagocytosis and release of ROS and proteolytic enzymes. Eosinophils, another type of polymorphonuclear leukocyte, are nondividing granulated cells which release four proteins: major basic protein, eosinophil cationic protein (ECP), eosinophil peroxidase, and eosinophil-derived murotoxin in response to allergens or parasitic infections. Eosinophils migrate into the lungs in response to chemokines such as IL-5 produced by T_H2 lymphocytes.

The alveoli contain about 10% lymphocytes, of which 50% are CD4$^+$ T-helper lymphocytes, 30% CD8$^+$ T-cytotoxic/suppressor cells, 5% B-lymphocytes, and 10 - 15% NK cells. T-Lymphocytes can be divided into two functional subpopulations: T_H1 and T_H2 cells. T_H1 lymphocytes are generally defined as cells which produce IFNγ and IL-2 and T_H2 as cells producing IL-4, -5, and -10; these subdivisions are not mutually exclusive and there may be some overlap in cytokine release. In addition, T_H1 lymphocytes are generally thought to mediate delayed-type hypersensitivity reactions, activate CD8$^+$ cytolytic T-lymphocytes, activate macrophages for microbicidal function, and induce IgG$_{2a}$ production. T_H2 lymphocytes release cytokines (e.g., IL-4) which participate in B-lymphocyte heavy chain class switch, resulting in production of IgG$_1$, IgA, and IgE antibodies. Thus, T_H2 lymphocytes are thought to participate in antibody-mediated hypersensitivity responses, including allergy. IL-10 from T_H2 lymphocytes can suppress the function of T_H1 lymphocytes whereas IL-12 from macrophages can suppress T_H2 lymphocyte function. With respect to immunologic defense of the lungs, T_H1 lymphocytes partake in immunity against most intracellular pathogens such as *Mycobacterium tuberculosis* (Cooper et al., 1993) while T_H2 lymphocytes are involved in responses to parasitic infections and in pulmonary allergic responses.

Immunoglobulins in the bronchoalveolar spaces of the lungs can function as opsonins by enhancing phagocytosis, in complement activation and bacterial killing, and in antagonizing viral infections. IgG is the predominant immunoglobulin in normal lungs with the strongest opsonic activity in the alveoli. However, secretory IgA can be increased in response to infection whereas IgE can be increased in response to allergens and parasitic infections.

EFFECT OF SELECTED XENOBIOTICS ON PULMONARY IMMUNITY

The studies cited in this section describe work evaluating changes in local pulmonary immunity as opposed to systemic immunity (i.e., the circulation or peripheral lymphoid organs). Studies in animal models have usually evaluated markers in bronchoalveolar lavage (BAL) fluid and cells, in lung homogenates, and in tissue sections whereas studies in humans have often evaluated markers in airway lining fluid obtained by nasal airway lavage, bronchial wash and BAL, and in bronchial biopsy specimens.

Asbestos

Asbestosis is associated with exposure to asbestos fibers and is characterized by a bilateral diffuse interstitial fibrosis of the lungs (American Thoracic Society, 1986). Occupational exposure to asbestos is also considered an important cause of malignant mesothelioma and bronchogenic carcinoma. Many reports have documented changes in cell-mediated (Tsang et al., 1988) and humoral immunity (Lange et al., 1974) in cells isolated from *peripheral* blood; few studies have examined changes in *pulmonary* immunity. This is important to note as changes in immune cells isolated from the periphery provide only an indirect indication of the immune status of the host lung. Natural killer cells are thought to be the first line defense against cancer cells or virally-infected cells (Santoli and Koprowski, 1979) since, unlike cytotoxic T-lymphocytes, NK cells can kill without recognition of foreign Class I proteins and they exist in great quantities within the interstitial lung lymphocyte population (Holt et al., 1985). Recently, an investigation examined the effect of inhaled asbestos upon interstitial NK numbers and function in C57Bl/6 mice exposed to chrysolite asbestos for 3 d and then sacrificed 7, 28 and 56 d later (Rosenthal et al., 1998). The ability of interstitial NK cells from exposed mice to lyse YAC-1 target cells *in vitro* was significantly suppressed 7 and 56 d post-exposure. There was also a significant decrease in the percentage of lung NK cells recovered from these mice. As there was no apparent effect upon the number of pulmonary T-lymphocytes isolated nor on the number of splenic NK cells isolated, this suggested a cell/organ specificity in impaired function. However, in asbestos-exposed patients, there were increases in BAL lymphocytes which were associated with less physiologic impairment than in patients with increased levels of PMN and eosinophils (Haslam et al., 1978). Reduced CD4:CD8 ratios in BAL fluid have also been described in asbestos-exposed workers and correlated with pleural thickening (Sprince et al., 1991).

Asbestos exposure has also been documented to cause changes in alveolar Type I and Type II cell populations. Moreover, in rats exposed to high airborne concentrations of asbestos, an increase in proliferation in bronchiolar epithelial cells and in cells in the alveolar duct regional and interstitial compartment was observed (BeruBe et al., 1996). In addition, asbestos exposure also caused an increase in the number of activated AM in humans (Rom et al., 1987) as well as in animals (Rom et al., 1991). These activated pulmonary macrophages are an abundant source of reactive oxygen/nitrogen species, many inflammatory mediators, as well as growth factors. Alveolar macrophages recovered by BAL from asbestos-exposed rodents have been shown to spontaneously release significant amounts of ROS during the progression of the disease (Rom et al., 1987). Production of ROS has also been shown in the lungs of animals exposed to asbestos by intratracheal (IT) instillation (Schapira et al., 1994). Inflammatory cells recovered from animals exposed to asbestos *in vivo* promote basal ROS release or prime cells for greater ROS generation after exposure to a second stimulus (Donaldson and Cullen, 1984). Increased oxidative stress induced by asbestos was also reflected by the presence of the potent antioxidant hemooxygenase in rats lungs 1 wk after a single IT exposure to chrysolite asbestos (Schapira et al., 1994) and by enhanced lipid peroxidation in the lungs of rats exposed to asbestos (Gulumian and Kilroe-Smith, 1987).

Increased production of reactive nitrogen species induced by asbestos has been documented by increases in levels of (a) nitrite (the stable oxidation product for NO) and nitrate (surrogate marker for peroxynitrite [ONOO·] formation) in AM recovered from rats exposed by inhalation (Quinlan et al., 1998), (b) steady state mRNA for inducible nitric oxide synthetase (iNOS) in lung homogenates from hosts receiving an IT injection (Quinlan et al., 1998), and (c) nitrotyrosine residues (marker of ONOO· formation) in AM and bronchiolar epithelial cells in the lungs of rats exposed to asbestos (Tanaka et al., 1998).

Detection of increased IL-1 and TNFα expression in asbestosis patients and in animal models is important as these cytokines serve many roles in T- and B-lymphocyte proliferation and activation, increases in arachidonic acid metabolism, increased inflammatory cell oxidative burst and degranulation, increased adhesion molecule expression (Driscoll, 1993), and upregulation of collagen and fibronectin gene expression in normal human fibroblasts exposed *in vitro* (Zhang et al., 1993). Increased mRNA expression and release of IL-1 and TNFα have been found in AM from patients with asbestosis (Zhang et al., 1993) or those with a history of chronic asbestos exposure (Perkins et al., 1993). Both of these cytokines also appear to be increased in animal models of asbestosis prior to development of inflammation and fibrosis (Driscoll et al., 1995a). Indeed, excess TNFα correlates with development of fibrosis in animal models of asbestosis (Lemaire and Ouellet, 1996). Similarly, IT asbestos instillation elicited increased IL-1 and TNFα from subsequently-isolated rat AM for at least 7 d (Driscoll et al., 1995b). The amounts of another interleukin, IL-6 (a cytokine which synergizes with TNFα and IL-1 and is important in lung fibroblast growth), has also been shown to be increased in BAL from patients with lung fibrosis and histories of long-term asbestos exposure (Simeonova et al., 1997).

Chemokines are inflammatory mediators which cause the directed migration of inflammatory cells. Several of these proteins, which include IL-8, MIP-2, MIP-1α and MIP-1β, and MCP-1, -2, and -3, have been implicated in the inflammatory response in asbestosis (Driscoll et al., 1993). In addition, AM from subjects with asbestosis release PMN chemotactic factors (Hayes et al., 1990). Activated AM recovered from asbestos-exposed individuals both evidenced more IL-8 mRNA and spontaneously released this interleukin associated with PMN infiltration (Broser et al., 1996). In addition, acute particle exposure to asbestos stimulated increased MIP-2 and CINC expression in the lungs of rats (Driscoll et al., 1995a). This is note-worthy since increased numbers of PMN are frequently found in animal models of asbestosis (Rom et al., 1991) and in humans occupationally exposed to asbestos (Robinson et al., 1986).

Growth factors that influence early events in G_0/G_1 are termed competence factors and include PDGF and fibronectin; progression factors (e.g., IL-1, IGF-1) induce cells to undergo DNA synthesis. Transforming growth factor β (TGFβ) is a potent chemotaxin for monocytes and PMN and increases production of collagen and fibronectin. Increased amounts of TGFβ have been detected by immunohisto-chemistry at sites of developing asbestotic lesions (Perdue and Brody, 1994). The AM from patients with asbestos exposure spontaneously release large amounts of fibronectin (Rom et al., 1987). Asbestos-stimulated pulmonary macrophages have been found to release fibroblast growth factors such as IGF-1 (Rom and Paakko, 1991) and TGF-α (Liu et al., 1996), agents with potent mitogenic activity expressed at sites of asbestos deposition in the lung (Kheradmand et al., 1994). PDGF genes are also unregulated in macrophages, epithelium, and interstitium next to alveolar duct bifurcations in asbestos-exposed animals (Liu et al., 1998).

Asbestos exposure has also been documented to upregulate early response genes which can participate as potent transcription factors. Increased levels of c-*jun* mRNA was found in lung homogenates of rats exposed by inhalation to asbestos (Quinlan et al., 1994). C-*jun* encodes one of the proteins that can dimerize to form activator protein-1 (AP-1) transcription factor that binds to promoter regions on any number of genes governing inflammation, proliferation, and apoptosis (Angel et al., 1991). Asbestos inhalation can also affect NF-κB transcription factor, as evidenced by an increased immunoreactivity of p65 (an NF-κB subunit) in rat lungs and its nuclear translocation in epithelial and mesothelial cells (Janssen et al., 1997).

Thus, asbestos exposure has been associated with changes in inflammatory cell number and function, in proliferation of epithelial cells, in the levels of inflammatory cytokines (TNFα, IL-1, IL-6), chemokines (IL-8, MIP-1, MIP-2, MCP), and growth factors (PDGF, TGFα,β), reactive oxygen/nitrogen species, and in early response gene regulation.

Silica

Silicosis is associated with exposure to crystalline silica (SiO_2) in occupations like stone cutting, quarrying, and mining (Craighead et al., 1988). Exposure induces a marked inflammatory response resulting in fibrotic lung disease. Following SiO_2 exposure, there is crystal deposition at the alveolar duct bifurcations, phagocytosis of particles (Brody et al., 1982), and a rapid influx of PMN and monocytes into the alveoli (Callis et al., 1985; Driscoll et al., 1990a). Phagocytosis by alveolar inflammatory cells may result in cell death or activation, with subsequent release of lysosomal enzymes (Callis et al., 1985) and lactate dehydrogenase (LDH) (Driscoll et al., 1990a), and generation of ROS (Vallyathan et al., 1992; Schapira et al., 1994) and NO (Blackford et al., 1994; Setoguchi et al., 1996) to elicit focal damage to Type I epithelial cells (Adamson and Bowden, 1984). In response to this damage, SiO_2 exposure is documented to cause replication of Type II epithelial cells at low doses and cytotoxicity at higher doses (Lesur et al., 1992). Silica-induced activation of AM, and the subsequent release of fibrogenic factors (Adamson et al., 1991) such as fibronectin (Driscoll et al., 1990a), appears to lead to pulmonary fibrosis (Reiser et al., 1983). As a reflection of this fibrotic response, animals exposed to SiO_2 also evidence impaired lung function (e.g. decreased lung compliance, vital capacity and diffusion capacity) (Begin et al., 1989).

Generation of ROS following SiO_2 exposure may be either directly from the crystal surface (Shi et al., 1988; Daniel et al., 1995) or indirectly through activation of inflammatory cells (Vallyathan et al., 1992; Antonini et al., 1994; Schapira et al., 1994). The AM lavaged from SiO_2-exposed rats evidenced increased zymosan- or phorbol ester-triggered chemiluminescence as early as 1 d post-exposure (Gossart et al., 1996). Also, SiO_2 exposure elicited significant pulmonary oxidative stress as well as increased lipid peroxidation and increased levels of antioxidant enzymes as early as 2 d post-instillation (Gossart et al., 1996). Rats exposed to SiO_2 through IT injection evidenced iNOS expression in their lungs and an excessive amount of NO in pulmonary lesions (Setoguchi et al., 1996). In another study, cells lavaged from exposed rats demonstrated increased iNOS mRNA 24 hr after instillation (Blackford et al., 1994). It is also likely that NO reacts with superoxide anion ($\cdot O_2^-$) to form the potent $ONOO\cdot$ free radical.

Silica also increases several inflammatory mediators such as IL-1 and TNFα. Silica can elicit the release of IL-1 from monocytes/macrophages following *in vitro* (Sari et al., 1993) or *in vivo* exposure (Struhar et al., 1989) or both. Silica exposure *in vivo* primed AM to release both IL-1 and TNFα following *in vitro* exposure to LPS (Driscoll et al., 1990b). Instillation of SiO_2 crystals into mice also elicited an increase in TNFα mRNA in lung tissue (Piguet et al., 1990). TNFα appears to participate in the development of fibrotic lesions; SiO_2-induced collagen deposition could be attenuated by anti-TNFα antibody yet significantly increased by continuous infusion of mouse recombinant TNFα. Infusion of a recombinant soluble TNFα receptor also prevented SiO_2-induced increases in lung hydroxyproline (Piguet and Vesin, 1994). In SiO_2-injected mice, there were significant levels of TNFα released from lavaged cells 28 d after exposure (Ohtsuka et al., 1995); other studies in rats noted increased TNFα in their BAL and from lavaged AM 3 d after SiO_2 instillation (Gossart et al., 1996). Finally, in a mouse inhalation model of SiO_2 exposure, a persistent overexpression of IL-1β and TNFα mRNA (measured by *in situ* hybridization) in alveolar cells, in aggregate lesions, and in mononuclear cells in BALT was noted (Davis et al., 1998).

Silica exposure also has been documented to increase the expression of chemotactic factors MIP-1 (which elicits both PMN and monocyte recruitment) and -2 (a potent chemotaxin for PMN). Driscoll et al (1993) noted increased levels of mRNA for both MIP-1α and -2 in lungs of rats IT-instilled with SiO_2 and that this effect was associated with the increased presence of PMN in the BAL. Other studies have also noted a transient increase in mRNA expression of PMN chemotaxins MIP-2 and KC in BAL cells from similarly-exposed rats (Yuen et al., 1996).

Silica exposure has also been shown to increase expression of ICAM-1 in a mouse model of SiO_2-induced lung injury (using IT instillation). Although introduction of anti-ICAM-1 antibody reduced PMN influx into the lungs (suggesting increased endothelial ICAM-1 expression) (Nario and Hubbard, 1996), the greatest increase in ICAM-1 expression (assessed by immunoperoxidase staining) was on alveolar/interstitial macrophages and Type II epithelial cells (Nario and Hubbard, 1997). Increased ICAM-1 on lavaged macrophages as well as increased levels of soluble ICAM-1 in the lavage fluid were also detected. As increased levels of cell-associated and soluble ICAM-1 were not detected in mice exposed to non-inflammatory titanium dioxide particles, increased levels of specific adhesion molecules may serve as a potential biomarker for exposure to toxic inorganic particles.

As with asbestos, exposure to SiO_2 induces changes in pulmonary inflammatory cell profiles, in epithelial cell proliferation, in the levels of inflammatory and chemotactic cytokines, in reactive oxygen/nitrogen species, and in adhesion molecule expression.

Oxidant Gases

Nitrogen Dioxide

Nitrogen dioxide (NO_2) is an oxidizing air pollutant found in outdoor air pollution produced by automobile exhaust and industrial sources and in indoor air pollution produced by unvented gas stoves or appliances. While indoor levels can reach up to 4 ppm in power plants, refineries, and ice skating rinks, outdoor levels usually do not exceed 0.5 ppm (Chauhan et al., 1998). Because NO_2 can react as a free radical, oxidant injury is suggested to be the principal mechanism by which NO_2 damages the lung. Nitrogen dioxide has also been postulated to adversely affect such lung defense mechanisms as mucociliary clearance, particle transport and detoxification by AM, and local and systemic immunity. Most studies evaluating the effect of NO_2 on local immunity have been conducted in mice evaluating their response to infection with bacteria or virus. Mice exposed to NO_2 either before or after an infectious challenge with *Klebsiella pneumonia* (Ehrlich, 1966) showed increased mortality. Pulmonary bactericidal activity against *Staphylococcus aureas* was also reduced after exposure to various NO_2 concentrations (Goldstein et al., 1973). Decreased intrapulmonary killing of *Mycoplasma pulmonis* related to a decrease in macrophage viability was also noted after exposure to NO_2 (Davis et al., 1992). In sum, pulmonary defense was more susceptible to the adverse effects of NO_2 when the lung was already infected with the organism. However, exposure to NO_2 prior to infection with murine cytomegalovirus (MCMV) reduced the inoculum of virus required to produce infection (Rose et al., 1988). Pulmonary macrophages were capable of ingesting the virus but could not eradicate it. Susceptibility to re-infection with MCMV was also increased after NO_2 exposure (Rose et al., 1989). The Authors suggested that NO_2 exposure damaged virus-specific immunity after a primary infection.

Human volunteers have also been exposed to NO_2 and AM were found to be impaired in their ability to phagocytose *Candida albicans* and to produce $\cdot O_2^-$ radical (Goldstein et al., 1973). In another study (Frampton et al., 1989) human volunteers were exposed to NO_2 continuously or intermittently. Continuous exposure elicited a decreased response of AM to inactivating influenza virus and an increase in IL-1 production in some individuals; intermittent exposure to NO_2 did not produce these effects. Many authors suggest that NO_2 may play a role in increasing susceptibility to respiratory virus infection; the effect of NO_2 on severity of illness could not be estimated. Repeated exposure to NO_2 (Sandstrom et al., 1992) have demonstrated a reduction in total macrophages, B-lymphocytes, NK cells, and CD4:CD8 ratios in the BAL. Exposure to NO_2 induced an increase in PMN and macrophages in the bronchial wash of smokers (Helleday et al., 1994). Chauhan et al. (1998) suggest that NO_2 may increase disease susceptibility by affecting both the upper airways (ciliated epithelium function, nasal filtering, etc) and lower airways (changes in cytokine production, changes in number and type of inflammatory cells, and increased production of free radical and arachidonic acid inflammatory mediators).

Nitrogen dioxide exposure may also change the early and late asthmatic response to inhaled allergen (Tunnicliffe et al., 1994). When 10 mild asthmatics were exposed to NO_2 and then challenged with allergen, there was a greater reduction in lung function in both early and late asthmatic responses. Nitrogen dioxide exposure may also increase IgE production. In sensitized rats, exposure to NO_2 resulted in higher levels of antigen-specific serum IgE, local IgA, IgG, and IgE antibody, and increased numbers of inflammatory cells in the lungs (Chauhan et al., 1998).

Ozone

Ozone (O_3) is another oxidizing air pollutant formed as a result of a series of complex chemical reactions between oxides of nitrogen, volatile organic compounds, and sunlight. Ambient levels seldom exceed 0.2 ppm O_3, but peak levels can reach 0.45 ppm O_3 in heavily-polluted areas (Krishna et al., 1998a). Short-term exposure of healthy subjects to an ambient concentration of O_3 has shown to result in an increase in PMN and in several biomarkers of inflammation, including levels of LDH (cytotoxicity), myeloperoxidase (PMN influx), C3a (complement activation), total protein, albumin (alveolar-capillary permeability), factor VII, IL-6, IL-8, and GM-CSF (Devlin et al., 1996) in the BAL. Other studies have shown an increase in mast cells, T-lymphocytes, and eosinophils in healthy subjects (Torres et al., 1997). Aires et al. (1993) reported in exposed humans an influx in PMN accompanied by an increase in ICAM-1 in the bronchial mucosa after a 4 hr exposure to 0.2 ppm O_3. An increase in another adhesion molecule, P-selectin, was also found in the microvascular endothelium of bronchial mucosa prior to PMN influx or plasma extravasation (Krishna et al., 1997a). The Authors suggest that upregulation of P-selectin may be one of the earliest events in this inflammatory response. Exposed (0.2 ppm O_3) humans also evidenced an increase in chemokines IL-8 and GROα and plasma extravasation 6 hr later (Krishna et al., 1998b). Ozone exposure in humans also resulted in a significant decrease in detection of substance P in the subepithelial sensory nerves suggesting a release of this neuropeptide into the airways (Krishna et al., 1997b). These studies also detected an increase in PMN and ciliated epithelial cells in BAL of these patients suggesting that O_3 exposure induced damage to bronchial epithelium and that the stimulated release of substance P from subepithelial sensory nerves contributed to PMN recruitment.

Ozone exposure to asthmatics elicited an inflammatory response similar to that seen in the airways of healthy subjects, except for a greater PMN influx (Scannell et al., 1996). After a short-term exposure to O_3 (0.2 ppm), evidence of plasma extravasation and activation of PMN and eosinophils was evident in the sputum (increase albumin, protein, MPO and ECP) 24 hr after exposure (Krishna et al., 1997c). Ozone has also been shown to potentiate response to inhaled allergen (Molfino et al., 1991). In one study, it was shown that O_3 could induce an acute inflammatory response and prime eosinophils in the nasal airways to a subsequent challenge with allergen (Peden et al., 1995). The Authors speculate that O_3 could exacerbate preexisting mucosal inflammation and prime eosinophils.

Thus, exposure to oxidant gases such as O_3 and NO_2 can affect pulmonary host defense by increasing susceptibility to microbial infection, by changing the profile of inflammatory cells in the lungs, and by altering the levels of inflammatory cytokines and chemokines as well as adhesion molecule expression.

Beryllium

Occupational exposure (by workers in ceramics manufacture, dental alloy manufacture, electronics, nuclear reactors, etc.) to beryllium (Be) by inhalation has been associated with an acute chemical pneumonitis and a cell-mediated granulomatous response (Type IV hypersensitivity) known as chronic beryllium disease (CBD). Studies in both patients and in animal models have documented a role for sensitized T-lymphocytes in the pathogenesis of the latter disease.

One of the first studies to demonstrate a role for an acquired pulmonary immune response demonstrated that T_H lymphocytes isolated from the BAL of a patient with beryllium disease proliferated in response to Be salts in an IL-2-dependent and MHC restricted-manner (Saltini et al., 1989). Investigators determined that the relevant immunogen in this disease was the hapten beryllium oxide (BeO) which conjugated with normal lung proteins to form a complete immunogen. Patients with beryllium disease had an elevated number of leukocytes and lymphocytes in their BAL which correlated with BAL beryllium lymphocyte transformation (proliferation) tests. The levels of these BAL constituents were also found to correspond with radiographic changes and changes in pulmonary function measurements (Newman et al., 1994). Other Investigators studied BAL cells from patients with CBD and found an increase in pulmonary lymphocytes as well. This group evaluated T-lymphocytes recovered from BAL of patients with CBD and found that beryllium sulfate ($BeSO_4$) stimulated production of both $TNF\alpha$ and IL-6 (Tinkle and Newman, 1997) and IL-2 and $IFN\gamma$ (Tinkle et al., 1997), but not of Il-4 or IL-7. In addition, it was noted that BAL levels of the α subunit of the soluble IL-2 receptor were elevated and that this too correlated with the degree of pulmonary lymphocytosis and clinical measure of disease severity. Detection of increased production of IL-2 and $IFN\gamma$ might suggest the involvement of T_H1 cells in the progression of this disease.

Investigators have also established a mouse model of CBD by inhalation of Be metal aerosols. Six months post-exposure, the lungs of Be-exposed mice had interstitial compact aggregates of lymphocytes, vacuolated macrophages and giant cells in the alveoli, PMN in alveoli and alveolar septa, multifocal interstitial granulomas, interstitial infiltrates of lymphocytes, plasma cells, monocytes, and macrophages, and minimal-to-mild fibrosis. The majority of lymphocytes were $CD4^+$ T-lymphocytes with some B-lymphocytes present, though lymphocytes from tracheobronchial lymph nodes, blood or spleen were not stimulated in response to the hapten $BeSO_4$ (Nikula et al., 1997).

Another mouse model of beryllium disease has been developed by IT instillation of the soluble salt (i.e., $BeSO_4$) or particle (i.e., BeO), with the cells present in the BAL subsequently evaluated by flow cytometry. The BAL from these hosts presented a significant increase in the numbers of lymphocytes following $BeSO_4$ treatment, with 1/3 being $\gamma\delta^+$ 2 wk after exposure followed by $CD4^+\alpha\beta^+$ at 4 wk post-exposure. Expression of MAC-1 on isolated macrophages was also elevated. Conversely, exposure to BeO induced histopathologic changes only 8 mo after (Huang et al., 1992).

Although CBD is initiated by hapten-specific T-lymphocytes, other mediator such as ROS may also participate. Investigators evaluated lung epithelial lining fluid from patients with CBD for levels of antioxidants as an indicator of oxidative stress. Although SOD and catalase levels/activites were not noted to be increased, there was measured increases in glutathione peroxidase protein and activity as well as higher levels of glutathione (Comhair et al., 1999).

Thus, the most significant biomarker of the Type IV hypersensitivity response in the lungs are an increase in the number and proliferation of antigen-reactive CD4[+] T-lymphocytes and the presence of T_H1 cytokines.

Occupational Asthma

Occupational asthma (OA) encompasses the respiratory diseases which are causally related to exposure in the working environment to airborne proteins, dusts, gases, vapors, or fumes and is manifested by episodic airway narrowing and/or by airway hyperresponsiveness (Newman Taylor, 1980). About 250 agents have been documented to induce OA in susceptibility individuals and can include high molecular weight (MW) substances (e.g., flour, grain dust, enzymes, animal urine/dander) and low MW chemicals (e.g., acid anhydrides, reactive dyes, diisocyanates, Western red cedar). Most of the high MW agents are large enough to be complete immunogens whereas low MW chemicals may act as haptens and covalently conjugate with self proteins to form complete immunogens. Both types of agents often induce asthma through IgE-mediated degranulation of mast cells and basophils. Consistent with other types of immediate-type hypersensitivity responses, one might anticipate finding an increase in the number of mast cells and detection of increased levels of IgE in the serum and BAL, increased number of pulmonary T- and B-lymphocytes and eosinophils, and increased detection of the chemotaxin IL-5, RANTES, and adhesion molecules in lung sections and BAL-associated cells. However, most cases of OA in humans are diagnosed by changes in pulmonary function tests rather than by detection of inflammatory cells and mediators. Recently, examination of induced sputum was used to evaluate changes in inflammatory cell profiles/proteins in OA subjects. For example, 10 patients diagnosed with OA (induced by grain, isocyanates, methylmethacrylates) were found to have increases in the numbers of sputum eosinophils and levels of eosinophil cationic protein (Lemiere et al., 1999). In bronchial biopsy specimens obtained from six patients with grain dust-induced asthma, there was a significant increase in the numbers of mast cells and PMN as well as an increase in levels of IL-8 in induced sputum (Park et al., 1998) Bronchial biopsy specimens taken from subjects with isocyanate-induced asthma evidenced an increase in IL-2 receptor-positive cells (suggestive of activated T-lymphocytes) and an increase in both the numbers of total and activated eosinophils (Bentley et al., 1992). Elevated levels of serum IgE were detected only in a small portion of subjects with asthma induced by isocyanate or plicatic acid (Western red cedar tree) (Chan-Yeung and Lam, 1986). Lastly, in a mouse model of TDI respiratory hypersensitivity, mononuclear inflammatory cell cuffing around the pulmonary vasculature but not around pulmonary airways was noted (Satoh et al., 1995).

Similar to lung diseases mediated by Type IV hypersensitivity, diseases (like OA) mediated by Type I mechanisms are characterized by increases in activated T-lymphocytes. In addition, there is evidence of increases in the numbers of eosinophils and PMN chemokines present in the lungs.

POTENTIAL   BIOMARKERS   REFLECTING   ALTERED
PULMONARY IMMUNITY FROM XENOBIOTIC EXPOSURE

Biomarkers have been defined as representative "signals in a continuum of events between causal exposure and resultant disease" (Schulte, 1989). In addition, biomarkers should participate in the biological events which play a role in this continuum between exposure and the progression to clinical disease. Biomarkers of altered pulmonary immune responses and the accompanying inflammatory response can be measured in different compartments and can include measures at the level of tissue, cells, proteins, mediators and nuclear transcription factors.

Perhaps the most common method for evaluating changes in immune/inflammatory status of the lungs is by noting changes in the number and type of cells in the airway, alveoli, or interstitium. Increases in PMN numbers usually suggest an acute (i.e., within hours) inflammatory response whereas increases in macrophages levels suggest a more chronic (i.e., over days) response. Changes in PMN, macrophages, and NK cells profiles would reflect modulation in the first-line innate pulmonary immune response whereas changes in numbers or functional subpopulations of lymphocytes would reflect an alteration in acquired immunity. Additionally, detection of increased numbers of lymphocytes would suggest an antigen-dependent response; increased numbers of both lymphocytes and eosinophils might indicate a pulmonary allergic response. Measuring the balance of functional subpopulations of T-lymphocytes (T_H1 vs T_H2) by evaluating cytokines or IgG isotype has allowed determinations to be made as to whether the inhaled toxicant is eliciting or may alter a humoral (T_H2) or cell-mediated (T_H1) immune response.

Since epithelial cells and resident macrophages in the lung are the first cells to be affected by inhaled toxicants, they may respond with an increase in activity reflected by changes in reactive oxygen/nitrogen species, and/or in inflammatory cytokines, chemokines or growth factors. Increased oxygen consumption and production of oxygen based free radicals (i.e., $\cdot OH$, $\cdot O_2^-$, HOCl) or oxidants (H_2O_2) are a reflection of stimulation and phagocytosis by macrophages and PMN. These mediators can be measured by electron spin resonance, by chemiluminescence or by interaction with substrates (e.g. cytochrome C reduction, dichlorohydrofluorescein oxidation). Detection of NO-derived reactive nitrogen intermediates may reflect a more complex inflammatory response since iNOS, the enzyme responsible for NO production, is upregulated by endotoxins or cytokines. Nitric oxide can be detected by interaction with substrates (e.g. Greiss reagent) to produce nitrites. Nitric oxide can also induce a number of covalent interactions with various biomolecules. One such modification yields 3-nitrotyrosine and detection of this adduct in proteins is often used to identify involvement of NO-derived oxidants in pulmonary responses. The conjugation of 3-nitrotyrosine with proteins can be analyzed either by immunohistochemistry or by an ELISA of lung tissue homogenates or isolated lung cells (Haddad et al., 1994). Western immunoblot analysis may also help characterize any proteins associated with nitrotyrosine. Evidence of increased NO production can also be evaluated by detecting increased iNOS mRNA or protein in lung tissue or lavage cells. Evidence of oxidative stress can be measured indirectly by lipid peroxidation (thiobarbuturic acid-reactive substances) or by increases in antioxidant enzymes (catalase, superoxide dismutase, glutathione peroxidase, glutathione, heme oxygenase).

Inflammatory cytokines (TNFα, IL-1 or IL-6) are a diverse group of inter-cellular signaling proteins that regulate local and system immune/inflammatory responses (Oppenheim and Ruscetti, 1997). Both TNFα and IL-1 are produced predominantly by activated macrophages and provide co-stimulatory signals for T-lymphocytes, promote B-lymphocyte proliferation, increase production of prosta-glandins and other cytokines/chemokines, increase adhesion molecule expression, induce epithelial cell proliferation, and activate fibroblasts to produced collagenase and cytokines. IL-6 is also produced by activated macrophages as well as by endo-thelial cells and fibroblasts. This interleukin has multiple biologic activities in a variety of cells, including synergizing with TNFα and IL-1 and inducing hepatic acute phase responses. TGFβ is produced by activated macrophages, lymphocytes, and platelets and promotes wound healing by stimulating fibroblasts. It also has anti-inflammatory activity by suppressing production of most lymphokines and monokines, inhibiting NK cell function. and decreasing proliferation of many cell types. The C-X-C family of chemokines include IL- 8, GROα and GROβ. The C-C family includes MIP-1α, MIP-1β, MIP-2, MCP-1,2,3, RANTES, and eotaxin. Although most of these cytokines attract many different types of leukocytes, in general, it is believed that MIP-2 is primarily a chemokine for neutrophils whereas RANTES, MCP-1, MIP-1α, and MIP-1β are primarily chemokines for monocytes and lymphocytes; at higher concentrations chemokines can also activate cellular effector functions. Both cytokines and chemokines can be detected in cell free fluid by activity (e.g. TNFα, IL-1) or by antigenicity in lung sections or lavaged cells by immunohistochemistry, by Western blot analysis or ELISA of tissue homogenates or by mRNA levels (*in situ* hybridization, Northern analysis). And, finally, measuring increased expression of adhesion molecules such as ICAM-1 on endo-thelial or epithelial cells or $\beta2$ integrins on leukocytes would reflect an ongoing inflammatory response involving trafficking of immune/inflammatory cells (DeLisser and Albelda, 1998).

Normal macrophages are generally able to ingest and kill many types of micro-organisms that gain access to the respiratory tract. Measuring the ability of these cells exposed *in situ* to eliminate pathogenic bacteria/virus is a longstanding biomarker of the potentiation of the toxicant to alter host resistance in man.

NF-κB regulates the expression of many genes involved in inflammatory responses in the lungs (e.g. TNFα, Il-1β, IL-6, IL-12, IL-8, MIP-1α, MCP-1, RANTES, iNOS, ICAM-1). In turn, activation of NF-κB has been shown in lung tissue and cells in response to inflammatory cytokines such as TNFα, IL-1β and in response to oxidants. Activator protein-1 (AP-1) is another important transcription factor composed of heterodimers of c-*fos*, c-*jun* or c-*jun*/c-*jun* homodimers. Activators of protein kinase C such as phorbol myristate acetate and ROS can active AP-1. Production of ROS (often induced by TNFα) can activate AP-1 through activation of the c-*jun* N-terminal kinases (JNKs) members of the MAP kinase superfamily. AP-1 also regulates expression of cytokines such as IL-6, -1β, -12, and TGFβ (Rahman and MacNee, 1998). Thus, detection of these activated transcription factors or kinases would signal initiation of an inflammatory response and suggest potential molecular mechanisms underlying xenobiotic induced lung injury.

CONCLUDING REMARKS

Table 1 provides a summary of possible biomarkers of pulmonary inflammation and immunity. Clearly, changes in a single biomarker should not be interpreted as an isolated event but, rather, as a component of a multiple cellular and molecular response. The significance of changes in these biomarkers should also be considered as part of a continuum in the response to exposure and physiologic outcome.

TABLE 1. BIOMARKERS OF PULMONARY IMMUNITY AND INFLAMMATION

Observed Pulmonary Response:	Response is Biomarker of:
Increased PMN in BAL or tissue	Acute inflammation
Increased macrophages in BAL or tissue	Chronic inflammation
Increased lymphocytes; lymphocyte proliferation in response to immunogen	Antigen driven immune response
Eosinophils	Allergic response
TNFα, IL-1, IL-6	Inflammatory cytokines
IL-6, TGFβ	Fibroblast activity
MIP-2, MIP-1α, IL-8 (human), MCP-1, RANTES, eotaxin	Cell trafficking
NO, iNOS protein and mRNA, nitrotyrosine residues; ROS in isolated cells lipid peroxidation antioxidant enzymes	Oxidative stress, inflammatory response
IL-4; IL-5, IgG_{2a} IFNγ, IL-2, IgG_1	T_H2 activity T_H1 activity
Pulmonary defense against pathogens	Host resistance
Early response genes and transcription factor activation	Initiation of immune/inflammatory response at the molecular level
Adhesion molecules	Cell trafficking; inflammatory response

REFERENCES

Adamson IYR, Bowden DH. Role of polymorphonuclear leukocytes in silica-induced pulmonary fibrosis. Am. J. Pathol., 1984;117:37-43.

Adamson IYR, Letourneau HL, Bowden DH. Comparison of alveolar and interstitial macrophages in fibroblast stimulation after silica and long or short asbestos. Lab. Invest., 1991;64:339-344.

Albelda SM, Smith CW, Ward PA. Adhesion molecules and inflammatory injury. FASEB J., 1994;8:504-512.

American Thoracic Society. Diagnosis of non-maligant diseases related to asbestos. Am. Rev. Respir. Dis., 1986;134:363-368.

Angel P, Karin M. The role of *Jun*, *Fos*, and the AP-1 complex in cell proliferation and transformation. Biochim. Biophys. Acta, 1991;1072:129-157.

Antonini JM, Van Dyke K, Ye Z, DiMatteo M, Reasor MJ. Introduction of luminol-dependent chemiluminescence as a method to study silica inflammation in the tissue and phagocytic cells of rat lung. Environ. Health Perspect., 1994;102:37-42.

Aris RM, Christian D, Hearne PQ, Kerr K, Finkbeiner WE, Balmes RJ. Ozone-induced airway inflammation in human subjects as determined by airway lavage and biopsy. Am. Rev. Respir. Dis., 1993;148:1363-72.

Bedard M, McClure CD, Schiller NL, Francoeur C, Cantin A, Denis M. Release of interleukin-8 and interleukin-6 and colony stimulating factors by upper airway epithelial cells: Implications for cystic fibrosis. Am. J. Respir. Cell Mol. Biol., 1993;9:455-462.

Begin R, Dufresne A, Cantin A, Possmayer F, Sebastien P. Quartz exposure, retention and early silicosis in sheep. Exp. Lung Res., 1989;15:409-428.

Bentley AM, Maestrelli P, Saetta M, Fabbri LM, Robinson DS, Bradley BL, Jeffery PK, Durham SR, Kay AM. Activated T lymphocytes and eosinophils in the bronchial mucosa in isocyanate-induced asthma. J. Allergy Clin. Immunol., 1992;89:821-829.

BeruBe KA, Quinlin TR, Moulton G, Hemenway D, O'Shaughnessy P, Vacek P, Mossman BT. Comparative proliferative and histopathologic changes in rat lungs after inhalation of chrysotile or crocidolite asbestos. Toxicol. Appl. Pharmacol., 1996;137:67-74.

Blackford JA, Antonini JM, Castranova V, Dey RD. Intratracheal instillation of silica up regulates inducible nitric oxide synthase gene expression and increases nitric oxide production in alveolar macrophages and neutrophils. Am. J. Respir. Cell Mol. Biol., 1994;11:426-431.

Brody AR, Roe MW, Evans JN, Davis, GS. Deposition and translocation of inhaled silica in rats. Quantification of particle distribution, macrophage participation and function. Lab. Invest., 1982;47:533-542.

Broser M, Zhang Y, Aston C, Harkin T, Rom WN. Elevated interleukin-8 in the alveoli of individuals with asbestos exposure. Int. Arch. Occup. Environ. Health, 1996;68:109-114.

Bullard DC, Qin L, Lorenzo I, Quinlin WM, Doyle NA, Bosse R, Vestweber D, Doerschuk CM, Beaudet AL. P-selectin/ICAM-1 double mutant mice: Acute emigration of neutrophils into the peritoneum is completely absent but is normal into pulmonary alveoli. J. Clin. Invest., 1995;95:1782-1788.

Callis AH, Sohnle PG, Mandel GS, Wiessner J, Mandel NS. Kinetics of inflammatory and fibrotic changes in a mouse model of silicosis. J. Lab. Clin. Med., 1985;105:547-533.

Chan-Yeung M, Lam S. Occupational asthma: State of the art. Am. Rev. Respir. Dis., 1986;133:688-703.

Chauhan AJ, Krishna MT, Frew AJ, Holgate ST. Exposure to nitrogen dioxide and respiratory disease risk. Rev. Environ. Health, 1998;13:73-90.

Comhair SAA, Lewis MJ, Bhathena PR, Hammel JP, Erzurum SC. Increased glutathione and glutathione peroxidase in lungs of individuals with chronic beryllium disease. Am. J. Respir. Crit. Care Med., 1999;159:1824-1829.

Cooper AM, Dalton DK, Stewart TA, Griffin JP, Russell DG, Orme IM. Disseminated tuberculosis in interferon-γ gene disrupted mice. J. Exp. Med., 1993;178:2243-2247.

Craighead JE, Kleinerman J, Abraham JL, GIbbs AR, Green FHY, Harley RA, Ruettner JR, Vallyathan NV, Juliano EB. Diseases associated with exposure to silica and non-fibrous silicate minerals. Arch. Pathol. Lab. Med., 1988;112:673-720.

Crapo JD, Young Sl, Fram EK, Pinkerton DE, Barry BE, Crapo RO. Morphometric characteristics of cells in the alveolar region of mammalian lungs. Am. Rev. Respir. Dis., 1983;128:S42-S46.

Daniel LN, Mao Y, Wang TCL, Markey CJ, Markey SP, Shi X, Saffiotti U. DNA strand breakage, thymine glycol production, and hydroxyl radical generation induced by different samples of crystalline silica *in vitro*. Environ. Res., 1995;71:60-73.

Davis GS, Pfeiffer LM, Hemenway DR. Persistent overexpression of interleukin-1β and tumor necrosis factor-α in murine silicosis. J. Environ. Pathol. Toxicol. Oncol., 1998;17:99-114.

Davis JK, Davidson MK, Schoeb TR, Lindsey JR. Decreased intrapulmonary killing of *Mycoplasma pulmonis* after short-term exposure to NO$_2$ is associated with damaged alveolar macrophages. Am. Rev. Respir. Dis., 1992;145:406-411.

DeLisser HM, Albelda SM. The function of cell adhesion molecules in lung inflammation: More questions than answers. Am. J. Respir. Cell Mol. Biol., 1998;19:533-536.

Devlin RB, McDonnell WF, Becker S, Madden MC, McGee MP, Perez R, Hatch G, House DE, Koren HS. Time-dependent changes of inflammatory mediators in lungs of humans exposed to 0.4 ppm ozone for 2 hrs: A comparison of mediators found in bronchoalveolar lavage fluid 1 and 18 hrs after exposure. Toxicol. Appl. Pharmacol., 1996;138:176-185.

Doerschuk CM, Beyers N, Coxson HO, Wiggs B, Hogg JC. Comparison of neutrophil and capillary diameters in relation to neutrophil sequestration in the lung. J. Appl. Physiol., 1993;74:3040-45.

Doerschuk CM, Winn RK, Coxson HO, Harlan JM. CD18-dependent and -independent mechanisms of neutrophil adherence in the pulmonary and systemic microcirculation of rabbits. J. Immunol., 1990;114:2327-33.

Donaldson K, Cullen RT. Chemiluminescence of asbestos-activated macrophages. Br. J. Exp. Pathol., 1984;65:81-90.

Driscoll KE, Hasenbein DG, Carter J, Poynter J, Asquith TN, Grant RA, Whitten J, Purdon MP, Takigiku T. Macrophage inflammatory protein 1 and 2: Expression by rat alveolar macrophages, fibroblasts, and epithelial cells and in rat lung after mineral dust exposure. Am. J. Respir. Cell Mol. Biol., 1993;8:311-318.

Driscoll KE, Hassenbein DG, Carter JM, Kunkel SL, Quinlan TR, Mossman BT. TNFα and increased chemokine expression in rat lung after particle exposure. Toxicol. Lett., 1995a;82:483-489.

Driscoll KE, Lindenschmidt RC, Maurer JK, Higgins JM, Ridder G. Pulmonary response to silica or TiO$_2$: Inflammatory cells, alveolar macrophage-derived cytokines, and histopathology. Am. J. Respir. Cell Mol. Biol., 1990b;2:381-390.

Driscoll KE, Maurer JK, Higgins J, Poynter J. Alvaeolar macrophage cytokine and growth factor production in a rat model of crocidolite-induced pulmonary inflammation and fibrosis. J. Toxicol. Environ. Health, 1995b;46:155-169.

Driscoll KE, Maurer JK, Lindenschmidt RC, Romberger D, Rennard SI, Crosby L. Respiratory tract responses to dust: Relationships between dust burden, lung injury, alveolar macrophage fibronectin release and the development of pulmonary fibrosis. Toxicol. Appl. Pharmacol., 1990a;106:88-101.

Driscoll KE. "*In Vitro* Evaluation of Mineral Cytotoxicity and Inflammatory Activity." In *Health Effects of Mineral Dusts. Reviews in Mineralogy. Vol. 28*, GD Guthrie, BT Mossman, eds. Washington, DC: Mineralogic Society of America, 1993, pp 489-512.

Dustin ML, Rothlein R, Bhan AF, Dinarello CA, Springer TA. A natural adherence molecule (ICAM-1): Induction by IL-1 and IFN-γ, tissue distribution, biochemistry and function. J. Immunol., 1986;137:245-254.

Ehrlich R. Effect of nitrogen dioxide on resistance to respiratory infection. Bacteriol. Rev., 1966;30:604-614.

Ferrick DA, Schrenzel MD, Mulvania T, Hsieh B, Felin WG, Lepper H. Differential production of interferon γ and interleukin-4 in response to TH1 and TH2 stimulating pathogens by gamma delta T cells *in vivo*. Nature, 1995;373:255-257.

Frampton MW, Smeglin AM, Roberts NJ, Finkelstein JN, Morrow PE, Utell MJ. Nitrogen dioxide exposure *in vivo* and human alveolar macrophage inactivation of influenza virus *in vitro*. Environ. Res., 1989;48:179-192.

Franke-Ullmann G, Pfortner C, Walter P, Steinmuller C, Lohmann-Matthes ML, Kobzik L. Characterization of murine lung interstitial macrophages in comparison with alveolar macrophages *in vitro*. J. Immunol., 1996;157:3097-3104.

Goldstein E, Eagle MC, Hoeprich PD. Effect of nitrogen dioxide on pulmonary bacterial defense mechanisms. Arch. Environ. Health, 1973;26:202-204.

Gossart S, Cambon C, Orfila C, Seguelas MH, Lepert JC, Rami J, Carre P. Pipy B. Reactive oxygen intermediates as regulators of TNFα production in rat lung inflammation induced by silica. J. Immunol., 1996;156:1540-1548.

Gulumian M, Kilroe-Smith TA. Crocidolite-induced lipid peroxidation in rat lung microsomes. Environ. Res., 1987;43:267-273.

Haddad IY, Pataki G, Hu P, Galliani C, Beckman JS, Manlon S. Quantitation of nitrotyrosine levels in lung sections of patients and animals with acute lung injury. J. Clin. Invest., 1994;94:2407-2413.

Haslam PL Lukoszek A, Merchant JA, Turner Warwick M. Lymphocyte responses to phytohemagglutinin in patients with asbestosis and pleural mesothelioma. Clin. Exp. Immunol., 1978;31:178-188.

Hayes AA, Venaille TJ, Rose AH, Musk AW, Robinson WS. Asbestos release of a human alveolar macrophage-derived neutrophil chemotactic factor. Exp. Lung Res/. 1990;16:121-130.

Helleday R, Sandstrom T, Stjernberg N. Differences in bronchoalveolar cell response to nitrogen dioxide between smokers and non-smokers. Eur. Respir. J., 1994;7:1213-1220.

Holt PG, Degebrodt A, Venaille T, O'Leary C, Krska K, Flexman J, Farrell H, Shellam G, Young P, Penhale J. Preparation of interstitial lung cells by enzymatic digestion of tissue slices: Preliminary characterization by morphology and performance in functional assays. Immunology, 1985;54:139-147.

Holtzman MJ. Arachidonic acid metabolism in airway epithelial cells. Ann. Rev. Physiol., 1992;54:303-329.

Huang H, Meyer KC, Kubai L, Auerbach R. An immune model of beryllium-induced pulmonary granulomata in mice. Histopathology, immune reactivity, and flow cytometric analysis of bronchoalveolar lavage-derived cells. Lab. Invest., 1992;67:138-146.

Janssen YMW, Driscoll KE, Howard B, Quinlan TR, Treadwell M, Barchowsky A, Mossman BT. Asbestos causes translocation of p65 protein and increased NF-κB DNA binding activity in rat lung epithelial and pleural mesothelial cells. Am. J. Pathol., 1997;151:389-401.

Kheradmand F, Folkesson HG, Shum L, Derynk R, Pytela R, Mathhay MA. Transforming growth factor-α enhances alveolar epithelial cell repair in a new *in vitro* model. Am. J. Physiol., 1994;267:L728-L738.

Krishna MT, Blomberg A, Biscione GL, Kelly F, Sandstrom T, Frew A, Holgate S. Short-term exposure to 0.12 ppm ozone upregulates P-selectin in normal human airways. Am. J. Respir. Crit. Care Med., 1997a;155:1798-1803.

Krishna MT, Chauhan AJ, Frew AJ and Holgate ST. Toxicological mechanisms underlying oxidant pollutant-induced airway injury. Rev. Environ. Health, 1998a;13:59-71.

Krishna MT, Madden J, Teran LM, Biscione GL, Lau LC, Withers NJ, Sandstrom T, Mudway I, Kelly FJ, Walls A, Frew AJ, Holgate ST. Effects of 0.2 ppm ozone on biomarkers of inflammation in BAL fluid and bronchial mucosa of healthy human subjects. Eur. Respir. J., 1998b;11:1294-1300.

Krishna MT, Newson E, Lau LCK, Salvi SS, Holgate ST, Frew AJ. Effect of 0.2 ppm ozone on lung function, exhaled nitric oxide and biomarkers of inflammation in sputum of mild atopic asthmatics. BSACI Annual Meeting, Nottingham, UK, July 1997, 1997c.

Krishna MT, Springall DR, Meng QH, Withers N, Macleod D, Biscione G, Frew A, Polak J, Holgate S. Effects of ozone on epithelium and sensory nerves in the bronchial mucosa of healthy humans. Am. J. Respir. Crit. Care Med., 1997b;156:943-950.

Lange A, Smolik R, Zatonki W, Szymanska J. Autoantibody and serum immunoglobulin levels in asbestos workers. Int. Arch. Arbeitsmed., 1974;32:313-325.

Lehnert BE. Pulmonary and thoracic macrophage subpopulations and clearance of particles from the lung. Environ. Health Perspect., 1992;97:17-46.

Lemaire I, Ouellet S. Distinctive profile of alveolar macrophage-derived cytokine release induced by fibrogenic and non-fibrogenic mineral dusts. J. Toxicol. Environ. Health, 1996;47:465-478.

Lemiere C. Pizzichini MMM, Balkissoon R, Clelland L, Efthimiadis A, O'Shaughnessy D, Colovich J, Hargreave FE. Diagnosing occupational asthma: Use of induced sputum. Eur. Respir. J., 1999;13:482-488.

Lesur O, Cantin AM, Tanswell AK, Melloni B, Beaulieu JF, Begin R. Silica exposure induces cytotoxicity and proliferative activity of type II pneumocytes. Exp. Lung Res., 1992;18:173-190.

Liu JY, Brass DM, Hoyle GW, Brody AR. TNFα receptor knockout mice are protected from the fibroproliferative effects of inhaled asbestos fibers Am. J. Pathol., 1998;153:1839-1847.

Liu JY, Morris GF, Lei WH, Corti M, Brody AR. Upregulated expression of transforming growth factor-α in the bronchiolar-alveolar duct regions of asbestos-exposed rats. Am. J. Pathol., 1996;149:205-215.

Macatonia SE, Knight SC, Edwards AJ, Griffiths S, Fryer P. Localization of antigen on lymph node dendritic cells after exposure to the contact sensitizer fluorescein isothiocyanate: Functional and morphological studies. J. Exp. Med., 1987;166:1654-1667.

Mariassy AT, Wheeldon EB. The pleura: A combined light microscopic, scanning transmission electron microscopy study in sheep. Exp. Lung Res., 1983;4:293-314.

Molfino NA, Wright SC, Katz I, Tarlo S, Silverman F, McClean PA, Szalai JP, Raizeme M, Slutsky AS, Samel N. Effect of low concentrations of inhaled allergen response in asthmatic patients. Lancet, 1991;338:199-203.

Muller WA. The role of PECAM-1 (CD31) in leukocyte migration: Studies *in vitro* and *in vivo*. J. Leukocyte Biol., 1995;57:523-528.

Mulligan MS, Polley MJ, Bayer RJ, Nunn MF, Paulson JC, Ward PA. Neutrophil-dependent acute lung injury: Requirement for P-selectin (GMP-140). J. Clin. Invest., 1992;90:1600-1607.

Mulligan MS, Vaporciyan AA, Warner RL, Jones ML, Foreman KE, Miyasaka M, Todd RF, Ward PA. Compartmentalized roles for leukocytic adhesion molecules in lung inflammatory injury. J. Immunol., 1995;154:1350-1363.

Nario RC, Hubbard AK. Localization of intercellular adhesion molecule-1 (ICAM-1) in the lungs of silica-exposed mice. Environ. Health Perspect., 1997;105:1183-1190.

Nario RC, Hubbard AK. Silica exposure increases expression of pulmonary intercellular adhesion molecule-1 (ICAM-1) in C57Bl/6 mice. J. Toxicol. Environ. Health, 1996;49:599-617.

Newman LS, Bobka C, Schumacher B, Daniloff E, Zhen B, Mroz MM, King TE, Jr. Compartmentalized immune response reflects clinical severity of beryllium disease. Am. J. Respir. Crit. Care Med., 1994;150:135-42.

Newman Taylor AJ. Occupational asthma. Thorax, 1980;35:241-245.

Nikula KJ, Swafford DS, Hoover MD, Tohulka MD, Finch GL. Chronic granulomatous pneumonia and lymphocytic responses induced by inhaled beryllium metal in A/J and C3H/HeJ mice. Toxicol. Pathol., 1997;25:2-12.

Ohtsuka Y, Munakata M, Ukita H, Takahashi T, Satoh A. Increased susceptibility to silicosis and TNFα production in C57Bl/6 mice. Am. J. Respir. Crit. Care Med., 1995;152:2144-2149.

Oppenheim JJ, Ruscetti FW. "Cytokines." In Medical Immunology, 9th Edition. DP Stites, AI Terr, TG Parslow, eds. Stamford, CT: Appleton & Lange, 1997, pp. 146-168.

Park HS, Jung KS, Hwang SC, Nahm DH, Yim HE. Neutrophil infiltration and release of IL-8 in airway mucosa from subjects with grain dust-induced occupational asthma. Clin. Exp. Allergy, 1998;28:724-730.

Peden DB, Setzer RW, Jr., Devlin RB. Ozone exposure has both a priming effect on allergen induced responses and an intrinsic inflammatory action in the nasal airways of perennially allergic asthmatics. Am. J. Respir. Crit. Care Med., 1995;151:1336-1345.

Perdue TD, Brody AR. Distribution of transforming growth factor-beta-1, fibronectin and smooth muscle actin in asbestos-induced pulmonary fibrosis in rats. J. Histochem. Cytochem., 1994;42:1061-1070.

Perkins RC, Scheule RK, Hamilton R, Gomes G, Friedman G, Holian A. Human alveolar macrophage cytokine release in response to in vitro and in vivo asbestos exposure. Exp. Lung Res., 1993;19:55-65.

Piguet PF, Collart MA, Crau GE, Sappino A-P, Vassalli P. Requirement of tumor necrosis factor for development of silica-induced pulmonary fibrosis. Nature, 1990;344:245-247.

Piguet PF, Vesin C. Treatment by human recombinant soluble TNF receptor of pulmonary fibrosis induced by bleomycin or silica in mice. Eur. Respir. J., 1994;7:515-518.

Pober JS, Bevilacqua MP, Mendrick DL, LaPierre LA, Fiers W, Gimbrone MA. Two distinct monokines, interleukin-1 and tumor necrosis factor, independently induce biosynthesis and transient expression of the same antigen on the surface of cultured human vascular endothelial cells. J. Immunol., 1986;136:1680-1687.

Quinlan TR, Berube KA, Hacker MP, Taatjes DJ, Timblin CR, Goldberg J, Kimberley P, O'Shaughnessy P, Hemenway D, Torino J, Jinenez LA, Mossman BT. Mechanisms of asbestos-induced nitric oxide production by rat alveolar macrophages in inhalation and in vitro models. Free Rad. Biol. Med., 1998;24:778-88.

Quinlan TR, Marsh JP, Janssen YMW, Leslie KO, Hemenway D, Vacek P, Mossman BT. Dose-responsive increases in pulmonary fibrosis after inhalation of asbestos. Am. J. Respir. Crit. Care Med., 1994;150:200-206.

Rahman I, MacNee W. Role of transcription factors in inflammatory lung diseases. Thorax, 1998;53:601-612.

Reiser KM, Hascher WM, Hesterberg TW, Last JA. Experimental silicosis. II. Long term effects of intratracheally-instilled quartz on collagen metabolism and morphologic characteristics of rat lungs. Am. J. Pathol., 1983;110:30-41.

Robinson BWS, Rose AH, James A, Whitaker D, Musk AW. Alveolitis of pulmonary asbestosis: Bronchoalveolar lavage of crocidolite and chrysotile exposed individuals. Chest, 1986;90:396-402.

Rom WN, Bitterman PB, Rennard SI, Cantin A. Crystal R. Characterization of the lower respiratory tract inflammation of nonsmoking individuals with interstitial lung disease associated with chronic inhalation of inorganic dusts. Am. Rev. Respir. Dis., 1987;136:1429-1434.

Rom WN, Paakko P. Activated alveolar macrophages express the insulin-like growth factor-I receptor. Am. J. Respir. Cell Mol. Biol., 1991;4:432-439.

Rom WN, Travis WD, Brody AR. Cellular and molecular basis of the asbestos-related disease. Am. Rev. Respir. Dis., 1991;143:408-422.

Rose RM, Fuglestad JM, Skornik WA, Hammer SM, Wofthal SF, Beck BD,Brain JC. The pathophysiology of enhanced susceptibility to murine cytomegalovirus respiratory infection during short-term exposure to 5 ppm nitrogen dioxide. Am. Rev. Respir. Dis., 1988;137:912-917.

Rose RM, Pinkston P, Skornik WA. Altered susceptibility to viral respiratory infection during short-term exposure to nitrogen dioxide. Res. Report Health Effects Inst., 1989;24:1-24.

Rosenthal GJ, Corsini EC, Simeonova P. Selected new developments in asbestos immunotoxicity. Environ. Health Perspect., 1998;106:159-169.

Rouzer C, Scott W, Hamill A, Cohn Z. Synthesis of leukotriene C and other arachidonic acid metabolites by mouse pulmonary macrophages. J. Exp. Med., 1982;155:720-733.

Sallusto F, Lanzavecchia A. Efficient presentation of soluble antigen by cultured human dendritic cells is maintained by granulocyte/macrophage colony stimulating factor plus interleukin-4 and downregulated by tumor necrosis factor α. J. Exp. Med., 1994;179:1109-1118.

Saltini C, Winestock K, Kirby M, Pinkston P, Crystal RG. Maintenance of alveolitis in patients with chronic beryllium disease by beryllium-specific helper T-cells. New Engl. J. Med., 1989;320:1103-1109.

Sandstrom T, Ledin MC, Thomasson L, Helleday R, Stjernberg N. Reductions in lymphocyte subpopulations after repeated exposure to 1.5 ppm nitrogen dioxide. Br. J. Ind. Med., 1992;49:850-854.

Santoli D, Koprowski H. Mechanims of activation of human natural killer cells against tumor and virus-infected cells. Immunol. Rev., 1979;44:125-163.

Sari M, Souvannavong V, Brown SC, Adam A. Silica induces apoptosis in macrophages and the release of IL-1α and IL-1β. J. Leukocyte Biol., 1993;54:407-413.

Satoh T, Kramarik JA, Tollerud DJ, Karol MH. A murine model for assessing the respiratory hypersensitivity potential of chemical allergens. Toxicol. Lett., 1995;78:57-66.

Scannell C, Chen L, Aris RM, Tager I, Christian D, Ferrando R, Welch B, Kelly T, Balmes JR. Greater ozone-induced inflammatory responses in subjects with asthma. Am. J. Respir. Crit. Care Med., 1996;154:24-29.

Schapira RM, Ghio AH, Effros RM, Morrisey J, Dawson CA, Hacker AD. Hydroxyl radicals are formed in the rat lung following asbestos instillation in vivo. Am. J. Respir. Cell Mol. Biol., 1994;10:573-579.

Schulte P. A conceptual framework for the validation and use of biological markers. Environ. Res., 1989;48:129-144.

Sertl K, Takemura T, Schachler E, Ferrans VJ, Kaliner MA, Shevach EM. Dendritic cells with antigen presenting capability reside in airway epithelium, lung parenchyma and visceral pleura. J. Exp. Med., 1986:163:436-451.

Setoguchi K, Takeya M, Akaike T, Suga M, Hattori R, Maeda H, Ando M, Takahashi K. Expression of inducible nitric oxide synthase and its involvement in pulmonary granulomatous inflammation in rats. Am. J. Pathol., 1996;149:2005-2022.

Shi X, Dalal NS, Vallyathan V. ESR evidence for the hydroxyl radical formation in aqueous suspension of quartz particles and its possible significance to lipid peroxidation in silicosis. J. Toxicol. Environ. Health, 1988;25:237-245.

Sibille Y, Reynolds HY. Macrophages and polymorphonuclear neutrophils in lung defense and injury. Am. Rev. Respir. Dis., 1990;141:471-501.

Simeonova PP, Toriumi W, Kommineni C, Erkan M, Munson AE, Rom WN, Luster MI. Molecular regulation of IL-6 activation by asbestos in lung epithelia cells. J. Immunol., 1997;159:3921-3928.

Sorokin SP, Brain JD. Pathways of clearance in mouse lungs exposed to iron oxide aerosols. Anat. Rec., 1975;181:581-625.

Sprince NL, Oliver LC, McLoud TC, Eisen EA. Christiani DC, Ginns LC. Asbestos exposure and asbestos related pleural and parenchymal disease. Association with immune imbalance. Am. Rev. Respir. Dis., 1991;143:822-828.

Springer TA. Adhesion receptors of the immune system. Nature, 1990;346:425-434.

Struhar DJ, Harbeck RJ, Gegen N, Kawada H, Mason RJ. Increased expression of Class II antigens of the major histocompatibility complex on alveolar macrophages and alveolar type II cells and interleukin (IL-1) secretion from alveolar macrophages in an animal model of silicosis. Clin. Exp. Immunol., 1989;77:281-284.

Tanaka S, Choe N, Hemenway DR, Zhu S, Matalon S, Kagan E. Asbestos inhalation induces reactive nitrogen species and nitrotyrosine formation in the lungs and pleura of the rat. J. Clin. Invest., 1998;102:445-454.

Tinkle SS, Kittle LA, Schumacher BA Newman LS. Beryllium-induced IL-1 and IFNγ in berylliosis. J. Immunol., 1997;158:518-526.

Tinkle SS, Newman LS. Beryllium-stimulated release of tumor necrosis factor-α, interleukin-6, and their soluble receptors in chronic beryllium disease. Am. J. Respir. Crit. Care Med., 1997;156:1884-1891.

Torres A, Utell MJ, Morrow PE, Voter KZ, Whitin JC, Cox C, Looney RJ, Speers Dm, Tsai Y, Frampton MW. Airway inflammation in smokers and nonsmokers with varying responsiveness to ozone. Am. J. Respir. Crit. Care Med., 1997;156:728-736.

Tsang PH, Chu FN, Fischbein A, Bekesi G. Impairments in functional subsets of T-suppressor (CD8) lymphocytes, monocytes, and natural killer cells among asbestos-exposed workers. Clin. Immunol. Immunopathol., 1988;47:323-332.

Tunnicliffe WS, Burge PS, Ayres JG. Effect of domestic concentrations of nitrogen dioxide on airway responses to inhaled allergen in asthmatic patients. Lancet, 1994;344:1733-1736.

Underdown BJ, Schiff JM. Strategic defense at the mucosal surface. Ann. Rev. Immunol., 1986;4:389-417.

Vallyathan V, Mega JF, Shi X, Dalal NS. Enhanced generation of free radicals from phagocytes induced by mineral dusts. Am. J. Respir. Cell Mol. Biol., 1992;6:404-413.

Wegner CD, Wolyniec WW, LaPlante AM, Marschman K, Lubbe K, Haynes N, Rothlein R, Letts LG. ICAM-1 contributes to pulmonary oxygen toxicity in mice: Role of leukocytes revised. Lung, 1992;170:267-279.

Widdicombe JG. Relationship between the composition of mucus, epithelial lining liquid and adhesion of microorganisms. Am. J. Crit. Care Med. 1995;151:2088-2093.

Yuen IS, Hartsky MA, Snajdr SI, Warheit DB. Time-course of chemotactic factor generation and neutrophil recruitment in the lungs of dust-exposed rats. Am. J. Respir. Cell Mol. Biol., 1996;15:268-274.

Zhang Y, Lee TC, Guillemin B, Yu MC, Rom WN. Enhanced IL-1β and tumor necrosis factor α release and messenger RNA expression in macrophages from idiopathic pulmonary fibrosis or after asbestos exposure. J. Immunol., 1993;150:4188-4196.

INDEX

A

Acquired immunity, 63, 79, 396, 445
Acute beryllium disease, 213, 215-216
Adaptation, 303, 305, 311-312, 316, 418
Adhesion molecule(s), 69, 75-76, 79, 122, 128-130, 132-135, 156, 246, 258, 319, 434-435, 438, 440, 442, 444, 446
Aerodynamic diameter, 86-87, 216
Aerosol(s), 85-86, 88-89, 91, 100, 225, 233, 256-257, 272, 280, 308, 310, 317, 320, 338-340, 359, 361, 372, 376, 391, 414, 443
Aggregate(s), 16, 33, 43, 74, 79, 95, 132, 217, 228, 285, 440, 443
Air pollutants, 1, 25, 108, 116, 122, 301-302, 306, 320, 370, 412, 414-415, 417, 421, 424-425, 427
Allele, 231-232
Allergy, 107, 115, 118, 120, 204, 255, 307, 314, 419-420, 422-423, 425, 436
Alloy, 214, 223, 251, 255, 275, 442
Alumina, 267, 269
Aluminum, 214, 267-271, 312
Alveolar duct, 2, 19, 21, 34-36, 45, 91, 355-356, 358-359, 362, 364, 437-439
Alveolar macrophage(s), 29, 41-42, 47, 62, 64, 94-98, 119, 135, 137, 140, 181, 186, 244-246, 249-251, 253, 256-258, 269-271, 273, 277, 279, 281-282, 285, 304, 340-341, 345, 347, 355, 357-359, 364, 376, 380, 400-401, 412-413, 427
Alveoli, 18, 34-37, 42, 45-46, 50, 95, 97-98, 119, 436, 439, 443, 445
Alveolitis, 112, 282, 321
Aminocarb, 201
Ammonia, 8, 337-338, 389, 390
Ammonium, 254, 269, 282, 285, 308, 337

Antibacterial, 30, 40, 285, 320, 340-341, 345, 373
Antibody/antibodies, 26, 29, 62-65, 70, 74, 79, 108-112, 115-121, 133-134, 136, 143, 154-158, 169, 181, 186, 200, 202, 217, 232, 248-249, 255, 272, 274, 307-310, 312-313, 321, 325, 340, 345, 355, 363, 379-380, 396-398, 402, 420-422, 424, 434, 436, 440-441
Antibody-forming cells, 71, 200, 203, 248
Antigen response, 274, 307-308, 312-313, 402, 420-421, 424
Antigen presentation, 66, 76, 117, 119, 157, 231, 274, 435
Anti-neutrophil antibody (ANCA), 154-158, 165, 167, 319
Arachidonic acid, 69, 112, 128, 279, 322, 397, 434, 438, 441
Arsenic, 241-247, 389
Asbestos, 32, 128, 131, 241, 353-362, 364, 377, 433, 436-440
Asthma, 25, 44, 50, 78, 97, 107-110, 112-118, 120-122, 127, 131, 160, 168, 182, 186, 203-204, 254, 267-268, 281, 307, 323, 326, 371, 375-376, 379, 395, 397-398, 402, 412, 419-424, 426, 444
Atopy, 116-118, 122, 423, 425
Autoantibodies, 155, 157-158, 207, 273, 353, 398
Autoimmune, 108, 154, 158-159, 169, 315, 396, 398
Autoimmunity, 154, 158, 165, 167, 169, 231, 315

B

Bacillus thuringiensis, 183-186, 190-192
Bacteria, 1, 20, 25, 41-42, 70, 128, 130, 153, 159-162, 182-183, 185-187, 189, 206, 250, 273, 314, 317, 319-320, 326, 340, 344, 354, 401, 412, 414-416, 418, 434, 441, 446

H

I

W

Wegener's granulomatosis, 154-157, 167

Welding, 248, 251, 254, 268, 283, 286, 303, 342-343

Woodburning devices, 369-370

Woodsmoke, 369-374, 381

X

Xenobiotic(s), 1, 7, 12-13, 26, 29, 61, 120, 153, 159, 165-166, 168, 241, 322, 433, 445-446

Z

Zinc, 242, 247-248, 258, 271, 278, 283-287, 389, 391